CHARACTERIZATION OF IMPURITIES AND DEGRADANTS USING MASS SPECTROMETRY

WILEY SERIES ON PHARMACEUTICAL SCIENCE AND BIOTECHNOLOGY: PRACTICES, APPLICATIONS, AND METHODS

Series Editor:

Mike S. Lee
Milestone Development Services

Mike S. Lee • *Integrated Strategies for Drug Discovery Using Mass Spectrometry*

Birendra Pramanik, Mike S. Lee, and Guodong Chen • *Characterization of Impurities and Degradants Using Mass Spectrometry*

Mike S. Lee and Mingshe Zhu • *Mass Spectrometry in Drug Metabolism and Disposition: Basic Principles and Applications*

CHARACTERIZATION OF IMPURITIES AND DEGRADANTS USING MASS SPECTROMETRY

Edited by

Birendra N. Pramanik
Mike S. Lee
Guodong Chen

WILEY

A JOHN WILEY & SONS, INC., PUBLICATION

Published by John Wiley & Sons, Inc., Hoboken, New Jersey
Published simultaneously in Canada

Library of Congress Cataloging-in-Publication Data:

Characterization of impurities and degradants using mass spectrometry /
edited by Birendra N. Pramanik, Mike S. Lee, Guodong Chen.
p. cm.
Includes index.
ISBN 978-0-470-38618-7 (cloth)
1. Drugs–Analysis. 2. Drugs–Spectra. 3. Mass spectrometry. 4. Contamination (Technology) I. Pramanik, Birendra N., 1944- II. Lee, Mike S., 1960- III. Chen, Guodong.
RS189.5.S65C53 2010
615'.1–dc22 2010023283

Printed in Singapore

eBook ISBN: 978-0-470-92136-4
oBook ISBN: 978-0-470-92137-1
ePub ISBN: 978-0-470-92297-2

10 9 8 7 6 5 4 3 2 1

CONTENTS

PREFACE

During the past decade, new formats for automated, high-throughput sample generation combined with a faster pace of drug development led to a shift in sample analysis requirements from a relatively pure sample type to a trace mixture. Mass spectrometry–based technologies played a significant role in this transition and assumed a critical role in pharmaceutical analysis throughout each stage of drug development ranging from drug discovery to manufacturing. A critical part of the development and support of a marketed product is the analysis of impurities and degradation products. Structural information on drug impurities can serve to accelerate the drug discovery–development cycle. The use of chromatographic methods such as high-performance liquid chromatography (HPLC) has long been a hallmark of impurity and degradant analysis. HPLC is often used to profile and classify molecules and work in concert with mass spectrometry to assist with the elucidation of structure. Identification of resulting impurities is based on direct comparison of the mass spectrometric fragmentation of the impurity with the parent drug tandem mass spectrometry (MS/MS) fragmentation patterns. The use of rapid and systematic strategies based on hyphenated analytical techniques such as liquid chromatography–mass spectrometry (LC-MS) profiling and liquid chromatography–tandem mass spectrometry (LC-MS/MS) substructural techniques has become a standard analytical platform for impurity identification activities. We are delighted to highlight current analytical approaches, industry practices, and modern strategies for the identification of impurities and degradants in drug development of both small-molecule pharmaceuticals and protein therapeutics. We provide an ensemble of analytical applications that require the combination of separation techniques and mass spectrometry methods that reflect achievements in impurity and degradant analysis.

We would like to acknowledge the special efforts of all the authors who have made significant contributions to this book. Special thanks go to the acquisitions and production editors at John Wiley & Sons, Inc. for their assistance.

Birendra N. Pramanik
Mike S. Lee
Guodong Chen

CONTRIBUTORS

Michael Ackerman, Bristol-Myers Squibb Company, Pennington, NJ

David W. Berberich, Covidien, St. Louis, MO

Guodong Chen, Bristol-Myers Squibb Company, Princeton, NJ

Hao Chen, Department of Chemistry and Biochemistry, Ohio University, Athens, OH

Himanshu S. Gadgil, Amgen Inc., Seattle, WA

Ming Gu, Cerno Bioscience, Danbury, CT

David M. Hambly, Amgen Inc., Seattle, WA

Tao Jiang, Covidien, St. Louis, MO

Brent Kleintop, Bristol-Myers Squibb Company, New Brunswick, NJ

Mike S. Lee, Milestone Development Services, Newtown, PA

Jiwen Li, Department of Chemistry and Biochemistry, Ohio University, Athens, OH

David Q. Liu, GlaxoSmithKline, King of Prussia, PA

Frances Liu, Novartis, East Hanover, NJ

Peiran Liu, Bristol-Myers Squibb Company, Pennington, NJ

Joseph McClurg, Covidien, St. Louis, MO

Frank Moser, Covidien, St. Louis, MO

Michael Motto, Novartis, East Hanover, NJ

Zheng Ouyang, Department of Biomedical Engineering, Purdue University, West Lafayette, IN

Changkang Pan, Novartis, East Hanover, NJ

Birendra N. Pramanik, Merck and Co., Kenilworth, NJ

Reb Russell, Bristol-Myers Squibb Company, Pennington, NJ

Ruth Waddell Smith, Department of Chemistry, Michigan State University, East Lansing, MI

Scott A. Smith, Department of Chemistry, Michigan State University, East Lansing, MI

Robert J. Strife, Procter & Gamble, Mason, OH

Mingjiang Sun, GlaxoSmithKline, King of Prussia, PA

Li Tao, Bristol-Myers Squibb Company, Pennington, NJ

Qinggang Wang, Bristol-Myers Squibb Company, New Brunswick, NJ

R. Randy Wilhelm, Covidien, St. Louis, MO

Lianming Wu, GlaxoSmithKline, King of Prussia, PA

Wei Wu, Bristol-Myers Squibb Company, Pennington, NJ

Yu Xia, Department of Chemistry, Purdue University, West Lafayette, IN

Gang Xue, Pfizer Inc., Groton, CT

Fa Zhang, Johnson & Johnson, Skillman, NJ

Yining Zhao, Pfizer Inc., Groton, CT

ACRONYMS*

ADCC	antibody-dependent cell-mediated cytotoxicity
ADME	adsorption, distribution, metabolism, excretion
AGC	automatic gain control
AGE	advanced glycation endproduct
AHOT	axial harmonic orbital trapping
ANDA	abbreviated new-drug application
APCI	atmospheric-pressure chemical ionization (DAPCI—desorption APCI)
API	atmospheric-pressure ionization
APTDI	atmospheric-pressure thermal desorption/ionization
ASAP	atmospheric solid analysis probe
AUC	analytical ultracentrifugation
CDC	complement-dependent cytotoxicity
CDR	complementarity-determining region
CE	capillary electrophoresis
Cf	continuous flow
CHO	Chinese hamster ovary
CI	chemical ionization; chemical impact
CID	collision-induced dissociation
CIT	cylindrical ion trap
CLND	chemiluminescent nitrogen detector
COM	center of mass
COSY	correlation spectroscopy
CV	coefficient of variation
CZE	capillary-zone electrophoresis
DAD	diode array detection
DAPPI	desorption atmospheric-pressure photoionization
DART	direct analysis in real time
DBDI	dielectric barrier discharge ionization
DE	delayed extraction
DEC	determination of elemental (de)composition
DEPT	distortionless enhancement by polarization transfer

*Partial list only; common terms (IR, HLC, GC, NMR, RF, etc.), proper names (FDA, NIST, etc.), and chemical compounds (SDS, TCA, etc.) omitted here.

DESI	desorption electrospray ionization (FD-DESI—fused-droplet DESI; MALDESI—matrix-assisted laser DESI)
DeSSI	desorption sonic spray ionization
DLI	direct liquid introduction
DOE	design of experiment(s)
DS	drug substance
ECD	electron capture dissociation
EESI	extractive electrospray ionization (ND-EESI—neutral desorption EESI)
EI	electron impact
EIC	electrospray ionization chromatography
ELDI	electrospray-assisted desorption/ionization
ESSI	electrosonic spray ionization
ETD	electron transfer dissociation
EU	enzyme unit
FAB	fast-atom bombardment
FFF	field flow fractionation
FIDI	field-induced droplet ionization
HAPGDI	helium atmospheric-pressure glow discharge ionization
HC/LC	heavy chain/light chain
HCD	higher-energy C-trap (*or* collision-induced) dissociation
HCP	host cell protein
HCV	hepatitis C virus
HF/LF	high field/low field
HIC	hydrophobic interaction chromatography
HMBC	heteronuclear multibond coherence
HTS	high-throughput screening
IAA	isotope abundance analysis
ICP	inductively coupled plasma
ICR	ion cyclotron resonance
IE/KE	internal energy/kinetic energy
IEC	ion exchange chromatography
IEF	isoelectric focusing
ILA	immunoligand assay
IMS	ion mobility spectrometry
JeDI	jet desorption ionization
LAESI	laser ablation electrospray ionization
LAL	limulus amebocyte lysate
LMC	liquid microjunction chromatography
LOD	limit of detection
LSIMS	liquid secondary ionization mass spectrometry
LTP	low-temperature plasma
LTQ	linear trap quadrupole
mAb	monoclonal antibody
MAGIC	monodisperse aerosol generation interface for chromatography

MALDI	matrix-assisted laser desorption/ionization
MCP	microchannel plate
MDD	maximum daily dose
MP	model protein
MPD	multiphoton dissociation
MPPSIRD	mass peak profiling from selected ion recording data
MSD	mass spectrometry detector
NCE	new chemical entity
NDA	new-drug application
NI/PI	negative ion/positive ion
NMP	next maximum projection
NOE	nuclear Overhauser effect
NOESY	nuclear Overhauser enhancement spectroscopy
OT	open tubular
OVAT	one variable at a time
PADI	plasma-assisted desorption/ionization
PAGE	polyacrylamide gel electrophoresis
PB	particle beam
PCA	principal-components analysis
PDAD	photodiode array detection
PDM	pharmaceutical development–manufacturing
PDMS	plasma desorption mass spectrometry
PET	positron emission tomography
PGM	profile generation model
PPC	practical peak capacity
PPIPPN	pulsed positive ion–pulsed negative ion
PTM	posttranslational modification
QIT	quadrupole ion trap
QMF	quadrupole mass filter
rFC	recombinant factor C
RIC	reconstructed ion chromatogram
RIT	rectilinear ion trap
RP	reversed phase
RRT	relative retention time
RSD	relative standard deviation
RT	retention time
SA	spectral accuracy
SEC	size exclusion chromatography
SFC	supercritical fluid chromatography
SID	surface-induced dissociation
SIMS	secondary-ion mass spectrometry
SMB	supersonic molecular beam
SPE	solid-phase extraction
SSP	surface sampling probe
SWIFT	stored waveform inverse Fourier transform

TDC	time-to-digital converter
TGA	thermogravimetric analysis
TIC	total-ion chromatogram
TOF	time of flight (oaTOF—orthogonal acceleration TOF; reTOF—reflectron TOF)
VOC	volatile organic compound
WBA	whole-body autoradiography

PART I

METHODOLOGY

CHAPTER 1

Introduction to Mass Spectrometry

SCOTT A. SMITH
Department of Chemistry, Michigan State University, East Lansing, MI 48824

RUTH WADDELL SMITH
Forensic Science Program, School of Criminal Justice, Michigan State University, East Lansing, MI 48824

YU XIA
Department of Chemistry, Purdue University, West Lafayette, IN 47907

ZHENG OUYANG
Weldon School of Biomedical Engineering, Purdue University, West Lafayette, IN 47907

1.1 HISTORY

Although mass spectrometry (MS) has aged by about one century, it has never ceased to evolve into an increasingly powerful and important technique for chemical analysis. The development of mass spectrometry can be folded into a few periods, where the capabilities of a particular discipline of science were advanced significantly and steadily due to the introduction of MS into that field. Those periods are, approximately, physics (1890s–1945), chemistry (1945–1975), materials science (1955–1990), and biology/medicine (1990–present) [1]. The history of MS shows that the technique has facilitated many significant scientific achievements, from the discovery of isotopes [2], to purifying the material for the first atomic bombs [3], to space exploration [4,5], to the mass analysis of whole red blood cells each weighing several tens of picograms [6]. The following is a short account of some of the notable feats that have transpired in this field.

Characterization of Impurities and Degradants Using Mass Spectrometry, First Edition.
Edited by Birendra N. Pramanik, Mike S. Lee, and Guodong Chen.

1.1.1 Atomic Physics

The technique now known as MS has its roots in atomic physics at the beginning of the twentieth century, when it was originally applied by physicists toward answering questions on the nature of atoms. Throughout much of the 1800s, the prevailing wisdom held that atoms were indivisible, that all atoms of a given element had the same mass, and that the masses of all elements were multiples of the mass of hydrogen [7–9]. Despite these beliefs, the interrogation of bulk elements through chemical means (gravimetric analyses) demonstrated that some atomic masses were, in fact, not unit integers of that of hydrogen (e.g., chlorine). Furthermore, for much of the century, relatively little was known of the nature and origins of electricity. Hence, the explanations for these phenomena awaited the discovery of electrons and isotopes through physical investigations.

Toward the end of the 1800s, many physicists were interested in unraveling the underlying principles of electricity. To study the properties of electric currents, they would create a potential difference between two electrodes in partially evacuated discharge tubes made of glass and containing various types of gas. Evidence for *cathode rays* (electron beams) was first observed by Plücker in 1859 when he noticed a green phosphorescence occurring on his discharge chamber at a position adjacent to the cathode [10]. In time, the investigations of other physicists led to an accumulation of clues about the nature of cathode rays, including observations that (1) they are directional, moving from the cathode to the anode, (2) they are energetic, as determined by observing platinum foil becoming white-hot when placed in their path, (3) they conduct negative charge, as determined by measurement with electrometers, (4) they are particles rather than waves, (5) their energy is proportional to the acceleration potential to which they are subjected, (6) they have dimensions that are smaller than those of atomic gases, as determined by considering their penetration depth through media of varying density, and (7) they may be derived from any atom through various means, including heat, X rays, or electrical discharge [10]. Thomson went on to develop the means for measuring electron mass in a discharge chamber evacuated to low pressure (see Figure 1.1) [11]. By applying a magnetic field (B) and an electric field (E), both at right angles to each other as well as to the direction of electron propagation, they could determine the electron velocity (v) by canceling out the deflections of the magnetic and electric forces (i.e., $|Bev - Ee| = 0$) such that the electrons travel in a straight line, yielding $v = E/B$. The ratio of electron mass to electron charge (m_e/z) could also be arrived at from experimental measurements as $(m_e/e) = (B^2 l/E\theta)$, where l is the distance traveled by an electron through a uniform electric field and θ is the angle through which electrons are deflected as they exit the electric field [11]. From this and other experiments, Thomson demonstrated that the mass of electrons are about $\frac{1}{1000}$th(0.001%) that of the proton (the mass of protons, the ionized form of the smallest known particles at the time, was by then known from electrolysis research) [11]. Thomson was awarded the 1906 Nobel Prize in Physics "in recognition of his theoretical and experimental investigations on the conduction of electricity by gases" [12].

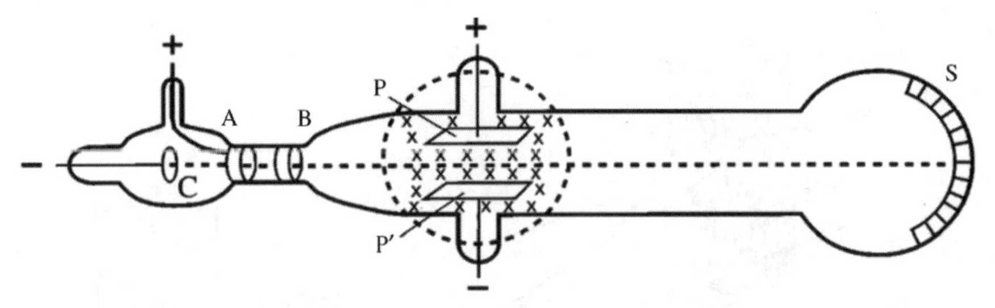

FIGURE 1.1 Thomson's apparatus for measuring electron mass-to-charge ratio (*m*/*z*). Components are as follows: (A, B) anodes with pinhole apertures to guide and narrow the beam; (C) cathode; (P, P′) electric field deflection electrodes; (S) detection screen. The magnetic field, when applied, was directed orthogonally to both the electron beam and the electric field (indicated by the tickmarks x). (Reprinted from Ref. 10, with permission of John Wiley & Sons, Inc.)

While progressing toward an understanding of electrons, physicists also became interested in understanding the positively charged particles (cations) that were present in discharges [13]. During studies of the effects of weak magnetic fields on cathode rays in 1886, Goldstein discovered positively charged *anode rays* that traveled in the opposite direction of electrons; unlike cathode rays, these anode rays were not susceptible to deflection by the weak magnetic fields used in Goldstein's experiments [14]. However, in 1898, Wein determined that anode rays in fact could be influenced by the presence of magnetic fields, provided the fields were relatively strong; with this knowledge, he determined that their masses were on the order of atoms rather than the substance of which cathode rays were composed [14]. Building on such early observations, Thomson created a device called the *parabolic mass spectrograph* (see Figure 1.2), in which he exposed anode rays to parallel magnetic and electric fields in such a way that, while propagating through the field region the rays were influenced vertically by the electric field and horizontally by the magnetic field, with the result that the ions impinged on a photographic plate positioned transverse to the direction of particle propagation [14]. The images on the plate were of parabolas, in which each particular parabola was specific for mass-to-charge ratio (*m*/*z*) and the occurrence of parabolic lines was attributed to distributions in kinetic energy [14]. Thomson's device was capable of identifying the presence of ionized gases, and he demonstrated its capabilities by acquiring a mass spectrograph of the mixture of gases constituting the atmosphere [14]. Notably, Thomson's atmospheric data showed the first instance of the rare isotope ^{22}Ne (neon-22) adjacent to the predominant ^{20}Ne; since he believed that stable elements could have only a single mass (a then widely held belief), he assumed that what was conventionally considered neon was actually a mixture of two elements, with that at mass 22 being previously unknown [2,14]. Shortly before this time, Rutherford and Soddy discovered nuclear transmutation, whereby fission products from radioactive elements produce as products chemically distinguishable elements of abnormal mass (i.e., *isotopes*) [15]; however, given the unusual nature of radioactive matter at the time of Thomson's observation, the link was not obvious that neon atoms could occur as distributions

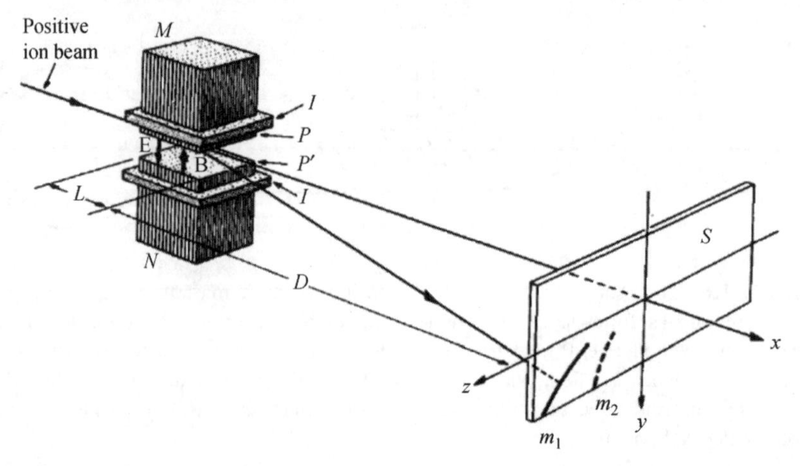

FIGURE 1.2 Ion separations on Thomson's parabolic mass spectrograph. Components are as follows: (I) insulator; (M, N) magnet poles; (P, P′) electric field deflection electrodes; (S) detection screen. The position of ion impact (shown here for two species labeled m_1 and m_2) on the screen was dependent on ion charge and kinetic energy, the electric and magnetic field strengths, and the dimensions of *L* and *D*. (Reprinted from Ref. 10, with permission of John Wiley & Sons, Inc.)

of varying mass. It wasn't until 1919, when Aston built an improved mass spectrograph and discovered the isotopes of dozens of elements, that isotope theory became widely accepted by the scientific community [16]. When he published the results of the measurements of the first 18 elements that he investigated, Aston demonstrated that they all were within $\frac{1}{1000}$th of whole-number units, with the exception of hydrogen, which has a very slight deviation from the whole-number trend [16]. For his efforts toward proving the existence of isotopes, Aston won the 1922 Nobel Prize in Chemistry.

The first breakthroughs in MS were made using equipment that required manual measurements of mass based on visual observation or the interpretation of photographic records that were prone to indicating disproportionate signal intensities based on the species analyzed [13]. These issues were resolved with the development of the first mass spectrometer, by Thomson, in 1912 [13,17]. Rather than detecting on an *image plane* under conditions of constant field strength (as in the mass spectrograph), in Thomson's mass spectrometer the field strengths to which the ions were exposed could be systematically varied while the ion intensities were acquired as electric current using an electrometer positioned behind a plate containing a parabolic slit [13]. This modification also removed a mass dependence on detection intensity, as a signal intensity bias existed on the photographic plates of the spectrograph that favored ions of lower mass, a feature that would be critically detrimental to accurate measurements of relative abundance [13].

As time passed, other physicists made improved mass spectrometers. In 1918, Dempster built a mass spectrometer featuring electronic detection and a 180° magnet capable of resolution values of around 100 (for atomic-range masses) [17]. Aston constructed several notable mass spectrometers; his first, in 1919, was a tandem-in-space *EB* configuration that featured energy correction (i.e., ions of a given *m/z* arrived at a single point on the detection plane regardless of the velocity distribution within the beam) that was capable of achieving a resolution of $\leq$130; later versions of a similar design achieved resolutions of 600 (in 1927) and 2000 (in 1942) [18]. In 1939, Nier produced a magnetic sector instrument that was much smaller than Dempster's (i.e., a few hundred pounds vs. 2000 lb) that was the basis for the design of all future magnetic sectors [19]. With isotope-based research taking off, various other teams also took up the challenge of creating better instruments and developing new applications.

1.1.2 Early Applications

Early applications of MS were centered on discovering isotopes and measuring their relative abundances. By 1935, all known elements in the periodic table had been evaluated for their isotopic compositions by MS [13]. As mass accuracy and precision improved, MS eventually supplanted gravimetric analysis as the predominant method for measuring atomic weights [18]. Another use for MS in the 1930s was for the dating of minerals (*geochronology*) by measuring the relative abundances of radioisotopes in a given sample; for example, by considering a sample's radioisotope ratios in the context of known rates of radioactive decay, the current age of Earth has been determined to be about 4.5 billion years [13]. Mass spectrometers may also lend themselves to separating radioisotopes in a preparative fashion, as in the case of uranium; an early attempt at such processing resulted in the separation and retrieval of some nanograms of the rare ^{235}U from the predominant ^{238}U—an amount sufficient to demonstrate for the first time that ^{235}U is the uranium isotope that readily undergoes fission reactions [19]. Interest in the use of fissile material in weapons ensued, and by spring 1945 hundreds of massive sector instruments ("calutrons") were operating in Oak Ridge, Tennessee to produce some of the ^{235}U used against the people of Hiroshima, Japan in World War II [3]. Although fairly quickly supplanted by the more efficient gas diffusion methods of ^{235}U purification, mass spectrometers nonetheless remained invaluable for enriched materials production for their use as leak detectors and for purity confirmation of the gas diffusion process [19]. It was also during this period that MS was applied toward another very different application—as a means of characterizing the molecular structure of hydrocarbons during crude oil processing [13].

1.1.3 Organic Structural Analysis

Driven by analytical demands from the petroleum and pharmaceutical industries for the characterization of refined petrochemicals and natural products, respectively, MS began to transition into its role as a powerful tool for molecular analysis. The early challenges of such applications were many, including sample introduction, hardware reliability, and spectral interpretation; the latter was particularly difficult as the fundamental rules of structural analysis took years to develop. The invention of electron ionization by Dempster in 1929 went a long way toward ensuring analytical reproducibility among different instruments, the basis for a community-wide effort toward developing a systematic approach for molecular structure interpretation. Rules were established to explain characteristic fragmentation patterns in mass spectra; an example was the "nitrogen rule," which could be applied to organics to interpret which peaks might contain nitrogen or alternatively to determine whether particular peaks corresponded to even- or odd-electron ions if the analyte is of a known composition. Mechanisms were derived to explain dissociation processes; well-known examples include those of metastable ions (where ion internal energy is sufficient for dissociation of an ionic system, yet the system does not fully fragment prior to detection, resulting in a broadened peak) [20] and the McLafferty rearrangement (intramolecular proton abstraction to a carbonyl oxygen from a γ-hydrogen) [21]. The structural analysis of hydrocarbons and other small organics was systematically delineated in McLafferty's seminal text *Interpretation of Mass Spectra* (ca. 1966 but updated as recently as 1993) [22,23]. As chemists became more confident in their spectral interpretation capabilities, the experiments they tried also increased in complexity; to meet these challenges, instrumentation became more sophisticated. Innovations such as tandem MS (MS/MS) [24] for stepwise fragmentation analysis and the coupling of gas chromatography with MS (GC-MS) [25] did much to improve the information attainable by MS as well as its applicability toward the analysis of complex mixtures. Insights into thermochemistry also began to be derived from MS. Ionization potentials for molecular ions and appearance energies for product ions could be determined through various methods, allowing the determination of chemical properties of isolated ionic systems [26].

1.1.4 The Biological Mass Spectrometry Revolution

By the early 1970s, MS was a mainstay in many analytical laboratories. In fact, the technique was also deemed essential outside the laboratory and off the planet as well, having been sent on the *Viking* space mission to Mars in 1976 [27]. Through the decade, commercialized versions became available for various platforms, including sectors, GC-MS (featuring quadrupole filters), time-of-flight (TOF), and Fourier transform ion cyclotron resonance (FT-ICR). The analysis of small organics had become relatively routine, and a major emphasis of research turned toward the problems of biology and the analysis of large, fragile biomolecules such as peptides and proteins. Although Biemann and coworkers had shown the potential for mass spectral sequencing of small peptides in 1959 [28], much was still to be done to improve the effectiveness of bioanalysis. Techniques that showed early promise in biomolecule analysis included desorption methods such as fast-atom bombardment (FAB) and liquid secondary ionization MS (LSIMS), where bombardment of a liquid sample matrix with high-energy neutral or charged particles (respectively) can facilitate the ejection of intact pseudomolecular ions; another technique applied to early protein analysis was plasma desorption MS (PDMS) [29], where bombardment of a surface-deposited sample by ^{252}Cf fission fragments could result in the expulsion of large ionized molecules that were predeposited on the surface. However, the glycerol matrix of FAB/LSIMS techniques can lead to high background, and the equipment for PD was limited to only a small number of laboratories. The advent of thermospray, the ionization of LC eluant in a heated vacuum interface, proved promising in that it allowed the online coupling of liquid chromatography to MS (i.e., LC-MS) for the analysis of nonvolatiles; however, thermospray is seldom employed today as its performance was surpassed by electrospray, a somewhat similar technique that was developed in the mid- to late 1980s by Fenn [30]. Fenn approached the issue of protein analysis by using a technique known as *electrospray ionization* (ESI) [31], whereby large biomolecular ions could be formed via the nebulization of an electrified liquid. Much headway was being made in the area of laser desorption in the 1980s, culminating with the mass analysis of very large intact biomolecular ions: Tanaka developed a method using UV-resonant metal nanoparticles to enable the intact ionization and volatilization of proteins, while Karas and Hillenkamp developed a similar technique which they termed, *matrix-assisted laser desorption ionization* (MALDI), wherein preformed ions reside in a solid matrix prior to their ejection by the UV photoexcitation and explosion of organic matrix crystals [32,33]. For their efforts toward establishing protein analysis by MS, Tanaka and Fenn shared the 2002 Nobel Prize in Chemistry.

Since the relatively recent establishment of proteomics (the study of protein structure and function) [34], other "omics" studies have also been developed using similar strategies, including metabolomics, lipidomics, glycomics, metallomics, and phosphoproteomics. Remarkable biological insights have resulted, including the protein sequencing of fossilized dinosaur remains [35]. Relatively recent contributions to instrumentation have included the successful introduction of a new ultra-high-resolution mass analyzer (the Orbitrap™, originally developed by Makarov at Thermo Fisher Scientific, Bremen, Germany) that can match the high-performance capabilities of FTICR for a fraction of the cost. The chemical imaging of tissues using MS shows promise for a future of highly enhanced medical and biological investigations [36]. New methods of ion activation have also been developed and applied toward biological problems, including electron capture dissociation (ECD) [37] and electron transfer dissociation (ETD) [38]; these two similar techniques are notable for their radical-directed dissociation mechanisms, which allow the analysis of proteins carrying posttranslational modifications (PTMs), whose locations would otherwise often be unidentified in analyses using conventional methods of activation [i.e., collision-induced dissociation (CID)]. The future of MS promises to resolve many more biological issues with ever-greater performance.

1.2 IONIZATION METHODS

Chemical analysis using MS is achieved by measuring the mass-to-charge ratios (*m/z*) of the charged forms of the analyte molecules. The first step in the mass analysis process is to generate the analyte as ionic species in the gas phase. A wide variety of ionization methods have been developed over the last several decades, which enabled the utilization of MS in different areas of chemical analysis. The main challenge in the development has always been preserving the molecular information while converting the analyte molecules from condensed phases into gas phase and making them charged. Soft ionization methods allow the preservation of the molecular structures in ions, which can be elucidated with the combination of the MS and MS/MS analysis. The energy deposition required for transferring analyte molecules into the gas phase and ionizing them can easily result in intense fragmentation of the molecules, as in certain desorption ionization methods, inductively coupled plasma (ICP), and electron impact (EI) ionization. This problem becomes much more severe when applying MS for the study of biompolymers such as peptides and proteins, whose volatility is low but whose structural information is highly valuable. Development of the electrospray ionization (ESI) and matrix assisted laser desorption/ionization (MALDI) provided the solution for this problem. Since the ionization methods have been comprehensively described in the literature, including the recent volume of *The Encyclopedia of Mass Spectrometry* (Vol. 6, *Ionization Methods*) [39], we have listed the characteristic features of the most commonly used ionization methods in Table 1.1. Their implementation with different types of instrumental setups for several applications is discussed later in this chapter.

1.3 MASS SPECTROMETER TYPES

Mass spectrometry is a discipline of analytical chemistry wherein the gas-phase ionic form of chemical species may be identified and characterized according to their mass and the number of elementary charges that they carry. There are several divisions of instrumental aspects of mass spectrometers including sample introduction, ion formation, ion transport, mass analysis, detection, vacuum systems, and software. In the following text we will introduce the reader to the principles of the various mass analyzers, providing a brief but comprehensive overview of the practical aspects of operation. This introduction is not meant to be exhaustive; lesser-used techniques or unlikely phenomena are mentioned only in passing or not at all. In the following sections we briefly describe the principal mass analyzers used in MS: magnetic sector (B), quadrupole mass filter (QMF), quadrupole ion trap (QIT), time-of-flight (TOF) analyzers, Fourier transform ion cyclotron resonance (FT-ICR), and Orbitrap.

1.3.1 Magnetic Sector Mass Spectrometers

The separation of ions in a strong electric or/and magnetic field constitutes the oldest form of mass spectrometric analysis, with roots dating back to the end of the nineteenth century. Under the influence of strong direct-current (DC) electric (E) and magnetic (B) fields, a gas-phase ion population may be made to undergo separations within an E field based on ion kinetic energy ($0.5\,mv^2$) or within a B field based on momentum (mv). Some founding innovators in the development of magnetic analyzers (and indeed MS) included Wein, Thomson, Aston, and Dempster.

TABLE 1.1 Characteristic Features of the Most Commonly Used Ionization Methods

Ionization	Pressure	Suitable Analytes	Ionic Species	m/z Range
Inductively coupled plasma (ICP)	> mTorr	All	Atomic ion, A^+	< 220
Electron impact (EI)	< mTorr	Volatile organic compounds (VOCs)	Molecular and fragment ions, $M^{+\cdot}$, $M^{-\cdot}$	< 1000
Chemical ionization (CI)	~ 1 Torr	VOCs	Molecular ions $(M + H)^+$, $(M - H)^{-\cdot}$ $M^{-\cdot}$, and adduct formation	< 1000
Atmospheric-pressure CI (APCI)	Atmospheric	VOCs, metabolites	Molecular ions $(M + H)^+$, $(M + H)^+$, $(M - H)^{-\cdot}$ $M^{-\cdot}$	< 1000
Electrospray ionization (ESI)	Atmospheric	Metabolites, lipids, peptides, proteins	Multiply protonated/deprotonated molecular ions $(M + nH)^{n+}$, $(M - nH)^{n-}$, and complexes	< 4000
Matrix-assisted laser desorption ionization (MALDI)	< 1 Torr	Metabolites, lipids, peptides, proteins	Singly protonated/deprotonated molecular ions $(M + H)^+$, $(M - H)^{-\cdot}$ and complexes	≤200,000

Early applications of sector mass analysis included investigations of fundamental atomic physics: for example, the existence of and the mass of electrons [11], in addition to the accurate determinations of the masses and natural abundances of isotopes [2,16]. Sector analyzers have also been used for the isotopic purification of ^{235}U for the first atomic bomb [3], as platforms for the study of much of the earliest tandem MS experiments [20,40], and for accurate determination of the age of materials based on isotope ratios (e.g., carbon dating) [17]. As understanding of ion trajectories and their impact on mass spectrometric performance matured, instruments evolved with increasing sophistication; in time, sector instruments achieved such sophistication as to allow achievable resolutions of up to 10^5 and single-digit part-per-million (ppm) mass accuracies. Today, sector analyzers have largely been supplanted by other mass spectrometer types, although they are still employed for some applications (e.g., ultra-accurate isotope ratio determinations) [18]. With the rate of development of sector instrumentation and applications in decline for some time, recent literature discussions on the matter are principally available in MS texts [41–43].

When an ion is exposed to a magnetic field occurring in a dimension perpendicular to the ion's trajectory, the ion experiences a force in a direction orthogonal to both B and the ion's velocity. The circular path that an ion takes through a homogenous magnetic field is dependent on a balance between centripetal and centrifugal forces, which can be described as

$$zevB = \frac{mv^2}{r} \tag{1.1}$$

where z is the number of elementary charges on an ion, e is the elementary charge (1.602×10^{-19} C), B is the magnetic field magnitude, m is the ion mass, v is the ion velocity, and r is the radius of the ion trajectory as it is deflected by the magnetic field. Often, B sector analyzers are referred to as "*momentum analyzers*", as can be seen by rearrangement of Eq. (1.1) to arrive at

$$r = \frac{mv}{zeB} \tag{1.2}$$

Hence, for ions of a given charge and a constant magnetic field strength, the degree of deflection that an ion incurs as it transits through a magnetic sector is dependent only on momentum (mv).

A magnetic sector mass spectrometer can consist simply of an ion source, an electromagnet, various slits to allow selective ion beam passage prior to and after the mass analyzer, a vacuum system, an electrometer, and a data processor. Ions are generated in a source and then accelerated (by way of an electric field) toward the entrance of the B sector analyzer, where they experience a deflection dependent on their mass, charge, and velocity. For sector MS, the extraction potential at the source is quite high (e.g., 10 keV) to maximize sensitivity by reducing beam broadening, and also to allow for ions to pass quickly through a sector as it is being scanned without significantly affecting resolution. Since ion sources do not produce monoenergetic ion beams, it is quite common to couple a magnetic sector mass analyzer in tandem with an electric sector analyzer such that the E sector can be made to select a range of ions having the same kinetic energy (E sectors are technically *energy analyzers* rather than mass analyzers). In such "double-focusing" geometries, correction is effected for both kinetic energy and angular dispersion in the electric and magnetic sectors, respectively. Kinetic energy correction is achieved in an E sector through a balance between centripetal and centrifugal forces, which is shown in the following equality:

$$zeV_{\text{accel}} = \frac{mv^2}{r} \tag{1.3}$$

which may be rearranged to consider ion trajectories along a circular arc:

$$r = \frac{mv^2}{zeV_{\text{accel}}} \tag{1.4}$$

Provided an ion beam is made monoenergetic by an E sector prior to mass analysis (as in the EB tandem configurations; e.g., see Figure 1.3), the m/z of the ions within the population may be determined by detection of ions either along an image plane or at a single point. For the former case, B is maintained at a constant value such that variation in ion m/z corresponds to variation in r, which results in ions arriving at different points along an image plane in an m/z-related manner (the image plane consists of either a photographic plate for early instruments or multicollector detectors for more modern ones). Such simultaneous broad-spectrum detection provides the highest sample efficiency, although achieving high resolution or sensitivity through such means requires stringent fabrication specifications for the detector [44]. Alternatively, given the means for scanning B, a tandem sector mass spectrometer may be operated in such a way that ions may be detected at a single position along the detection plane (e.g., at an electron multiplier behind a narrow slit). Such fixed-point detection is typically limited to a scan rate of 100 ms per decade

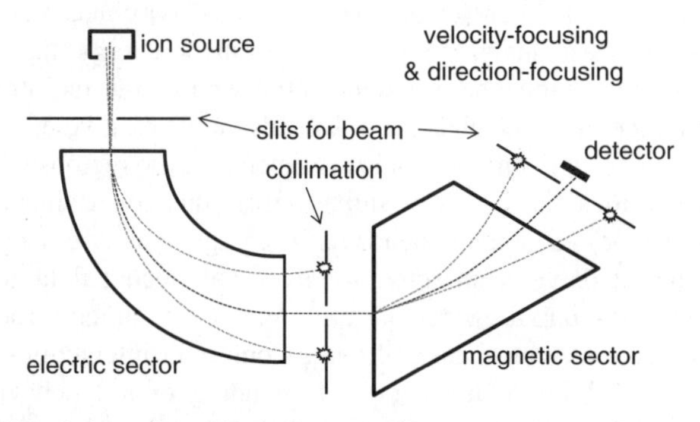

FIGURE 1.3 Depiction of an *EB* dual sector mass spectrometer. Ions are detected as a function of their deflections in the electric and magnetic sectors.

(e.g., from 100 to 1000 *m*/*z*), as higher scan rates can degrade resolution [42,43]. Additionally, the fact that *B* is scanned quadratically to achieve a linear correlation with *m*/*z*, and hence that *m*/*z*-dependent sensitivity and inaccurate relative abundances can occur, must be considered.

1.3.2 Quadrupole Mass Filter and Quadrupole Ion Trap Mass Spectrometers

Quadrupole mass analyzers separate ions through controlled ion motion in a dynamic quadrupolar electric field. First introduced by Paul and Steinwedel in 1953 [45], quadrupole mass analysis is performed on two types of mass analyzer: quadrupole mass filters (QMFs) and quadrupole ion traps (QITs). Common analytical traits of quadrupole mass spectrometers include "unit" resolution (i.e., differentiation of singly charged isotopes), mass-to-charge ratio (*m*/*z*) ranges of >1000 Th (Thomson; unit measuring *m*/*z*), and specific chemical structural information provided through tandem MS. The fundamental basis for ion stability is essentially the same for both analyzer types, yet some differences exist in geometry and the waveforms applied in order to produce mass spectra. The following information is intended only to convey the major principles of operation and their consequences on performance. For further and deeper discussions of the concepts associated with quadrupole MS, the interested reader is encouraged to explore several detailed reviews [46,47].

An *electric field* occurs when there is a potential difference between two objects. It is the nature of an electric field to store electrical potential energy, and ions in such a field may occupy any Cartesian coordinate position provided their kinetic energies match or surpass the electric potential energy (pseudopotential) associated with that position. A *quadrupole field* provides a linear restoring force as a function of the square of an ion's displacement from the field center. Hence, the form of the force *F* on an ion moving away from the trap center in a trapping dimension of a QIT is in accordance with Hooke's law for harmonic oscillation [48]

$$F(u) = -C \cdot u \tag{1.5}$$

for *u* is displacement in a dimension of ion motion and *C* is a constant. Given that ions enter a quadrupole mass analyzer with nonzero kinetic energy, they will undergo sinusoidal oscillation within the pseudopotential well of the radiofrequency (RF) field. The magnitude of ion displacement depends on the relative magnitudes of the ion and field energies, and ion position is restricted to those regions of the field with potentials that the ions can match or surpass given their own kinetic energy. The position and trajectory of an ion depends on its charge, mass, velocity, and starting position, and the repulsive or attractive forces of the electric field and other ions. Either the kinetic or internal energy of an ion may be modified through collisions between the ion and background gas or through Coulombic interactions between like- or oppositely charged ions. Given an understanding of ion behavior within an electrodynamic quadrupole field, an analyst can use a quadrupole mass spectrometer to manipulate and mass-selectively detect ions as mass spectra.

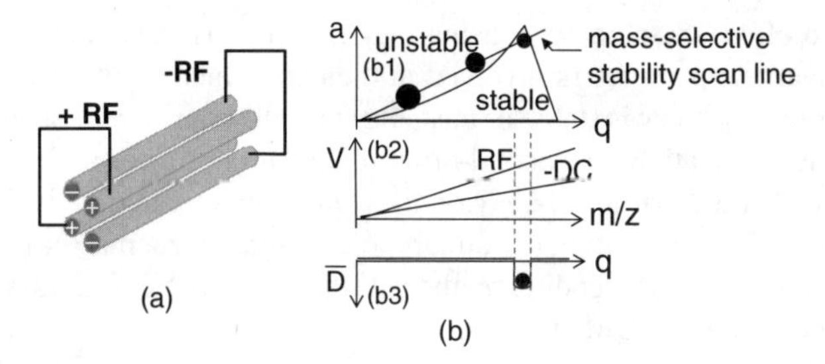

FIGURE 1.4 Depiction of some aspects of quadrupole mass filters: (a) isometric view of a quadrupole mass filter, where the electrodes are paired across the axis of the analyzer; (b) some principles involved in mass analysis — plot (b1) indicates relative positions of three ions in the *a,q* space of the Mathieu stability diagram; plot (b2) indicates the potentials applied as functions of ion seqular frequency; plot (b3) depicts the pseudopotential well depth, where the *m/z* of interest is shown in the only stable region.

In a QMF and, by analogy, QITs, an electric field occurs between two pairs of parallel electrodes, with each pair short-circuited together and situated opposite each other and equidistant about a central axis (see Figure 1.4). In ideal geometries, electrodes are of a hyperbolic form so as to provide the purest quadrupole electric potential (ϕ_2), which can be described mathematically for any point (x,y) in the cross-sectional plane of a QMF (for example) as [47]

$$\phi_2(x,y) = A_2\frac{(x^2-y^2)}{r_0^2} + C; \quad x^2 \le r_0^2; \quad y^2 \le r_0^2 \tag{1.6}$$

where x and y are displacements from the QMF center in their respective dimensions, r_0 is the inscribed radius of the QMF, A_2 is the amplitude of the applied potential, and C is a constant added to the potential to account for any "float" voltage applied equivalently to all electrodes, which is relevant for instances beyond the frame of reference of the quadrupole (e.g., the transport of ions into or out of the device). The lack of cross-terms between the Cartesian coordinates (e.g., xy) means that, in a quadrupole field, ion motion in each dimension is independent of the fields or motion in orthogonal directions; this feature makes it much easier to consider aspects of ion motion and manipulation in comparison to higher-order multipoles. The amplitude of the applied waveform A_2 is of the form

$$A_2 = \pm(U - V\cos(\Omega t)) \tag{1.7}$$

where U is the DC potential and V is the RF potential that oscillates at the angular frequency Ω and the plus/minus sign designates that the two rod pairs are of opposite sign. For QMFs, the force on an ion depends on its position within the electrodynamic field; at any given moment, an ion is simultaneously accelerated in two dimensions—attraction in one dimension and repulsion in an orthogonal dimension. For ions of stable trajectory, the potential on the electrode pairs will always reverse and attain sufficient amplitude to redirect ion trajectories before they discharge on an electrode's surface.

In order to effect mass analysis using quadrupole mass analyzers, relationships between the various parameters involved in the experiment and the state of the ion (i.e., whether its trajectory is stable or unstable) must be considered. Such is provided by the Mathieu equation, a second-order differential function that allows the prediction of charged particle behavior in a quadrupole electric field [47]. With the Mathieu function, ion motion in a quadrupole mass analyzer may be determined as either stable or unstable, depending on the values of two *stability factors*, namely, a_u and q_u (shown here for a QMF)

$$a_u = \frac{8zeU}{mr_0^2\Omega^2} \tag{1.8}$$

$$q_u = \frac{4zeV}{mr_0^2\Omega^2} \tag{1.9}$$

where the subscripted u in a_u and q_u denotes the dimension (x or y), U is the DC potential (zero-to-peak), V is the RF potential (zero-to-peak), z is the number of elementary charges on an ion, e is the elementary charge (1.602×10^{-19} C), m is the ion mass [in daltons (Da)], r_0 is the inscribed radius [in centimeters (cm)], and Ω is the RF angular frequency [in radians per second (rad/s)]. Because the potentials applied to the x and y pairs of an ideal QMF or 2D QIT (and by analogy 3D QITs) are 180° out of phase, at any given instant the y-dimension parameters a_y and q_y are of equal magnitude but opposite in sign to a_x and q_x; that is, $a_x = -a_y$ and $q_x = -q_y$. The relationship between the a and q terms may be represented graphically with a Mathieu stability diagram (Figure 1.5). The boundaries of the stability diagram represent the

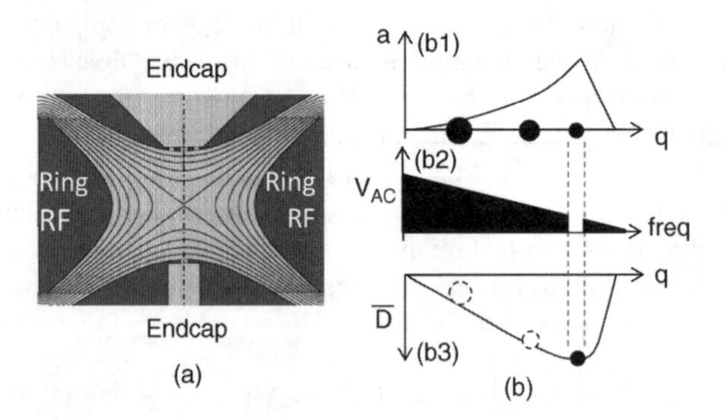

FIGURE 1.5 Depiction of some aspects of quadrupole ion traps. (a) cross section of a quadrupole ion trap, where electrodes are solid and equipotential field lines are indicated (image modified from Ref. 194, with permission of Elsevier); (b) some principles involved in mass-selective isolation: plot (b1) indicates relative positions of three ions along the q axis of the Mathieu stability diagram; plot (b2) indicates the applied waveform that resonantly accelerates and ejects all ions except those of the m/z of interest (which coincide with the waveform notch); plot (b3) depict the pseudopotential well depth, where the m/z of interest is shown in the deepest region.

set of a_u,q_u values at which an ion's trajectory transitions from stable to unstable. Although there are multiple regions of stability defined by the Mathieu function, only the region known as *region 1* is typically considered, as this region is the least demanding in terms of voltage required for ion trajectory stability (in terms of both DC and RF). The bounds of region 1 are defined as those at which the term β equals 0 or 1 for both the x and y dimensions.

Ion motion within quadrupole ion traps (QITs) are described by the Mathieu function in a manner similar to that for the QMF. However, the way in which QMFs and QITs perform mass analysis are different, owing to differences in the dimensionality of their electric fields. While QMFs can trap ions in the x/y plane, they cannot do so along the ion optical axis. In contrast, QITs have either RF or DC trapping potentials in the z dimension (for 3D and 2D traps, respectively). Geometrically speaking, a 2D trap can be created from a QMF by simply installing thin lenses at the ends of the QMF and applying DC stopping potentials to them. A 3D trap, which features RF trapping potentials in three dimensions, is typically is constructed of a toroidal *ring electrode* with two *endcap electrodes* that cover the openings at the top and bottom of the toroid.

There are several possible ways to create a mass spectrum with a QMF, but the device is usually operated in *mass-selective stability* mode, whereby a scanline is chosen on the Mathieu stability diagram, which is characterized by a constant a/q ratio. Through the course of an analytical scan, the DC and RF potentials are ramped such that only a narrow m/z range will be allowed passage through the QMF per unit time. Calibration of the a/q ramp with the detector timing allows mass spectra to be produced by plotting detected ion current versus time.

As with QMFs, there are multiple ways to perform mass analysis on a QIT. However, the most typical is that of a *mass-selective instability* [49] scan with resonant ejection; this mode features a scanline that lies along only the q axis of the stability diagram (no DC components). As ions oscillate within a trapping RF field, their travel is characterized by their secular frequencies

$$\omega_{u,n} = \pm\left(n + \tfrac{1}{2}\beta_u\right)\Omega \tag{1.10}$$

for $\omega_{u,n}$ is the secular frequency in the u dimension, n is the order of the fundamental secular frequency (typically $n = 0$), Ω is the fundamental frequency, and β_u is approximated as

$$\beta_u = \sqrt{a_u + \tfrac{1}{2} q_u^2} \tag{1.11}$$

for $q_u < 0.4$ [47]. During QIT mass analysis, $\omega_{u,0}$ is simplified as follows:

$$\omega_{u,0} = \frac{q_u \Omega}{2\sqrt{2}} \tag{1.12}$$

During QIT mass analysis, as the RF is ramped, ions acquire different secular frequencies. A low-voltage supplemental alternating-current (AC) waveform is applied to the electrodes in the dimension intended for ejection at a frequency corresponding to the desired q value at which ions are to be ejected (often between 0.7 and 0.9). As ions are scanned through this q-value, they become destabilized and are ejected through holes in an electrode for subsequent detection. By performing resonant ejection, one can scan an ion population through this "hole" of instability with the result of better resolution than can be obtained by RF-only scanning of the population through the stability boundary, which is subject to inherent instability in the mass spectrometer electronics in addition to a shallow D_u (which can cause frequency spreading of a given m/z) [50,51].

Should a quadrupole analyzer have nonlinear fields (nonquadrupolar contributions), the representation of the Mathieu stability diagram becomes overlain with internal points and lines of nonlinear resonance at which ions may be ejected or made to undergo undesired excitation [47]. In practice, QMFs and QITs are often subject to nonlinear resonances, particularly those devices that are of a simplified geometry (and hence have nonideal trapping fields). However, the addition of nonlinear resonances can also be utilized to advantage, as has been demonstrated with the rectilinear ion trap (RIT), where ions scanned through the nonlinear point at $q_x = 0.81$ were shown to have remarkably sharper peak shapes than at any other resonant frequency (Figure 1.4); this occurrence is attributed to the proximity of the RIT β_x value to the known octapolar nonlinear resonance point of $\beta = 0.7$ [50].

Finally, another way to determine trapped ion trajectory stability is to consider the pseudopotential well. The pseudopotential well is the representation of the strength of the electric field applied to the trap electrodes. At any instant, a trapped ion in a quadrupole field is at once stable in one dimension (figuratively, near the bottom of a potential "well") yet unstable in another (figuratively, near the top of a potential "hill"). In order to maintain stability, several conditions of energy must be met, including: (1) the ion kinetic energy KE_{ion} cannot exceed the pseudopotential well depth in a particular dimension D_u; (2) the RF frequency, which is 180° out of phase between electrode pairs, must alternate at a rate sufficient to ensure that the ions do not remain too long on the potential hill in the unstable dimension so as to escape the device. Paul has described the mechanical analog of the quadrupole pseudopotential as a "rotating saddle" that inverts with every 90° of phase. The Dehmelt approximation to the quadrupole pseudopotential describes the dependence of D_u on the q parameter and the RF potential in a 2D QIT [52]:

$$D_{x,y} \approx \frac{qV_{RF}}{4} \tag{1.13}$$

Consideration of D_u is important, as an ion with a kinetic energy higher than that of D_u will not be contained by that field. This is of practical consequence for matters such as ion injection, ion ejection, and various forms of ion manipulation.

1.3.3 Time-of-Flight Mass Spectrometers

The premise of separations by time-of-flight MS (TOF-MS) is that a mixture of ions of varying *m/z* yet the same kinetic energy will separate over time owing to differences in velocity. These differences in velocity are measured in terms of the time that it takes ions to reach a detector surface after being pulsed into a flight tube (e.g., see Figure 1.6a). The technique was first demonstrated in 1948 by Cameron and Eggers, whose instrument, the "ion velocitron," was just capable of baseline resolution between N_2^+ (28 Th) and CF_2Cl^+ (85 Th) using a 320 V acceleration potential [53]. Despite the straightforward concept of mass analysis by TOF, the field took a long time to come to its present state of maturity. Issues with minimizing distributions of kinetic energy, spatial coordinates, and angular distributions have taken decades to overcome. Spectral quality was also notably hampered by the lack of fast detection electronics. In order to achieve high-quality spectra, TOF mass spectrometers must be constructed to and operated with very high standards, including high-precision mechanical tolerances and even strict control of the flight tube material and temperature to avoid thermal expansion and the consequent variation in mass calibration. Much progress in TOF development occurred in the 1980s, when the 'biological revolution' in MS spurred interest in the development of instruments capable of analyzing high-mass samples. The coupling of MALDI ion sources with TOF analyzers became very common, as the unlimited upper mass range of TOF meshed well with the high-mass ions and pulsed-ion generation characteristic of MALDI. Through it all, TOF mass spectrometers have evolved to take their place as

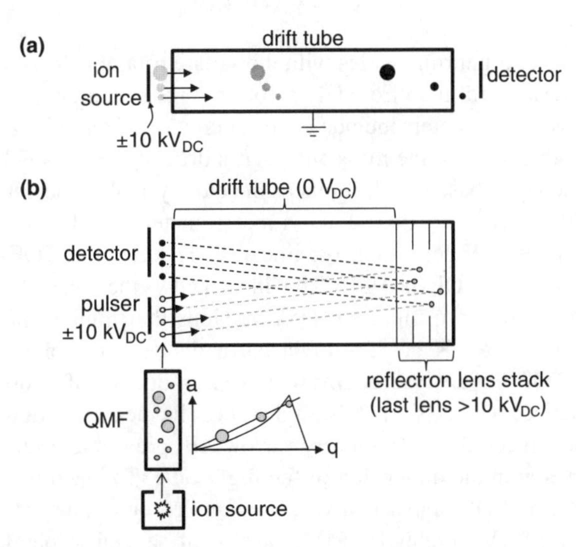

FIGURE 1.6 Two depictions of time-of-flight mass spectrometers: (a) a linear TOF arrangement where the filled circles represent ions of different masses and the arrows represent their velocity magnitudes; (b) a quadrupole TOF (Q-TOF) in which an ion beam is purified in *m/z* first by a quadrupole mass filter and the ions are subsequently pulsed orthogonally into a reflectron TOF.

popular instruments valued for their high resolution, mass accuracy, and sensitivity. Today, commercial TOF-based instruments are capable of providing resolutions of >10,000, mass accuracies of a few ppm, and the capability for performing MS/MS. The interested reader is referred to a few recent reviews for further details on TOF-MS than those provided here [54–59].

The TOF concept is observed in the equivalence of kinetic energy with an ion's acceleration through an electric field

$$\tfrac{1}{2}mv^2 = zeV_{\text{accel}} \tag{1.14}$$

where m is ion mass (in kg), v is ion velocity (in m/s), z is the number of elementary charges on an ion (unitless), e is the elementary charge (1.602×10^{-19} C), and V_{accel} is the potential through which the ions are accelerated (in volts). Rearrangement of this identity yields the equivalence of the mass-to-charge ratio m/z

$$\frac{m}{z} = 2zeV_{\text{accel}}\frac{1}{v^2} \tag{1.15}$$

which demonstrates that m/z scales with the inverse square of velocity. Stated another way, the TOF for an ion may be derived by substituting velocity with distance over time ($v = d/t$). By doing so and rearranging Eq. (1.15) to solve for t, we obtain

$$t = \sqrt{\left(\frac{m}{z}\right)\left(\frac{d^2}{2eV_{\text{accel}}}\right)} \tag{1.16}$$

which indicates that flight time scales with the square root of m/z, an equality that is perhaps more meaningful than Eq. (1.15) from the measurement perspective.

All TOF mass spectrometers include an ion source, a gating mechanism to initiate or stop ion introduction into the mass analyzer, a drift region in which ions of near-equivalent KE separate based on differences in velocity, a plane detector, and a high-vacuum system. The category of TOF mass spectrometers may be further subdivided into three classes (see Figure 1.6a, b): linear TOF, reflectron TOF (reTOF), and orthogonal acceleration TOF (oaTOF). The linear TOF is the first and hence "classic" design, while the latter two represent later versions that implement performance-enhancing design features. Regardless of the particular design employed or method of ion generation, TOF-MS analysis begins with the introduction of a population of ions into the region adjacent to the start of the flight tube. The ideal conditions in this region are such that the ion population has a very narrow KE spread and occupies a region of space that is narrow in the dimension of the flight tube. To inject ions into the drift tube, a high-potential DC signal is applied (example conditions: rise time = 25 ns; <1 μs duration; 1.5 kV potential) [54] at a rate of several kilohertz (kHz), thus triggering the initiation of ion flight into the field-free drift tube (no further acceleration occurs in this region). Within the drift tube (often 1–2 m in length), ions will travel with kinetics defined by their acceleration potential [typically a few kiloelectronvolts (keV) to several tens of keV] with KE spreads within a several tens of millielectronvolts (meV) or better (depending on the source type and instrument quality). Within the drift region, ion separation is based on differences in velocity, with ions of low *m*/*z* traveling with the greatest speed. Detection of ions occurs at a planar detector, which records signal intensity versus the time since the injection pulse was triggered.

The optimization of TOF construction and operation has proved critical to achieving high-quality data in terms of resolution, sensitivity, and duty cycle. Three important advances made at the TOF source region were those of time lag focusing (for gas-phase ions) [60], delayed extraction (DE, a time lag focus analog for linear MALDI-TOF) [61,62], and the advent of orthogonal acceleration TOFs. The principle of timelag focusing as well as delayed extraction is centered on correcting space and velocity distributions at the analyzer source, that is, correction of the kinetic energy disparity arising from ions initially moving in a wide distribution of velocities and directions in the moments shortly before injection into the drift tube. To minimize this directional disparity, rather than triggering ion injection into the drift tube immediately after an ion population enters the source region, a tunable time delay (μs timescale) may be employed to allow the ions to expand over a wider distance (a few mm) in the dimension of the flight tube. Then, when a potential is applied to accelerate ions into the drift region, the ion population will have expanded across a greater distance along the dimension of the flight tube; the end result is such that the ions nearer the rear of the source volume will be accelerated for a longer time than will the ions nearer to the drift tube. Provided an appropriate time delay is employed, the "lagging" ions of a particular *m*/*z* range can be made to "catch up" with the leading ions and hence enhance spectral resolution. Another way in which spatial distributions in the analyzer source are minimized is through *orthogonal acceleration*. Mass spectrometers that feature orthogonal acceleration have the ionization occur in a region distinct from the analyzer, forming a beam that will ultimately travel transverse to the flight tube axis. Often, these beams are collisionally damped at moderate pressures to minimize kinetic energy distributions as well as beamwidth (including spatial distribution in the dimension of the flight tube), which may be further narrowed by slits whose smallest dimension is parallel with the flight tube axis. Hence, oaTOFs allow the sampling of a quasiplanar beam. Because the ions have a velocity component transverse to the flight tube, the detector may need to be shifted so that the center of the TOF source and the detector are not co-axial. It should also be noted here that the use of RF multipoles in oaTOFs permits the storage of the ion beam (via trapping) during the period in the TOF cycle when ions are not pulsed into the flight tube; hence, oaTOF designs are one way in which TOFs achieve a high duty cycle (between 5% and 100%) [59].

In addition to instrumental advances at the TOF source, TOF performance was also improved by correcting dispersions at one or multiple points within the ion flight path using a reflectron (or ion mirror). The idea for the reflectron originated from Mamyrin in the early 1970s when he was reminiscing on childhood games in which the objective was to see who could throw a ball the highest [58]; in such a game, the highest-thrown ball traveled with the greatest initial and final velocity, yet it also necessarily spent the most time decelerating at the apex of its arc before its return trip to Earth. From this concept was born the TOF equivalent, the "ion mirror," a stack of grids or annular disks to which is applied static DC potentials such that ions may penetrate the field to varying depths dependent on their kinetic energy. As the mirror potential is ultimately higher than all ion kinetic energies, the result is that each ion entering it will decelerate, arrest its forward motion, and then reaccelerate away from the mirror, eventually achieving the same velocity at which it entered the mirror. Since the highest-KE ions penetrate the mirror to the greatest depths, they must necessarily spend the most time in the mirror; in effect, this allows lower-velocity ions of the same m/z to travel a shorter distance through the mirror before returning to the drift region on the way to the detector (or another ion mirror). The end result is a narrowing in flight time distributions for ions of a given m/z (higher resolution; e.g., while commercial linear TOFs may provide $R \sim 1000$, commercial reTOFs are capable of $R = 7{,}500–20{,}000$) [43]. The reflectron is also a convenient way to effectively lengthen the drift tube length while allowing an instrument to have a smaller footprint.

The detector and data acquisition elements of TOFs have been subjected to marked improvements coinciding with the development of fast digital electronics. Since TOF analysis requires ions to travel at several tens of keV for optimal resolution, sensitivity, and duty cycle, the signal acquisition must accommodate time resolution on a very short timescale (ns range); should detection time resolution be slower than the arrival time distribution of a peak, the spectral resolution is then limited by the detector. Hence, conventional TOF detectors [microchannel plates or (MCPs); essentially a thin planar detector with thousands of continuous-dynode channels on the scale of tens of μm in diameter] [63] as well as signal processing circuits [time-to-digital converters (TDCs)] must characteristically have ultralow intrinsic signal delay. Fortunately, most MCPs and TDCs used in modern instrumentation are capable of operation with sub-nanoscale time resolution.

1.3.4 Fourier Transform Ion Cyclotron Resonance Mass Spectrometers

The history of ion cyclotron resonance MS (ICR-MS) begins in the 1930s when Lawrence and coworkers developed the theory and early instrumentation for cyclotrons [64,65]. Cyclotrons are instruments capable of trapping and manipulating ions under high vacuum in a strong magnetic field, in which ions move with a circular trajectory on a plane perpendicular to the magnetic field (see Figure 1.7a). Cyclotron research has long focused on aspects of atomic physics that require ions to be accelerated to very high energies. For example, cyclotrons are used for the bombardment of atomic ions to induce transmutation—the formation of one element as a fission product of a heavier element; such research allows the production of rare unstable isotopes suitable for scientific or medical interests. As early as 1949, the potential for measuring mass using cyclotron principles was recognized by Hipple, Sommer, and Thomas, who constructed the first cyclotron mass spectrometer (termed the *omegatron*). In the omegatron, ions were induced by a low-voltage resonant AC frequency to gradually expand their orbital paths along a plane orthogonal to a

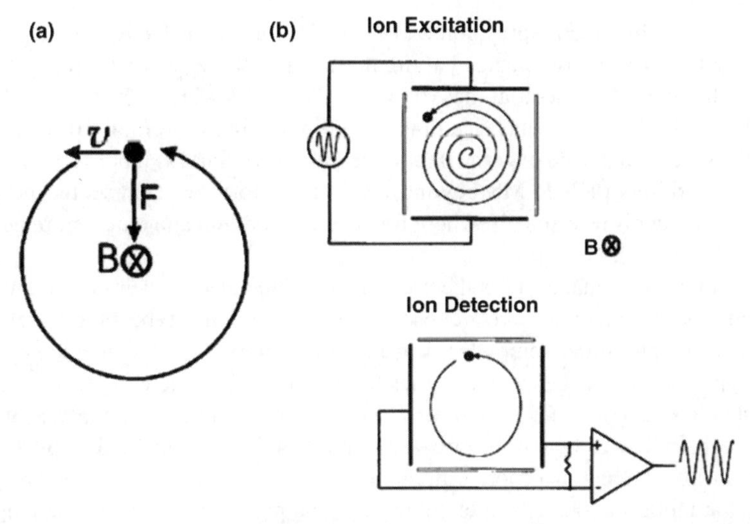

FIGURE 1.7 (a) If an ion in motion (indicated by arrow with velocity vector *v*) experiences a magnetic field in a dimension orthogonal to the ion motion (as indicated by the tickmark x), the ion will experience an inward force that causes circular motion in the plane transverse to the magnetic field lines; (b) depiction of ion excitation to the detection radius followed by transient image current detection (both diagrams reprinted from Ref. 69, with permission of John Wiley & Sons, Inc.).

magnetic field; this resulted in the radii of the excited ions' orbits expanding until impact occurred with an electrometer electrode [66]. The earliest efforts were focused on studying low-mass particles and their properties, with results including the baseline resolution of deuterium and diatomic hydrogen ions, which differ in mass by just less than 0.002 Da [66,67]. With the incorporation of *image current* detection in 1965 by Wobschall, ions could be detected nondestructively via the differential amplification of current flowing through a circuit connecting opposing electrodes as ion packets passed by them one at a time; hence, image currents allowed higher signal intensity and lower sample consumption [68]. Many early ICR-MS applications involved ion–molecule reaction studies, but the technique was hindered for broader applicability by the necessity of scanning the magnetic field over several tens of minutes to acquire a mass spectrum at unit resolution [69,70]. In 1974, Comisarow and Marshall published their conception of an ICR mass spectrometer that was capable of the simultaneous detection of multiple *m/z*, a development that allowed ICR mass spectrometers to acquire spectra in $\frac{1}{100}$th the time of conventional methods [71,72]. To perform the Fourier transform technique, all ions comprised within a broad *m/z* range were excited to the detection radius and simultaneously detected as a complex signal in the time domain, which could then be processed by Fourier transform (FT) for conversion to the frequency domain and then subsequent conversion into the *m/z* domain via application of a calibration function. Today, FT-ICR MS is used for many applications requiring the ultimate mass spectrometric performance in terms of resolution and mass accuracy; one good example is the detection of protein ions generated by electrospray ionization (ESI), for which FT-ICR is capable of resolving the isotopes of species having masses >100 kDa while allowing assignment of the accurate molecular mass to within 3 Da [73,74]. Routine figures of merit include single-digit ppm mass accuracies and resolving powers of 10^6–10^4 (for a typical 1 s mass spectrum over the *m*/*z* range up to 1000 Th) [73]. For further insight into the workings of ICR MS beyond the introductory level presented here, the interested reader is referred elsewhere for several excellent and much more detailed reviews [41,69,70,73–75].

Ion cyclotron resonance mass spectrometers are built around their magnet, with the geometry of the analytical cell chosen depending on the type of magnet used; permanent magnets and some electromagnets operate with cubic geometry so that the magnet poles might face each other to produce a homogenous field, while the highest field strengths are achieved with solenoidal superconducting magnets that are typically paired with coaxially oriented open-ended cylindrical cells [69,73]. Ion cyclotron instruments incorporate a balance of attractive and repulsive forces that causes ion motion to describe a circular path on a plane orthogonal to the magnetic field. Magnetically induced ion motion in this plane is due to a balance of centripetal (inward-drawing) and centrifugal (outward-drawing) forces, as can be seen in the following identity:

$$qvB = \frac{mv^2}{r} \tag{1.17}$$

Here q is the product of the elementary charge (e) and the number of elementary charges (z), B is the magnitude of the magnetic field, m is ion mass, v is ion velocity, and r is the orbital radius. By application of a low-amplitude RF excitation signal in one dimension of the orbital plane, ion trajectories can be made to spiral outward as the ions gain kinetic energy (see Figure 1.7b). When ions reach the detection orbit, the excitation signal is turned off and the ions travel along the detection orbital radius. By substituting angular frequency ($\omega = v/r$) into Eq. (1.17), the cyclotron frequency (ω_c) may be characterized as

$$\omega_c = \frac{q}{m}B \tag{1.18}$$

A remarkable quality of the cyclotron frequency is that it is independent of ion kinetic energy—where two ions of the same *m*/*z* to have different energies, the end result would be only that the one of higher energy (and hence orbital radius) would travel nearer to the detection electrodes and create a relatively larger detection signal. Hence, with no need for energy focusing, ICRs have an inherent advantage over other mass analyzers in their capability to produce spectra with high resolution.

In addition to ion cyclotron motion in the cyclotron plane (the x–y plane), ions within an ICR cell typically also experience two other types of field-induced motion. The first is motion in the axial (z) dimension, which is due to a low-voltage electrostatic trapping field that is maintained for the duration of an MS experiment. Hence, as ions precess in their cyclotron orbits, they will also oscillate in the axial dimension to a small extent. The other motion an ion experiences is that of *magnetron motion*, a consequence of the magnetic and electric fields acting on ions in orthogonal directions. Magnetron motion arises because the radial electrodes (one pair each of excitation and detection electrodes) are grounded while the z-dimension "trapping" electrodes are of equal potential; hence, the midpoint between the trapping plates have some nonzero potential and a net force is directed radially outward toward ground [69]. Consequently, magnetron motion causes a minor localized spiraling trajectory in ion motion as the ions travel on the cyclotron orbits, which are of much higher magnitude. Magnetron motion is typically of a much lower magnitude and frequency relative to axial motion, which in turn is of a much lower magnitude and frequency relative to cyclotron motion [41]. Since it is generally undesirable, magnetron motion is effectively controlled by minimizing the axial ion velocity.

A unique aspect of FT-MS is that it allows detection of all ions over a broad m/z range via image current, whereby the differential amplification of current in the circuit that links the two detection electrodes results in a time-domain signal representative of the cyclotron frequency. By detecting ions in a nondestructive manner, each ion may be detected many times, enabling enhanced signal-to-noise ratio (S/N) as well as reduced sample consumption. In order to allow detection of all ions pseudosimultaneously, they must all be at about the same radius within the ICR; such a condition is made possible through stored waveform inverse Fourier transform (SWIFT). Developed by Marshall and coworkers, SWIFT involves the calculation of a waveform in the frequency domain followed by its inverse Fourier transform into the time domain, with the result that all ions of interest will be excited to the same radial amplitude through application of the SWIFT waveform [76]. Previous methods of broadband excitation, such as chirping (a rapid scan through all frequencies), lead to undesirable dispersions in orbital radius, which makes accurate relative abundances difficult to determine. The resultant multifrequency image current signal is processed by Fourier transform from the time domain into the frequency domain, from which a conversion to m/z may be done on the basis of preexisting calibration data [71,72].

1.3.5 Orbitrap Mass Spectrometers

The Orbitrap mass analyzer is an "evolved" version of the Kingdon ion trap that contains ions in an electric field between a coaxial pair of electrodes. The Kingdon trap, originally conceived in 1923 by K. H. Kingdon [77], consists of an electrostatic trapping potential defined by the logarithmic field between a cylinder and a wire that it surrounds; ion confinement is achieved when the electrostatic attraction between ions and the central electrode (centripetal force) is countered by the ion motion in the direction tangential to, and hence away from, the center electrode (centrifugal force). The balanced relationship between centripetal and centrifugal forces in the Kingdon trap may be observed as [77]

$$qV = \frac{1}{2}mv^2 \cdot \left(\frac{R}{r}\right) \tag{1.19}$$

where q is the product of the quantity of elementary charges and the elementary charge ($q = z \cdot e$, for z is a unitless integer and e is 1.602×10^{-19} C), V is the electric potential, the ion kinetic energy $\frac{1}{2}mv^2$ consists of mass (m) and velocity (v) components, R is the radius of the outer electrode, and r is the radius of the inner electrode. As a consequence of the centripetal and centrifugal forces being balanced, an ion that is stably trapped will assume an elliptical orbit about the wire with its eccentricity (degree of elliptical vs. circular orbit character) depending on the field potential, the ion kinetic energy, and the ion's angle of approach. To contain ions in the axial dimension of a Kingdon trap, repelling potentials are applied to electrodes located at both ends of the cylinder.

Kingdon traps have been used for a variety of purposes, including the study of ions by optical spectroscopy and development of the orbitron ion pump [78,79]. The first application of a Kingdon trap as a mass analyzer occurred in 1981 when R. D. Knight created a modified-geometry Kingdon trap that had no endcap electrodes and an outer electrode that was tapered at both ends and bisected at its middle, a geometry that provided an approximately quadrupolar field in the axial dimension [80]. Knight used his trap to monitor the image current produced on the bisected outer electrodes by the near-harmonic oscillation of ions in the axial dimension, allowing him to perform mass analysis on plasma species generated by laser ablation of various metals [80]. However, field nonidealities in this simple cylinder-about-a-wire geometry caused an undesired interdependence of ion motion between the axial dimension and the radial plane, inhibiting truly harmonic oscillations (and hence spectral quality) and prompting the development of an "ideal" electrode geometry (i.e., the Orbitrap) to provide a purer axial quadrupole field; of anecdotal interest, this is the opposite trend of the developments in the RF ion trap community, which began with ideal geometries (of hyperbolic shape) and later branched out to simplified versions [e.g., the cylindrical ion trap (CIT) [81] or the rectilinear ion trap (RIT) [50]] when it was determined that performance for those devices was not severely worsened by implementing simpler fabrication procedures.

In 2000, Makarov created a "Knight-style Kingdon trap" (the Orbitrap), which featured an electrode geometry allowing a near-ideal quadrupole field along the axial dimension [82]. Rather than a cylinder-about-a-wire geometry faced with endcap plates (as for the Kingdon trap), the Orbitrap geometry consists of an outer "barrel-like" electrode, which is tapered near its ends, in addition to a "spindle-like" center electrode, which is also tapered at its ends yet broadened at its middle (see Figure 1.8). With its outer electrode actually split into two isolated electrodes at the equatorial axis ($z = 0$), the Orbitrap is able to perform differentially amplified image current detection using the outer electrode pair in lieu of the endcap plates of previous designs. The quadrologarithmic potential of the Orbitrap, $U(r,z)$, is the sum of a quadrupolar field with that of a cylindrical capacitor, and is defined as [82]

$$U(r,z) = \frac{k}{2}\left[\left(z^2 - \frac{r^2}{2}\right) + R_m^2 \cdot \ln\left(\frac{r}{R_m}\right)\right] + C \qquad (1.20)$$

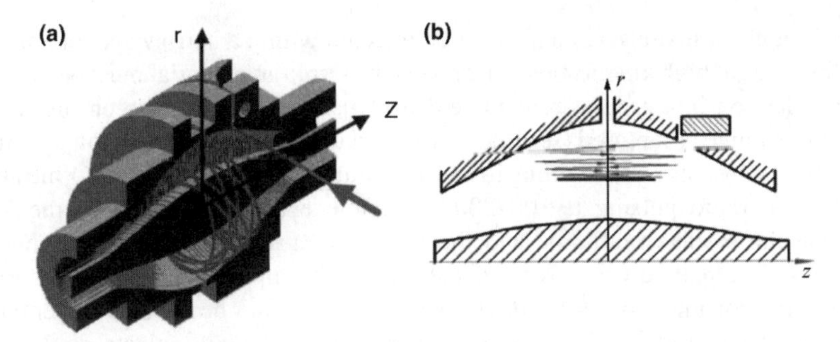

FIGURE 1.8 (a) Isometric cutaway view of the Orbitrap mass analyzer; (b) cutaway view of axial ion trajectory at injection and during electrodynamic squeezing (both diagrams reprinted from Ref. 78, with permission of John Wiley & Sons, Inc.).

where r and z are cylindrical coordinates, k is field curvature, R_m is the characteristic radius, and C is a constant. An important consequence of Orbitrap geometry is that trapped ion motion along the axial (z) dimension is independent of ion kinetic energy, the angle of the orbit with respect to the z axis, and spatial spread of the ion population in the radial (r) plane (as indicated by the absence of rz cross-terms in the potential). Detection on an Orbitrap analyzer occurs through differential amplification of the image current induced in the two outer electrode halves, which are bisected at the $z = 0$ plane. The shape of the Orbitrap electric field is such that ions moving axially are trapped in a quadrupolar potential well and hence their axial motion is sinusoidal in nature with a characteristic frequency dependent on m/z, as described in the following relationship:

$$\omega = \sqrt{k\frac{z}{m}}$$

Here ω is ion axial frequency and k is a constant of the axial restoring force. It is interesting to note that resolution ($R = m/\Delta m$) for an Orbitrap decays with an inverse-root relationship with m/z; hence, for all except the highest-field-strength FT-ICRs and for analyses of practical timescales (~1 s), the Orbitrap has superior performance at and above m/z of moderate value (i.e., 2000 Th) [79]. To measure mass-to-charge ratio (m/z) on the Orbitrap, the frequency of ion motion in the z dimension is measured as an image current by the split outer electrodes and then fast-Fourier-transformed into a frequency spectrum, from which a mass spectrum can be obtained after accounting for m/z calibration. Since the image current method of detection does not destroy ions, they are each detected many times during a given analysis, and hence less sample may be consumed per spectrum relative to "destructive" techniques.

Ions must be introduced into an Orbitrap mass analyzer with stable trajectories of narrow distribution along the z axis to ensure the highest spectral quality. To ensure that this occurs, several conditions must be met: (1) the Orbitrap must be operated at high vacuum (2×10^{-10} mbar) to avoid energy and trajectory-dispersive ion–molecule collisions, (2) ions must be injected within a narrow spatiotemporal window (i.e., at high kinetic energy), and (3) the Orbitrap potential must be ramped during ejection in a process termed "electrodynamic squeezing" to ensure that a wide mass range is captured (see Figure 1.8b). Narrow spatiotemporal ion injection is facilitated by collisional cooling in a storage multipole just outside the Orbitrap in addition to rapid pulsing (~100–200 ns; kinetic energies >1 keV) of the ions between the storage mutlipole and the Orbitrap entrance [78]. Electrodynamic squeezing is a method whereby the center electrode amplitude is increased during the period of ion injection (~20–100 μs) so that the ions may be injected tangentially to the outer electrode surface yet avoid discharge on the outer electrode through constriction of their orbital radii [82]. An additional requirement of injection is that the ions be introduced at a position offset from $z = 0$ such that they may be induced into harmonic oscillations via their introduction on one side of the quadratic potential well.

The most impressive attributes of the Orbitrap are its resolution and mass accuracy capabilities, which rival those of FT-ICR for many practical applications. Resolution on an FT mass spectrometer depends on the total acquisition time, but typically mass spectrometers are expected to provide about one spectrum per second for practical reasons (e.g., online chromatography) [75]. Orbitrap data that last for one second have been shown to yield a resolution of >100,000 for *m/z* 400 [83]. The Orbitrap is currently capable of acquiring data for up to 1.8 s before the transient signal fades away as a result of packet decoherence [79]. The limitation to extended data acquisition times on the Orbitrap is that the axial ion phase will become decoherent and hence meaningful signal will not be achieved. The mass accuracy on an Orbitrap is generally <5 ppm if externally calibrated or <2 ppm with internal calibration; optimized mass accuracies of 0.2 ppm have been demonstrated [79]. The dynamic range of mass accuracy has been shown to be about 1 : 5000 [83], meaning that an ion species present at a very low signal will be ascribed the same *m/z* as that same species were it to be present as a population of 5000 × greater abundance; this is a testament to the low susceptibility of the Orbitrap to space charge influences.

In commercial embodiments of Orbitrap mass spectrometers, the analyzer is located behind a two-dimensional quadrupole ion trap (2D QIT) [84]. The presence of the 2D QIT allows for a variety of advantages that cannot be achieved by the Orbitrap alone. Most notably, tandem MS (MS/MS) may be performed "upstream" in the 2D QIT prior to introduction of isolated species or product ions for high-resolution/accurate mass analysis in the Orbitrap. Fittingly, automated workflows have been developed to allow simultaneous analyses on both analyzers for the purpose of high-throughput analyses for which the LTQ-Orbitrap™ (where LTQ = linear trap quadrupole) is employed.

1.4 TANDEM MASS SPECTROMETRY

Tandem MS is the practice of using multiple stages of ion manipulations, including *m/z* isolation and fragmentation. Tandem MS is also referred to as "MS/MS" for two stages of mass analysis or "MS^n," where n is the number of stages of mass analysis. The benefits to MS^n include the following: (1) increased signal-to-noise by reduction in chemical background peaks, (2) the capacity for purifying a mixed ion population to provide an isolated reagent species for reactions, and (3) enhanced chemical specificity in terms of structural analysis when employing strategies of fragmentation.

The isolation and fragmentation of analyte ions for the purpose of structural analysis is the most common application of MS^n. A mass spectrometrist can use an ion activation technique to provide detailed structural characterization of ions by allowing the attribution of both molecular and fragment peaks to a particular chemical species. For example, single-stage mass analysis of the protonated form of caffeine [molar mass = 194.2 Da] will provide only the m/z of the intact precursor. Since unknown chemical species are seldom identified with such limited information, it can be extremely difficult for the analyst to determine the most probable structure without knowledge of the fragment masses and an interpretation of the intramolecular connectivity. This is particularly true for any chemical system with more than a few atoms when measured on an instrument that cannot accurately determine the exact mass (from which the elemental composition can be derived) to good precision (such is typically the case for ion traps). However, if the analyst were to first isolate the unknown species and then induce fragmentation, clues about the ion structure will be provided according to the m/z and relative abundances of the product such that relatively low-resolution instruments can suffice for accurate structural identification.

There are a variety of means by which ion activation may be achieved, including interactions of the analyte ion with neutral molecules, surfaces, electrons, photons, electric fields, or other ions. The most commonly applied method for activation is *collision-induced dissociation* (CID), in which ion–molecule collisions result in characteristic fragmentation of the precursor ion. The technique known as *surface-induced dissociation* (SID) [50,85] involves the collision of precursors with a surface such that the resultant product ions are similar to those arrived at by CID. Electron-induced dissociation techniques, such as electron capture dissociation (ECD) [37] and electron transfer dissociation (ETD) [38,86], involve interactions between ions and free electrons or between ions of opposite polarity, respectively, which leads to radical-driven fragmentation which yields product ion information that is often complementary to CID. Depending on the specific technique applied, photoionization techniques can involve photoabsorption to provide thermodynamically favored product ions [e.g., infrared multiphoton dissociation (IRMPD) [87]] or even field ionization by high-power ($>10^{14}$ W/cm^2) femtosecond lasers [88]. For brevity and relevance to the audience of this book, the following discussion of ion activation is focused on three techniques that are commercially available: CID, ECD, and ETD.

1.4.1 Ion Isolation

Prior to target ion fragmentation, it is common to perform an isolation step so that no interfering species remain to hamper spectral interpretation. Most mass spectrometers of the trapping type (QITs and ICRs) are capable of performing *tandem-in-time* MS^n as sequential operations in a single analyzer. Quadrupole ion trap and FT-ICR isolation is achieved by application of notched broadband AC waveforms, which permits the ejection of all ions except those that are desired to maintain stable trajectories [73]. The notable exception for on trap MS/MS capability in is the Orbitrap, which cannot practically implement MS/MS for structural analysis without destabilizing precisely controlled ion trajectories; however, a workaround to this problem has been developed by coupling the Orbitrap with an upfront ion trap to perform preselection and any necessary fragmentation [84]. The remaining analyzer types (magnetic sectors, quadrupole filters, and TOFs) require *tandem-in-space* configurations, where two or more analyzers are used; the typical setup involves a first analyzer performing the target ion selection and a second analyzer performing the analysis, with an intermediate "collision cell" in place to effect the activation. Magnetic sectors can perform target ion isolation by adjustment of field strengths such that only the m/z of interest can be passed through a narrow slit and onward to a collision cell or/and the next stage of mass analysis. Isolation in quadrupole filters is achieved by application of constant a (DC) and u (RF) values to the rods such that only a particular range of m/z maintains stable trajectories. TOF-MS/MS can be accomplished by placing a collision cell in the ion flight path and gating the cell open or shut depending on which ions are intended to pass into it and fragment; TOF mass spectrometers are also often arranged as Q-TOF platforms so that precursor isolation is effected within the quadrupole.

1.4.2 Ion-Molecule Collisions and Collision-Induced Dissociation

As MS is never performed in a perfect vacuum (except for some computer simulations), the presence of background gases will affect the instrument performance and in certain cases can assist in particular processes (i.e., collisional activation, thermal cooling, or directional velocity damping). Collisional processes between an ion and neutral commonly occur in all except the highest-vacuum instruments, with the frequency of collisions being determined by the *mean free path*, (λ) [41]

$$\lambda = \frac{kT}{\sqrt{2}P\sigma} \tag{1.21}$$

where λ is in units of centimeters, k is the Boltzmann constant ($k = 1.381 \times 10^{-21}$ J/K), T is temperature (in kelvins), P is pressure (in pascals; 1 Torr = 133.3 Pa), and σ is the collision cross section (in m^2) with an area represented by $\sigma = \pi(r_i + r_n)^2$ where r_i and r_n are the molecular radii of the participating ion and neutral, respectively. The collision between an ion and a neutral in free space has the center-of-mass (COM) energy E_{COM}

$$E_{COM} = \frac{1}{2}\frac{m_i m_n}{m_i + m_n} v_R^2 \tag{1.22}$$

where m_i is the ion mass, m_n is the neutral mass, and v is the relative velocity of the two particles. When accelerated through an electric or magnetic field, $\mathrm{KE_{ion}}$ often becomes considerably greater than $\mathrm{KE_{neutral}}$, which is probably at thermal energy ($\frac{3}{2}kT \approx 0.04$ eV for $T = 293.15$ K). In such an instance, the energy brought to the collision is practically wholly brought by the ion while that brought by the neutral can be ignored as follows

$$E_\mathrm{L} = \tfrac{1}{2} m_\mathrm{i} v_R^2 \tag{1.23}$$

where E_L, the "laboratory energy" involved in the experiment, is defined as the kinetic energy of the incident ion. Substitution of the v_R^2 term from Eq. (1.23) in Eq. (1.22) yields the following formula:

$$E_\mathrm{COM} = E_\mathrm{L} \frac{m_\mathrm{n}}{m_\mathrm{i} + m_\mathrm{n}} \tag{1.24}$$

Hence, the maximum energy available for uptake by the ion is essentially proportional to the ion kinetic energy times the quotient of the neutral mass over the summed mass of the neutral and ion. For example, Eq. (1.24) can be used to compare the energy available for uptake by an ion in the case that it undergoes collisions with He [nominal mass = 4 Da (Where u is the universal mass unit, used to express both atomic and molecular masses)] versus Ar (nominal mass = 40 Da). For an ion of *m/z* 100 with, say, 20 eV KE, a collision with helium will result in as much as $(20 \times 4/104) = 0.77$ eV being taken up as ion internal energy $\mathrm{IE_{ion}}$. Compare this with the collision of the ion with argon: $20 \times 40/140 \simeq 5.71$ eV. With a greater relative energy uptake per collision, fewer collisions are necessary to achieve fragmentation when argon is the neutral, which exemplifies why higher-mass neutrals are often used in reaction cells in which an ion has a very limited probability of participating in multiple collisions. Two practical consequences of the choice of buffer gas can be considered in terms of analytical resolution and transfer efficiency. Consider the case of a quadrupole where the RF ramp of mass-selective instability mode is occurring, and ions are being sequentially ejected according to their *m/z*. Suppose that a population of ions of the same *m/z* are experiencing resonant excitation; they are perhaps 10 cycles from be ejected, enough time for a collision to occur with a background neutral gas atom or molecule. Naturally, the ions will eject over a distribution of times owing to their differences in position, trajectory, and velocity. However, if some ions also undergo a collision with a He atom shortly before ejection, they may perhaps lose enough KE (by conversion to $\mathrm{IE_{ion}}$) to eject a cycle or two later than expected, causing a bit broader distribution of peak width. However, if the analyte ions instead collide with N_2, they could assume a still broader distribution of position/trajectory/velocity and subsequently eject over an even longer time period (poorer resolution than the He case). On beam-type instruments, a collision gas must be chosen that is of sufficient mass to promote efficient fragmentation on the timescale of ion passage through the collision cell, yet of a mass not so high as to cause excessive beam dispersion and loss of transfer efficiency (this is, of course, dependent on the incident ion kinetic energy).

1.4.3 Electron Capture Dissociation and Electron Transfer Dissociation

Electron capture dissociation (ECD) was developed in the McLafferty group in 1998 following an accidental discovery that low-energy electrons ($< 10\,\text{eV}$) captured by multiply charged peptide ions trapped in an ICR cell can induce unique fragmentation of the peptide ions [89,90]. A general chemical equation for the ECD process can be written as follows:

$$[\text{M}+n\text{H}]^{n+} + \text{e}^- \rightarrow ([\text{M}+n\text{H}]^{(n-1)\cdot +})_{\text{transient}} \rightarrow \text{fragments}$$

Since charge reduction of the precursor ions is involved, multiply charged (very often protonated) analyte cations formed by electrospray ionization (ESI) are employed in order to detect fragment ions in the ECD process. Low-energy electrons are produced with a filament-based electron gun and injected into the ICR cell at relative low energy for reaction with analyte cations [91]. Distinct from the collisional activation process, in which part of the kinetic energy is transfered to the internal energy for activation, the recombination energy between the electron and the cation causes the subsequent excitation and fragmentation. Since an odd-electron species is formed with the electron capture, ion fragmentation in ECD is majorly driven by radical chemistry [92]. Taking the multiply protonated peptide ions as an example, the normally strong N–C_α backbone bonds are cleaved homolytically in ECD, producing a complementary pair of *c* and *z* fragments as indicated in Figure 1.9a, while *b*- and *y*-type fragment ions are typically observed in CID as a result of the heterolytic cleavage of the amide bonds as shown in Figure 1.9b [93]. Note that the peptide fragmentation nomenclature follows the definition proposed by Roepstorff and Fohlman [94].

There are several unique features of ECD for the analysis of biomolecules, especially for peptides and proteins. As compared to CID, sequence analysis by ECD tends to have a smaller dependence on protein sequence and gives good sequence coverage by inducing widespread cleavages along the backbone with the exception of cleavages *N*-terminal to proline residues (owing to the residue's cyclical structure). Such nonspecific fragmentation is especially useful for identification of unknown peptides and proteins, where extensive fragmentation is required to facilitate identification of the molecular structure. Since disulfide linkages can be cleaved on electron capture, ECD has also been proved useful in characterizing peptides and proteins that contain disulfide bonds [95]. Another very attractive feature of ECD is that labile post-translational modifications (PTMs) are often preserved, such as glycosylation, phosphorylation, and sulfonation, whereas in CID they are preferentially cleaved. This feature allows not only sequence identification but also pinpointing of the location of the modification [96].

Because of the requirement of storing cations and electrons in overlapping space simultaneously, implementation of ECD is generally limited to Fourier transform ion cyclotron resonance mass spectrometers, which have the capacity to stably trap electrons. In 2004, Hunt's group demonstrated that ion/ion electron transfer gave rise

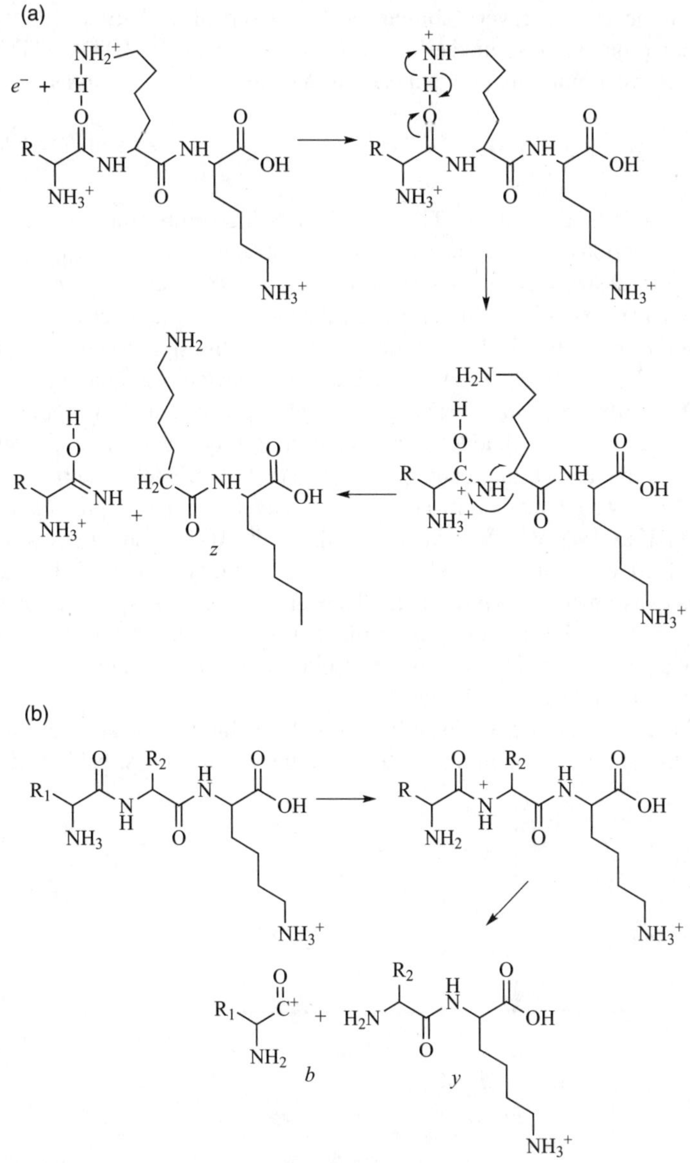

(a)

NH_2^+
e^- + H
O
R NH NH O OH
NH_3^+ O
NH_3^+

+
NH
H
O
R NH NH O OH
NH_3^+ O
NH_3^+

NH_2
H
O
R NH + H_2C NH O OH
NH_3^+ O
z

H_2N
H
O
R C + NH NH O OH
NH_3^+ O
NH_3^+

(b)

O R_2 H O
R_1 N N OH
NH_3 H O
NH_3^+

O R_2 H O
R + N N OH
NH_2 H O
NH_3^+

O
R_1 C^+ + H_2N R_2 H N O OH
NH_2 O
b *y*
NH_3^+

FIGURE 1.9 (a) Fragmentation scheme for production of *c*- and *z*-type ions after reaction of a low-energy electron with a multiply protonated peptide; (b) fragmentation scheme for the production of *b*- and *y*-type ions by CID of a multiply protonated peptide (both schemes reprinted from Ref. 93, with permission of the National Academy of Sciences USA).

to dissociation behavior very similar to that observed with ECD [93,97]. This dissociation process was termed *electron transfer dissociation* (ETD). ETD involves reactions between multiply charged cations and singly charged anions:

$$[M+nH]^{n+} + Y^- \rightarrow ([M+nH]^{(n-1).+})_{\text{transient}} + Y \rightarrow \text{fragments} + Y$$

The reagent anions selected for ETD typically have high probabilities for transferring electrons to cations; what's more, such reagent anions should also have a low probability of abstracting protons from the cations [98]. Currently, the most commonly used reagents include fluoranthene and azobenzene, the negative ions of which can be produced by chemical ionization (CI) [93] or atmospheric-pressure chemical ionization (APCI) [99]. ETD can be conducted on conventional quadruple ion traps, which are readily accessible; because of a higher pressure in QITs, good reaction efficiencies are typically obtained in relatively short times (tens of ms), allowing the coupling of the activation technique with online liquid chromatography [93]. Figure 1.10 shows the tandem mass spectra obtained from a phosphopeptide eluted during a *n*HPLC/MS/MS experiment [100]. In the ETD tandem mass spectrum (Figure 1.10b), every single backbone cleavage product is observed. The sequence can be easily assigned as RKpSILHTIR. The CID tandem mass spectrum of the same peptide (Figure 1.10a), however, is dominated by a single peak corresponding to the loss of a phosphoric acid moiety. No peptide backbone cleavage is observed, and the sequence identification is impossible.

Both ECD and ETD are still in early stages of development by any standards. This is exemplified by the ongoing debates on the mechanisms behind ECD and

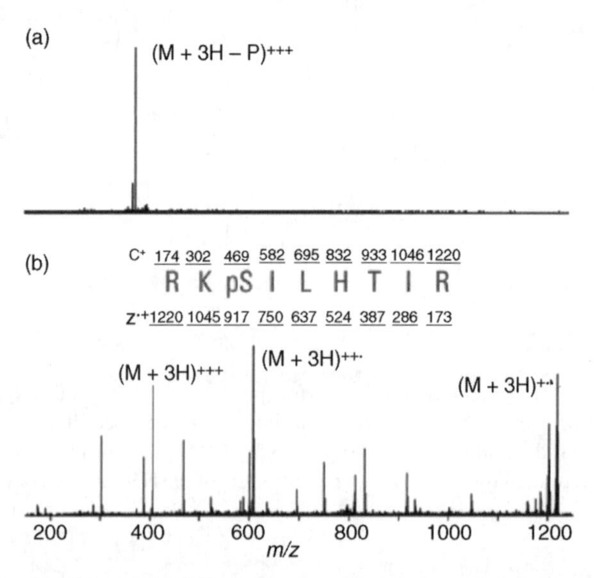

FIGURE 1.10 (a) CID and (b) ETD tandem mass spectra recorded for the phosphopeptide RKpSILHTIR (Reprinted from Ref. 100, with permission of the American Chemical Society).

ETD [101,102]. Nevertheless, the unique features of ECD and ETD guarantee them as a new category of tandem mass spectrometric technique with potentials in a variety of applications. With the launch of several commercial tandem mass spectrometers that are capable of ECD or ETD, a much wider impact of these techniques on the analysis of complex mixtures can be foreseen.

1.5 SEPARATION TECHNIQUES COUPLED TO MASS SPECTROMETRY

1.5.1 Gas Chromatography–Mass Spectrometry

During the late 1950s and early 1960s, significant advances in gas chromatography (GC) were made, allowing the technique to become one of the most prominently used in modern analytical chemistry. Early work by Martin and Synge used two liquid phases (liquid–liquid partition chromatography) to separate acetyl derivatives of amino acids [103]. Although theoretical advantages of replacing one of the liquid phases with a gaseous phase were noted, the practical application of gas–liquid partition chromatography was not reported until 1952 [104]. The same group also reported gas–liquid partition chromatography for the separation of ammonia and methylamines, as well as for the separation of volatile aliphatic amines [105,106]. Perhaps one of the greatest advances in GC instrumentation occurred in the late 1950s when Golay developed open tubular (capillary) columns [107]. Although not realized at the time, the development of these columns would make coupling of GC with MS considerably simpler in later years.

Although GC instrumentation was still in its infancy in the late 1950s, the first GC-MS instruments were actually reported around this time [25,108,109]. The "hyphenation" of the two techniques was an important advance in exploiting the benefits and minimizing the limitations of each instrument. While gas chromatography is highly efficient for the separation of complex mixtures, definitive identification of sample components is not possible with only the retention time from the chromatogram. In MS, definitive identification is possible, based on the fragmentation pattern of the molecule. Thus, MS offers highly sensitive detection, as well as structural information for each sample component separated by GC.

In one of the first GC-MS instruments reported, a conventional gas–liquid partition gas chromatograph was coupled with a time-of-flight (TOF) mass spectrometer [108]. The gas chromatograph contained four columns, the ends of which fed into a block that also contained two exit lines: one feeding into the mass spectrometer and the second feeding into a thermal conductivity detector that was used to continuously monitor the effluent. The mass spectrometer was slightly modified, replacing the oil diffusion pump with a mercury diffusion pump and the water-cooled baffle with a Freon −22 refrigerated baffle, in order to reduce the instrument background levels. Mass spectra were visualized in real time on an oscilloscope. With this early instrument, spectra were collected at the rate of 2000 scans/s, scanning the range *m/z* 1–6000.

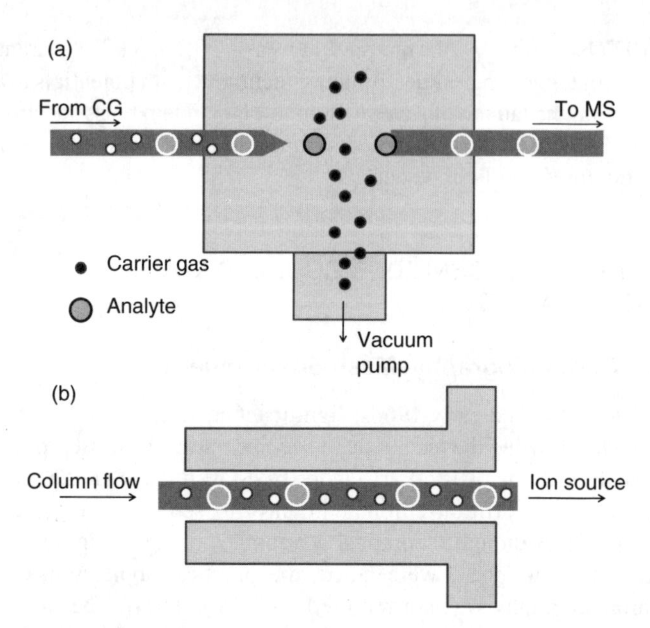

FIGURE 1.11 Coupling of GC to mass spectrometers via (a) a jet separator or (b) direct introduction (reprinted from Ref. 112, with permission of John Wiley & Sons, Inc.).

Initial problems in coupling the two techniques arose, in part, from the use of packed columns in GC. In such columns, flow rates are in the order of 10–20 mL/min, which are not compatible with the MS vacuum system [110]. Hence, ways to reduce the flow rate from the GC column before entering the mass spectrometer were needed. A number of different interfaces were developed to alleviate this problem, but only the more commonly employed ones are discussed here.

In molecular (or jet) separators (Figure 1.11a), introduced by Ryhage in 1964, the effluent from the GC feeds into an evacuated chamber through a restricted capillary [111]. An expanding, supersonic jet, composed of carrier gas and sample molecules, is formed at the tip. Lower-molecular-weight compounds diverge from the jet and are lost, mainly as a result of collisions. The higher-molecular-weight compounds form the core of the jet that is subsequently sampled into the mass spectrometer [112]. Not only is flow rate reduced, but sample components are enriched prior to entering the mass spectrometer. However, transport efficiency is low (~ 40%), and the separator discriminates against lower-molecular-weight sample components [112].

An alternative to molecular separators is the open-split interface that splits the flow, thereby reducing the total gas flow entering the mass spectrometer. The interface uses a T-connector that contains a restrictor tube [113,114]. The GC column feeds into one end of the restrictor, and a second length of fused silica feeds from the restrictor into the mass spectrometer. The length and diameter of this second piece of fused silica is chosen such that a compatible flow rate is delivered to the mass spectrometer. A flow of helium into the T-connector removes excess carrier gas flow, and the connector is heated to prevent condensation of the separated components [113]. Although no sample enrichment is possible, this interface allows switching of columns without the need to vent the mass spectrometer. Additionally, the interface is compatible with a wide range of GC flow rates. However, transport efficiency into the mass spectrometer was low (20–50%), which can have a detrimental effect on the sensitivity achievable [113,115].

In modern GC-MS instruments, the need for molecular separators or flow splitting is eliminated with the use of wall-coated open tubular (capillary) columns. Flow rates for these columns are generally less than 2 mL/min, which is more compatible with the vacuum system of the mass spectrometer [113]. As a result, the GC column can be fed directly into the ion source, via a heated transfer line, with no restrictions necessary (Figure 1.11b). As with the open-split interface, there is no sample enrichment, but this is not necessary as there is 100% transport efficiency from the column into the ion source. Despite the simplicity of this approach, there are some disadvantages. Since all the effluent directly enters the ion source, there is an increased risk of contamination and, since the end of the column is under high vacuum, the mass spectrometer must be vented in order to change the column [110]. Nonetheless, the vast majority of GC-MS applications today use this direct coupling interface.

Considering commercial GC-MS instruments, numerous ionization and mass analyzer combinations are available. Most commonly, EI, positive CI, and negative CI are offered, with one manufacturer offering a pulsed positive ion–pulsed negative ion (PPIPNI) CI source. The single quadrupole, triple quadrupole, and ion trap (linear or quadrupole) are among the more common mass analyzers used in benchtop GC-MS instruments. However, hyphenated instruments using time-of-flight mass analyzers and magnetic sector analyzers are readily available commercially.

1.5.2 Liquid Chromatography–Mass Spectrometry

Liquid chromatography (LC) is often used to analyze samples that are not amenable to GC analysis, namely, nonvolatile or thermally labile compounds. As it is beyond the scope of this section to discuss LC theory and principles, interested readers are directed to many of the excellent texts available [113,116]. In the early 1970s, research efforts began to focus on coupling LC with MS. As stated previously, MS offers definitive identification of samples, and is a highly sensitive detector. Additionally, an LC-MS system allows separation and detection of samples that are not readily analyzed by GC.

However, interfacing LC with MS proved more difficult than for GC, for a number of reasons. Flow rates in LC are in the range 0.1–10 mL/min, although more typically 1–2 mL/min for conventional LC [112,113]. Considering methanol as an example, a liquid flow rate of 1 mL/min corresponds to a gas flow rate of 593 mL/min at atmospheric pressure [112], which is too high to be introduced directly into the mass spectrometer. Furthermore, the mobile-phase composition is seldom compatible with MS because of the presence of various nonvolatile additives. Since the late 1970s, a number of different interfaces have been developed and commercialized [117]. These interfaces are based mainly on (1) removing the liquid mobile phase from the effluent, (2) splitting the effluent flow, or (3) ionizing the effluent at atmospheric pressure, prior to entering the mass spectrometer. This section is intended as only a very brief overview of the more common interfaces that were developed and those that are in current use; interfaces and their development are discussed in greater detail in the literature, and readers are directed to these resources for further details [112,116].

The moving-belt interface developed by McFadden et al. was the first interface to be commercialized in 1977 [117,118]. The interface was based on an earlier design by Scott et al. and involved removing the mobile phase prior to entering the mass spectrometer [119]. The column effluent was deposited on a ribbon composed of stainless steel or polyimide that was continuously moving. The ribbon passed under heaters and through vacuum locks to evaporate the mobile-phase solvent. The belt entered the mass spectrometer and the remaining sample residue was vaporized via a flash vaporizer directly attached to the ion source. Because of the speed and efficiency of vaporization, the sample entered the ion source with little decomposition and was subsequently ionized by EI or CI. On leaving the ion source, the belt passed over another heater to clean off any remaining residues and minimize carryover into the next sample. On the basis of this design, sample components had to be volatile (to some extent) to be desorbed and vaporized from the belt. Although the moving-belt interface could accommodate flow rates of 1–2 mL/min [113], obtaining reproducible spectra was dependent on depositing a uniform layer of sample on the belt. This was more difficult to achieve when water was present in the mobile phase, which gave rise to droplets, rather than a film, on the belt. In addition, the mechanical nature of the moving belt rendered it more prone to breakage and therefore, less robust [117].

Early work by Tal'roze et al. used a capillary to introduce the column effluent directly into the ion source at flow rates less than 1 μL/min [120]. Volatile analytes were subsequently ionized via EI. Baldwin and McLafferty reported the direct introduction of liquids into a CI source via a 2-mm capillary drawn into a fine tip at the end [121]. The capillary was introduced into the ion source using a normal sample probe, and the solvent was used as the reagent gas for CI. Later, a narrow capillary [75 μm i.d. (inner diameter)] was used to restrict effluent flow into the ion source [122]. However, problems occurred when the liquid evaporated in the capillary as a result of the high vacuum in the ion source. Attempts to overcome this problem by restricting the capillary had limited effectiveness as blocking of the capillary then became problematic. In 1980, Melera replaced restricted capillaries with a diaphragm that allowed the formation of a stable liquid jet [123]. Pinholes in the diaphragm restricted the effluent flow entering a desolvation chamber, where the effluent was then nebulized. As before, the mobile-phase liquid was used as the reagent gas for chemical ionization of the analytes. This design, known as the *direct liquid introduction* (DLI) *interface*, was commercialized in 1980, soon after commercialization of the moving-belt interface [117]. Niesson published a two-part review of the DLI interface, with the first part focusing on instrumental aspects and the second part on MS and applications [124,125]. The DLI interface could accommodate flow rates in the range 50–100 μL/min [112] meaning that, in order to maintain sensitivity, the interface was best used with at least microbore columns, which were not conventionally used in routine applications. In addition, pinholes in the diaphragm were still prone to clogging [112]. With further advances in different interface designs, the DLI interface is now considered obsolete [112].

Willoughby et al. described the monodisperse aerosol generation interface for chromatography (known as "MAGIC") in 1984, which became the basis for the particle beam (PB) interface [116,126]. Aspects of interfacing LC with MS using a PB interface have been reviewed by Cappiello [127]. Similar to the moving-belt interface, the PB interface also removes the mobile phase solvent prior to ionization. The effluent is nebulized into a desolvation chamber, the outer walls of which are heated to 50–70°C [116]. The resulting droplets are dispersed, by means of a perpendicular flow of helium, and desolvated. The less volatile analyte components coagulate to form small particles, with diameters ranging from 50 to 300 nm [116]. The mixture passes through a narrow nozzle into a two-stage momentum separator. In the first stage, the pressure is typically 10 kPa. In the resulting expanding jet, the lower-mass components (mobile phase and nebulizing gas) diverge and are pumped away. The heavier mass analyte particles are sampled through a skimmer into the second pumped region (typically 30 kPa). Again, diverging particles are pumped away, leaving a beam of particles enriched in analyte that is introduced into the ion source through a transfer line. Transferred particles strike the walls of the ion source, which is heated to approximately 250°C [116]; hence, flash vaporization or disintegration of particles occurs and subsequent ionization is achieved using EI or CI. One of the main advantages of this interface is the ability to generate EI spectra from samples analyzed by LC, enabling comparison with library databases. However, the performance of the interface varies with mobile-phase composition. For example, the presence of water in the mobile phase detrimentally affects sensitivity since the high surface tension and boiling point of water prevents formation of completely solvent-free particles [127]. As such, the PB interface is now also considered obsolete [116].

Until this point in the development of LC-MS, ionization was mainly achieved by EI or CI, meaning that the analysis of nonvolatile or thermally labile sample components was limited. Although fast-atom bombardment (FAB) ionization had been reported with a moving-belt interface [112], it wasn't until the development of flow FAB interfaces in the mid 1980s [128–130] that the analysis of nonvolatile and thermally labile samples was more readily realized. In FAB, the sample is mixed with a nonvolatile matrix (typically glycerol), applied to the end of a probe, and then bombarded with a beam of fast atoms [114]. Surface atoms are ionized, and the ions are then focused and accelerated toward the mass analyzer [114]. Thus, in developing the FAB interface, the additional concern of adding the matrix to the sample had to be addressed. Technological advances in the development of the flow–FAB interfaces, as well as applications, have been reviewed [131].

Ito et al. developed the frit FAB interface in 1985 [130] and Caprioli et al. reported the continuous-flow (Cf) FAB interface in 1986 [128]. In each case, the FAB probe is modified to accommodate a capillary through which the effluent, mixed with the FAB matrix, flows. The matrix can be mixed with the effluent pre- or postcolumn. In frit FAB, a porous stainless-steel frit is positioned at the end of the capillary, while in Cf

FAB, the capillary is directed toward a FAB target. The mobile phase evaporates at the frit or target, leaving a thin film of sample and matrix, which is then subjected to FAB [116]. In 1988, de Wit et al. introduced the coaxial CfFAB interface to enable coupling of packed microcapillary or open tubular (OT) LC columns with MS [129]. In their design, two separate capillary columns were used to deliver the column effluent and FAB matrix to the probe tip, where mixing occurred [129]. As a result, independent control of flow rates was possible [132]. Irrespective of design, typical flow rates for flow FAB interfaces are in the 5–15 μL/min range. Thus, microbore columns must be used or, if conventional packed columns are used, the effluent flow must be split, which affects sensitivity [116].

The development and commercialization of atmospheric-pressure ionization (API) sources in the mid- to late 1980s widened the range of sample types that could be analyzed by LC-MS [117]. In fact, many of the previously discussed interfaces are obsolete, with commercial LC-MS systems using API sources as interfaces. In these sources, the effluent is ionized at atmospheric pressure and the resulting ions are continuously sampled into the mass spectrometer for separation and detection. Today, two API sources dominate in commercially available LC-MS instruments: electrospray ionization (ESI) and atmospheric-pressure chemical ionization (APCI). These sources have been discussed elsewhere in this chapter and thus are not described again here. Examples of interfaces for ESI- and APCI-LC-MS are given in Figures 1.12 and 1.13, respectively.

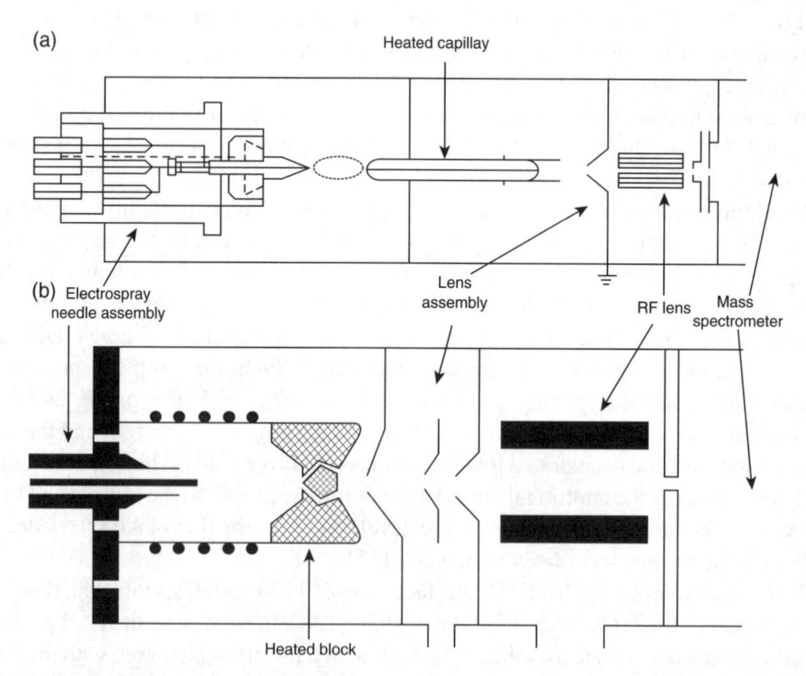

FIGURE 1.12 Interfaces for electrospray LC-MS: (a) heated capillary and (b) heated block designs (reprinted from Ref. 117, with permission of John Wiley & Sons, Inc.).

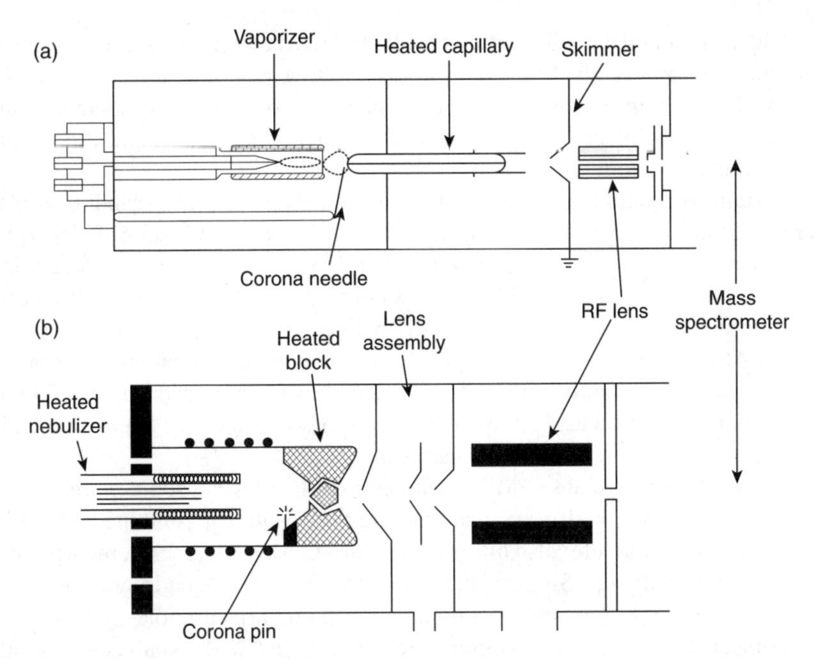

FIGURE 1.13 Interfaces for atmospheric-pressure chemical ionization LC-MS: (a) Thermofinnigan, Hemel Hempstead, UK; (b) Micromass UK Ltd., Manchester, UK (reprinted from Ref. 117, with permission of Wiley & Sons, Inc.).

Research in the late 1960s and early 1970s by Dole and coworkers [133–135] introduced the concept of electrospray, in which molecular beams of macromolecules were generated in vacuo. In 1984, Yamashita and Fenn applied this concept to create molecular beams of analytes sufficiently small to be mass analyzed using a conventional quadrupole analyzer [136]. Later work by Fenn's group described the use of the electrospray source as an interface in LC-MS [137]. It was with these developments in ESI that attention returned to investigating the use of APCI as an interface in LC-MS, even though the concept of APCI was first reported in the literature in the early 1970s [138].

Atmospheric-pressure chemical ionization is used for the ionization of low- to medium-polarity compounds while higher-molecular-weight, polar compounds are ionized by ESI. In terms of ionization, APCI typically generates singly charged ions of the molecular species, with little fragmentation. Electrospray ionization is also considered a mild ionization process but, in contrast to APCI, large molecules with numerous ionizable sites will form multiply charged ions. This not only increase sensitivity but also enables the analysis of compounds with molecular weights beyond the working mass range of the spectrometer [114].

For APCI, flow rates range from 100–200 μL/min to 1–2 mL/min, while for ESI, flow rates range from 1 μL/min to 1 mL/min [139]. Further modifications have resulted in ESI sources that operate at significantly lower flow rates. Emmett and Caprioli described a micro-ESI source that accommodated flow rates in the range

300–800 nL/min [140], while Wilm and Mann described the nano-ESI, which used flow rates as low as 20 nL/min [141]. Lower flow rates are not detrimental to sensitivity, which, in ESI, is dependent on concentration rather than sample volume introduced. In fact, lower flow rates can be advantageous in certain applications owing to less sample consumption.

In certain applications, such as the identification of potential drug candidates in the pharmaceutical industry, it is often desirable to use a combination of ionization techniques for full analyte characterization. This impetus was the driving force behind the development of combined ESI-APCI sources. Gallagher et al. reported a combined source for high-throughput LC-MS analyses [142]. Within a single analysis, ESI and APCI scans were alternately collected with polarity switching. Nowadays, several instrument manufacturers offer dual-source LC-MS instruments, allowing sample characterization using two different ionization techniques while eliminating the need to change and re-optimize hardware.

In terms of commercial LC-MS instruments, ESI and APCI sources are the most commonly available, and many manufacturers also offer a combined ESI/APCI source. Each manufacturer also offers a range of LC-MS instruments, incorporating different mass analyzers. Single-quadrupole, ion trap, triple-quadrupole, and time-of-flight (TOF) analyzers are most common, although quadrupole-TOF (Q-TOF) and Fourier transform ion cyclotron resonance (FT-ICR) analyzers are also available.

1.5.3 Capillary Electrophoresis–Mass Spectrometry

Originating with the initial work of Jorgenson and Lukacs, capillary electrophoresis (CE) has evolved to become a sensitive and efficient separation technique [143,144]. Samples are carried through a narrowbore fused-silica capillary (10–100 cm long, 25–100 μm i.d.) in a flow of buffer, and are separated under the influence of an applied electric field [113]. Separated components can be detected inline using optical detectors (e.g., UV absorbance or fluorescence) to generate an electropherogram, which is a plot of detector response versus migration time. In CE, separation of charged and uncharged species in a single analysis is possible, and today, the term *capillary electrophoresis* is applied to a range of different separation modes, all based on the same principles, but chosen according to sample type. For example, in capillary zone electrophoresis (CZE), initially described by Jorgenson and Lukacs [143,144], separation is based on differences in hydrodynamic radius/charge ratios [145]. In capillary gel electrophoresis, a molecular sieve is used to separate samples according to size, while in capillary isoelectric focusing, amphoteric samples are separated based on differences in isoelectric points [113]. Several texts and articles are available that discuss the history, development, and applications of capillary electrophoresis [113,146–150].

Because of the relatively short, narrowbore capillary columns used, flow rates in CE are in the nanoliter range, meaning that coupling to MS is more amenable than is coupling LC. Additionally, due to the low flow rates, highly sensitive detectors are desirable. This led research efforts to focus on coupling CE with MS with

Olivares et al. first reporting online MS detection for CZE in 1987 [151]. Similar to chromatographic techniques, using MS as a detector offers definitive identification of sample components, as well as the desired sensitivity. More recently, a number of review articles have been published that discuss the development of interfaces that have enabled coupling of CE with MS [145,152,153]. Hence, this section focuses only on the common interface that uses an electrospray ionization source to couple the two instruments.

Using an ESI source, interface configurations can be subdivided into two classes: sheath flow interfaces and sheathless interfaces [152]. In the former, the voltage necessary for ESI is applied to the CE buffer indirectly by means of a sheath liquid while in the latter, the voltage is applied directly to the buffer. With sheath flow interfaces, the buffer composition can be altered to be more compatible with subsequent analysis and detection techniques (i.e., ESI and MS); however, with the addition of the sheath liquid, the buffer and hence, sample components, are diluted, which can have detrimental effects on sensitivity. The sheathless interface configuration can, therefore, offer greater sensitivity since there is no dilution effect, but it is more limited in the choice of compatible CE buffers.

Considering the sheath flow interface (Figure 1.14), two configurations are common: the coaxial sheath flow and the liquid junction configurations. In the coaxial configuration, the CE capillary is surrounded by a wider-diameter, outer tube. The sheath liquid is introduced, either hydrodynamically or by external

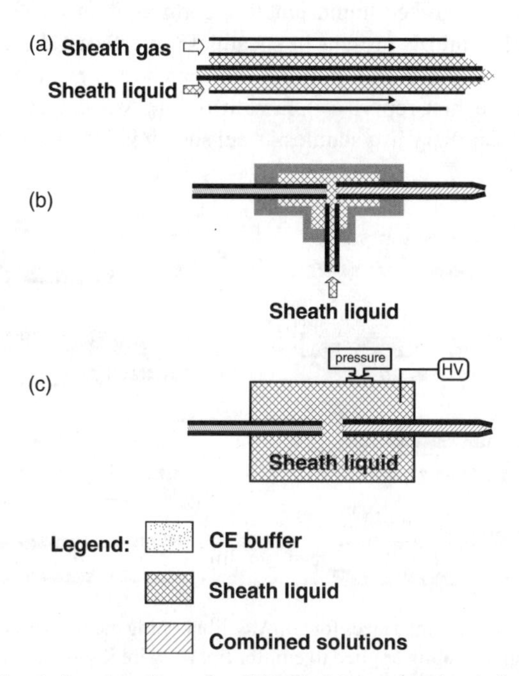

FIGURE 1.14 Sheath flow interfaces for CE-MS: (a) coaxial sheath flow with sheath gas; (b) liquid junction; (c) pressurized liquid junction (reprinted from Ref. 195, with permission of Elsevier).

pumping, and flows through this tube. The CE buffer and sheath liquid mix at the Taylor cone, at the tip of the electrospray emitter. Since mixing only occurs as the separated components leave the capillary, the separation process is undisturbed. A further modification to this configuration involves the addition of a second outer tube that encompasses the previous assembly. This tube accommodates a flow of sheath gas, which may be necessary to stabilize the spray and facilitate droplet formation, particularly at higher flow rates [152,153].

The liquid junction configuration is similar in principle, except for the position at which the buffer and sheath liquid are mixed. The end of the capillary and ESI emitter are housed in a T-junction, with a narrow gap (typically 20–50 μm) between the two [152]. The sheath liquid is introduced orthogonally into the gap. There is less dilution in this configuration such that sensitivity is not as detrimentally affected as in the coaxial configuration. However, with the orthogonal introduction of liquid, band broadening can occur as the separated components leave the capillary, which can lead to poorer resolution. This effect can be overcome using the pressurized liquid junction configuration that was described by Fanali [154]. In essence similar to the liquid junction, the main differences are the larger gap between the capillary and ESI emitter (100 μm) and the fact that the junction is contained within a pressurized reservoir of sheath liquid [152,154]. With the applied pressure, band broadening is minimized, leading to improved resolution and separation efficiency [152]. In addition, although the CE buffer is still diluted with sheath liquid, the dilution factor is not as great as with the coaxial and unpressurized liquid junction configurations [152].

Since no sheath liquid is present in sheathless interfaces, the necessary voltage must be applied directly to the CE buffer, which can be achieved in a number of ways (Figure 1.15). In the first report of CE with online MS detection, Olivares et al. terminated the CE capillary in a stainless-steel sheath [151]. A potential was applied

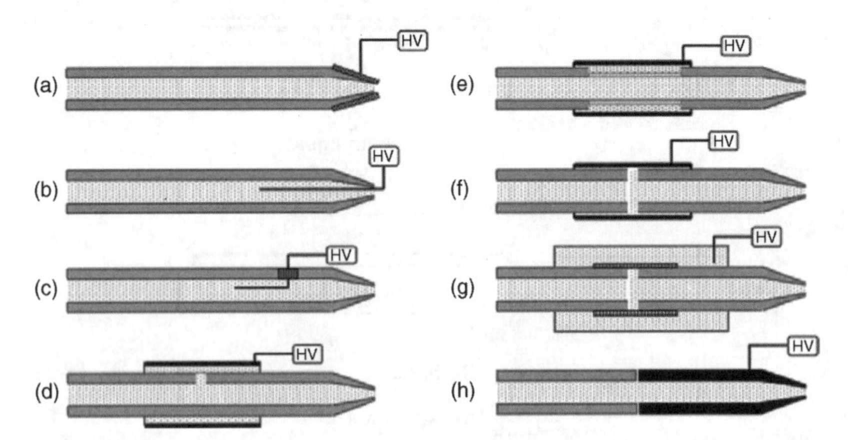

FIGURE 1.15 Sheathless interfaces for CE-MS, illustrating methods used to create electrical contact: (a) conductive coating applied to emitter tip; (b) wire inserted at tip; (c) wire inserted through hole; (d) split-flow interface with metal sheath; (e) porous, etched capillary walls in metal sleeve; (f) junction with metal sleeve; (g) microdialysis junction; (h) junction with conductive emitter tip (reprinted from Ref. 195, with permission of Elsevier).

to the sheath, which was used as the CZE cathode, as well as the electrospray needle. In later work, silver vapor was deposited onto the capillary and ESI needle to improve the electrical contact [155]. Other options for sheathless interfaces include coating the capillary end with a metal [153]. The sprayer tip can also be metal-coated or fabricated from a conductive material (e.g., metal or polymer), with the capillary end positioned in direct contact with the tip [153]. Alternatively, a wire electrode can be inserted into the capillary, either through the end of the capillary or via a small hole drilled near the end of the capillary [152,153].

While ESI sources constitute the most common interfaces for CE-MS, the use of other ionization sources as interfaces has been reported, which can widen the range of sample types that can be separated and detected by CE-MS. Inductively coupled plasma (ICP) sources are widely used as interfaces for applications involving elemental speciation and the development, and applications of CE-ICP-MS were reviewed by Kannamkumarath et al. in 2002, by Michalke in 2005, and by Álvarez-Llamas et al., also in 2005 [156–158].

Tanaka et al. reported an APCI interface for CE-MS [159]. A commercial APCI manifold was modified to incorporate a stainless–steel tube that surrounded the CE capillary and accommodated a coaxial flow of sheath liquid. A nebulizing gas was also introduced. The steel tube, containing the capillary, was then positioned in the APCI nozzle for subsequent ionization of the sample [159]. As mentioned in Section 1.5.1, APCI offers ionization of less-polar compounds that are not amenable to ESI. In addition, CE–atmospheric-pressure photoionization–MS (CE-APPI-MS) has been reported in the literature, by modifying an APPI source originally intended for LC-MS [160]. A wider range of CE buffers can be used with APPI sources; nonvolatile buffers did not cause the analyte signal suppression observed using ESI [160].

In a review of CE-MS developments and applications, Schmitt-Koplin and Englmann reported on trends in mass analyzers used in the hyphenated instruments [161]. Between 1987 and 2001, the majority of CE-MS publications described the use of single-quadrupole or triple-quadrupole mass analyzers. Tandem quadrupole, ion trap, time-of-flight (TOF), Fourier transform ion cyclotron resonance (FT-ICR), and sector field analyzers were also reported, albeit in fewer publications. With the exception of the tandem quadrupole, publications using each type of mass analyzer increased between 2001 and 2004. The greatest increase was observed for the ion trap, followed by the single quadrupole, then triple quadrupole, and TOF. By 2004, less than 10 publications reported the use of either FT-ICR or sector field analyzers [161].

1.5.4 Ion Mobility Spectrometry–Mass Spectrometry

Ion mobility spectrometry (IMS) separates ions on the basis of differences in differential mobility through a weak, uniform electric field, in the presence of a counterflow of drift gas [162,163]. Mobility depends on collision cross section and thus, ions are separated according to size-to-charge ratio [164]. The technique was initially termed *plasma chromatography* and in fact, one of the first papers on IMS described a hyphenated GC-IMS-MS instrument [165]. Samples were separated by GC, detected by IMS or further separated on the basis of mobility, then mass-separated and detected by MS. A ^{63}Ni source was initially used as the ion source, limiting IMS to the analysis of volatile samples in the vapor phase [163]. However, today, atmospheric-pressure ionization sources such as ESI and MALDI are also used, thus increasing the range of sample types amenable to analysis by IMS [163]. For further details on the history, theory, and principles of IMS, readers are directed to the excellent text by Eiceman and Karpas that covers these aspects in detail [162].

Coupling IMS with MS is advantageous because of differences in separation mode between the two techniques. In IMS, ions are separated according to size-to-charge ratio while in MS, separation is based on mass-to-charge ratio. Hence, isomers and conformers can be separated by IMS, which is not possible by MS. However, definitive identification of analytes is possible by MS but not by IMS alone since collision cross section is not sufficiently specific. In the following section, interface designs and IMS/MS instrument configurations are discussed; however, this discussion is certainly not exhaustive, and readers are directed to several review articles and texts that are available in the literature for further details [162–164].

In coupling IMS and MS, differences in operating pressures have to be considered as well as differences in electrical potentials of the two instruments and the interface. Two interface designs are commonly employed. In the first, a pinhole orifice (20–50 μm), or skimmer cone with orifice diameter 100 μm, is used and ions are transferred directly from the IMS drift tube to the high vacuum of the mass spectrometer [162]. The second design uses two skimmer cones, or one skimmer cone and a large pinhole membrane, to transfer ions from the drift tube through two differentially pumped regions, into the mass spectrometer [162]. To ensure that ions can pass from the drift tube, through the interface, and into the mass spectrometer, it is necessary to increase the potential of the interface with respect to the mass analyzer (particularly for quadrupoles) and increase the potential of the drift tube with respect to the interface. A series of ion lenses are also incorporated into the interface to focus the ion beam into the mass spectrometer.

One of the early commercially available IMS/MS instruments, the Alpha II PC/MS, incorporated a quadrupole mass analyzer and the drift tube contained two ion gates to enable different modes of data acquisition (Figure 1.16) [164]. The entrance gate was positioned at the beginning of the drift tube, allowing ions to enter, while the exit gate was positioned at the end of the tube, allowing ions to leave the drift tube and enter the mass spectrometer. With both gates open, ions in the drift tube continually entered the mass analyzer, which was scanned to generate a mass spectrum of all ions in the sample. By pulsing the entrance gate and keeping the exit gate open, ions entered the drift tube (similar to operation in standalone IMS systems) and passed into the mass analyzer, which was operated in total-ion monitoring mode, enabling all ions to be detected, albeit with no mass separation. By gating both gates with suitable delays, ions of a selected mobility entered the mass analyzer, which was tuned to a specific *m/z* for ion identification. Throughout the 1970s and 1980s, most of the research conducted used instruments similar to the aforementioned design, and MS was essentially seen as a selective detector for the ion mobility spectrometer [164].

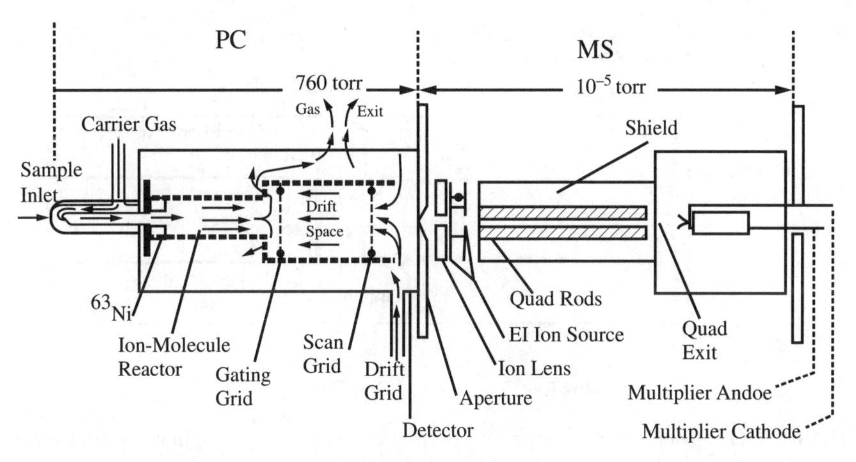

FIGURE 1.16 Alpha II plasma chromatography/mass spectrometry interface (reprinted from Ref. 196, with permission of Elsevier).

A later IMS–quadrupole MS instrument, manufactured by PCP Inc., included an EI source in the mass spectrometer, thus allowing ionization of neutral molecules that entered from the drift tube [162]. Dwivedi et al. reported the use of an IMS–quadrupole MS system for chiral separation of pharmaceuticals, carbohydrates, and amino acids that was achieved through addition of a chiral additive to the drift gas [166]. Subsequent separation in the drift tube was based not only on size and charge but also on stereospecific interactions with the chiral gas. Further details and applications of IMS–quadrupole MS have been reviewed and are available in the literature [162,163].

Into the 1990s, developments in instrument design continued to generate and sustain interest in IMS/MS. Guevremont et al. modified a commercially available IMS/MS instrument, replacing the ^{63}Ni source with an ESI source and replacing the quadrupole with a time-of-flight mass analyzer [167]. After passing through an orifice interface (25 μm diameter) between the drift tube and mass spectrometer, ions passed through a series of four lenses and were accelerated into the flight region by application of simultaneous pulses to two grids (Figure 1.17). In terms of data acquisition, it was possible to "gate" individual IMS peaks; that is, data were acquired only when an IMS peak of interest entered the mass analyzer [167].

In TOF-MS, spectra are collected in the microsecond range, while ion mobility spectra are collected in the millisecond range. This means that, when the two are coupled, hundreds of mass spectra can be collected for each ion in the ion mobility spectrum [163]. The advantages of this aspect were highlighted by Clemmer and coworkers for the gas-phase separation of protease digests [168]. As the mass analyzer continuously scanned throughout the duration of the ion mobility cycle, the sample could be characterized in terms of ion mobility, *m/z*, and intensity. More recently, IMS-TOF-MS has been used further in the separation of protein–peptide mixtures [169,170], and amphetamine-type drugs [171], as well as for the characterization of oligosaccharides [172], among other applications [163].

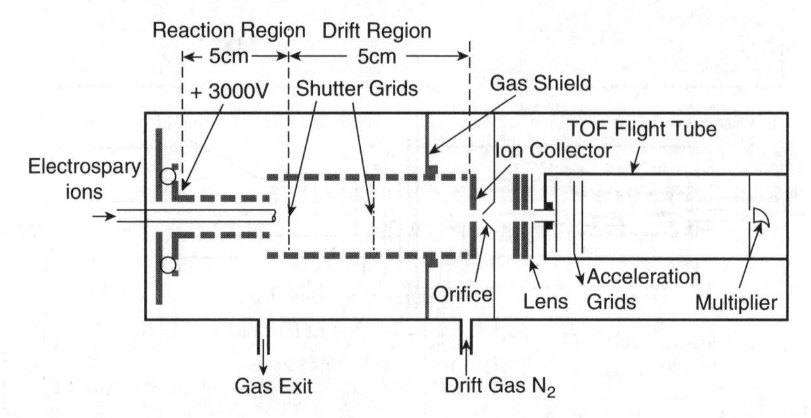

FIGURE 1.17 Coupling of ion mobility spectrometry with time-of-flight mass spectrometry (reprinted from Ref. 167, with permission of the American Chemical Society).

Clowers and Hill described coupling IMS with a quadrupole ion trap (QIT) MS [173]. The drift tube contained two Bradbury–Nielson gates. Opening of the second gate was delayed with respect to the first to allow ions of a specific mobility to enter the trap. The process was repeated to accumulate sufficient population of the specific mobility ion in the trap for MS^n experiments [173]. A quadrupole ion trap was also used by Creaser et al. in their tandem ion trap/ion mobility spectrometer [174]. In this design, the ion trap could be used in the conventional sense as a mass analyzer or alternatively, as an ion source for IMS.

Although not as common as the mass analyzers discussed above, IMS has been coupled with Fourier transform ion cyclotron resonance mass spectrometers [163]. Tang et al. incorporated a two-gate system and coupled the two instruments via a flared inlet capillary interface [175]. Thus, separation occurred in the ion mobility spectrometer and ions of specific mobility were selected for subsequent mass analysis. Using this system, separation of two peptides, as well as separation of two isomeric phosphopeptides, was demonstrated [175].

1.6 PROSPECTS FOR MASS SPECTROMETRY

Mass spectrometry is indeed a powerful tool for chemical analysis that has been widely applied in a variety of areas related to scientific research, industrial production, and governmental regulations. An even wider range of applications of MS can be foreseen for applications such as production quality control, food safety regulation, disease diagnosis or personal healthcare, and security — applications that are highly dependent on the availability of MS analysis systems with acceptable size and cost and a minimum requirement of skills for operation. After over a century of development, the mass spectrometers themselves, as in-laboratory instruments, have become more and more automated and robust, although the cost of ownership and maintenance is still far beyond what is acceptable for use at pharmacies, grocery

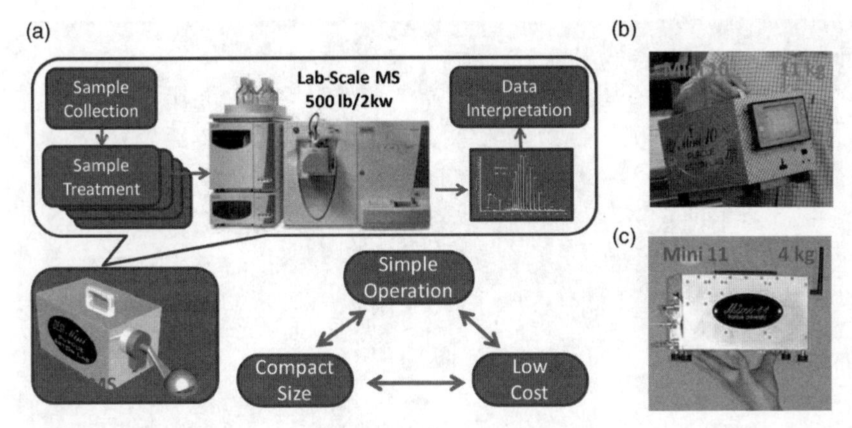

FIGURE 1.18 (a) Conceptual schematic of a miniaturized mass spectrometry analytical system; (b) Mini 10 and (c) Mini 11 handheld ion trap mass spectrometers (illustrations reprinted from Ref. 176, with permission from *Annual Reviews*).

stores, or households. Current mass spectrometric chemical analysis procedures are also highly dependent on sample type, which ultimately requires the personnel performing the analysis to have the knowledge and experience for the sample preparation and data interpretation. These are the limiting factors that prevent the current MS analysis systems to be employed outside of analytical laboratories. Since the 1990s, some major efforts have been put into the development of miniature mass spectrometers of all types [176] (see Figure 1.18). Handheld mass spectrometers weighing only 5 kg have been developed with the capability of coupling in vacuo as well as atmospheric-pressure ionization sources [177,178]. A discontinuous atmospheric-pressure interface (DAPI) [179] was developed, which opens periodically for ion introduction. It allowed the transfer of ions generated in air, such as by ESI or APCI, into the vacuum manifold for mass analysis without using complicated differential pumping stages or high-capacity pumping systems. The capabilities of these small instruments have been demonstrated with the analysis of volatile organic compounds (VOCs) in air and water as well as nonvolatile compounds, including peptides and proteins in samples in condensed phases. Such miniature instrumentation represents the state of the art in MS technologies, exemplifying how MS systems can be developed with much smaller size, lower weight, and lower cost, which ultimately economically justifies the development of specialized MS systems targeting narrow ranges of applications.

With the emerging possibilities of miniature mass spectrometers, finding solutions for simplifying the sample preparation is becoming highly desirable. Ambient ionization methods, which aim at the direct sampling of analytes from complex sample matrices, have been developed and applied for MS analysis with minimal or no sample pre-treatment. Starting with desorption electrospray ionization (DESI) [180] and direct analysis in real time (DART) [181], more than 20 ambient ionization methods [182–184] have been developed since 2004. The majority of these methods use charged droplets, lasers, or plasma to desorb and ionize the

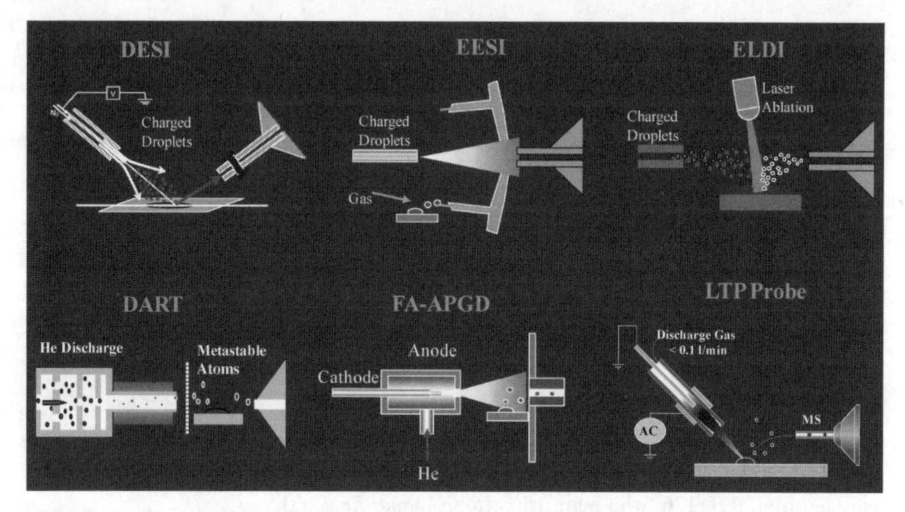

FIGURE 1.19 Some ambient ionization methods. DESI = desorption electrospray ionization [180]; EESI = extractive electrospray ionization [197,198]; ELDI = electrospray-assisted laser desorption/ionization [199]; DART = direct analysis in real time, FA-APGD = flow afterglow atmospheric-pressure glow discharge [200]; LTP = low-temperature plasma [183].

analytes from samples in condensed phases. The processes of desorption and ionization can be simultaneous or sequential (Figure 1.19), depending on the individual methods, but all techniques generally require little time for gas-phase ions to be generated for MS analysis. The direct analysis of complex samples has been demonstrated using these methods, including the analysis of nonvolatile organics, such as explosives, from various surfaces [185], steroids and metabolites in urine [186,187], ingredients in drug tablets [188], fatty acids from olive oil [189], and lipids from raw tissue [190]. MS imaging of raw tissues has also been developed with DESI for lipid [191] and drug distributions [192]. Most of the ambient ionization methods are implemented for qualitative or semiquantitative analysis, since sample treatments, including the addition of internal standards, are minimized to simplify the procedure. A new method of generating ions directly from paper loaded with samples was developed by applying a high voltage to the paper wetted with solution; this method of analysis has been applied for therapeutic drug monitoring with direct analysis of dried blood spots (DBS) and highly quantitative information can be obtained over the therapeutic ranges of various drugs, including imatinib (Gleevec) and atenolol [193].

The combination of ambient ionization methods and miniature mass spectrometers potentially will provide a unique instrumentation platform for the development of MS analysis systems of small size, low cost, more importantly, with simplified analysis procedures. It is foreseen that a wide variety of miniature MS systems, each targeting specialized applications, will be developed to allow personnel inexperienced in analytical chemistry to be able to use them outside analytical laboratories. Knowledge and experience accumulated in chromatography, reactive ionization, and selective derivatization in real time will be used in the development of sampling devices that have enhanced specificity and sensitivity for target analytes in complex samples. The first adoption of such miniature MS systems will likely occur with applications in areas such as clinical diagnosis or therapy, where the need for complex mixture analysis by MS has been demonstrated and simple analysis procedures are mandated by governmental regulations.

REFERENCES

1. Cooks, R. G.; Ouyang, Z. (2002), The American Society for Mass Spectrometry and Allied Topics, Orlando, FL.
2. Aston, F. W. (1922), Nobel Prize acceptance speech, Stockholm, Sweden.
3. Parkins, W. E. (2005), *Phys. Today 58*, 45–51.
4. Palmer, P. T.; Limero, T. F. (2001), *J. Am. Soc. Mass Spectrom. 12*, 656–675.
5. Nemes, P.; Vertes, A. (2007), *Anal. Chem. 79*, 8098–8106.
6. Nie, Z.; Cui, F.; Tzeng, Y. K.; Chang, H. C.; Chu, M.; Lin, H. C.; Chen, C. H.; Lin, H. H.; Yu, A. L. (2007), *Anal. Chem. 79*, 7401–7407.
7. Dalton, J. (1808), *A New System of Chemical Philosophy*; Henderson & Spalding; London.
8. Prout, W. (1815), *Ann. Philos. 6*, 321–330.
9. Prout, W. (1816), *Ann. Philos. 7*, 111–113.
10. Griffiths, I. W. (1997), *Rapid Commun. Mass Spectrom. 11*, 3–16.
11. Thomson, J. J. (1897), *Philos. Mag. 5*, 293–316.
12. `http://nobelprize.org`; (2009).
13. Grayson, M.; Brenna, J. T.; Busch, K. L.; Caprioli, R. M.; Cotter, R. J.; Grigsby, R. D.; Judson, C. M.; Ramanathan, R.; Siuzdak, G.; Story, M. S.; Thomas, J. J.; WIlloughby, R. C.; Yergey, A. L. (2002), *Measuring Mass: From Positive Rays to Protein*, American Society for Mass Spectrometry, Santa Fe, NM.
14. Thomson, J. J. (1913), *Proc. Roy. Soc. Lond. A (Containing Papers of a Mathematical and Physical Character) 89*, 1.
15. Downard, K. M. (2007), *Mass Spectrom. Rev. 26*, 713–723.
16. Aston, F. W. (1920), *Nature 105*, 617–619.
17. Budzikiewicz, H.; Grigsby, R. D. (2006), *Mass Spectrom. Rev. 25*, 146–157.
18. De Laeter, J. R. (2009), *Mass Spectrom. Rev. 28*, 2–19.
19. Nier, A. O. (1989), *J. Chem. Educ. 66*, 385–388.
20. Cooks, R. G.; Beynon, J. H.; Caprioli, R. M.; Lester, G. R. (1973), *Metastable Ions*, Elsevier, Amsterdam.
21. McLafferty, F. W. (1959), *Anal. Chem. 31*, 82–87.
22. McLafferty, F. W. (1966), *Interpretation of Mass Spectra*, Benjamin.
23. McLafferty, F. W.; Turucek, F. (1993), *Interpretation of Mass Spectra*, 4th ed., University Science Books, Sausalito, CA.

24. Busch, K. L.; Glish, G. L.; McLuckey, S. A. (1988), *Mass Spectrometry/Mass Spectrometry: Techniques and Applications of Tandem Mass Spectrometry*; VCH Publishers, New York.
25. Gohlke, R. S.; McLafferty, F. W. (1993), *J. Am. Soc. Mass Spectrom. 4*, 367.
26. Bouchoux, G. (2007), *Mass Spectrom. Rev. 26*, 775–835.
27. Biemann, K.; Oro, J.; Toulmin, P.; Orgel, L. E.; Nier, A. O.; Anderson, D. M.; Simmonds, P. G.; Flory, D.; Diaz, A. V.; Rushneck, D. R.; Biller, J. A. (1976), *Science 194*, 72–76.
28. Biemann, K.; Gapp, F.; Seibl, J. (1959), *J. Am. Chem. Soc. 81*, 2274–2275.
29. Sundqvist, B.; Macfarlane, R. D. (1985), *Mass Spectrom. Rev. 4*, 421–460.
30. Gelpi, E. (2009), *J. Mass Spectrom. 44*, 1137–1161.
31. Fenn, J. B.; Mann, M.; Meng, C. K.; Wong, S. F.; Whitehouse, C. M. (1989), *Science 246*, 64–71.
32. Karas, M.; Bachmann, D.; Bahr, U.; Hillenkamp, F. (1987), *Int. J. Mass Spectrom. Ion Process. 78*, 53–68.
33. Tanaka, K.; Waki, H.; Ido, Y.; Akita, S.; Yoshida, Y.; Matsuo, T. (1988), *Rapid Commun. Mass Spectrom. 2*, 151–153.
34. Yates, J. R. (1998), *J. Mass Spectrom. 33*, 1–19.
35. Asara, J. M.; Schweitzer, M. H.; Freimark, L. M.; Phillips, M.; Cantley, L. C. (2007), *Science 316*, 280–285.
36. Franck, J.; Arafah, K.; Elayed, M.; Bonnel, D.; Vergara, D.; Jacquet, A.; Vinatier, D.; Wisztorski, M.; Day, R.; Fournier, I.; Salzet, M. (2009), *Molec. Cell. Proteom. 8*, 2023–2033.
37. Zubarev, R. A.; Kelleher, N. L.; McLafferty, F. W. (1998), *J. Am. Chem. Soc. 120*, 3265–3266.
38. Syka, J. E. P.; Coon, J. J.; Schroeder, M. J.; Shabanowitz, J.; Hunt, D. F. (2004), *Proc. Nat. Acad. Sci. 101*, 9528–9533.
39. Gross, M. L.; Caprioli, R. M., eds. (2007), *The Encyclopedia of Mass Spectrometry*, Vol. 6, *Ionization methods*, Elsevier, Burlington, MA.
40. McLafferty, F. W. (1980), *Acct. Chem. Res. 13*, 33–39.
41. de Hoffmann, E.; Stroobant, V. (2001), *Mass Spectrometry: Principles and Applications*, 2nd ed., Wiley, New York.
42. Downard, K. (2004), *Mass Spectrometry: A Foundation Course*, Royal Society of Chemistry, Cambridge, U.K.
43. Watson, J. T.; Sparkman, O. D. (2004), *Introduction to Mass Spectrometry: Instrumentation, Applications, and Strategies for Data Interpretation*, 4th ed., Wiley, Hoboken, NJ.
44. Burgoyne, T. W.; Hieftje, G. M. (1996), *Mass Spectrom. Rev. 15*, 241–259.
45. Paul, W.; Steinwedel, H. (1953), *Z. Naturforsch. A (Journal of Physical Sciences) 8*, 448–450.
46. March, R. E. (1997), *J. Mass Spectrom. 32*, 351–369.
47. March, R. E.; Todd, J. F. J. (2005), *Quadrupole Ion Trap Mass Spectrometry*, 2nd ed., Wiley, Hoboken, NJ.
48. Paul, W. (1990), *Rev. Modern Phys. 62*, 531–540.

49. Stafford, G. C.; Kelley, P. E.; Syka, J. E. P.; Reynolds, W. E.; Todd, J. F. J. (1984), *Int. J. Mass Spectrom. Ion Process. 60*, 85–98.
50. Ouyang, Z.; Wu, G. X.; Song, Y. S.; Li, H. Y.; Plass, W. R.; Cooks, R. G. (2004), *Anal. Chem. 76*, 4595–4605.
51. Tabert, A. M.; Goodwin, M. P.; Cooks, R. G. (2006), *J. Am. Soc. Mass Spectrom. 17*, 56–59.
52. Douglas, D. J.; Frank, A. J.; Mao, D. M. (2005), *Mass Spectrom. Rev. 24*, 1–29.
53. Cameron, A. E.; Eggers, D. F. (1948), *Rev. Sci. Instrum. 19*, 605–607.
54. Short, R. T. (1997), *Physica Scripta T71*, 46–49.
55. Cotter, R. J. (1999), *Anal. Chem. 71*, 445A–451A.
56. Guilhaus, M.; Selby, D.; Mlynski, V. (2000), *Mass Spectrom. Rev. 19*, 65–107.
57. Chernushevich, I. V.; Loboda, A. V.; Thomson, B. A. (2001), *J. Mass Spectrom. 36*, 849–865.
58. Mamyrin, B. A. (2001), *Int. J. Mass Spectrom. 206*, 251–266.
59. Ens, W.; Standing, K. G. (2005), *Biol. Mass Spectrom. 402*, 49–78.
60. Wiley , W. C.; McLaren, I. H. (1955) *Rev. Sci. Instrum. 26*, 1150–1157.
61. Colby, S. M.; King, T. B.; Reilly, J. P. (1994), *Rapid Commun. Mass Spectrom. 8*, 865–868.
62. Brown, R. S.; Lennon, J. J. (1995), *Anal. Chem. 67*, 1998–2003.
63. Koppenaal, D. W.; Barinaga, C. J.; Denton, M. B.; Sperline, R. P.; Hieftje, G. M.; Schilling, G. D.; Andrade, F. J.; Barnes, J. H. (2005), *Anal. Chem. 77*, 418A–427A.
64. Lawrence, E. O.; Edlefson, N. E. (1930), *Science 72*, 367–377.
65. Lawrence, E. O.; Livingston, M. S. (1932), *Phys. Rev. 40*, 19–35.
66. Hipple, J. A.; Sommer, H.; Thomas, H. A. (1949), *Phys. Rev. 76*, 1877–1878.
67. Sommer, H.; Thomas, H. A.; Hipple, J. A. (1951), *Phys. Rev. 82*, 697–702.
68. Wobschall, D. (1965), *Rev. Sci. Instrum. 36*, 466–475.
69. Amster, I. J. (1996), *J. Mass Spectrom. 31*, 1325–1337.
70. Marshall, A. G.; Hendrickson, C. L. (2002), *Int. J. Mass Spectrom. 215*, 59–75.
71. Comisarow, M. B.; Marshall, A. G. (1974), *Chem. Phys. Lett. 25*, 282–283.
72. Comisarow, M. B.; Marshall, A. G. (1974), *Chem. Phys. Lett. 26*, 489–490.
73. Marshall, A. G.; Hendrickson, C. L.; Jackson, G. S. (1998), *Mass Spectrom. Rev. 17*, 1–35.
74. Marshall, A. G. (2000), *Int. J. Mass Spectrom. 200*, 331–356.
75. Marshall, A. G.; Hendrickson, C. L. (2008), *Annu. Rev. Anal. Chem. 1*, 579–599.
76. Marshall, A. G.; Wang, T. C. L.; Ricca, T. L. (1985), *J. Am. Chem. Soc. 107*, 7893–7897.
77. Kingdon, K. H. (1923), *Phys. Rev. 21*, 408–418.
78. Hu, Q. Z.; Noll, R. J.; Li, H. Y.; Makarov, A.; Hardman, M.; Cooks, R. G. (2005), *J. Mass Spectrom. 40*, 430–443.
79. Perry, R. H.; Cooks, R. G.; Noll, R. J. (2008), *Mass Spectrom. Rev. 27*, 661–699.
80. Knight, R. D. (1981), *Appl. Phys. Lett. 38*, 221–223.
81. Wells, J. M.; Badman, E. R.; Cooks, R. G. (1998), *Anal. Chem. 70*, 438–444.
82. Makarov, A. (2000), *Anal. Chem. 72*, 1156–1162.
83. Scigelova, M.; Makarov, A. (2006), *Proteomics* 16–21.

84. Makarov, A.; Denisov, E.; Kholomeev, A.; Baischun, W.; Lange, O.; Strupat, K.; Horning, S. (2006), *Anal. Chem. 78*, 2113–2120.
85. Cooks, R. G.; Terwilliger, D. T.; Ast, T.; Beynon, J. H.; Keough, T. (1975), *J. Am.Chem. Soc. 97*, 1583–1585.
86. Coon, J. J.; Syka, J. E. P.; Schwartz, J. C.; Shabanowitz, J.; Hunt, D. F. (2004), *Int. J. Mass Spectrom. 236*, 33–42.
87. Brodbelt, J. S.; Wilson, J. J. (2009), *Mass Spectrom. Rev. 28*, 390–424.
88. Kalcic, C. L.; Gunaratne, T. C.; Jonest, A. D.; Dantus, M.; Reid, G. E. (2009), *J. Am. Chem. Soc. 131*, 940–942.
89. Guan, Z.; Kelleher, N. L.; O'Connor, P. B.; Aaserud, D. J.; Little, D. P.; McLafferty, F. W. (1996), *Int. J. Mass Spectrom. Ion Processes, 157/158*, 357–364.
90. Zubarev, R. A.; Kelleher, N. L.; McLafferty, F. W. (1998), *J. Am. Chem. Soc. 120*, 3265–3266.
91. Zubarev, R. A. (2003), *Mass Spectrom. Rev. 22*, 57–77.
92. Leymarie, N.; Costello, C. E.; O'Connor, P. B. (2003), *J. Am. Chem. Soc. 125*, 8949–8958.
93. Syka, J. E. P.; Coon, J. J.; Schroeder, M. J.; Shabanowitz, J.; Hunt, D. F. (2004), *Proc. Natl. Acad. Sci. USA 101*, 9528–9533.
94. Roepstorff, P.; Fohlman, J. (1984), *Biomed. Mass Spectrom. 11*, 601–601.
95. Zubarev, R. A.; Kruger, N. A.; Fridrikkson, E. K.; Lewis, M. A.; Horn, D. M.; Carpenter, B. A.; McLafferty, F. W. (1999), *J. Am. Chem. Soc. 121*, 2857–2862.
96. Cooper, H. J.; Hakansson, K.; Marshall, A. G. (2005), *Mass Spectrom. Rev. 24*, 201–222.
97. Coon, J. J.; Syka, J. E. P.; Schwartz, J. C.; Shabanowitz, J.; Hunt, D. F. (2004), *Int. J. Mass Spectrom.236*, 33–42.
98. Gunawardena, H. P.; He, M.; Chrisman, P. A.; Pitteri, S. J.; Hogan, J. M.; Hodges, B. D. M.; McLuckey, S. A. (2005), *J. Am. Chem. Soc. 127*, 12627–12639.
99. Liang, X. R.; Xia, Y.; McLuckey, S. A. (2006), *Anal. Chem. 78*, 3208–3212.
100. Coon, J. J. (2009), *Anal. Chem. 81*, 3208–3215.
101. Zubarev, R. A.; Haselmann, K. F.; Budnik, B.; Kjeldsen, F.; Jensen, F. (2002), *Eur. J. Mass Spectrom. 8*, 337–349.
102. Syrstad, E. A.; Turecek, F. (2005), *J. Am. Soc. Mass Spectrom. 16*, 208–224.
103. Martin, A. J. P.; Synge, R. L. M. (1941), *Biochem. J. 35*, 1358.
104. James, A. T.; Martin, A. J. P. (1952), *Biochem. J. 50*, 679.
105. James, A. T. (1952), *Biochem. J. 52*, 242.
106. James, A. T.; Martin, A. J. P.; Smith, G. H. (1952), *Biochem. J. 52*, 238.
107. Golay, M. J. E. (1958), in Coates, V. J.; Noebels, H. J.; Fagerson, I. S., eds., *Gas Chromatography*, Academic Press, New York, *1958*, pp. 1–13.
108. Gohlke, R. S. (1959), *Anal. Chem. 31*, 535.
109. Holmes, J. C.; Morrell, F. A. (1957), *Appl. Spectrosc. 11*, 86.
110. Niesson, W. M. A. (2001), *Current Practice of Gas Chromatography—Mass Spectrometry*, Marcel Dekker, New York.
111. Ryhage, R., *Anal. Chem.* (1964) *36*, 759.
112. Abian, J. (1999), *J. Mass Spectrom. 34*, 157.
113. Braithwaite, A. J.; Smith, F. J. (1996), *Chromatographic Methods*, 5th ed., Blackie, Glasgow.

114. De Hoffmann, E.; Stroobant, V. (2002), *Mass Spectrometry Principles and Applications*, 2nd ed., Wiley, Chichester, UK.
115. Hendrich, L. H. (1995), in Grob, R. L., ed., *Modern Practice of Gas Chromatography*, 3rd ed., Wiley, New York, pp. 265–321.
116. Niesson, W. M. A. (2006), *Liquid Chromatography—Mass Spectrometry*, 3rd ed., CRC Press, Boca Raton, FL.
117. Ardrey, B. (2003), *Liquid Chromatography—Mass Spectrometry: An Introduction*, Wiley, Chichester, UK.
118. McFadden, W. H.; Schwartz, H. L.; Evans, S. (1976), *J. Chromatogr. 122*, 389.
119. Scott, R. P. W.; Scott, C. G.; Munroe, M.; Hess, J., Jr., (1994), *J. Chromatogr. 99*, 395.
120. Tal'roze, V. L.; Karpov, G. V.; Gorodetskii, I. G.; Skurat, V. E. (1968), *Russ. J. Phys. Chem. 42*, 1658.
121. Baldwin, M. A.; McLafferty, F. W. (1973), *Org. Mass Spectrom. 7*, 1111.
122. Arpino, B.; Baldwin, M. A.; McLafferty, F. W. (1974), *Biomed. Mass Spectrom. 1*, 80.
123. Melera, A. (1980), *Adv. Mass Spectrom. 8B*, 1597.
124. Niesson, W. M. A. (1986), *Chromatographia 21*, 277.
125. Niesson, W. M. A. (1986), *Chromatographia 21*, 342.
126. Willoughby, R. C.; Browner, R. F. (1984), *Anal. Chem. 56*, 2626.
127. Cappiello, A. (1996), *Mass Spectrom. Rev. 15*, 283.
128. Caprioli, R. M.; Fan, T. (1986) *Anal. Chem, 58*, 2949.
129. De Wit, J. S. M.; Deterding, L. J.; Moseley, M. A.; Tomer, K. B.; Jorgenson, J. W.; Caprioli, R. M. (1988), *Rapid Commun. Mass Spectrom. 2*, 100.
130. Ito, Y.; Takeuchi, T.; Ishii, D.; Goto, M. (1985), *J. Chromatogr. 346*, 161.
131. Caprioli, R. M.; Suter, M. J.-F. (1992), *Int. J. Mass Spectrom. Ion Process. 118/119*, 449.
132. Moseley, M. A.; Deterding, L. J.; De Wit, J. S. M.; Tomer, K. B.; Kennedy, R. T.; Bragg, N.; Jorgenson, *J. W.* (1989), *Anal. Chem. 61*, 1577.
133. Clegg, G. A.; Dole, M. (1971), *Biopolymers 10*, 821.
134. Dole, M.; Mack, L. L.; Hines, R. L.; Mobley, R. C.; Ferguson, L. D.; Alice, M. B. (1968), *J. Chem. Phys. 49*, 2240.
135. Mack, L. L.; Kralik, P.; Rheude, A.; Dole, M. (1970), *J. Chem. Phys. 52*, 4977.
136. Yamashita, M.; Fenn, J. B. (1984), *J. Phys. Chem. 88*, 4451.
137. Whitehouse, C. M.; Dreyer, R. N.; Yamashita, M.; Fenn, J. B. (1985), *Anal. Chem. 57*, 675.
138. Horning, E. C.; Horning, M. G.; Carroll, D. I.; Dzidic, I.; Stillwell, R. N. (1973), *Anal. Chem. 45*, 936.
139. Cody, R. B. (2002), in Pramanik, B. N., Ganguly, A. K., Gross, M. L.eds., *Applied Electrospray Mass Spectrometry*, Marcel Dekker, New York, pp. 1–105.
140. Emmett, M. R.; Caprioli, R. M. (1994), *J. Am. Soc. Mass Spectrom. 5*, 605.
141. Wilm, M.; Mann, M. (1996), *Anal. Chem. 68*, 1.
142. Gallagher, R. T.; Balogh, M. P.; Davey, P.; Jackson, M. R.; Sinclair, I.; Southern, L. J. (2003), *Anal. Chem. 75*, 973.
143. Jorgenson, J. W.; Lukacs, K. D. (1981), *Anal. Chem. 53*, 1298.
144. Jorgenson, J. W.; Lukacs, K. D. (1983), *Science 222*, 266.

145. Klampfl, C. W. (2006), *Electrophoresis 27*, 3.
146. Baker, D. R. (1995), *Capillary Electrophoresis*, Wiley, Chichester, UK.
147. Camilleri, P. (1993), *Capillary Electrophoresis Theory and Practice*, CRC Press, Boca Raton, FL.
148. Schmitt-Koplin, P. (2008), *Capillary Electrophoresis Methods and Protocol*, Humana Press, Totowai, NJ.
149. Shintani, H.; Polonský, J. (1997), *Handbook of Capillary Electrophoresis Applications*, Chapman & Hall, London.
150. Weinberger, R. (1993), *Practical Capillary Electrophoresis*, Academic Press, San Diego, CA.
151. Olivares, J. A.; Nguyen, N. T.; Yonker, C. R.; Smith, R. D. (1987), *Anal. Chem. 59*, 1230.
152. Maxwell, E. J.; Chen, D. D. Y. (2008), *Anal. Chim. Acta 627*, 25.
153. Schmitt-Koplin, P.; Frommberger, M. (2003), *Electrophoresis 24*, 3837.
154. Fanali, S.; D'Orazio, G.; Foret, F.; Kleparnik, K.; Aturki, Z. (2006), *Electrophoresis 27*, 4666.
155. Smith, R. D.; Barinaga, C. J.; Udseth, H. R. (1988). *Anal. Chem. 60*, 1948.
156. Álvarez-Llamas, G.; del Rosario Fernández de laCampa, M.; Sanz-Medel, A. (2005), *Trends Anal. Chem. 24*, 28.
157. Kannamkumarath, S. S.; Wrobel, K.; Wrobel, K.; B'Hymer, C.; Caruso, J. A. (2002), *J. Chromatogr. A 975*, 245.
158. Michalke, B. (2005), *Electrophoresis 26*, 1584.
159. Tanaka, Y.; Otsuka, K.; Terabe, S. (2003), *J. Pharm. Biomed. Anal. 30*, 1889.
160. Mol, R.; de Jong, G. J.; Somsen, *G. W.* (2005), *Electrophoresis 26*, 146.
161. Schmitt-Koplin, P.; Englmann, M. (2005), *Electrophoresis 226*, 1209.
162. Eiceman, G. A.; Karpas, Z. (2005), *Ion Mobility Spectrometry*, 2nd ed., CRC Press, Boca Raton, FL.
163. Kanu, A. B.; Dwivedi, P.; Tam, M.; Matz, L.; Hill, H. H., Jr., (2008), *J. Mass Spectrom. 43*, 1.
164. Collins, D. C.; Lee, M. L. (2002), *Anal. Bioanal. Chem. 372*, 66.
165. Cohen, M. J.; Karasek, F. W. (1970), *J. Chromatogr. Sci. 8*, 330.
166. Dwivedi, P.; Wu, C.; Matz, L. M.; Clowers, B. H.; Siems, W. F.; Hill, H. H., Jr., (2006), *Anal. Chem.*, *78*, 8200.
167. Guevremont, R.; Siu, K. W. M.; Wang, J.; Ding, L. (1997), *Anal. Chem. 69*, 3959.
168. Valentine, S. J.; Counterman, A. E.; Hoaglund, C. S.; Reilly, J. P.; Clemmer, D. E. (1998), *J. Am. Soc. Mass Spectrom. 9*, 1213.
169. Hoaglund, C. S.; Valentine, S. J.; Sporleder, C. R.; Reilly, J. P.; Clemmer, D. E. (1998), *Anal. Chem. 70*, 2236.
170. Ruotolo, B. T.; Gillig, K. J.; Stone, E. G.; Russell, D. H.; Fuhrer, K.; Gonin, M.; Schultz, J. A. (2002), *Int. J. Mass Spectrom. 219*, 253.
171. Steiner, W. E.; Clowers, B. H.; Fuhrer, K.; Gonin, M.; Matz, L. M.; Siems, W. F.; Schultz, A. J.; Hill, H. H., Jr., (2001), *Rapid Commun. Mass Spectrom. 15*, 2221.
172. Liu, Y.; Clemmer, D. E. (1997), *Anal. Chem. 69*, 2504.
173. Clowers, B. H.; Hill, H. H., Jr., (2005), *Anal. Chem. 77*, 5877.

174. Creaser, C. S.; Benyezzar, M.; Griffiths, J. R.; Stygall, J. W. (2000), *Anal. Chem. 72*, 2724.
175. Tang, X.; Bruce, J. E.; Hill, *H. H., Jr.* (2007), *Rapid Commun. Mass Spectrom. 21*, 1115.
176. Ouyang, Z.; Cooks, R. G. (2009), *Annu. Rev. Anal. Chem. 2*, 187 214.
177. Gao, L.; Song, Q. Y.; Patterson, G. E.; Cooks, R. G.; Ouyang, Z. (2006), *Anal. Chem. 78*, 5994–6002.
178. Gao, L.; Sugiarto, A.; Harper, J. D.; Cooks, R. G.; Ouyang, Z. (2008), *Anal. Chem. 80*, 7198–7205.
179. Gao, L.; Cooks, R. G.; Ouyang, Z. (2008), *Anal. Chem. 80*, 4026–4032.
180. Takats, Z.; Wiseman, J. M.; Gologan, B.; Cooks, R. G. (2004), *Science 306*, 471–473.
181. Cody, R. B.; Laramee, J. A.; Durst, H. D. (2005), *Anal. Chem. 77*, 2297–2302.
182. Venter, A.; Nefliu, M.; Cooks, R. G. (2008), *Trends Anal. Chem. 27*, 284–290.
183. Harper, J. D.; Charipar, N. A.; Mulligan, C. C.; Zhang, X. R.; Cooks, R. G.; Ouyang, Z. (2008), *Anal. Chem. 80*, 9097–9104.
184. Na, N.; Zhao, M. X.; Zhang, S. C.; Yang, C. D.; Zhang, X. R. (2007), *J. Am. Soc. Mass Spectrom. 18*, 1859–1862.
185. Takats, Z.; Cotte-Rodriguez, I.; Talaty, N.; Chen, H. W.; Cooks, R. G. (2005), *Chem. Commun., 15*, 1950–1952.
186. Huang, G.; Chen, H.; Zhang, X.; Cooks, R. G.; Ouyang, Z. (2007), *Anal. Chem. 79*, 8327–8332.
187. Chen, H. W.; Pan, Z. Z.; Talaty, N.; Raftery, D.; Cooks, R. G. (2006), *Rapid Commun. Mass Spectrom. 20*, 1577–1584.
188. Chen, H. W.; Talaty, N. N.; Takats, Z.; Cooks, R. G. (2005), *Anal. Chem. 77*, 6915–6927.
189. Garcia-Reyes, J. F.; Mazzoti, F.; Harper, J. D.; Charipar, N. A.; Oradu, S.; Ouyang, Z.; Sindona, G.; Cooks, R. G. (2009), *Rapid Commun. Mass Spectrom. 23*, 3057–3062.
190. Wiseman, J. M.; Puolitaival, S. M.; Takats, Z.; Cooks, R. G.; Caprioli, R. M. (2005), *Angew. Chem. Int. Ed. 44*, 7094–7097.
191. Wiseman, J. M.; Ifa, D. R.; Song, Q. Y.; Cooks, R. G. (2006), *Angew. Chem. Int. Ed. 45*, 7188–7192.
192. Wiseman, J. M.; Ifa, D. R.; Zhu, Y. X.; Kissinger, C. B.; Manicke, N. E.; Kissinger, P. T.; Cooks, R. G. (2008), *Proc. Natl. Acad. Sci. USA 105*, 18120–18125.
193. Wang, H.; Liu, J.; Cooks, R. G.; Ouyang, Z. (2008), *Angew. Chem. Int. Ed. 49*, 877–880.
194. Plass, W. R.; Li, H. Y.; Cooks, R. G. (2003), *Int. J. Mass Spectrom., 228*, 237–267.
195. Maxwell, E. J.; Chen, D. D. Y. (2008), *Anal. Chim. Acta 627*, 25.
196. Karpas, Z.; Stimac, R. M.; Rappoport, Z. (1988), *Int. J. Mass Spectrom. Ion Process. 83*, 163–175.
197. Chen, H.; Venter, A.; Cooks, R. G. (2006), *Chem. Commun.* 2042–2044.
198. Chen, H. W.; Venter, A.; Cooks, R. G. (2006), *Chem. Commun.* 2042–2044.
199. Shiea, J.; Huang, M.-Z.; HSu, H.-J.; Lee, C.-Y.; Yuan, C.-H.; Beech, I.; Sunner, J. (2005), *Rapid Commun. Mass Spectrom. 19*, 3701–3704.
200. Andrade, F. J.; Wetzel, W. C.; Webb, M. R.; Gamez, G.; Ray, S. J.; Hieftje, G. M. (2008), *Anal. Chem. 80*, 2654–2663.

CHAPTER 2

LC Method Development and Strategies

GANG XUE

Pfizer Global R&D, Eastern Point Road, Groton, CT 06340

YINING ZHAO

Pfizer Global R&D, 50 Pequot Avenue, New London, CT 06320

2.1 INTRODUCTION

During the production of active pharmaceutical ingredients (APIs), many opportunities for the generation of impurities may arise [1]. The safety of the drug depends not only on the toxicological properties of the API itself but also on the impurities that it contains. For this reason, accurate assessment of impurity profiles of API is one of the most important fields of activity in pharmaceutical analysis. Spectroscopic analyses such as mass spectroscopy (MS), infrared spectroscopy (IR), and nuclear magnetic resonance (NMR) provide accurate structure characterization of the impurities. However, because of the mixed nature of the samples, appropriate chromatographic separation of each component is often required before spectroscopic analysis to ensure the accurate spectral acquisition.

A number of chromatographic separation techniques are widely adopted for chemical analysis, such as thin-layer chromatography (TLC) [2], gas chromatography (GC) [3,4], high-performance liquid chromatography (HPLC) [5,6], supercritical fluid chromatography (SFC) [7,8], and capillary electrophoresis (CE) [9–11]. Selection of the optimal separation technology is the first step in sample analysis, which has to be determined by the compounds' physiochemical properties such as boiling point, polarity, and molecular weight. For most small-molecule druglike compounds, HPLC is by far the most reliable and efficient separation technique, which serves as the main workhorse for impurity analysis.

Unfortunately, although HPLC has been extensively practiced for over three decades, most scientists still tend to develop the separation methods by trial and

Characterization of Impurities and Degradants Using Mass Spectrometry, First Edition.
Edited by Birendra N. Pramanik, Mike S. Lee, and Guodong Chen.

error, which is often time-consuming and labor-intensive. This chapter summarizes a more efficient systematic screening optimization approach, which would greatly reduce or even eliminate the method development time.

2.2 COLUMN, pH, AND SOLVENT SCREENING

2.2.1 Resolution: Goal of Separation

The quality of the chromatography separation is measured by resolution (R_s) of two adjacent peaks, which is defined as follows:

$$R_s = \frac{2(t_2 - t_1)}{w_1 + w_2} \tag{2.1}$$

Here t_1 and t_2 are the retention times of the first and second adjacent peaks while w_1 and w_2 are their corresponding baseline bandwidths. The greater the R_s, the greater the distance between the two peaks (i.e., the better the separation). Chromatography peaks of similar heights would correspond to a resolution of 1.5 when baseline-separated, which is typically the goal of HPLC method development for accurate quantitation. However, a minimum resolution of 2 is required to ensure baseline separation of peaks with distinct heights, for example the 1% impurity eluting next to the main peak.

Equation (2.2) illustrates the correlation between the resolution and experimental conditions:

$$R_s = \frac{\sqrt{N}}{4} \frac{\alpha - 1}{\alpha} \frac{k'}{1 + k'} \tag{2.2}$$

Here, N is column theoretic plates, k' is the average retention factor for the two peaks, and α is the selectivity, or the ratio of the two retention factors. Obviously, increases in N, α, and k' all lead to improved R_s. Among the three factors, the N increase contributes least to the resolution improvement as shown by the least steep $f(N)$ curve in Figure 2.1. The resolution increases very significantly with k' initially. But the increase flattens out very quickly as k' exceeds 5 [curve $f(k)$]. The selectivity α, instead, remains the most significant contributor to the continuous improvement in resolution [curve $f(\alpha)$]. In summary, this means that a minimum retention factor of approximately 1–2 is required for good resolution, but strong retention will not further enhance the separation; the column efficiency is not effective for improving separation, but the selectivity is. Hence, effective method development should focus on the selectivity.

2.2.2 Screening: Systematic Approach to Seeking Selectivity

The conventional trial-and-error approach starts with "random" initial condition (column and mobile phase) followed by step-by-step one-variable-at-a-time (OVAT)

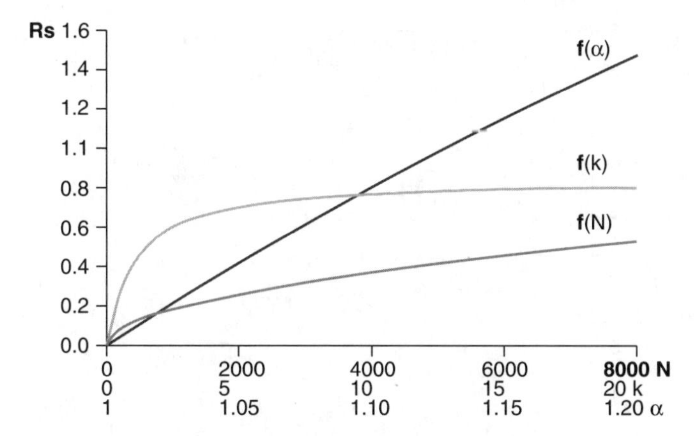

FIGURE 2.1 Contribution of theoretical plates, retention factor, and selectivity to resolution.

optimization as shown in Figure 2.2a. The key drawback of the OVAT stepwise approach is slow and labor intensive. Depending on the "random" initial condition, this approach may lead to some local optimal condition such as the condition point *A* or *B* in Figure 2.2a, while the optimal condition is at point *C*. Instead, systematic design of experiments (DOE) will allow mapping the entire chromatography space with limited number of experiments in one shot as shown in Figure 2.2b. Statistical analysis of the DOE results will elucidate the global optimal conditions. Such DOE experiments in HPLC method development are executed as screening.

More than a dozen of the experimental parameters that affect retention can be included in the screening. But the factorial combination of all parameters will lead to an unmanageable number of experiments. Practical screening should start with the key parameters relating to the variable of focus, the selectivity. In reversed-phase HPLC, three key parameters will impact selectivity most: pH, stationary phase, and organic modifier.

Figure 2.3 shows the log D plot of five analytes with respect to pH. As a result of the pK_a difference, the log D of the five compounds respond very differently to the change in pH. On the other hand, the log D is a measure of the hydrophobicity of each compound, which is directly proportional to the retention factor k'. For this reason, as pH changes, the selectivity, which is the k' ratio of two adjacent peaks, may change significantly. The overlaid log D plots such as that in Figure 2.3 allows very intuitive selection of the appropriate pH range where the compounds show difference in log D values and are thus easily separated. Another important characteristic is the flatness of the log D plots. The flatter the curve range, the less sensitive retention is to pH change; that is the more robust it is. For these reasons, pH 1–3 is most suitable for robustly separating the five compounds in Figure 2.3. When all compound structures are not available, empirical retention data at different pH levels can be acquired and extrapolated to show such trends.

Stationary phase is another parameter that would change the selectivity drastically. HPLC columns made with different material, bonding chemistry, carbon load, or

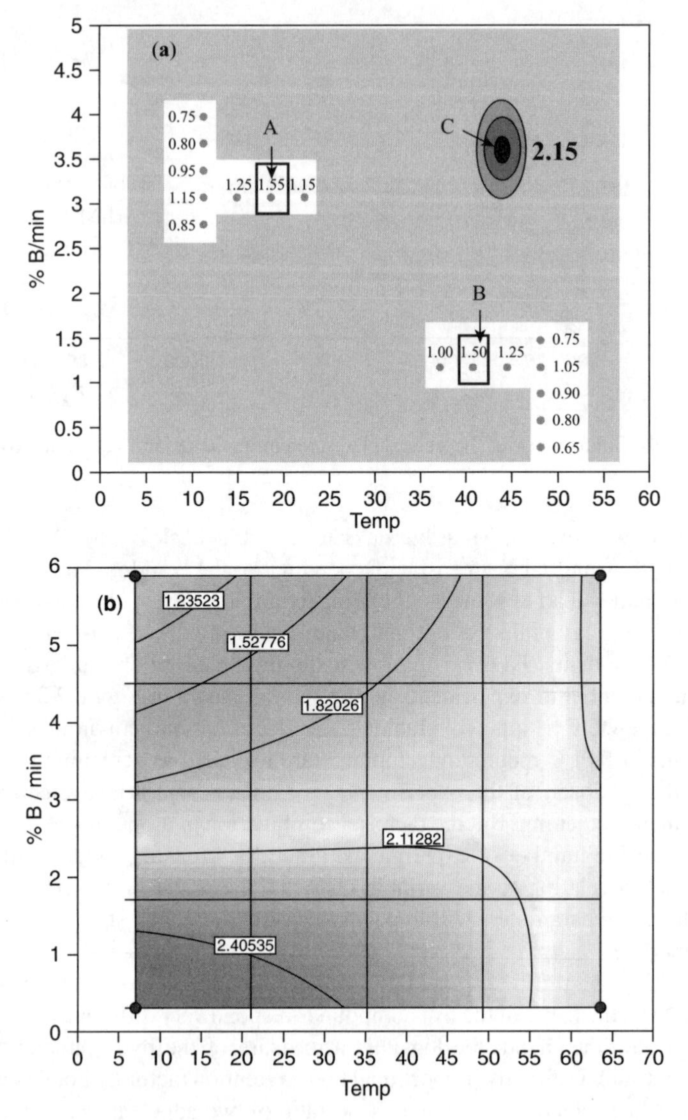

FIGURE 2.2 Comparison of conventional OVAT and systematic screening approaches of method development: (a) OVAT approach; (b) DOE screening approach (adapted from Ref. 26).

surface coverage will show difference in types and strengths of interaction with various compounds. Figure 2.4 illustrates the selectivity and retentivity characterization of some common HPLC columns on the market. The diversely scattered column distribution in the two-dimensional space indicates the different retention behavior expected when analyzing the same compounds. Finding the right column chemistry thus is one of the most effective means to optimize the selectivity.

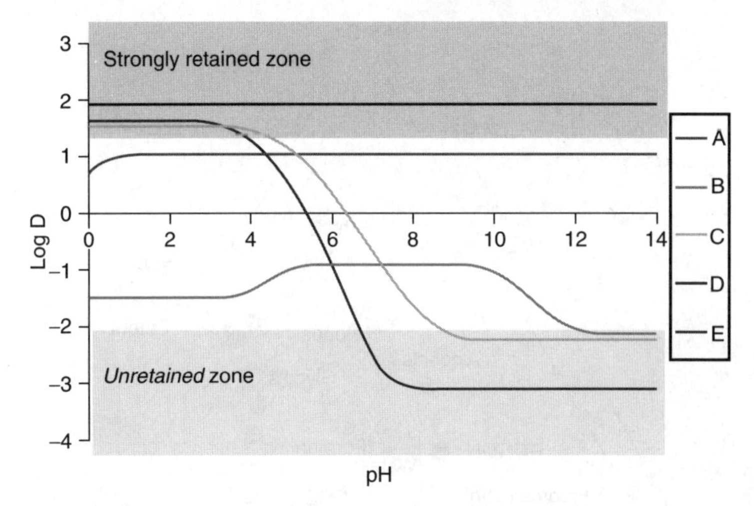

FIGURE 2.3 Effect of pH on log D of different compounds.

The primary effect of organic modifier in reversed-phase HPLC is the decrease in retention factor k' with the increase in solvent strength ($B\%$ or ϕ) as shown in Eq. (2.3).

$$\log k' = \log k'_x - S\phi \tag{2.3}$$

Correspondingly, the selectivity changes with respect to $B\%$ as Follows:

$$\log \alpha = \log \frac{k'_2}{k'_1} = \log \frac{k'_{x2}}{k'_{x1}} - \Delta S\phi \tag{2.4}$$

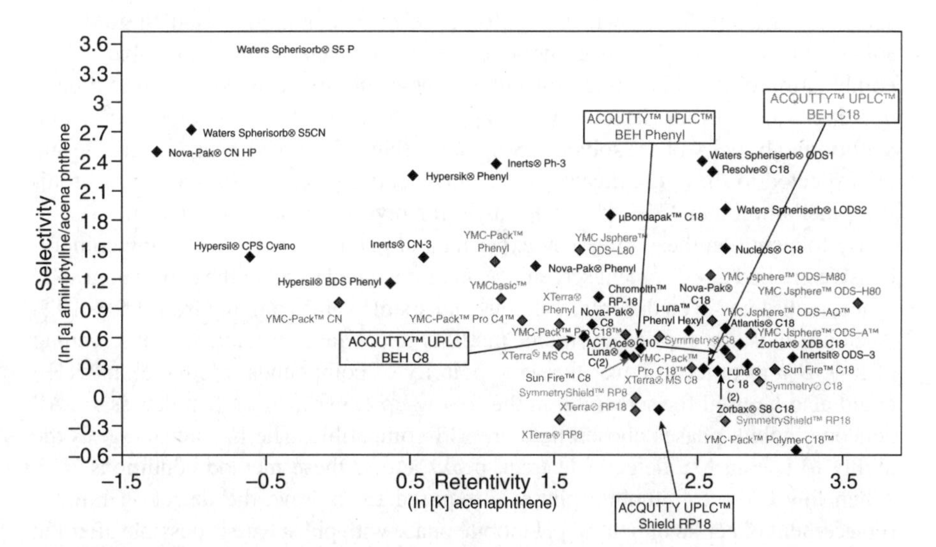

FIGURE 2.4 Waters reversed-phase column selectivity chart (courtesy of Waters).

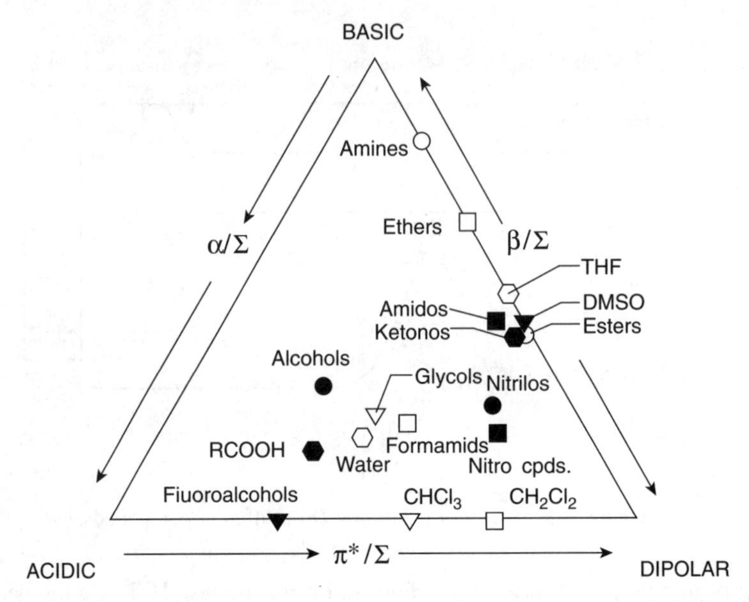

FIGURE 2.5 Solvent selectivity triangle (reprinted with permission from Ref. 12; copyright 1993 Elsevier).

Here the k'_x and S are constants specific to each compound in different solvents. With the same solvent, the logarithm of selectivity (α, or k'_2/k'_1) changes linearly to solvent strength with a negative slope of ΔS; that is, the lower the $B\%$, the greater the selectivity. However, lower $B\%$ means slower separation and broader peaks. Alternatively, a change in organic solvent type may vary both k'_{x2}/k'_{x1}, and ΔS thus yields very different selectivity without compromising the speed. The solvent–selectivity triangle as shown in Figure 2.5 is most commonly used to guide the solvent selection [12]. Since the interaction between solute and solvent is a combination of three key intermolecular forces—dipole, dispersion, and hydrogen bonding—the solvent selectivity is expected to depend on the dipole moment, acidity, and basicity of the solvents. Solvents fall in different parts of the triangle and are expected to show significantly different selectivity, values. In practice, acetonitrile, methanol, and THF are used mostly for reversed-phase (RP) HPLC.

By focusing on these three key experimental parameters, the following comprehensive method development screening can be designed to cover the chromatography space, including seven different columns and six pH values ranging from 1.9 to 10.5. Acetonitrile is first applied as organic modifier to conduct generic gradient screening of 5–95$B\%$ in order to cover the wide polarity of compounds. Methanol and THF could also be used for screening in the follow-up experiments when necessary. All aqueous mobile phases chosen here are MS-compatible. The key advantage is the ability to use an MS detector to track peaks across these method conditions [13]. When low-UV-cutoff mobile phase is required to improve the detection limit, a replacement of acidic or neutral pH mobile phase with phosphate is possible after the screening and before the further optimization.

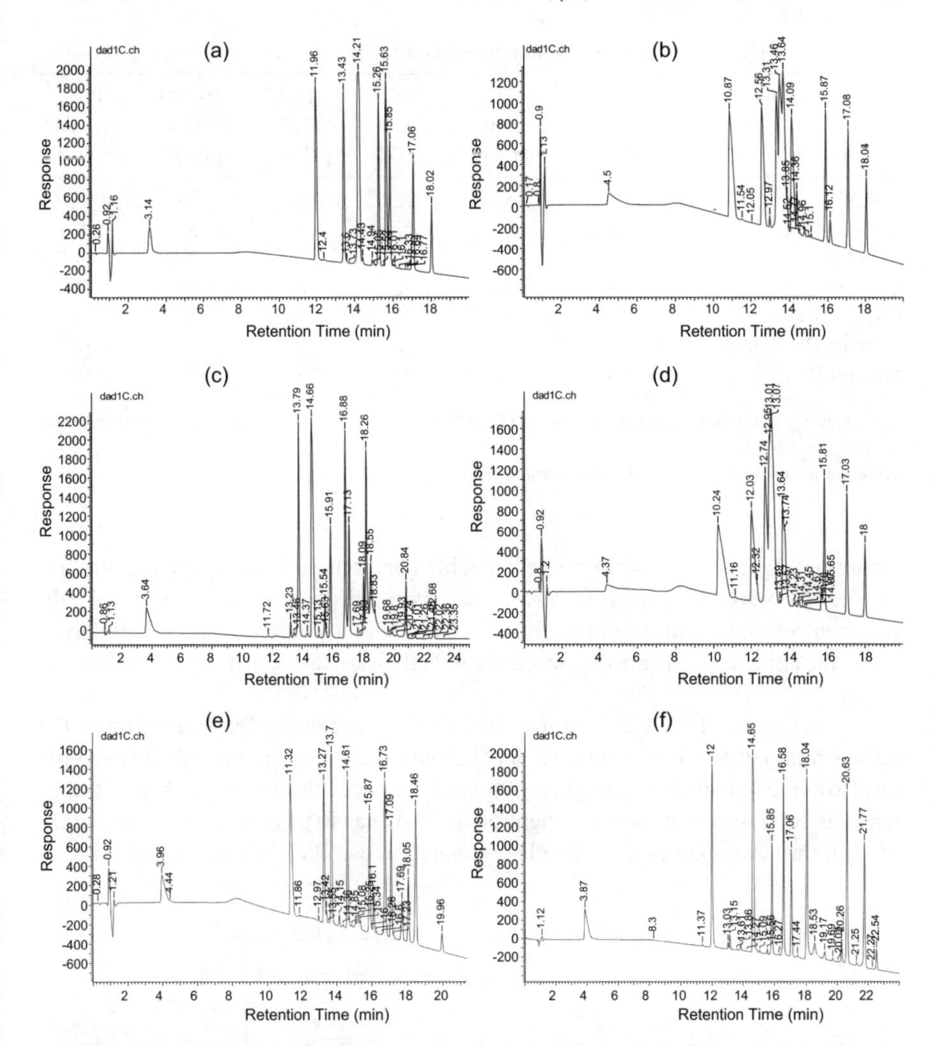

FIGURE 2.6 Plots showing pH screening of a drug compound with 10 spiked in impurity standards on XBridge C8 with conditions listed in Table 2.1: (a) pH 1.9; (b) pH 2.6; (c) pH 3.2; (d) pH 4.5; (e) pH 6.8; (f) pH 10.5.

One example of a drug purity method screening is shown in Figure 2.6. To better illustrate the peak profile, the synthesized standards of 10 key impurities are spiked into the API sample. Although it is impossible to clearly show all 36 chromatograms in one page, the separation of the spiked sample at six different pH values, all with the XBridge C8 column and acetonitrile, are shown here. As expected, the resolution and peak shape vary significantly as pH changes. Peak tracking with MS allows more quantitative analysis of the separation as plotted in Figure 2.7. The retention pH correlation plot provides data very similar to those obtained from the overlaid log D plot in Figure 2.3. The retention times of some nonpolar analytes are relatively flat

TABLE 2.1 HPLC Method Development Screening[a]

Column[b]	0.1% TFA, pH 1.9	0.1% Formic, pH 2.6	0.1% AcOH, pH 3.2	10 mM NH_4Ac, pH 4.5	10 mM NH_4Ac, pH 6.8	0.1% NH_4OH, pH 10.5
SB-CN	X	X	X	X	X	—
Ace Phenyl	X	X	X	X	X	—
Halo C8	X	X	X	X	X	—
Ace C8	X	X	X	X	X	—
XBridge C8	X	X	X	X	X	X
XBridge RP Shield C18	X	X	X	X	X	—
Atlantis T3	X	X	X	X	X	—

[a]All screening methods use a generic gradient of 5–95% organic in 15 min with flow rate 0.84 mL/min and equilibration time 4 min.
[b]All columns are 3 × 100 mm with 3 μm particle size.

(impurities 7, 9, and 10), while some ionizable compounds show curved pH titration profiles. However, no structure information is required here, which is obviously important when the purpose of method development is to characterize the impurity. Often, the empirical data are also more accurate than are theoretically predicted log *D* values.

A closer look at Figure 2.7 suggests that pH 7–9 is probably the best pH range for the separation of all components, not only because the retention times spread out most but also as the retention plots begin to flatten out. A visual check of peak shapes further confirms such selection. Similar retention analysis has to be conducted for the other six columns. An optimal column–pH combination can then be determined.

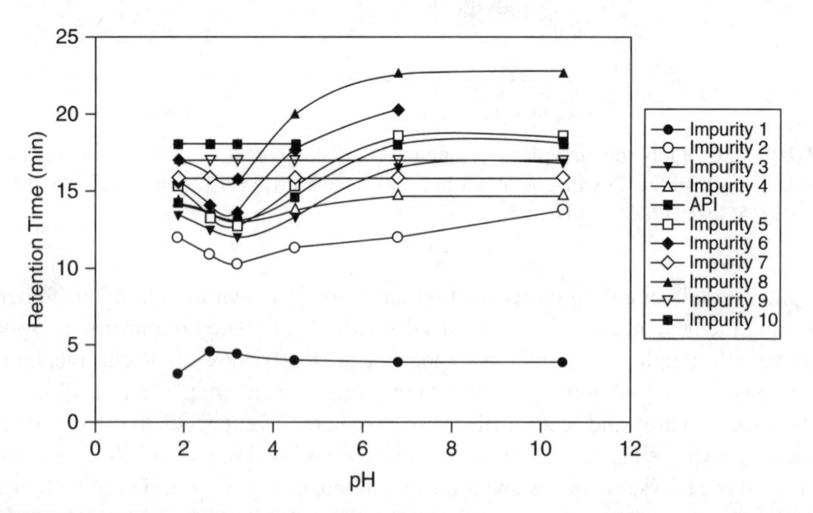

FIGURE 2.7 Effect of pH on the retention of a drug compound and its associated impurities on XBridge column with conditions listed in Table 2.1.

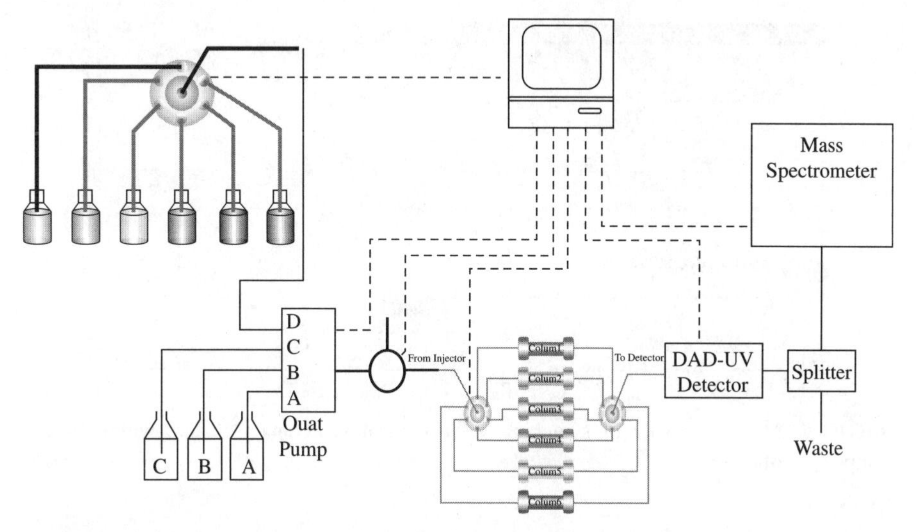

FIGURE 2.8 Schematics of a HPLC screening system with integrated six-port solvent selection valve and six-column switching valve (adapted with permission from Ref. 13; copyright 2004 Elsevier).

2.2.3 Screening Instrumentation and Controlling Software

Standard HPLC instruments can be directly used for method screening execution. However, since most HPLC instruments support only up to two columns and four mobile phase inputs, manual exchange of columns and mobile phases are required to support the screening experimental design. Switching valves can be easily integrated to the HPLC instruments to automate the process: the low-pressure solvent selection valves and the high-pressure column switching valves (Figure 2.8). Table 2.2 lists a number of switching valves available on the market.

The software control options of these valves vary by both chromatography data system and valve models. Most valves can be mixed with different HPLC instruments via contact closure communication. For example, in Waters Empower software, the four channels of contact closure can be programmed to output binary code with ON

TABLE 2.2 Commercial HPLC Switching Valves

Solvent Selection Valves (Low Pressure)			Column Switching Valves (High Pressure)		
Vendor	Model	Number of Ports	Vendor	Model	Number of Ports
Waters	Solvent Selection	6	Waters	Column Selection	6
Agilent	12PS/13PT	12	Agilent	6PS	12
Gilson	ValveMate II	6	Gilson	ValveMate II	6
VICI	ChemInert	6	VICI	MultiPS	6
			Chiralizer	LC Spiderling	10

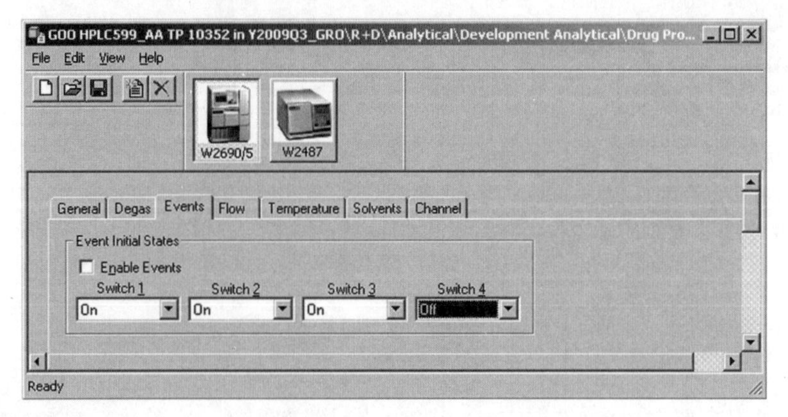

FIGURE 2.9 Screenshot of switching valve control via contact closure within Waters Empower software.

corresponding to 0 and OFF for 1; Thus "ON–ON–ON–OFF" indicates valve position 1 (binary code 0001) as shown in Figure 2.9. Agilent ChemStation offers more integrated software control as shown in Figure 2.10. Unfortunately, this simple valve control supports Agilent valves only.

In addition, a number of specialized screening software applications have become available, such as the Agilent ChemStation Method Scouting Wizard and ACD AutoChrom and ChromSword by LC Resources. Along with the supported switching valves, these applications greatly streamline the screening experimental design and allow the screening operation to be fully automated.

Finally, systematic screening explores the chromatography space and manages to pinpoint the suitable column, pH, and organic modifier with very little labor. However, the screening itself may not yield the optimal separation method. Instead, systematic screening points to the space close to the "sweet spot." A follow-up wave of optimization is often required to fine-tune the method to maximize the resolution and minimize the separation time.

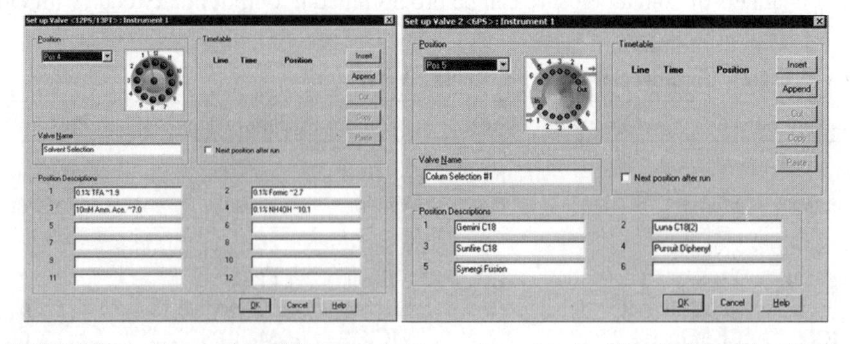

FIGURE 2.10 Screenshot of direct switching valve control within Agilent ChemStation software.

2.3 GRADIENT AND TEMPERATURE OPTIMIZATION

In addition to the aforementioned three key parameters, two more factors could be adjusted to improve the resolution: gradient and temperature. As shown in Eq. (2.4), when compounds show variation in S (i.e., $\Delta S \neq 0$), the solvent strength ϕ can be adjusted to achieve better selectivity. The other and often neglected factor, temperature, affects the retention factor (k') through the change in phase equilibration constant. As a general rule of thumb, a 1°C increase in temperature usually will decrease k' by approximately 1–2% [14]. The following quadratic equation gives a good estimation of their relationship:

$$\ln k' = a + \frac{b}{T} + \frac{c}{T^2} + \ln \beta \tag{2.5}$$

Here T is the temperature and β is the phase ratio, that is, the volume ratio of stationary phase and mobile phase in the column. When compared to stationary phase, pH, and organic modifier, the impact of these two parameters to the separation is limited. However, once the first three parameters are determined, gradient and temperature can be "tweaked" to effectively fine-tune the resolution and can often significantly reduces the runtime. In addition, as shown in Eqs. (2.3) and (2.5), the relationship between the retention factor and these two parameters are so well defined that very accurate predictions can be achieved with a very small set design of experiments (DOE). A minimum of two gradients and three temperatures is required to fit to the linear [Eq. (2.3)] and quadratic [Eq. (2.5)] equations, respectively. An example of such six-experiment injection DOE is shown in Figure 2.11. This DOE includes two 5–95% acetonitrile linear gradients with gradient times of 10 and 45 min respectively, and three separation temperatures of 30°C, 48°C, and 65°C. As shown in the six chromatograms, although not too many peaks switched retention order between methods, the resolution and peak distribution across the separation window did vary from method to method.

While manual retention time fitting to Eqs. (2.3) and (2.5) is possible with Excel, commercial simulation software such as DryLab and the ACD LC simulator can greatly simplify the data analysis. The temperature and gradient simulation of the DOE results obtained with LC simulator is shown in Figure 2.12. Part (b) is the simulated chromatogram, and part (a) is the resolution map, that is, the minimum resolution of any two peaks at a certain temperature and gradient condition. The optimal conditions (warm color zone) with maximized resolution are clearly illustrated. As the cursor is moved to the optimal condition, the simulated chromatogram is updated to reflect the virtual method change. For this example, 30–60°C and 30–40B% results in a very good separation with the resolution of any two peaks exceeds 8. However, the simulated chromatogram shows separation time exceeds 45 min. Instead, 50°C and 90B% shows that a minimum resolution of ~5 and the separation time can be cut down to less than 12 min. With the limited six experiments, the entire temperature and gradient space can be mapped and optimal condition selected, without the laborious trial and error. Usually, one additional final conformational

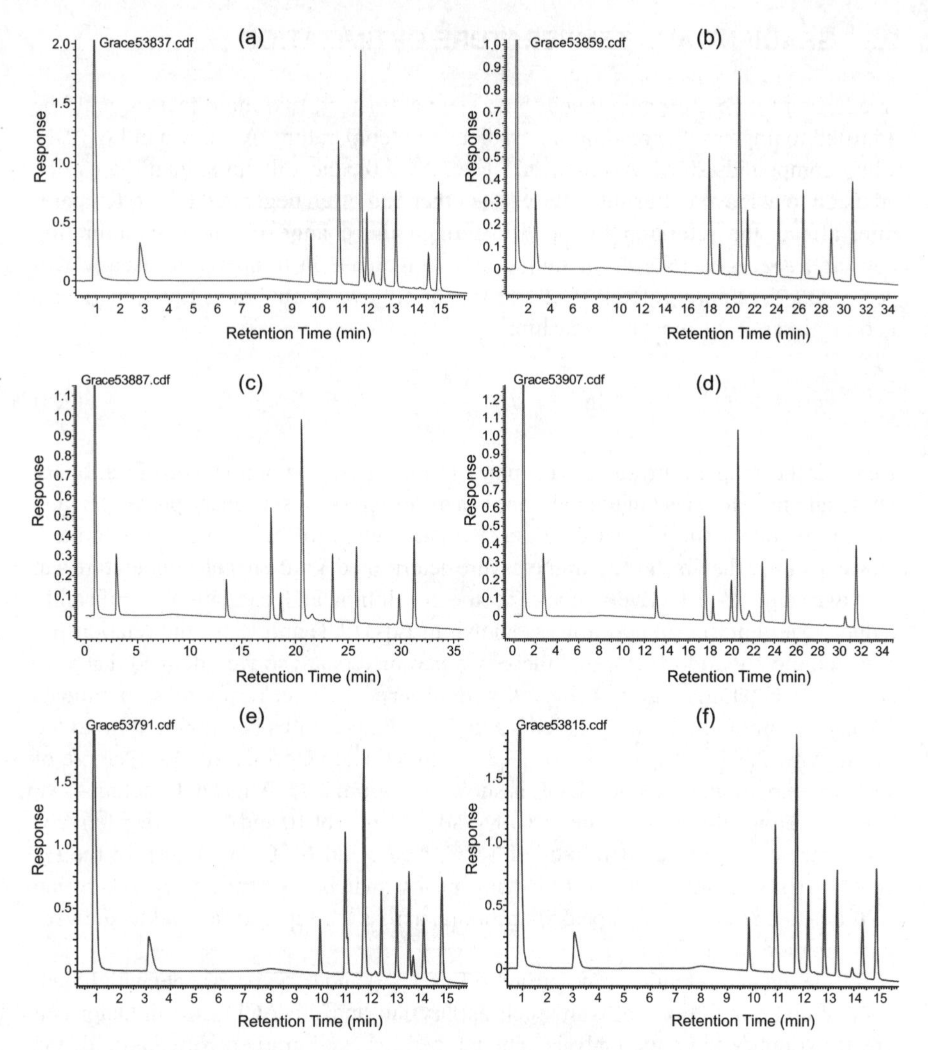

FIGURE 2.11 Design of experiments for temperature and gradient optimization of a drug impurity HPLC method optimization. All experiments were run with a gradient of 5–95% methanol with 10 mM ammonium acetate on BEH C18 (2.1 × 100 mm) at a flow rate of 0.4 mL/min.

run is necessary to confirm the predicted separation, which completes the method development cycle.

2.4 ORTHOGONAL SCREENING

The column, pH, and solvent screening combined with temperature and gradient optimization provides a systematic and scientific approach to achieve a method

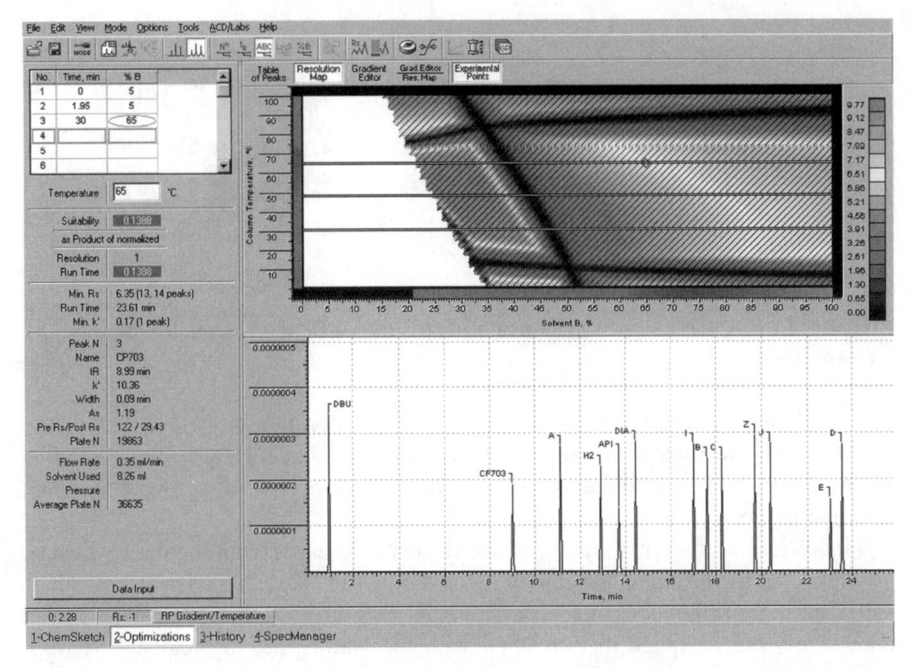

FIGURE 2.12 ACD LC simulator gradient and temperature prediction for the DOE data in Figure 2.11: (a) resolution map; (b) simulated chromatograms.

suitable to separate all impurities from the API. This strategy significantly reduces the cycle time required for method development compared to the conventional OVAT approach. The method is then validated and transferred for batch release, stability test, and so on. The key drawback, though, lies in the first wave of screening. Although the screening can be narrowed down to the three key method parameters, 36 injections are still a considerable number of experiments and can take about 2 days to complete because of the necessary factorial experimental design. For impurity and degradation characterization, where a single separation method is not necessary, orthogonal screening may help further reduce or even eliminate the method development [13,15–19].

2.4.1 Method Orthogonality

Figure 2.13 contains the conceptual plot of orthogonal HPLC methods. The *X* and *Y* axes of the plots are the retention times of all analytes in the same sample separated by two different methods. In plot (a), the two separations appear very similar with the same peak retention order. The correlation plot shows every component close to the diagonal. A simple squared linear regression (r^2) of the retention times (or other retention indices such as retention factors or normalized retention times) is a good measure of the method orthogonality. The closer r^2 is to 0, the more orthogonal the two methods are. Highly correlated separations such as in methods 1 and 2 have r^2

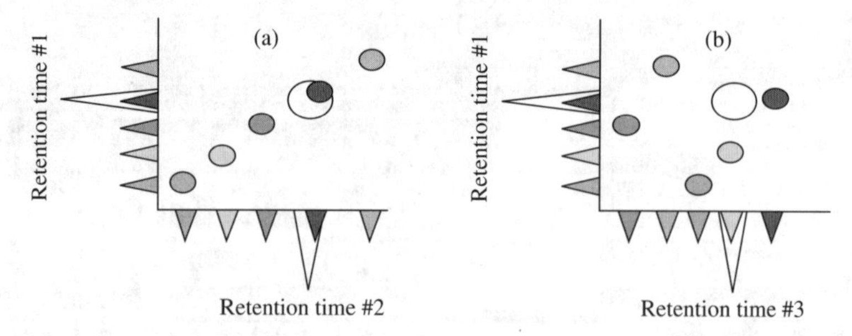

FIGURE 2.13 Conceptual illustration of method orthogonality: (a) correlated methods; (b) orthogonal methods.

close to 1, which means least orthogonal. Two peaks coelute in method 1 (red and white peaks). The same peak pair coelute in method 2 as well. Hence, analysis of the sample using the second nonorthogonal method does not offer any additional information about the mixture.

Figure 2.13b shows methods 1 and 3 with very different retention orders, which is also reflected by the slight correlation in the plot with r^2 close to 0. Thus, methods 1 and 3 are more orthogonal methods. Although method 3 also has one pair of peaks unresolved, the red peak that coeluted with the white peak in method 1 is well separated from other peaks. The new coeluting peak (yellow) has been separated in method 1. These two methods provide complementary information about the content of the mixture; thus they are orthogonal methods. Collectively, every analyte is resolved by at least one method even though a single method could not separate all components. When a sufficient number of orthogonal methods are grouped together, the generic orthogonal method set could separate most of the samples, thus eliminating the need for method development.

2.4.2 Selection of Orthogonal Methods

The selection of orthogonal methods requires a representative set of test compounds, followed by screening these compounds with different column, pH, and solvent as described earlier in this section. In our screening performed in late 2004, we included 46 different compounds of varying hydrophobicity, pK_a, size, and shape [20]. The test mixture contained a number of APIs with their precursors, process impurities, and degradants, including diastereomers, which reflected the complexity of real drug samples. In total, 27 HPLC columns comprising both the latest and established phases at the time of screening (all with 3 μm particle size) were assessed for orthogonality with the six MS-compatible buffers from pH 1.9 to 10.5, listed in Table 2.1. The screening was repeated with acetonitrile and methanol as organic modifiers.

While the method orthogonality (r^2) can be easily calculated for each method pair, the analysis and comparison of orthogonality of the entire screening method set can be overwhelming. Instead, principal components analysis (PCA) can be applied to reduce multidimensional chromatographic data as rapid qualitative

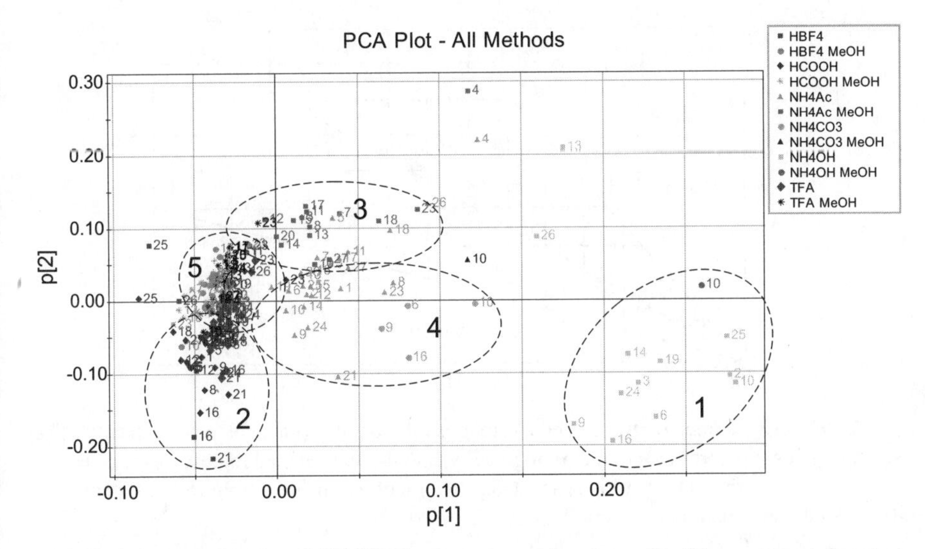

FIGURE 2.14 PCA plot of 27 HPLC column screening data with 46 component drug test mixture. The aqueous mobile phases include HBF_4, TFA, HCOOH, NH_4Ac, NH_4CO_3, and NH_4OH. Acetonitrile and methanol are used as organic modifiers. The numbers in the plot represent the 27 different columns, while the mobile phases are marked by the legend. (Reprinted from Ref. 20)

preprocessing [21,22]. Figure 2.14 illustrates the PCA result of the entire screening data set. The top ranked two principal components account for about 75% of all variation. From this plot, the greatest differences between points would be expected to provide the highest separation orthogonality. Not surprisingly, many of the method data points clustered in the left middle section of the space demonstrate their similarity, due to their close column chemistry and buffer pH. However, there exist some methods distant from the major cluster, For example, the data points in the lower right corner. These methods feature high pH 0.1% ammonium hydroxide, which confirms the strong impact of pH on selectivity. All the screening data can be divided into approximately five distinct "regions" as marked by the ovals in the plot, which cover most of the methods in the column screening. One method can be selected from each region to give the five-method orthogonal set.

The selection process then requires the quantitative comparison of r^2 for maximized orthogonality. However, poor separation with broad peaks and significant co-elution can bias the assessment with superficial low r^2 when compared with high-efficiency separations. For this reason, a second attribute, practical peak capacity (PPC), needs to be assessed for the combined chromatography efficiency as first proposed by Liu et al. [23]. The PPC is the combined peak capacity from two methods corrected by their orthogonality as defined in the following equation:

$$N_p = N_1 \times N_2 - \tfrac{1}{2}(N_2^2 \tan\gamma + N_1^2 \tan\alpha) \tag{2.6}$$

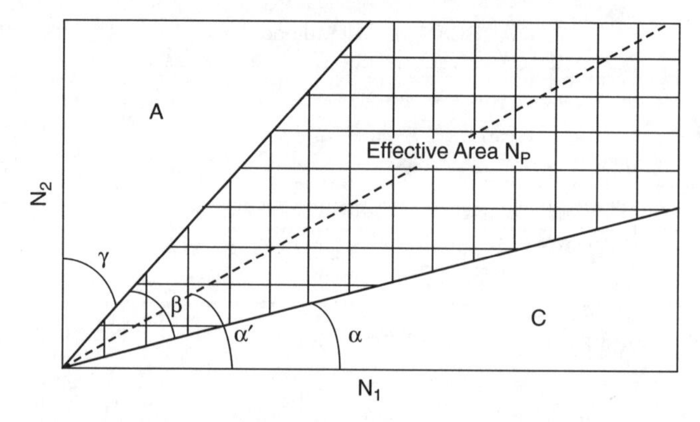

FIGURE 2.15 Effect of the method orthogonality on the practical peak capacity. The spreading angle β correlates to the orthogonality, and the shaded area represents the practical peak capacity in the 2D retention space. (Reprinted with permission from Ref. 23; copyright 1995 American Chemical Society.)

Here N_1 and N_2 are the peak capacities of methods 1 and 2, respectively, and N_p is that of the PPC, while

$$\alpha = \tan^{-1}\frac{N_2}{N_1} - \frac{\beta}{2} \tag{2.7}$$

$$\gamma = 90 - (\alpha + \beta) \tag{2.8}$$

as defined in Figure 2.15. Here β is the spread angle corresponding to $\cos^{-1}$ (r^2). The N_p thus represents the shaded area of the retention plane. Part of the space shown in white becomes unavailable because of the partially correlated two methods yield some redundant information. The most orthogonal methods give the widest the spread angle β, and thus maximize the shaded area N_p. Of course, N_p also depends largely on the individual method efficiency, N_1 and N_2.

With these assessments, the following five methods stand out as the most orthogonal with the best separation efficiency as listed in Table 2.3. The entire screening, including the blank injections for each method, takes about 5 h to complete.

2.4.3 Impurity Orthogonal Screening

A real-world example is shown in Figure 2.16, which is acquired from screening one experimental drug sample with the orthogonal method set. Because a significant number of peaks switch peak order across the orthogonal methods, MS data are required to track the peaks by their molecular weights. A total of six chemical entities are identified in this drug sample (A–E). None of the five methods is able to resolve all six analytes. Compounds B and C coeluted in M1; A and E coeluted in M4 and M5; C and E coeluted in M2; and B, D, E, and F all merged in M3 with poor tailing. However, since all six analytes are well resolved and clearly identified in at least one of

TABLE 2.3 HPLC Orthogonal Screening[a]

Method	Column[b]	0.1% TFA, pH 1.9	0.1% Formic, pH 2.6	10 mM NH_4Ac, pH 6.8	0.1% NH_4OH, pH 10.5
M1	Synergi Fusion	X	—	—	—
M2	Luna C18 (2)	—	X	—	—
M3	Pursuit diphyenyl	—	—	X[c]	—
M4	Sunfire C18	—	—	X	—
M5	Gemini C18	—	—	—	X

[a]All screening methods use a generic gradient of 5–95% acetonitrile (except method M3[c]) in 18 min with flow rate 0.84 mL/min and equilibration time 4 min.
[b]All columns are 3 × 100 mm with 3 μm particle size.
[c]With methanol as organic modifier.

the five methods, no critical impurity content information is missing. This collective illustration of comprehensive impurity profile without method development is clearly one of the most appealing advantages of the generic orthogonal screening.

Since 46 diverse drug molecules were included in the design and selection of such screening, these orthogonal methods are applicable to the analyses of a wide variety of drug samples with no need to change methods or develop a specific method for each sample. It is good idea, however, to continue monitoring the new columns

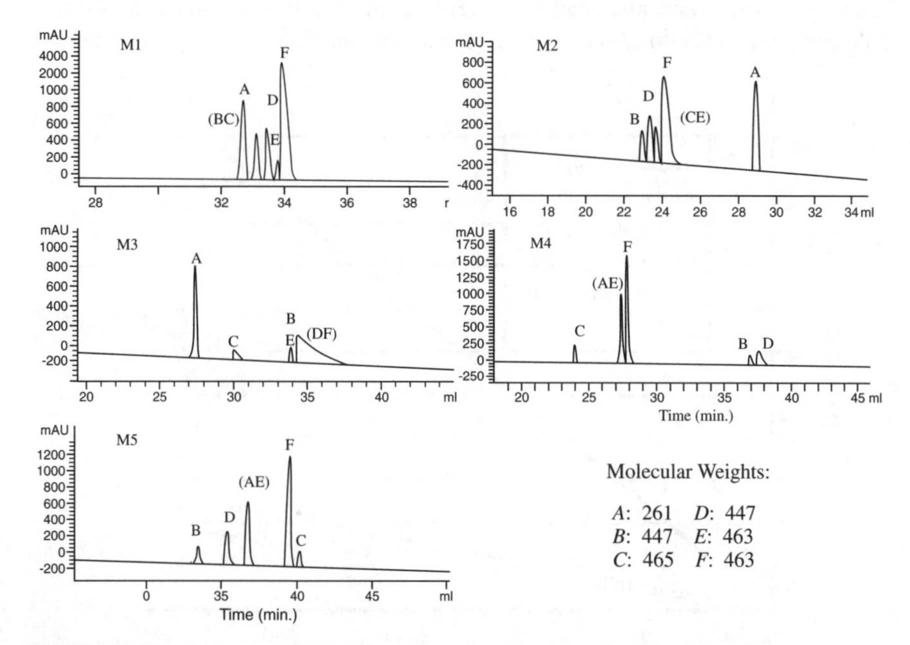

FIGURE 2.16 Orthogonal HPLC screening of a spiked drug purity sample with the methods listed in Table 2.3. Coeluting analytes are indicated in parentheses. The molecular weights determined for each analyte are (A) 261 Da; (B) 447 Da; (C) 465 Da; (D) 447 Da; (E) 463 Da; (F) 463 Da. (Adapted with permission from Ref. 13; copyright 2004 by Elsevier.)

introduced into the marketplace and add the data into the analysis in order to take advantage of the advances in column chemistry.

2.5 HIGH-EFFICIENCY SEPARATION

The more recent introduction of UHPLC (ultra-high-performance liquid chromatography) instruments such as Waters UPLC, Thermo Accela, and Agilent 1290, allows the utilization of sub-2 μm columns in routine drug analysis. The profound impact of UHPLC will be discussed in more detail in a later chapter. But very briefly, UHPLC allows very high-efficiency separation within very short analysis time. As shown in Eq. (2.1), the column efficiency or plate number (N) is one of the three terms that contribute to the resolution. Although the increase in N is reflected by only a square-root improvement in resolution (see Figure 2.1), it is most predictable and comes "free" with the UHPLC and <2-μm columns. The increase in efficiency gives more tolerance for less-than-perfect selectivity α to achieve the resolution goal of 2.

A recent analysis of the abovementioned column screening dataset using the next maximum projection (NMP) method [24] shows that for HPLC separation with 10-cm and 3-μm columns, which typically give 15,000 plates, about five orthogonal methods are required to separate a mixture of 35 compounds as shown in Figure 2.17. When analyzed with UHPLC and <2-μm columns, the average efficiency increases to about 50,000 plates and only three orthogonal methods

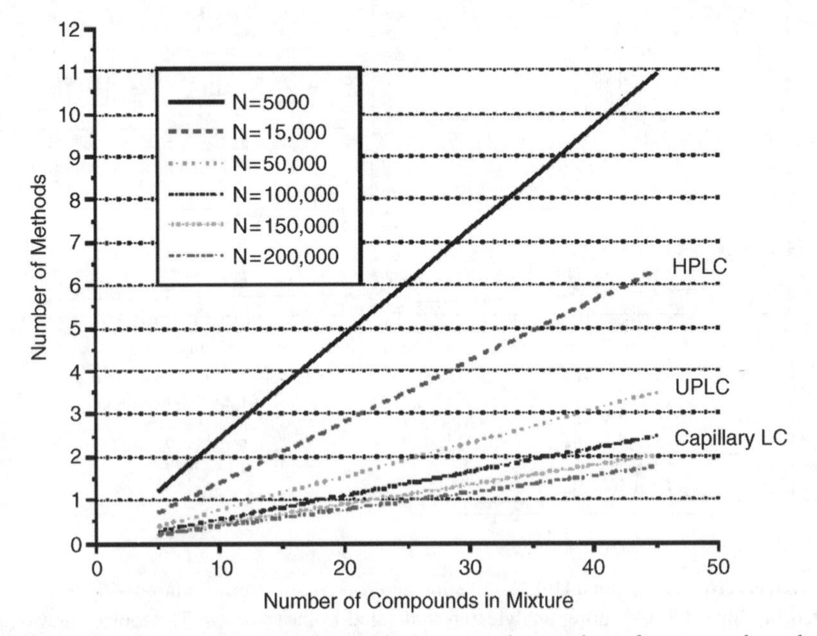

FIGURE 2.17 NMP analysis: relationship between the number of compounds and the number of methods (reprinted from Ref. 24).

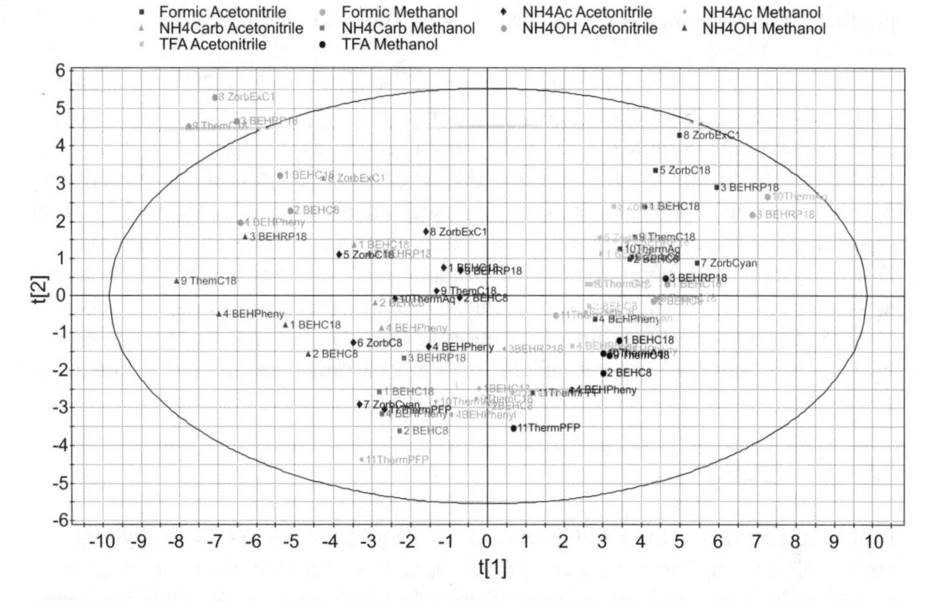

FIGURE 2.18 PCA plot of 20 UHPLC column screening data with 46 component drug test mixture. The aqueous mobile phases include TFA, HCOOH, NH_4Ac, NH_4CO_3, and NH_4OH. Acetonitrile and methanol are used as organic modifiers. The numbers in the plot represent the 20 different columns, while the mobile phases are labeled by letter.

are needed. Here the 35-compound resolution doesn't assume these compounds to be as evenly spaced as in the peak capacity calculation. Instead, the efficiency is determined from empirical data of the 46-component test mixture including difficult diastereomer pairs.

A subsequent column screening was conducted with 20 available <2-μm columns. A similar PCA analysis shows interesting V-shape method distribution (Figure 2.18). A selection of a three-orthogonal-method "region" at the vertices of the triangle followed by correlation coefficient and PPC calculation yields the UHPLC version of orthogonal screening as listed in Table 2.4. With the faster

TABLE 2.4 UHPLC Orthogonal Screening[a]

Method	Column[b]	0.1% Formic, pH 2.6	10 mM NH_4Ac, pH 6.8	0.1% NH_4OH, pH 10.5
M1	BEH Shield RP18	X[c]	—	—
M2	BEH C18	—	X[d]	—
M3	BEH Phenyl	—	—	X[d]

[a]All screening methods use a generic gradient of 5–95% modifier in 10.5 min with flow rate 0.4 mL/min and equilibration time 2.5 min.
[b]All columns are 2.1 × 100 mm with 1.7 μm particle size.
[c]With acetonitrile as organic modifier.
[d]With methodal as organic modifier.

gradient and fewer methods, the screening can now be completed within 2 h with sample and blank.

An interesting observation from Figure 2.18 is that when 200,000 plates is achieved, one single method can separate any 25 of the 46 compounds, which is about the average complexity for most sample mixtures. Lestremau et al. demonstrated an effective plate number of 162,000 when eight columns of 25 cm each were daisy-chained together [14,25]. This work was done with the tradeoff of ~ 2 h of separation time and operation at 80°C to reduce the backpressure on a HPLC system. However, with the progression of high-efficiency <2-μm, emerging 1-μm columns and enhancements in UHPLC instruments, it is anticipated that routine analysis with 100,000–200,000 plates within a 1-h timeframe could be achieved in the near future. By then, the "universal separation" dream of any chromatographer may come true.

2.6 CONCLUSIONS

Instead of optimizing the HPLC methods one variable at a time (OVAT) by trial and error, well-designed screening can provide a more systematic and efficient alternative to reduce the method development cycle time. Among all chromatography parameters, selectivity has most impact on the resolution and should be focused during method development. The two-wave approach, column and pH screening followed by temperature and gradient optimization, permits quick exploration for optimal selectivity. High-efficiency UHPLC separation with increased theoretical plates, however, can allow more tolerance in selectivity, and thus further expedite the method development. For impurity profiling and structure characterization, a generic orthogonal method set can collectively illustrate comprehensive impurity contents, thus eliminating the need to develop a sample-specific HPLC method.

REFERENCES

1. Ryan, T. W. (1998), *Anal. Lett. 31*, 2447.
2. Wiltshire, H.; Nedderman, A. (2008), in Venn, R. F., ed., *Principles and Practice of Bioanalysis*, 2nd ed., CRC Press, Boca Raton, FL, pp. 141–150.
3. Cowan, D. A. (2008), *Essays in Biochemistry*, Vol 44, Portland Press Ltd., pp. 139–148.
4. Maurer, H. H. (2008), *Handbook of Analytical Separations*, Vol. 6, Elsevier, B.V., pp. 425–445.
5. Thompson, R.; LoBrutto, R. (2007), *HPLC Pharm. Sci.* 641.
6. Bynum, K. C. (2007), *Sep. Sci. Technol. 8*, 297.
7. Smith, J.; Wikfors, R.; Fogelman, K.; Berger, T. A. (2004), *Abstracts of Papers*, 227th ACS Natl. Meeting, Anaheim, CA, March 28–April 1, 2004.
8. Berger, T. A.; Smith, J.; Fogelman, K.; Kruluts, K. (2002), *Am. Lab. 34*, 14.
9. Marsh, A.; Broderick, M.; Altria, K.; Power, J.; Donegan, S.; Clark, B. (2008), *Meth. Molec. Biol. 384*, 205.

10. Altria, K.; Marsh, A.; Sanger-van de Griend, C. (2006), *Electrophoresis 27*, 2263.
11. Altria, K. D.; Chen, A. B.; Clohs, L. (2001), *LC-GC Eur. (Liquid Chromatography–Gas Chromatography Europe) 14*, 736.
12. Snyder, L. R.; Carr, P. W.; Rutan, S. C. (1993), *J. Chromatogr. A 656*, 537.
13. Xue, G.; Bendick, A. D.; Chen, R.; Sekulic, S. S. (2004), *J. Chromatogr. A 1050*, 159.
14. Lestremau, F.; Cooper, A.; Szucs, R.; David, F.; Sandra, P. (2006), *J. Chromatogr. A 1109*, 191.
15. Demond, W.; Kenley, R. A.; Italien, J. L.; Lokensgard, D.; Weilersbacher, G.; Herman, K. (2000), AAPS (American Association of Pharmaceutical Scientists) *PharmSciTech. 1*, np.
16. Van Gyseghem, E.; Crosiers, I.; Gourvenec, S.; Massart, D. L.; Vander Heyden, Y. (2004), *J. Chromatogr. A 1026*, 117.
17. Van Gyseghem, E.; Jimidar, M.; Sneyers, R.; Redlich, D.; Verhoeven, E.; Massart, D. L.; Vander Heyden, Y. (2005), *J. Chromatogr. A 1074*, 117.
18. Wang, X.; Li, W.; Rasmussen, H. T. (2005), *J. Chromatogr. A 1083*, 58.
19. Pellett, J.; Lukulay, P.; Mao, Y.; Bowen, W.; Reed, R.; Ma, M.; Munger, R. C.; Dolan, J. W.; Wrisley, L.; Medwid, K.; Toltl, N. P.; Chan, C. C.; Skibic, M.; Biswas, K.; Wells, K. A.; Snyder, L. R.; (2006), *J. Chromatogr. A 1101*, 122.
20. Ferguson, P.; Whitlock, M. (2005), *Pfizer Internal Report.*
21. Massart, D. L.; Vander Heyden, Y. (2005), *LC-GC Eur. 18*, 84.
22. Euerby, M. R.; Petersson, P. (2006), The use of principal component analysis for the characterization of reversed-phase liquid chromatographic stationary phases, in Kromidas, S., ed., *HPLC Made to Measure*, Wiley-VCH Verlag GmbH & Co. KGaA, Weinheim, Germany, pp. 264–279.
23. Liu, Z.; Patterson, D. G. Jr.; Lee, M. L. (1995), *Anal. Chem. 67*, 3840.
24. Bemish, R. (2007), *Pfizer Internal Report.*
25. Lestremau, F.; de Villiers, A.; Lynen, F.; Cooper, A.; Szucs, R.; Sandra, P. (2007), *J. Chromatogr. A 1138*, 120.
26. Snyers, R.; Janssens, W.; Huybrechts, T.; Vrielynck, S.; Somers, I. (2007), Conference oral presentation at HPLC 2007, Ghent, Belgium, June 17–21.

CHAPTER 3

Rapid Analysis of Drug-Related Substances using Desorption Electrospray Ionization and Direct Analysis in Real Time Ionization Mass Spectrometry

HAO CHEN and JIWEN LI

Center for Intelligent Chemical Instrumentation, Department of Chemistry and Biochemistry, Clippinger Laboratories, Ohio University, Athens, OH 45701

3.1 INTRODUCTION

Mass spectrometry (MS) has advanced enormously in more recent years. The advent of new methods of ion production, novel mass analyzers, and new tools for data processing has made it possible to analyze almost all chemical entities, ranging from small organic compounds, to large biological molecules, to whole living cells and tissues. Today mass spectrometry has become one of the most powerful and popular modern physiochemical methods for studying the details of elemental and molecular processes in nature and plays an increasing significant role in the chemical and life sciences because of its unparalleled capability of providing information on molecular weight, chemical structures, isotopic content, and even molecular dynamics.

Ambient mass spectrometry (MS) [1,2] is a conceptually new innovation in the field, initiated with the introduction of desorption electrospray ionization (DESI) [3] by Cooks and coworkers and direct analysis in real-time (DART) [4] ionization mass spectrometry by Cody and coworkers. The strength of this new family of technologies stems from their capacity for the direct analysis of ordinary objects in the open atmosphere of the laboratory or in their natural environment of samples, bypassing most elements of the analytical system and transferring ions into the mass

Characterization of Impurities and Degradants Using Mass Spectrometry, First Edition.
Edited by Birendra N. Pramanik, Mike S. Lee, and Guodong Chen.

spectrometer with little or no sample preparation [2]. This new field of MS has developed rapidly, and a number of ambient ionization methods have been developed, including desorption atmospheric-pressure chemical ionization (DAPCI) [5,6], electrospray-assisted laser desorption/ionization (ELDI) [7], matrix-assisted laser desorption electrospray ionization (MALDESI) [8], extractive electrospray ionization (EESI) [9], atmospheric solid analysis probe (ASAP) [10] jet desorption/ionization (JeDI) [11], desorption sonic spray ionization (DeSSI) [12], field-induced droplet ionization (FIDI) [13], desorption atmospheric-pressure photoionization (DAPPI) [14], plasma-assisted desorption/ionization (PADI) [15], dielectric barrier discharge ionization (DBDI) [16], liquid microjunction surface sampling probe method (LMJ-SSP) [17], atmospheric-pressure thermal desorption/ionization (APTDI) [18], surface sampling probe (SSP) [19], fused-droplet electrospray ionization (FD-ESI) [20], helium atmospheric-pressure glow discharge ionization (HAPGDI) [21], neutral desorption extractive electrospray ionization (ND-EESI) [22], laser ablation electrospray ionization (LAESI) [23], and low-temperature plasma (LTP) for ambient desorption/ionization [24]. Figure 3.1 shows the cartoon illustration of several ambient ionization techniques, which integrate sampling process with ionization (i.e., directly desorb and ionize samples from a surface in the condensed phase into gaseous ions for MS detection). Mechanistically, these new methods involve desorption or ionization by either spray, heat, plasma, high electric field, or laser impact. All these techniques have shown that ambient MS can be used as a rapid tool to provide efficient desorption and ionization and hence to allow mass spectrometric characterization of target compounds. The general analytical aspects of

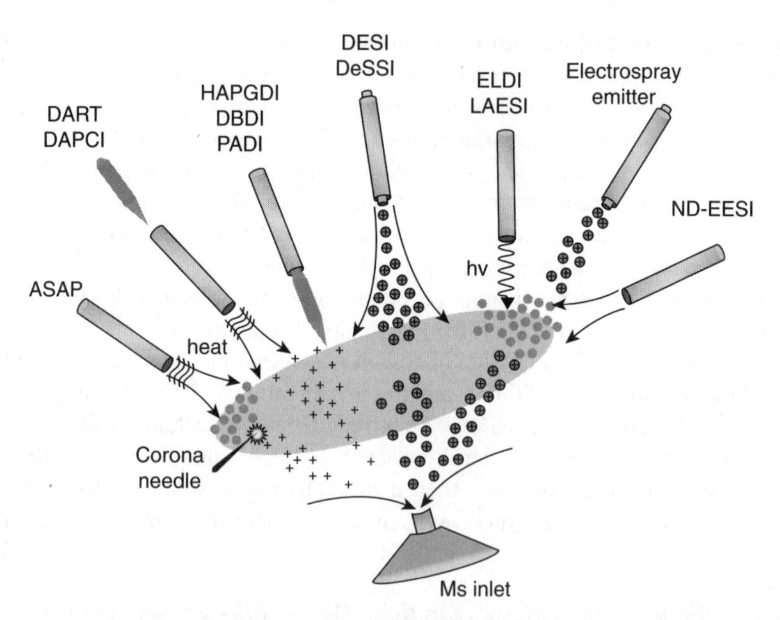

FIGURE 3.1 Techniques used in ambient desorption/ionization (reprinted with permission from Ref. 2; copyright 2008 by Elsevier Ltd.).

ambient mass spectrometry have been reviewed [2,25]. Although DESI, the first ambient ionization method, was developed in less than 7 years, there have been many reports published in the literature with regard to the wide applications of ambient mass spectrometry, covering the detection and analysis of both small molecules (e.g., explosives) to macromolecules (e.g., proteins and polymers). In this chapter, we introduce only the DESI and DART analysis of drug-related substances including impurities. First, we start with the description of the experimental apparatus and ionization mechanisms of DESI and DART along with their general analytical performances; then we present some novel applications of these methods for drug analysis in complicated biological systems (e.g., in urine, plasma and skin), high-throughput analysis [26,27], chemical imaging of drugs in tissues by DESI [28,29], and near-instantaneous chemical profiling of living animals by DART [30].

3.2 IONIZATION APPARATUS, MECHANISMS, AND GENERAL PERFORMANCE

3.2.1 Desorption Electrospray Ionization (DESI)

As shown in Figure 3.2 [3], the DESI experiment, in its simplest form, uses a fine spray of charged droplets (with diameters of <10 μm [2]) to desorb and ionize small organic molecules or large biomoleucles deposited on a surface of interest. Typically, the charged droplets used in DESI is generated by electrosonic spray ionization (ESSI) [31] of an appropriate solvent such as a mixture of methanol and water

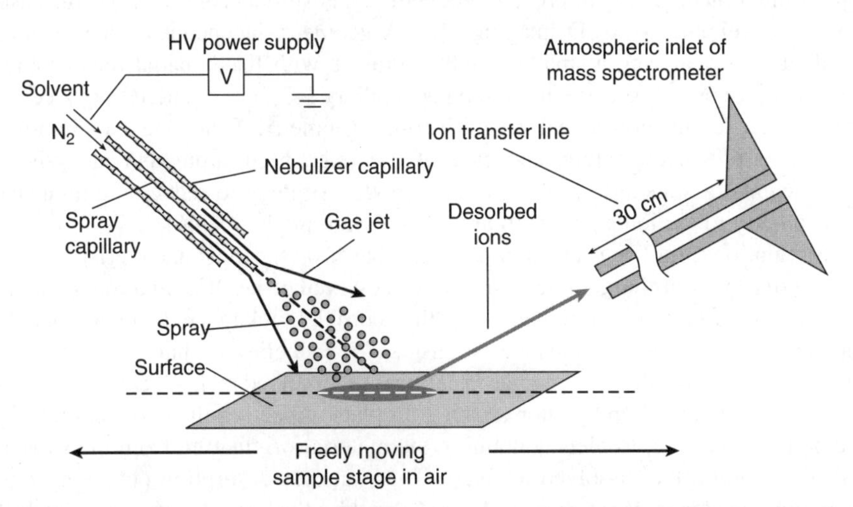

FIGURE 3.2 Schematic of a typical DESI experiment. The sample solution was deposited from solution and dried onto a PTFE surface, and methanol–water (1 : 1 containing 1% acetic acid or 0.1% aqueous acetic acid solution) was sprayed at a flow rate of 3–15 μL/min under the influence of a high (4 kV) voltage. (Reprinted with permission from Ref. 3; copyright 2004 by the American Association for the Advancement of Science.)

containing a small amount of acetic acid, to which an electrical potential of several kilovolts (e.g., 4–5 kV) is applied and high-pressure nebulizing gas is used to assist desolvation. The resulting high-momentum sprayed droplets (velocities typically in excess of 100 m/s) is directed to the analyte on the surface for desorption and ionization. The desorbed ions are sampled by a mass spectrometer equipped with an atmospheric-pressure interface. Depending on the capabilities of the mass spectrometer, tandem MS, selected ion monitoring, exact mass, and other types of measurements are possible [32], which are valuable characteristics in helping address the complexity of the sample mixtures as samples are usually analyzed without any preseparation from their natural matrices [33]. DESI has been implemented using various mass spectrometers, including triple quadrupoles [34], linear ion traps [35], Orbitrap [36], quadrupole time-of-flight (Q-TOF) [37], ion mobility/TOF and ion mobility/QTOF hybrids [38], Q-traps [39], and Fourier transform ion cyclotron resonance (FT-ICR) instruments [40] as well as fieldable ion trap mass spectrometers. In DESI, surface materials can be glass slides, paper, copper foil, Teflon, or other substances. In particular, one study shows that polymethylmethacrylate (PMMA) was found to give the best performance in terms of sensitivity for the analysis of pharmaceuticals and metabolites [41]. Commercial DESI ion sources are available from Prosolia Inc., USA.

It has been known [2] that the signal intensity of DESI spectra strongly depends on the various geometric factors, including the angle and the distance between the DESI sprayer and the surface and between the surface and the MS inlet. These geometric factors often have to be optimized to obtain a good signal and precise control in the x, y, z dimensions for the sample and for the sprayer together with angular control of the spray direction to greatly improve reproducibility. Positional control is also the basis for the development of 2D imaging [42]. A geometry-independent DESI source configuration featuring a small, airtight enclosure with fixed spatial relationships between the sprayer, surface and sampling capillary [43] was reported to reduce the dependence of the ion signal on the various geometric factors and to improve ionization efficiency, safety, and ease of use (e.g., high-throughput analysis of the contents of standard 96-well plates). Another strategy to reduce the influence of geometry to the ion signal is to use transmission mode [44–46].

Simulations [2,47] and anecdotal evidence based on spectral characteristics (e.g., strong similarity in charge state distributions observed in the DESI and conventional ESI spectra of protein samples) indicate that a droplet pickup mechanism probably operates under most circumstances. Instead, it is believed that the surface is prewetted by initial droplets. Surface analytes are dissolved in this localized solvent layer and are picked up by later-arriving droplets impacting the surface, creating numerous offspring droplets containing the material originating from the solvent layer, including the dissolved analytes. Thus, analyte desorption occurs by momentum transfer in the form of charged droplets that are then ionized by ESI mechanisms. Other possible mechanisms previously proposed [3] include chemical sputtering [48] and gas-phase ionization process through proton transfer or other ion–molecule reactions. Chemical sputtering involves charge transfer between a gas-phase ion and a molecular species on the surface with enough momentum transfer to lead to desorption of the surface-derived ions. Charge transfer can involve electron, proton, or other ion exchange, and the process is known from studies of ion–surface collision phenomena under vacuum [48]. For instance, direct ionization of carotenoids from fruit skin is probably accounted for by this mechanism [3]. A third suggested mechanism is volatilization/desorption of neutral species from the surface followed by gas-phase ionization through proton transfer or other ion–molecule reactions, as evidenced by the temperature effect, with a wide variety of nonvolatile compounds (heavy terpenoids, carbohydrates, peptides) showing high ionization efficiency at elevated surface temperatures [3].

In addition to being used regularly for solid sample analysis from surfaces, DESI has more recently been extended to allow the direct analysis of liquid samples or liquid films [39,44,45,49–51]. Interestingly, high-mass proteins (e.g., BSA with MW 66 kDa) apparently can be more easily desorbed and ionized from solution than from dried samples on surface, probably because there is less aggregation in solution than in the solid form [39]. Also, it has been demonstrated that liquid sample DESI can be used for online coupling of electrochemistry [39,52] and microfluidics [51] with mass spectrometry.

Reactive DESI [46,53–58] is a further development in DESI that exploits the potential for coupling specific ion–molecule reactions [59–67] with the ionization event and so greatly improves the selectivity and efficiency with which compounds with specific functionalities are detected. It involves the use of a spray solution that contains specific reagents intended to allow particular ionic reactions during the sampling process. Compared with traditional methods [68] employing solution phase derivatization followed by ESI, the online derivation in reactive DESI is much faster (typically taking seconds for one sample analysis), thereby speeding up the analytical process.

In terms of analytical performance in general, the limits of detection (LODs) in DESI are typically an order of magnitude greater than those in the corresponding ESI experiments [2]. A typical LOD for small molecules, such as explosives, is in the low-femtomole range [69], and biopolymers, such as peptides and proteins, approach similar levels [3]. Relative standard deviations (RSDs) below 5% have been reported, with the use of a particularly suitable surface, a porous polytetrafluoroethylene (PTFE) surface, which shows minimal cross-contamination between samples and improved sensitivity and signal stability.

3.2.2 Direct Analysis in Real Time (DART)

Like DESI, DART also operates in open air under ambient conditions. The basic DART source [4] (Figure 3.3) consists of a tube divided into three chambers through which a gas such as nitrogen or helium flows. The first discharge chamber contains a ground counterelectrode and a needle electrode to which an electrical potential of several kilovolts (1–5 kV) initiates an electrical discharge producing ions, electrons, and excited-state species in a plasma. It is believed that the electronic or vibronic excited state species (metastable helium atoms or nitrogen molecules) are the working reagent in DART. In a second chamber, a second perforated electrode can be biased to

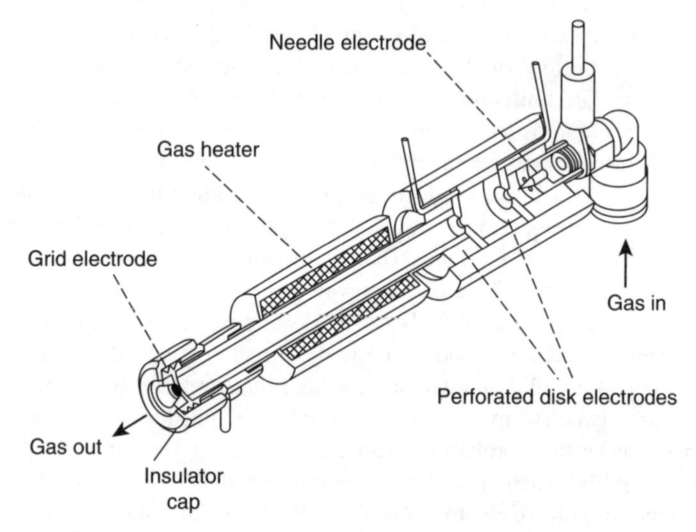

FIGURE 3.3 Cutaway view of the DART source (reprinted with permission from Ref. 4; copyright 2005 by the American Chemical Society).

remove ions from the gas stream. The gas flow then passes through a third region that can be heated; the gas temperature was found to be adjustable from room temperature up to 250°C. It was found that heating the gas aids the desorption of some analyte materials. Gas exiting through a third perforated electrode or grid can be aimed directly toward the mass spectrometer orifice, or the gas flow can be reflected off a sample surface and into the mass spectrometer. The DART position was adjustable on an *x,y,z* stage over a wide range of angles and distances. However, the exact positioning, distance, and angle of DART with respect to the sample surface and the mass spectrometer are not critical. A typical DART–sample–orifice distance was 5–25 mm. However, the polarity of the second perforated electrode and the grid electrode was critical. The potentials of these electrodes are biased to positive potentials (e.g., 100 and 250 V, respectively) for positive-ion detection and to negative potentials for negative-ion detection. The DART ion source is commercially available and a trademark of JEOL USA, Inc. The DART ion source usually is coupled with TOF instrument. Accurate mass of the first monoisotopic peak of the analyte, and the relative abundances of the peaks in the isotopic clusters provided reliable information for identification [70]. Other instruments such as triple-quadrupole [71], fieldable [32], and hybrid quadrupole TOF (Q-TOF) [27] mass spectrometers [32] were also used in conjunction with DART ion sources.

Ionization is performed by reaction of electronic or vibrionic excited-state metastable species [4,72–74] with reagent molecules and analytes. DART can be used for the analysis of gases, liquids, or solids, and is a useful tool for small-molecule analysis, and not a technique for the analysis of large biomolecules such as proteins [70]. The ionization mechanisms in the DART source are complex and do not follow a single process. Different ionization mechanisms occur depending on the nature of the carrier gas, analyte concentration, and polarity of ions. Two of the proposed mechanisms are penning ionization [75,76] and proton transfer from water clusters [77,78]. Penning ionization occurs when a metastable atom transfers energy to an analyte (M), resulting in the formation of a molecular ion $M^{+\cdot}$. This process will take place if the analyte molecule M has an ionization energy less than the internal energy of the metastable atom. For the proton transfer from water clusters to occur in producing an $[M + H]^+$, the analyte (M) must have a higher proton affinity than the ionized water cluster. The ionization mechanism of negative-ion direct analysis in real time (NI-DART) has been investigated using various organic compounds [79] and the NI-DART-generated ionization products were found to be similar to negative-ion–atmospheric-pressure photoionization (NI-APPI). It is suggested that four ionization processes, including electron capture (EC), dissociative EC, proton transfer, and anion attachment are involved in the NI-DART. An important feature for DART ionization is that, in comparison with electrospray ionization, alkali metal cation attachment is never observed [4], simplifying the interpretation of mass spectra of unknown compounds. Another significant feature of DART is that no memory effects or sample carryover are observed, favoring high-throughput analysis [4]. Unlike DESI, the plasma-based DART technique can also access nonpolar compounds, and their mass spectra are simple (no solvent clusters, multiply charged or alkali metal adducts).

3.3 DRUG ANALYSIS IN BIOLOGICAL MATRICES USING DESI AND DART

As reported in the literature, DESI has become very successful in the fast analysis of a variety of different analytes including pharmaceuticals [37,80,81], metabolites [41,82], drugs of abuse [83,84], explosives [54,55,59], chemical warfare agents [85], polymers [86], bacteria [87], nature products [88], and even intact tissues [89,90] as well as thin-layer chromatography plates [91]. Likewise, DART has demonstrated success in the ionization of hundreds of chemicals, including chemical agents and their signatures, pharmaceuticals [4,92,93], metabolites, amino acids, peptides, oligosaccharides, synthetic organics, organometallics, explosives, toxic industrial chemicals, counterfeit drugs [94], bacterial fatty-acid methyl esters [95], flavors and fragrances [96], and also planar chromatography [97]. This section focuses on the discussion of DESI/DART analysis of drug-related substances in complicated biological matrices.

It is well known that two key bottlenecks in pharmaceutical bioanalysis are sample cleanup and chromatographic separation [71]. Although multiple approaches have been developed since the 1990s to either shorten or multiplex these steps, they remain the rate-limiting steps as absorption–distribution–metabolism–excretion (ADME) property screening is being routinely incorporated into the drug discovery process. Therefore, ambient ionization methods such as DESI and DART will be advantageous in expediting the pharmaceutical bioanalysis because sample cleanup and chromatographic separations are not needed as a result of their capability of direct examining biological samples.

3.3.1 DESI Application

One of the remarkable applications, of DESI is in vivo sampling of living tissue surfaces [3]. In the experiment, an aqueous–alcohol DESI spray was directed onto the finger of a person who had taken 10 mg of the over-the-counter antihistamine loratadine. About 40 min after ingestion of the tablet, the molecule became detectable directly on the skin, as shown in Figure 3.4. In another study [5], a thin layer of ibuprofen gel containing 5% w/w of the active ingredient was applied to the surface of a human finger. The gel was gently massaged until absorbed by the skin. Using the DESI technique, one could readily detect the drug at the point of application 20 min after applying the gel. Figure 3.5a shows the negative DESI mass spectrum in which the base peak in the spectrum is deprotonated ibuprofen (*m/z* 205), and its assignment was confirmed with collision-induced dissociation (CID, Figure 3.5b).

Drugs of abuse and their metabolites in complicated biological matrices such as urine can also be analyzed by DESI-MS [98]. The acquired DESI spectrum of a urine sample containing benzodiazepines (the structures of diazepam and its metabolites are illustrated in Figure 3.6) is displayed in Figure 3.7. The identified ions in the spectrum include the $[M+H]^+$ ions of demethylated (*N*-desmethyldiazepam at *m/z* 271), demethylated and hydroxylated (oxazepam at *m/z* 287), hydroxylated (temazepam at *m/z* 301), and a dihydroxylated diazepam metabolite (hydroxytemazepam at *m/z* 317; parahydroxytemazepam). The assignments were confirmed by recording the product ion spectra of each ion and by comparing these spectra to those of the corresponding standards and literature and were also in agreement with separate GC-MS experiments. High sensitivity was achieved in the analysis and good signal intensity was detected even after a 100 fold dilution of the samples. Interestingly, selectivity in the DESI-MS measurements for different kinds of analytes could be increased further by optimizing the spray solvent composition; the use of an entirely

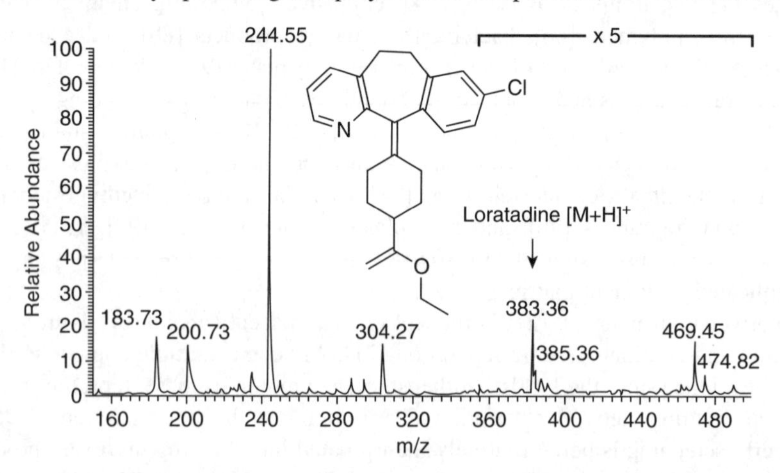

FIGURE 3.4 DESI spectrum recorded by spraying methanol–water onto the finger of a person 50 min after taking 10 mg of the over-the-counter antihistamine loratadine (*m/z* 383/385) (reprinted with permission from Ref. 3; copyright 2004 by the American Association for the Advancement of Science).

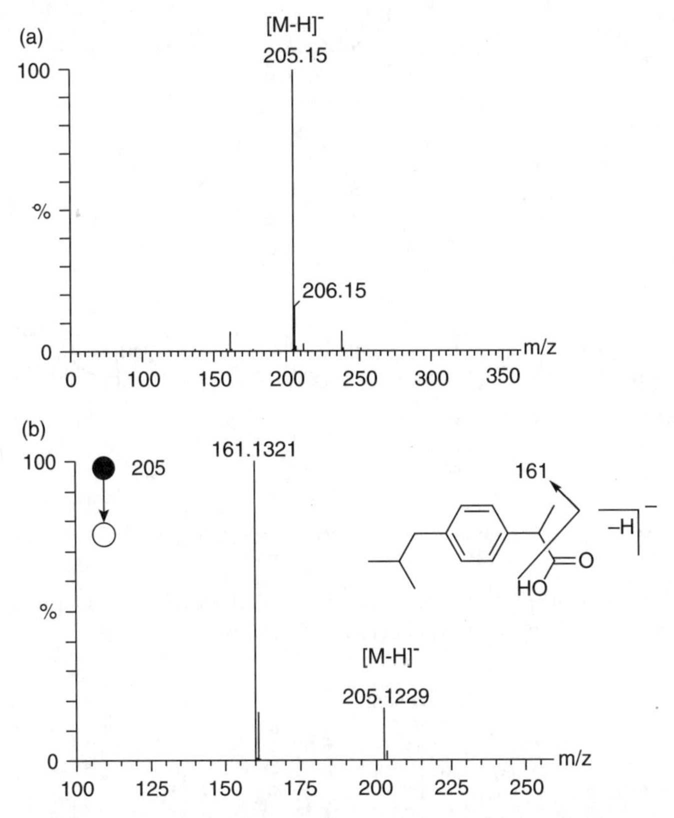

FIGURE 3.5 (a) Negative-ion DESI-MS spectrum of ibuprofen gel desorbed off skin 20 min after applying the gel; (b) MS/MS spectrum obtained for the deprotonated ibuprofen ion (reprinted with permission from Ref. 5; copyright 2006 by John Wiley & Sons, Ltd.).

aqueous solvent enhanced the signal of polar analytes, such as the benzodiazepines, whereas the use of a spray solvent with a high organic content increased the signal of less polar analytes, such as codeine and morphine. Choosing an appropriate spray solvent is critical to DESI ionization. In 2010, it was shown that using organic solvent such as acetonitrile can enable online desalting for the direct analysis of urine glucose [46].

3.3.2 DART Application

Pharmaceuticals in the form of tablets and capsules, such as prescription drugs, over-the-counter supplements, veterinary medicines, and confiscated illicit drugs, have been successfully revealed by using DART without breaching or opening the tablet in any way. The rapid detection of acetaminophen and oxycodone in a painkiller is one example (Figure 3.8) [4]. These active ingredients were detected within seconds without crushing, breaching, or extracting the tablet.

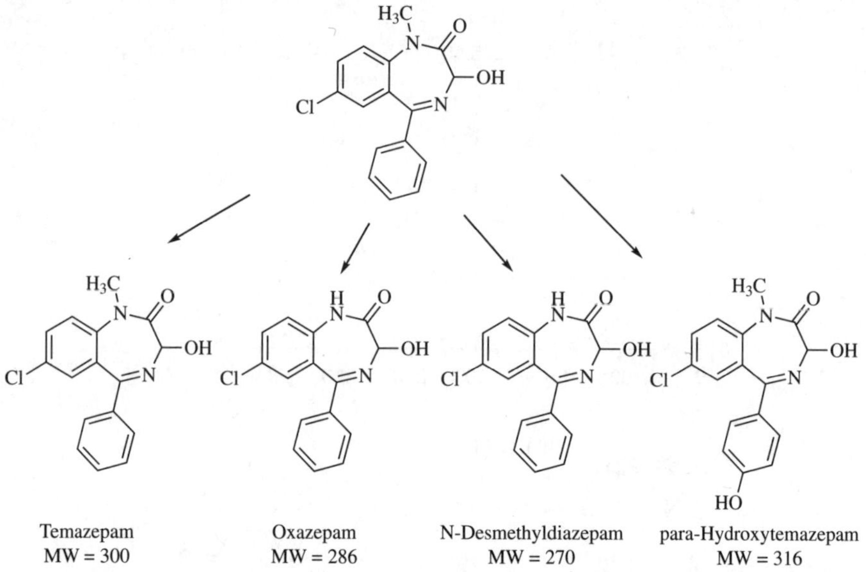

Temazepam
MW = 300

Oxazepam
MW = 286

N-Desmethyldiazepam
MW = 270

para-Hydroxytemazepam
MW = 316

FIGURE 3.6 Structures of diazepam and metabolites (reprinted with permission from Ref. 98; copyright 2007 by the Royal Society of Chemistry).

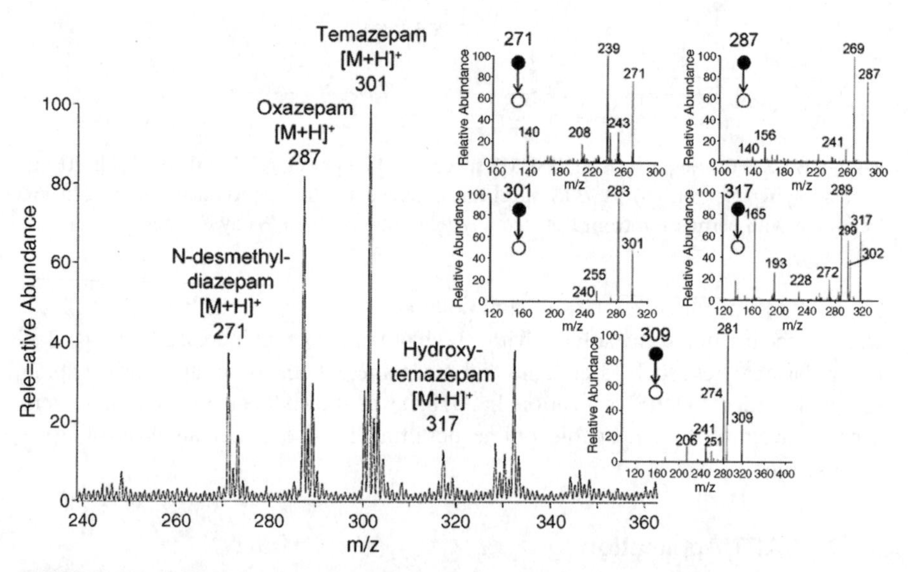

FIGURE 3.7 DESI mass spectrum of a benzodiazepine containing urine sample showing the presence of the diazepam metabolites. The insets show the product ion MS/MS spectra of *m/z* 271 ($[M + H]^+$ of *N*-desmethyldiazepam), *m/z* 287 ($[M + H]^+$ of oxazepam), *m/z* 301 ($[M + H]^+$ of temazepam), *m/z* 317 ($[M + H]^+$ of hydroxytemazepam), and *m/z* 309 ($[M + H]^+$ of alprazolam). The spray solvent was water–formic acid (100 : 0.1%) at 3 μL/min. Teflon was used as the sampling surface. (Reprinted with permission from Ref. 98; copyright 2007 by the Royal Society of Chemistry.)

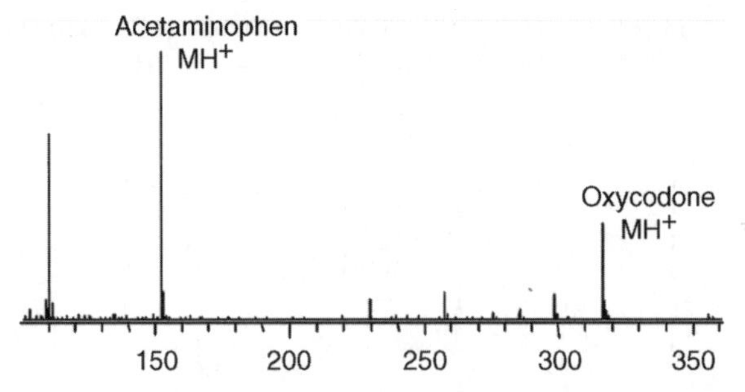

FIGURE 3.8 Rapid detection of acetaminophen and oxycodone in an intact painkiller tablet (reprinted with permission from Ref. 4; copyright 2005 by the American Chemical Society).

Bodily fluids including blood, saliva, and urine are also suitable for DART analysis without sample preparation. In these experiments, a glass rod was dipped into the fluid and placed in front of the DART ionizing beam. Endogenous substances such as amino acids, urea, uric acid, and creatinine as well as exogenous substances such as over-the-counter medicines, prescription drugs, and caffeine were detected. DART can provide rapid and easy drug screening in time-critical situations. For example, ranitidine is an over-the-counter medication used to prevent and treat symptoms of heartburn associated with acid indigestion and sour stomach. A person ingested a 300 mg dose, and a urine sample was analyzed with DART 5 h later. As shown in Figure 3.9, an excellent-quality mass spectrum was obtained within 30 s, which clearly shows the presence of ranitidine and its metabolites desmethylranitidine and ranitidine *N*-oxide [4]. Screening of cocaine and its metabolites in human urine samples by DART coupled to time-of-flight mass spectrometry after online preconcentration utilizing microextraction by packed sorbent was also demonstrated [70].

In order to gain high reproducibility, an automated DART ionization source coupled with a LEAP Technologies autosampler has been developed and evaluated for quantitative bioanalysis [71]. As shown in Figure 3.10, a total of nine injections of benzoylecgonine in unextracted rat plasma were made onto the system and the coefficient of variance (CV) of the peak height was about 3.1%, which is sufficient for

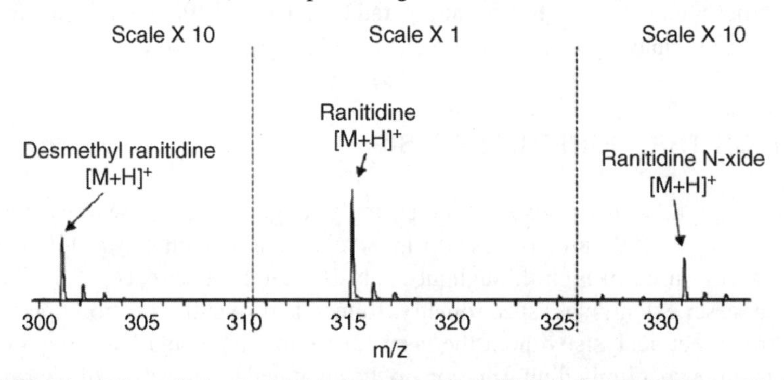

FIGURE 3.9 Urinanalysis obtained in less than 30 s with DART. Ranitidine and its metabolites were detected in raw, untreated urine 5 h after ingestion of the pharmaceutical. (Reprinted with permission from Ref. 4; copyright 2005 by the American Chemical Society.)

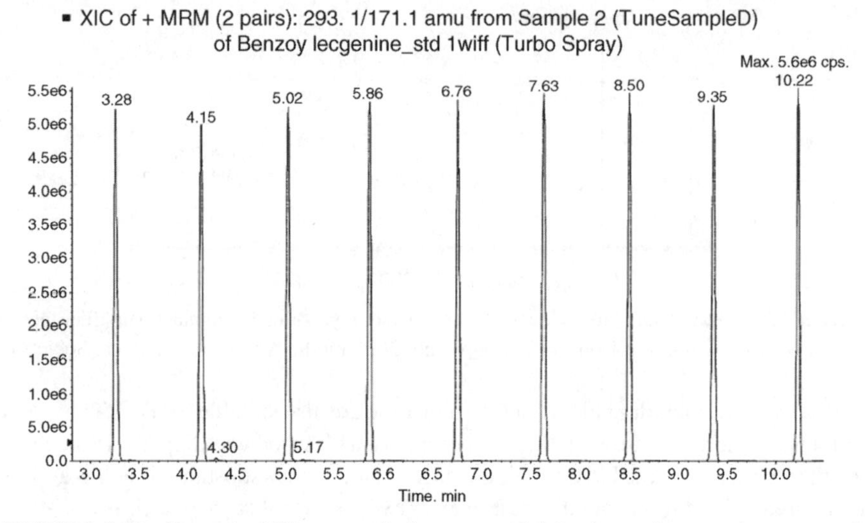

FIGURE 3.10 Reproducibility test of nine repeated injections of a rat plasma sample containing 1 μM benzoylecgonine. The %CV of the peak height is 3.1%. (Reprinted with permission from Ref. 71; copyright 2009 by the American Chemical Society.)

quantitative bioanalytical work. Precision and accuracy were also tested for multiple test compounds over a dynamic range of four orders of magnitude. The system has been used to analyze biological samples from both in vivo pharmacokinetic studies and in vitro microsomal/S9 stability studies, and the results generated were similar to those obtained with conventional LC-MS/MS methods. As a demonstration, an oral mouse PK study [71] at 25 mg/kg was conducted with a Millennium proprietary compound (compound B). PK samples at seven timepoints were collected with three animals per timepoint. Plasma samples were analyzed using the conventional LC-MS/MS method, as well as the DART-MS/MS system. It turns out that the mean percentage difference of the concentrations measured with these two methods ranged from −4.7% to 16.4%, demonstrating a good correlation between the two methods. These experiments suggest that the automated DART system has significant potential for high-throughput bioanalysis and real-time bioanalysis.

3.4 HIGH-THROUGHPUT ANALYSIS

High-throughput measurements are increasingly sought in many research areas such as drug discovery [99], proteomics [100], and combinatorial chemistry [101]. Techniques widely used in high-throughput analysis include spectroscopic techniques, and, to a lesser extent, mass spectrometry. Raman and near-IR spectroscopies allow high-throughput analysis without the need for sample manipulation. They provide useful if somewhat limited information on the chemical composition of pharmaceuticals, in a nondestructive fashion [102,103]. Ambient mass spectrometry methods such as DESI and DART are also compatible with high-throughout analysis.

In the DESI experiments for fast scanning of multiple samples, a variable-speed moving belt was built for sampling and used to provide rapid qualitative and semi-quantitative information on drug constituents in tablets. The sample transport system was successful in placing sample at the appropriate positions with respect to the DESI source and allowing good-quality mass spectra to be recorded. Sampling rates as high as 3 samples/s were achieved in the ambient environment, with RSD of 2–8%. Impurities and components present at levels as low as ~0.1% can be identified. As an example, a series of 16 mass spectra for 16 Claritin tablets acquired at a speed of 0.76 samples/s is shown in Figure 3.11a. All the spectra have

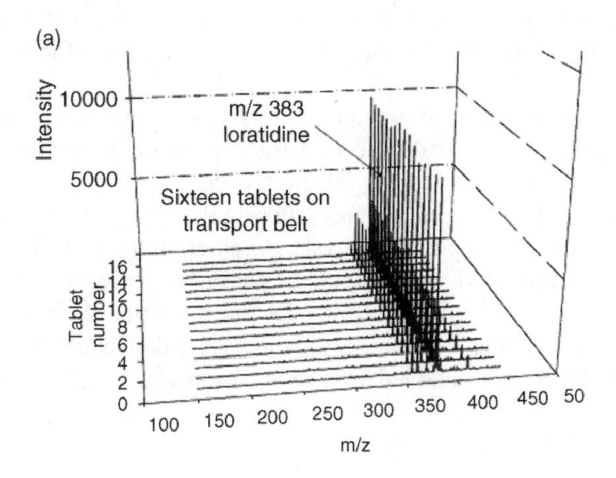

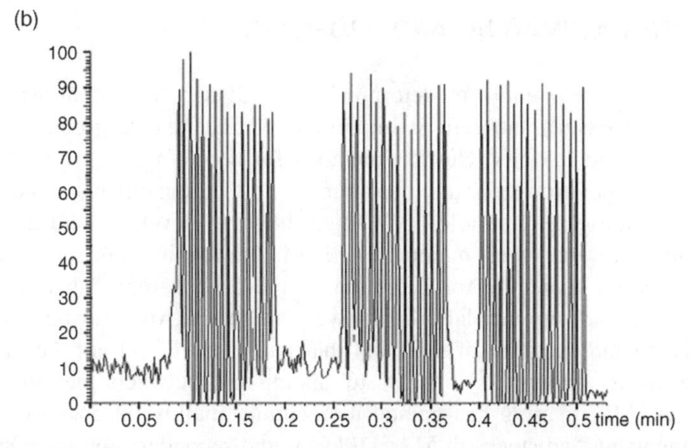

FIGURE 3.11 (a) Individual mass spectra of the 16 Claritin tablets on a moving belt at a speed of 0.76 sample/s. The main peak observed is at *m/z* 383, and the chloride signature is also observed at 385. All the peaks at *m/z* 383 for all tablets have almost identical intensities with a standard deviation of ~2%. (b) Ion chromatogram of high-throughput DESI (2.67 samples/s). By monitoring of ions at *m/z* 383, three sets of 16 Claritin tablets were investigated using the moving belt as sample transportation system. (Reprinted with permission from Ref. 26; copyright 2005 by the American Chemical Society.)

almost the same relative ion intensities for the peak *m/z* 383 with 4.8% RSD, and these peaks are well resolved. Other typical data recorded at 2.67 samples/s operating frequency are shown in Figure 3.11b, for three sets of 16 tablets. All 16 tablets remain resolved at this analysis speed, but the RSD of the ion abundance of the base peak *m/z* 383 in each set of tablets increased to 7.8%. The increasing RSD of the signal intensity with increasing sampling frequency is associated with the smaller data acquisition time and also to a less reproducible sampling position, since the moving belt also undergoes more lateral movements at higher speed. A more stable moving-belt system is in need to allow considerably higher sample throughput. Another factor that limits performance at high sampling speeds is sample carryover, a common effect in mass spectrometry. This was determined to be less than 0.1 s for most tablets, by measuring the time required for the signal intensity to drop from 100% to 10%.

With the appealing features of no memory effects and simplicity in operation, DART-MS is a significant and invaluable tool for high-throughput analysis for drugs and their metabolites as well. In this regard, the automated DART ionization source coupled with the LEAP Technologies autosampler [71] mentioned above would be ideal for high-throughput bioanalysis. Another study using DART for metabolomic fingerprinting [104] in serum samples has been performed. Again, in that study, each DART run required only 1.2 min, during which time more than 1500 different spectral features were observed in a time-dependent fashion. A repeatability of 4.1–4.5% was obtained for the total-ion signal using a manual sampling arm.

3.5 CHEMICAL IMAGING AND PROFILING

The information gained from determining the distribution of drugs and their metabolites in tissues and cells is important for understanding and predicting a drug's action and toxicity. Current methods for acquiring drug distributions in tissues include positron emission tomography (PET), magnetic resonance imaging (MRI), autoradiography, whole-body autoradiography (WBA), and fluorescence microscopy. However, their major limitation is the requirement for a radioactive label or reporter molecule for detection of the drug compound, rendering the procedure time-consuming, labor-intensive, and costly. Also, for both WBA and PET, it is the radioactivity of the label that is measured and not the intact drug molecule itself, making it difficult to distinguish between the drug and its metabolites. MS imaging has gained momentum, mainly because of continued improvements and advances in MALDI [105] and secondary-ion mass spectrometry (SIMS) [106], as these two techniques offer high spatial resolution to enable applications on the cellular or subcellular levels. However, the challenge for small-molecule imaging by MALDI is that matrix ions and mixed analyte–matrix clusters crowd the low-mass range, limiting confident detection of analyte ions of <750 Da. An alternative method to MALDI for small-molecule detection is DESI, which also allows for direct analysis [107] and imaging [90] of biological tissues

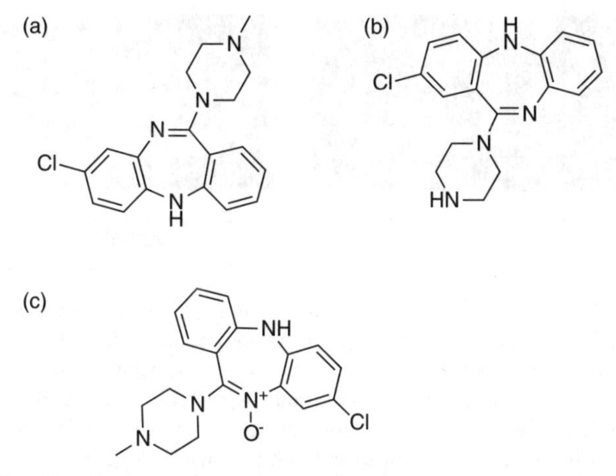

FIGURE 3.12 Chemical structures of (a) clozapine, (b) *N*-desmethylclozapine, and (c) clozapine-*N*-oxide (reprinted with permission from Ref. 28; copyright 2008 by the National Academy of Sciences of the USA).

and other surfaces [108] without using a chemical matrix for ionization. In studies of model pharmaceutical compounds, the DESI signal response was shown to be linear ($R^2 = 0.996$), accurate (relative error ±7%), and precise (relative standard deviation 7%) when analyzing neat solutions deposited on hydrophobic surfaces [109]. DESI-MS imaging provides information on the spatial distribution of molecules at or near the surface with a lateral resolution that was reported to be ~250 μm [110].

In a DESI chemical imaging of tissues in 2008, clozapine (Figure 3.12a), an atypical antipsychotic developed in the 1960s targeted toward treatment of schizophrenia, was chosen as a drug for the study. The main active metabolites of clozapine are desmethylclozapine (Figure 3.12b) and *N*-clozapine-*N*-oxide (Figure 3.12c). Clozapine distribution in brain, liver, lung, and spleen has been described as including a tendency for slower clearance from lung tissue than from brain [111]. Clozapine is found in relatively high density in the cerebral cortex, caudate putamen, nucleus accumbens, olfactory tubercle, and substantia nigra pars reticulate in both rat and monkey brains [112] but with less frequency in other areas of the brain, such as the cerebellum and striatum. DESI-MS imaging was applied to visualize the distribution of clozapine in rat brain sections after the animals had been dosed and euthanized. Figure 3.13a, b shows the optical image and the DESI image, respectively, from a rat brain section taken from an animal that had been dosed at 50 mg/kg via oral gavage and the brain removed 30 min after dose. The tissue was imaged in the MS/MS mode, and the DESI product ion spectra resulting from fragmentation of clozapine at *m/z* 327.1 $[M+H]^+$ into *m/z* 270.1 were recorded with a pixel size of 245 × 245 μm. The distribution of clozapine in the tissue section showed relatively high levels in areas of the brain corresponding to cortical regions, which is in general agreement with studies in monkeys [113].

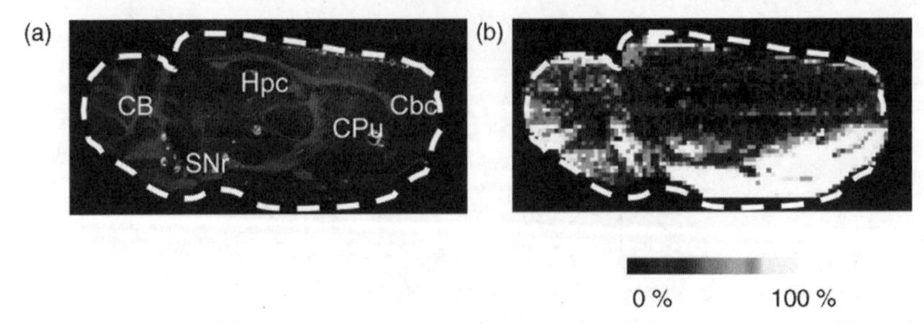

FIGURE 3.13 Optical image of a 22 × 11-mm^2 sagittal rat brain section and the corresponding selected ion image: (a) optical image of a sagittal rat brain section taken from animal 992 (0.5 h after dose) (CB, cerebellum; Cbc, cerebral cortex; Cpu, caudate–putamen; Hpc, hippocampus; SNr, substantia nigra); (b) DESI mass spectral image of clozapine in the brain section recorded in MS/MS mode. The image of the fragment ion at *m/z* 270.1 is shown by using false colors in raw pixel format. (Reprinted with permission from Ref. 28; copyright 2008 by the National Academy of Sciences of the USA.)

The corresponding images for clozapine, desmethylclozapine, and the endogenous phosphatidylcholine (PC 16 : 0/16 : 0) taken from full-scan MS spectra for lung tissue sections are shown in Figure 3.14. Clozapine (Figure 3.14b), desmethylclozapine (Figure 3.14c), and the PC (Figure 3.14d) display relatively homogeneous distributions across the entire lung tissue section. To confirm the results obtained by DESI-MS imaging, the more routine LC-MS/MS method, performed after homogenization and extraction of the equivalent tissues from the same rat collected at the same time, was used to quantify the tissue and plasma concentrations of clozapine for comparison. The DESI-MS imaging data were found to correlate well with and the corresponding LC-MS/MS results, both confirming the presence of clozapine in the lung. As illustrated in Figure 3.14e, the DESI results show a linear relationship to the plasma concentrations with $R^2 = 0.9669$, as do the LC-MS/MS responses ($R^2 = 0.9838$). In addition, the relative concentration–time profile for clozapine in the lung, as recorded by using DESI in full-scan MS mode, indicates that the highest concentration found in the lung is 30 min after dose and that the terminal half-life ($t_{1/2}$) ranges between 0.75 and 1.5 h. This is also in good agreement with the LC-MS/MS results.

Another study for chemical imaging of whole-body thin-tissue sections of mice intravenously dosed with propranolol was reported using DESI-MS/MS in comparison with WBA [29]. Figure 3.15a depicts the scanned optical image of a whole-body thin-tissue section of a mouse dosed with 7.5 mg/kg propranolol and euthanized after 20 min. Figure 3.15b is a spatial distribution plot of the SRM ion current for propranolol (*m/z* 260 → 116) obtained from the same section using DESI-MS/MS. On the basis of the predicted metabolic pathway of propanolol, several additional transitions were monitored in the same experiment but were not very successful due to low signal-to-noise ratio. Nevertheless, in comparison to parts (a) and (b) of Figure 3.15, part (c) shows the scanned optical image of a whole-body tissue section

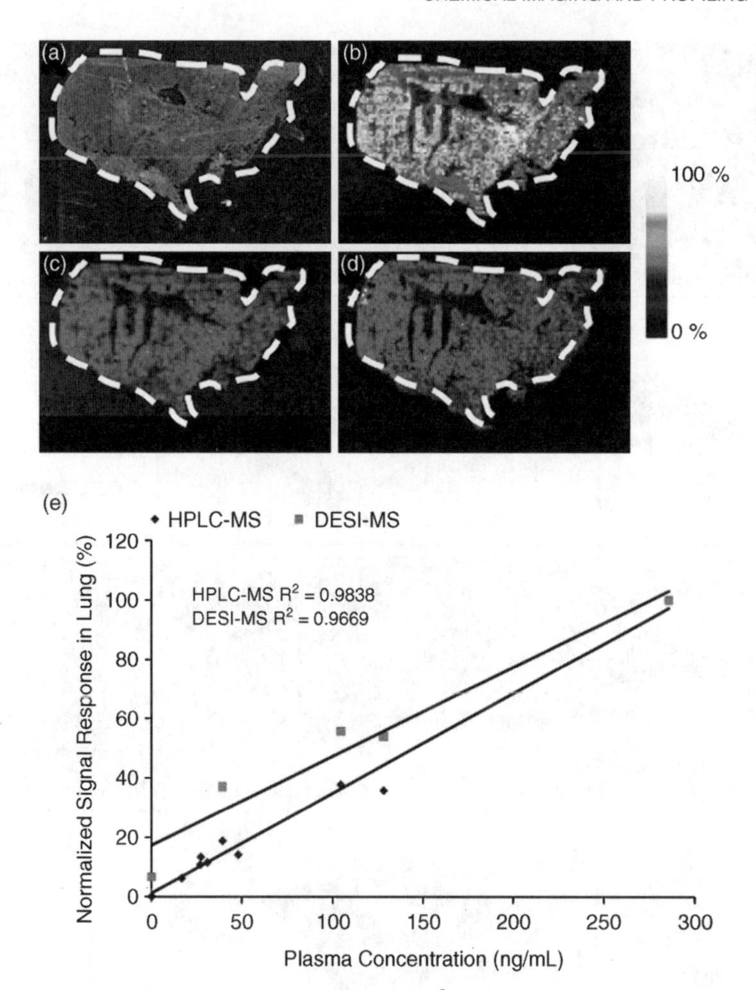

FIGURE 3.14 Optical image of a $30.6 \times 16.1\text{-mm}^2$ lung tissue section and the corresponding selected ion images, each having 132×70 pixels and shown in false colors in raw pixel format: (a) optical image; (b) image of clozapine at *m/z* 327.1; (c) image of desmethylclozapine at *m/z* 313.1; (d) image of sodiated phosphatidylcholine (16 : 0/16 : 0) at *m/z* 756.4; (e) DESI-MS imaging and LC-MS/MS results of *D*. The signal responses in each method were normalized to the maximum response in each experiment. The normalized signal responses were then plotted against the clozapine plasma concentrations as determined by LC-MS/MS. (Reprinted with permission from Ref. 28; copyright 2008 by the National Academy of Sciences of the USA.)

of a mouse dosed intravenously with 7.5 mg/kg [^{3}H]propranolol and euthanized 20 min after dose and its corresponding WBA image (Figure 3.15d). Visual comparison of the DESI-MS/MS image (Figure 3.15b) with the WBA image (Figure 3.15d) reveals that both methods confidently detected propranolol in the brain, lung, stomach, and kidney regions.

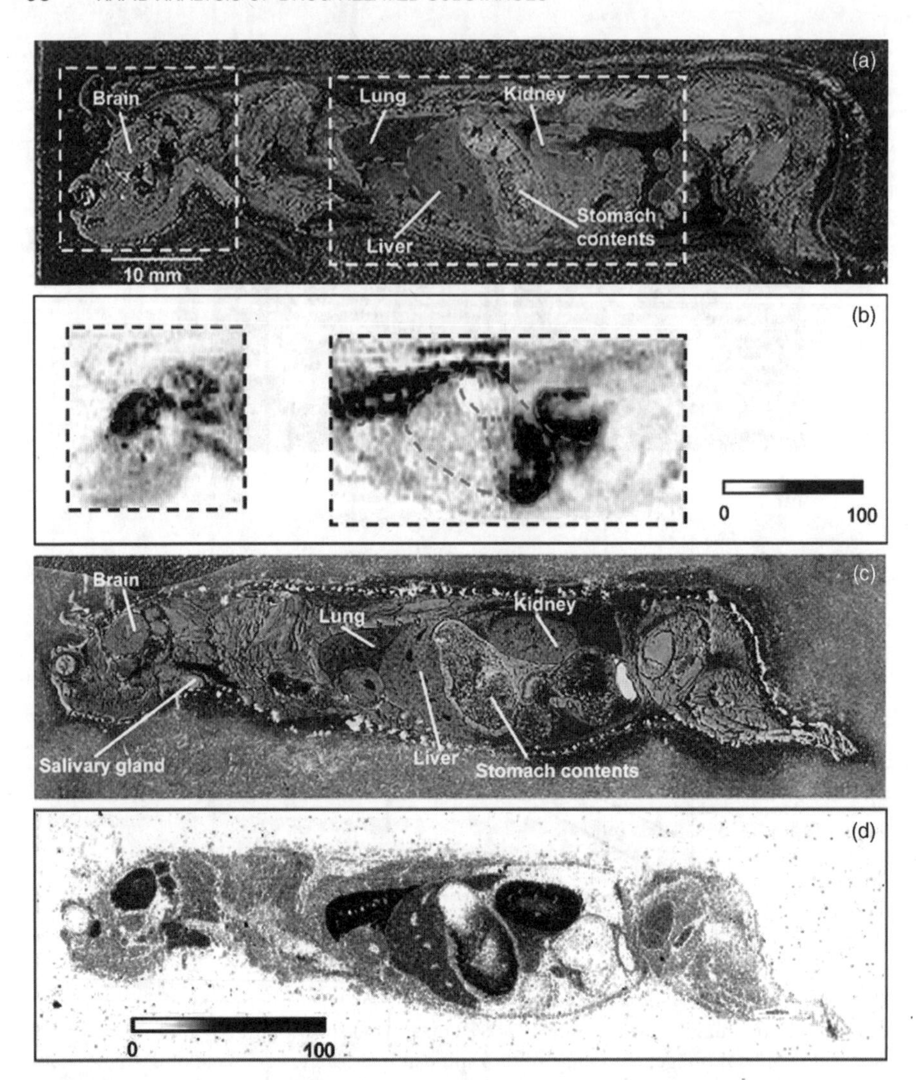

FIGURE 3.15 (a) Scanned optical image of a 40-μm-thick sagittal whole-body tissue section of a mouse dosed intravenously with 7.5 mg/kg propranolol and euthanized 20 min after dose; (b) distribution of propranolol in 20 × 20 mm and 38 × 20 mm areas measured by DESI-MS/MS (SRM: m/z 260 → 116); (c) scanned optical image of a 40- μm-thick sagittal whole-body tissue section of a mouse dosed intravenously with 7.5 mg/kg [^{3}H]propranolol and euthanized 20 min after dose; (d) autoradioluminograph of [^{3}H]propranolol-related material in the tissue section presented in (c) (reprinted with permission from Ref. 29; copyright 2008 by the American Chemical Society).

Analogous to DESI chemical imaging of tissues, DART has been explored for novel chemical profiling, for example, in the analysis of pheromones and other surface molecules from an living fly [30]. In mammals and insects, pheromones strongly influence social behaviors such as aggression and mate recognition. In

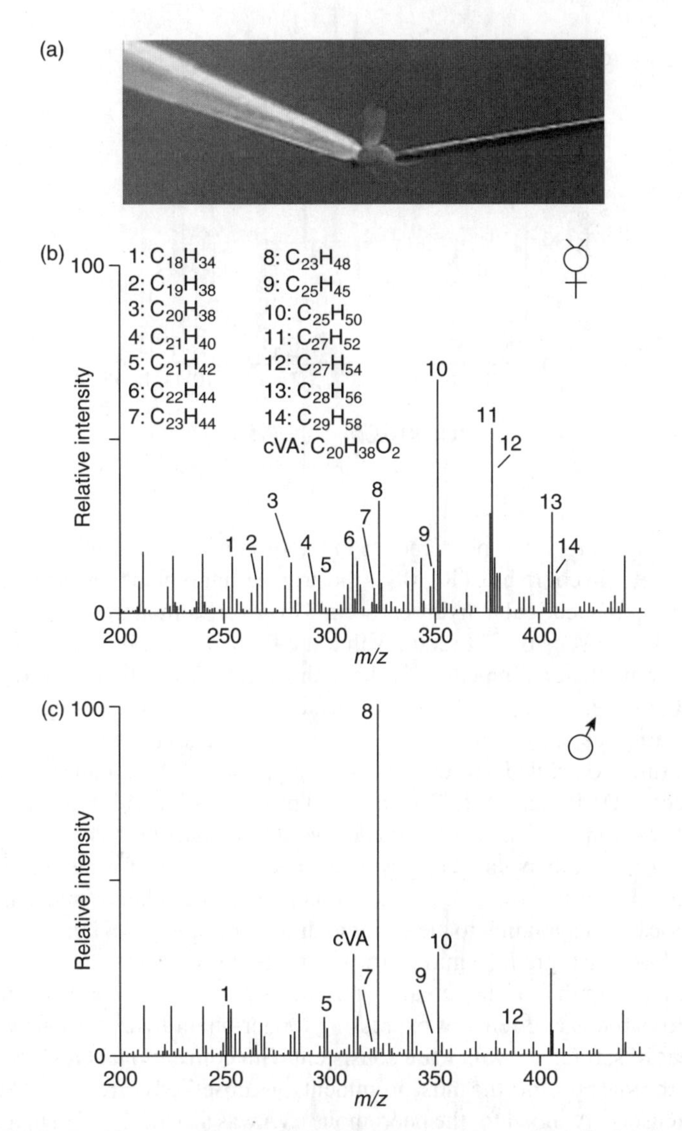

FIGURE 3.16 Positive mass spectra from DART mass spectral analysis of virgin male and virgin female flies. (a) To profile cuticular compounds from live *D. melanogaster*, the fly was held by a vacuum applied through a pipette tip and probed with a metal pin. The averaged positive-ion mass spectra obtained from DART mass spectral analysis revealed profile differences between a virgin female (b) and male (c). (d) A histogram of the relative intensities of each of the identified hydrocarbon species showed that predominantly longer-chain hydrocarbons were detected on the female fly cuticle. (Reprinted with permission from Ref. 30; copyright 2008 by the National Academy of Sciences of the USA.)

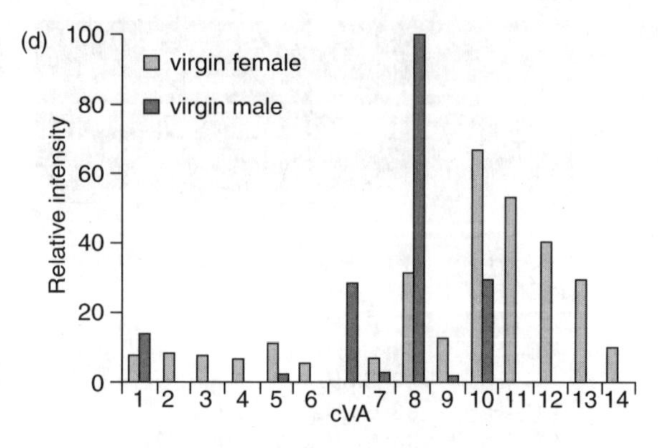

FIGURE 3.16 (*Continued*)

Drosophila melanogaster, pheromones in the form of cuticular hydrocarbons play prominent roles in courtship. GC-MS is the primary analytical tool currently used to study *Drosophila* cuticular hydrocarbons. However, although GC-MS is highly reproducible and sensitive, it requires that the fly be placed in a lethal solution of organic solvent, thereby impeding further behavioral studies. In one study, cuticular hydrocarbons were sampled from the surface of a restrained (using vacuum for immobilization as shown in Figure 3.16a), awake behaving fly by using several brief, carefully controlled depressions of the abdomen with a small steel probe that was subject to DART analysis. The chemical profiles of individually isolated wild-type *D. melanogaster* males and females were first analyzed. The averaged mass spectra from a socially isolated virgin female and virgin male fly measured with this method are shown in Figure 3.16b, c, respectively. The relative intensities of the detected ions corresponding to the major hydrocarbon species in females and males are plotted as a histogram (Figure 3.16d). The results show that the male profile is less complex than that of the female, with males expressing shorter-chain unsaturated hydrocarbons and females expressing longer-chain hydrocarbons with varying degrees of saturation, which are consistent with conventional GC-MS measurements. In the male profile, the most prominent ions observed correspond to tricosene and pentacosene. A signal for the pheromone cVA was also detected in males but not in females. In females, the ions with the greatest intensity corresponded to pentacosene and heptacosadiene, although the intensity ratios of these two hydrocarbons relative to each other varied between flies. In addition, DART MS was used to address whether there are spatial differences in the expression of cuticular hydrocarbons by comparing the profiles of individual male flies probed on the lateral thorax versus the anal–genital region. The signal for cVA was consistently greater from the anal–genital region than the thorax. This study provides a novel case of near-instantaneous analysis of an individual animal's chemical profile in parallel with behavioral studies and could be extended to other models of pheromone-mediated behavior.

3.6 FUTURE PERSPECTIVES

On a final note, the ambient mass spectrometric techniques such as DESI and DART have opened up a new subfield of mass spectrometry and have found extensive novel analytical applications. This chapter summarizes their applications in the rapid identification of drug-related substances in complicated biosystems, high-throughput analysis, chemical imaging in tissues, and chemical profiling of living animals. The direct sample analysis capability without sample cleanup and chromatographic separation makes these methods an alternative as well as a challenge to traditional LC-MS or GC-MS. With better understanding in the mechanisms and ion chemistry occurring during the ionization and the innovative instrumentation, there is no doubt that one will continue to see new exciting development in this area in the near future.

REFERENCES

1. Cooks, R. G.; Ouyang, Z.; Takats, Z.; Wiseman, J. M. (2006), Ambient mass spectrometry, *Science 311*, 1566.
2. Venter, A.; Nefliu, M.; Cooks, R. G. (2008), Ambient desorption ionization mass spectrometry, *Trends Anal. Chem. 27*, 284.
3. Takats, Z.; Wiseman, J. M.; Gologan, B.; Cooks, R. G. (2004), Mass spectrometry sampling under ambient conditions with desorption electrospray ionization, *Science 306*, 471.
4. Cody, R. B.; Laramee, J. A.; Durst, H. D. (2005), Versatile new ion source for the analysis of materials in open air under ambient conditions, *Anal. Chem. 77*, 2297.
5. Williams, J. P.; Patel, V. J.; Holland, R.; Scrivens, J. H. (2006), The use of recently described ionisation techniques for the rapid analysis of some common drugs and samples of biological origin, *Rapid Commun. Mass Spectrom. 20*, 1447.
6. Song, Y.; Cooks, R. G. (2006), Atmospheric pressure ion/molecule reactions for the selective detection of nitroaromatic explosives using acetonitrile and air as reagents, *Rapid Commun. Mass Spectrom. 20*, 3130.
7. Shiea, J.; Huang, M. Z.; Hsu, H. J.; Lee, C. Y.; Yuan, C. H.; Beech, I.; Sunner, J. (2005), Electrospray-assisted laser desorption/ionization mass spectrometry for direct ambient analysis of solids, *Rapid Commun. Mass Spectrom. 19*, 3701.
8. Sampson, J. S.; Hawkridge, A. M.; Muddiman, D. C. (2007), Direct characterization of intact polypeptides by matrix-assisted laser desorption electrospray ionization quadrupole Fourier transform ion cyclotron resonance mass spectrometry, *Rapid Commun. Mass Spectrom. 21*, 1150.
9. Chen, H.; Venter, A.; Cooks, R. G. (2006), Extractive electrospray ionization for direct analysis of undiluted urine, milk and other complex mixtures without sample preparation, *Chem. Commun.* 2042.
10. McEwen, C. N.; McKay, R. G.; Larsen, B. S. (2005), Analysis of solids, liquids, and biological tissues using solids probe introduction at atmospheric pressure on commercial LC/MS instruments, *Anal. Chem. 77*, 7826.
11. Takats, Z.; Katona, M.; Czuczy, N.; Skoumal, R. (2006), *Proc. 54th ASMS Conf. Mass Spectrometry and Allied Topics*, Seattle, WA.

12. Haddad, R.; Sparrapan, R.; Eberlin, M. N. (2006), Desorption sonic spray ionization for (high) voltage-free ambient mass spectrometry, *Rapid Commun. Mass Spectrom. 20*, 2901.
13. Grimm, R. L.; Beauchamp, J. L. (2005), Dynamics of field-induced droplet ionization: Time-resolved studies of distortion, jetting, and progeny formation from charged and neutral methanol droplets exposed to strong electric fields, *J. Phys. Chem. B 109*, 8244.
14. Haapala, M.; Pol, J.; Saarela, V.; Arvola, V.; Kotiaho, T.; Ketola, R. A.; Franssila, S.; Kauppila, T. J.; Kostiainen, R. (2007), Desorption atmospheric pressure photoionization, *Anal. Chem. 79*, 7867.
15. Ratcliffe, L. V.; Rutten, F. J. M.; Barrett, D. A.; Whitmore, T.; Seymour, D.; Greenwood, C.; Aranda-Gonzalvo, Y.; Robinson, S.; McCoustra, M. (2007), Surface analysis under ambient conditions using plasma-assisted desorption/ionization mass spectrometry, *Anal. Chem. 79*, 6094.
16. Na, N.; Zhang, C.; Zhao, M.; Zhang, S.; Yang, C.; Fang, X.; Zhang, X. (2007), Direct detection of explosives on solid surfaces by mass spectrometry with an ambient ion source based on dielectric barrier discharge, *J. Mass Spectrom. 42*, 1079.
17. Van Berkel, G. J.; Kertesz, V.; Koeplinger, K. A.; Vavrek, M.; Kong, A. T. (2008), Liquid microjunction surface sampling probe electrospray mass spectrometry for detection of drugs and metabolites in thin tissue sections, *J. Mass Spectrom. 43*, 500.
18. Chen, H.; Ouyang, Z.; Cooks, R. G. (2006), Thermal production and reactions of organic ions at atmospheric pressure, *Angew. Chem. Int. Ed. 45*, 3656.
19. Ford, M. J.; Berkel, G. J. V. (2004), An improved thin-layer chromatography/mass spectrometry coupling using a surface sampling probe electrospray ion trap system, *Rapid Commun. Mass Spectrom. 18*, 1303.
20. Shieh, I.-F.; Lee, C.-Y.; Shiea, J. (2005), Eliminating the interferences from TRIS buffer and SDS in protein analysis by fused-droplet electrospray ionization mass spectrometry, *J. Proteome Res. 4*, 606.
21. Andrade, F. J.; Ray, S. J.; Webb, M. R.; Hieftje, G. M. (2007), *Proc. 55th ASMS Conf. Mass Spectrometry Allied Topics*. Indianpolis, Indiana, USA.
22. Chen, H.; Yang, S.; Wortmann, A.; Zenobi, R. (2007), Neutral desorption sampling of living objects for rapid analysis by extractive electrospray ionization mass spectrometry, *Angew. Chem. Int. Ed. 46*, 7591.
23. Nemes, P.; Vertes, A. (2007), Laser ablation electrospray ionization for atmospheric pressure, in vivo, and imaging mass spectrometry, *Anal. Chem. 79*, 8098.
24. Harper, J. D.; Charipar, N. A.; Mulligan, C. C.; Zhang, X.; Cooks, R. G.; Ouyang, Z. (2008), Low-temperature plasma probe for ambient desorption ionization, *Anal. Chem. 80*, 9097.
25. Takats, Z.; Wiseman, J. M.; Cooks, R. G. (2005), Ambient mass spectrometry using desorption electrospray ionization (DESI): Instrumentation, mechanisms and applications in forensics, chemistry, and biology, *J. Mass Spectrom. 40*, 1261.
26. Chen, H.; Talaty, N. N.; Takats, Z.; Cooks, R. G. (2005), Desorption electrospray ionization mass spectrometry for high-throughput analysis of pharmaceutical samples in the ambient environment, *Anal. Chem. 77*, 6915.

27. Zhou, M.; McDonald, J. F.; Fernández, F. M. (2010), Optimization of a direct analysis in real time/time-of-flight mass spectrometry method for rapid serum metabolomic fingerprinting, *J. Am. Soc. Mass Spectrom. 21*, 68.
28. Wiseman, J. M.; Ifab, D. R.; Zhu, Y.; Kissinger, C. B.; Manickeb, N. E.; Kissinger, P. T.; Cooks, R. G. (2008), Desorption electrospray ionization mass spectrometry: Imaging drugs and metabolites in tissues, *Proc. Natl. Acad. Sci. USA 105*, 18120–18125.
29. Kertesz, V.; Van Berkel, G. J.; Vavrek, M.; Koeplinger, K. A.; Schneider, B. B.; Covey, T. R. (2008), Comparison of drug distribution images from whole-body thin tissue sections obtained using desorption electrospray ionization tandem mass spectrometry and autoradiography, *Anal. Chem. 80*, 5168.
30. Yew, J. Y.; Cody, R. B.; Kravitz, E. A. (2008), Cuticular hydrocarbon analysis of an awake behaving fly using direct analysis in real-time time-of-flight mass spectrometry, *Proc. Natl. Acad. Sci. USA 105*, 7135–7140.
31. Takáts, Z.; Wiseman, J. M.; Gologan, B.; Cooks, R. G. (2004), Electro-sonic spray ionization—a gentle technique or generating folded proteins and protein complexes in the gas phase and studying ion-molecule reactions at atmospheric pressure, *Anal. Chem. 76*, 4050.
32. Wells, J. M.; Roth, M. J.; Keil, A. D.; Grossenbacher, J. W.; Justes, D. R.; Patterson, G. E.; Barket, D. J. (2008), Implementation of DART and DESI ionization on a fieldable mass spectrometer, *J. Am. Soc. Mass Spectrom. 19*, 1419.
33. Williams, J. P.; Scrivens, J. H. (2005), Rapid accurate mass desorption electrospray ionisation tandem mass spectrometry of pharmaceutical samples, *Rapid Commun. Mass Spectrom. 19*, 3643.
34. Shin, Y. S.; Drolet, B.; Mayer, R.; Dolence, K.; Basile, F. (2007), Desorption electrospray ionization-mass spectrometry of proteins, *Anal. Chem. 79*, 3514.
35. Myung, S.; Wiseman, J. M.; Valentine, S. J.; Takats, Z.; Cooks, R. G.; Clemmer, D. E. (2006), Coupling desorption electrospray ionization with ion mobility/mass spectrometry for analysis of protein structure: Evidence for desorption of folded and denatured states, *J. Phys. Chem. B 110*, 5045.
36. Hu, Q.; Talaty, N.; Noll, R. J.; Cooks, R. G. (2006), Desorption electrospray ionization using an Orbitrap mass spectrometer: exact mass measurements on drugs and peptides, *Rapid Commun. Mass Spectrom. 20*, 3403–3408.
37. Weston, D. J.; Bateman, R.; Wilson, I. D.; Wood, T. R.; Creaser, C. S. (2005), Direct analysis of pharmaceutical drug formulations using ion mobility spectrometry/quadrupole-time-of-flight mass spectrometry combined with desorption electrospray ionization, *Anal. Chem. 77*, 7572.
38. Kauppila, T. J.; Talaty, N.; Salo, P. K.; Kotiaho, T.; Kostiainen, R.; Cooks, R. G. (2006), New surfaces for desorption electrospray ionization mass spectrometry: Porous silicon and ultra-thin layer chromatography plates, *Rapid Commun. Mass Spectrom. 20*, 2143–2150.
39. Miao, Z.; Chen, H. (2009), Direct analysis of liquid samples by desorption electrospray ionization-mass spectrometry (DESI-MS), *J. Am. Soc. Mass Spectrom. 20*, 10.
40. Bereman, M. S.; Nyadong, L.; Fernandez, F. M.; Muddiman, D. C. (2006), Direct high-resolution peptide and protein analysis by desorption electrospray ionization Fourier transform ion cyclotron resonance mass spectrometry, *Rapid Commun. Mass Spectrom. 20*, 3409.

41. Kauppila, T.; Wiseman, J. M.; Ketola, R. A.; Kotiaho, T.; Cooks, R. G.; Kostiainen, R. (2006), Desorption electrospray ionization mass spectrometry for the analysis of pharmaceuticals and metabolites, *Rapid Commun. Mass Spectrom. 20*, 387.
42. Demian, J. M. W.; Ifa, R.; Song, Q.; Cooks, R. G. (2007), Development of capabilities for imaging mass spectrometry under ambient conditions with desorption electrospray ionization (DESI), *Int. J. Mass Spectrom. 259*, 8.
43. Venter, A.; Cooks, R. G. (2007), Desorption electrospray ionization in a small pressure-tight enclosure, *Anal. Chem. 79*, 6398.
44. Chipuk, J. E.; Brodbelt, J. S. (2008), Transmission mode desorption electrospray ionization, *J. Am. Soc. Mass Spectrom. 19*, 1612.
45. Chipuk, J. E.; Brodbelt, J. S. (2009), The influence of material and mesh characteristics on transmission mode desorption electrospray ionization, *J. Am. Soc. Mass Spectrom. 20*, 584.
46. Zhang, Y.; Chen, H. (2010), Detection of saccharides by reactive desorption electrospray ionization (DESI) using modified phenylboronic acids, *Int. J. Mass Spectrom. 289*, 98.
47. Costa, A. B.; Cooks, R. G. (2008), Simulated splashes: Elucidating the mechanism of desorption electrospray ionization mass spectrometry, *Chem. Phys. Lett. 464*, 1.
48. Cooks, R. G.; Jo, S. C.; Green, J. (2004), Collisions of organic ions at surfaces, *Appl. Surface Sci. 231–232*, 13.
49. Mulligan, C. C.; MacMillan, D. K.; Noll, R. J.; Cooks, R. G. (2007), Fast analysis of high-energy compounds and agricultural chemicals in water with desorption electrospray ionization mass spectrometry, *Rapid Commun. Mass Spectrom. 21*, 3729.
50. Miao, Z.; Chen, H. (2008), Analysis of continuous-flow liquid samples by desorption electrospray ionization-mass spectrometry (DESI-MS), *Proc. 56th Annual American Society for Mass Spectrometry Conf. Mass Spectrometry*, Denver.
51. Ma, X.; Zhao, M.; Lin, Z.; Zhang, S.; Yang, C.; Zhang, X. (2008), Versatile platform employing desorption electrospray ionization mass spectrometry for high-throughput analysis, *Anal. Chem. 80*, 6131.
52. Li, J.; Dewald, H.; Chen, H. (2009), Online coupling of electrochemical reactions with liquid sample desorption electrospray ionization mass spectrometry, *Anal. Chem. 81*, 9716.
53. Chen, H.; Talaty, N. N.; Takats, Z.; Cooks, R. G. (2005), *Anal. Chem. 77*, 6915.
54. Cotte-Rodriguez, I.; Chen, H.; Cooks, R. G. (2006), Rapid trace detection of triacetone triperoxide (TATP) by complexation reactions during desorption electrospray ionization, *Chem. Commun. 2006*, 953.
55. Cotte-Rodriguez, I.; Hernandez-Soto, H.; Chen, H.; Cooks, R. G. (2008), In situ trace detection of peroxide explosives by desorption electrospray ionization and desorption atmospheric pressure chemical ionization, *Anal. Chem. 80*, 1512.
56. Chen, H.; Cotte-Rodriguez, I.; Cooks, R. G. (2006), *Chem. Commun. 2006*, 597.
57. Huang, G.; Chen, H.; Zhang, X.; Cooks, R. G.; Ouyang, Z. (2007), Rapid screening of anabolic steroids in urine by reactive desorption electrospray ionization, *Anal. Chem. 79*, 8327.
58. Nyadong, L.; Green, M. D.; De Jesus, V. R.; Newton, P. N.; Fernandez, F. M. (2007), Reactive desorption electrospray ionization linear ion trap mass spectrometry of latest-generation counterfeit antimalarials via noncovalent complex formation, *Anal. Chem. 79*, 2150.

59. Gunawardena, H. P.; O'Hair, R. A. J.; McLuckey, S. A. (2006), Selective disulfide bond cleavage in gold(I) cationized polypeptide ions formed via gas-phase ion/ion cation switching, *J. Proteome Res. 5*, 2087.
60. Schroder, D.; Schwarz, H. (2008), Gas-phase activation of methane by ligated transition-metal cations, *Proc. Natl. Acad. Sci. USA 105*, 18114.
61. Nibbering, N. M. M. (1990), Gas-phase ion/molecule reactions as studied by fourier transform ion cyclotron resonance, *Acct. Chem. Res. 23*, 279.
62. Gronert, S. (2001), Mass spectrometric studies of organic ion/molecule reactions, *Chem. Rev. 101*, 329.
63. Cooks, R. G.; Chen, H.; Eberlin, M. N.; Zheng, X.; Tao, W. A. (2001), Polar acetalization and transacetalization in the gas phase: The Eberlin reaction, *Chem. Rev. 106*, 188.
64. Bowers, M. T.; Marshall, A. G.; McLafferty, F. W. (1996), Mass spectrometry: Recent advances and future directions, *J. Phys. Chem. 100*, 12897.
65. DePuy, C. H.; Grabowski, J. J.; Bierbaum, V. M. (1982), Chemical reactions of anions in the gas phase, *Science 218*, 955.
66. Kenttamaa, H. I.; Cooks, R. G. (1989), Identification of protonated β-hydroxycarbonyl compounds by reactive collisions in tandem mass spectrometry, *J. Am. Chem. Soc. 111*, 4122.
67. Eberlin, M. N. (2004), Gas-phase polar cycloadditions, *Int. J. Mass Spectrom. 235*, 263.
68. Naven, T. J. P.; Harvey, D. J. (1996), *Rapid Commun. Mass Spectrom. 10*, 829.
69. Takats, Z.; Cotte-Rodriguez, I.; Talaty, N.; Chen, H.; Cooks, R. G. (2005), Direct, trace level detection of explosives on ambient surfaces by desorption electrospray ionization mass spectrometry, *Chem. Commun. 2005*, 1950.
70. Jagerdeoa, E.; Abdel-Rehim, M. (2009), Screening of cocaine and its metabolites in human urine samples by direct analysis in real-time source coupled to time-of-flight mass spectrometry after online preconcentration utilizing microextraction by packed sorbent, *J. Am. Soc. Mass Spectrom. 20*, 891.
71. Yu, S.; Crawford, E.; Tice, J.; Musselman, B.; Wu, J.-T. (2009), Bioanalysis without sample cleanup or chromatography: The evaluation and initial implementation of direct analysis in real time ionization mass spectrometry for the quantification of drugs in biological matrixes, *Anal. Chem. 81*, 193.
72. Cody, R. B.; Larameee, J. A. (2005), US Patent 6, 949, 741.
73. Laramee, J. A.; Cody, R. B. (2006), US Patent 7,112,785.
74. Jones, R. W.; Cody, R. B.; McClelland, J. F. (2006), Differentiating writing inks using direct analysis in real time mass spectrometry, *J. Forensic Sci. 51*, 915.
75. Penning, F. M. (1927), Uber Ionization durch metastable Atome, *Naturwissenschaften 15*, 818.
76. Baldwin, K. G. H. (2005), Metastable helium: Atom optics with nano-grenades, *Contemp. Phys. 46*, 105.
77. Laramee, J. A.; Cody, R. B.; Niles, J. M.; Durst, H. D. (2007), *Forensic Application of DART Mass Spectrometry in Forensic Analysis on the Cutting Edge*, Wiley, Hoboken, NJ, p. 175.
78. Laramee, J. A.; Cody, R. B. (2007), Chemi-ionization and direct analysis in real time mass spectrometry, in Gross, M. L.; Caprioli, R. M., eds., *The Encyclopedia of*

Mass Spectrometry: Molecular Ionization Methods, Vol VI. Elsevier, Amsterdam p. 377.

79. Song, L.; Dykstra, A. B.; Yao, H.; Bartmess, J. E. (2009), Ionization mechanism of negative ion-direct analysis in real time: A comparative study with negative ion-atmospheric pressure photoionization, *J. Am. Soc. Mass Spectrom. 20*, 42.
80. Chen, H.; Talaty, N.; Takats, Z.; Cooks, R. G. (2005), Desorption electrospray ionization mass spectrometry for high-throughput analysis of pharmaceutical samples in the ambient environment, *Anal. Chem. 77*, 6915.
81. Williams, J. P.; Scrivens, J. H. (2005), Rapid accurate mass desorption electrospray ionisation tandem mass spectrometry of pharmaceutical samples, *Rapid Commun. Mass Spectrom. 19*, 3643–3650.
82. Jackson, A. U.; Werner, S. R.; Talaty, N.; Song, Y.; Campbell, K.; Cooks, R. G.; Morgan, J. A. (2008), Targeted metabolomic analysis of Escherichia coli by desorption electrospray ionization and extractive electrospray ionization mass spectrometry, *Anal. Biochem. 375*, 272.
83. Rodriguez-Cruz, S. E. (2006), Rapid analysis of controlled substances using desorption electrospray ionization mass spectrometry, *Rapid Commun. Mass Spectrom. 20*, 53.
84. Leuthold, L. A.; Mandscheff, J. F.; Fathi, M.; Giroud, C.; Augsburger, M.; Varesio, E.; Hopfgartner, G. (2006), Desorption electrospray ionization mass spectrometry: Direct toxicological screening and analysis of illicit ecstasy tablets, *Rapid Commun. Mass Spectrom. 20*, 103.
85. D'Agostino, P. A.; Hancock, J. R.; Chenier, C. L.; Lepage, C. R. (2006), Liquid chromatography electrospray tandem mass spectrometric and desorption electrospray ionization tandem mass spectrometric analysis of chemical warfare agents in office media typically collected during a forensic investigation, *J. Chromatogr. A 1110*, 86.
86. Nefliu, M.; Venter, A.; Cooks, R. G. (2006), Desorption electrospray ionization and electrosonic spray ionization for solid- and solution-phase analysis of industrial polymers, *Chem. Commun. 2006*, 888.
87. Song, Y.; Talaty, N.; Tao, A. W.; Pan, Z.; Cooks, R. G. (2007), Rapid ambient mass spectrometric profiling of intact, untreated bacteria using desorption electrospray ionization, *Chem. Commun. 2007*, 61.
88. Talaty, N.; Takats, Z.; Cooks, R. G. (2005), Rapid in-situ detection of alkaloids in plant tissue under ambient conditions using desorption electrospray ionization, *Analyst 130*, 1624.
89. Eberlin, L.S.; Dill, D.L.; Golby, A. J.; Ligon, K. L.; Wiseman, J. M.; Cooks, R. G. (2010), Discrimination of Human Astrocytoma Subtypes by lipid analysis using desorption electrospray ionization imaging mass spectrometry, *Angew. Chem. Int. Ed. 49*, 5953.
90. Wiseman, J. M.; Ifa, D. R.; Song, Q.; Cooks, R. G. (2006), Tissue imaging at atmospheric pressure using desorption electrospray ionization (DESI) mass spectrometry, *Angew. Chem. Int. Ed. 45*, 7188.
91. Van Berkel, G. J.; Ford, M. J.; Deibel, M. A. (2005), Thin-layer chromatography and mass spectrometry coupled using desorption electrospray ionization, *Anal. Chem. 77*, 1207.
92. Petucci, C.; Diffendal, J.; Kaufman, D.; Mekonnen, B.; Terefenko, G.; Musselman, B. (2007), Direct analysis in real time for reaction monitoring in drug discovery, *Anal. Chem. 79*, 5064.

93. Zhao, Y.; Lam, M.; Wu, D.; Mak, R. (2008), Quantification of small molecules in plasma with direct analysis in real time tandem mass spectrometry without sample preparation and liquid chromatographic separation, *Rapid Commun. Mass Spectrom. 22*, 3217.

94. Fernandez, F. M.; Cody, R. B.; Green, M. D.; Hampton, C. Y.; McGready, R.; Sengaloundeth, S.; White, N. J.; Newton, P. N. (2006), Characterization of solid counterfeit drug samples by desorption electrospray ionization and direct analysis-in-real-time coupled to time-of-flight mass spectrometry, *Chem. Med. Chem. 1*, 702.

95. Pierce, C. Y.; Barr, J. R.; Cody, R. B.; Massung, R. F.; Woolfitt, A. R.; Moura, H.; Thompson, H. A.; Fernandez, F. M. (2007), Ambient generation of fatty acid methyl ester ions from bacterial whole cells by direct analysis in real time (DART) mass spectrometry, *Chem. Commun. 8*, 807.

96. Haefliger, O. P.; Jeckelmann, N. (2007), Direct mass spectrometric analysis of flavors and fragrances in real applications using DART, *Rapid Commun. Mass Spectrom. 21*, 1361.

97. Morlock, G.; Ueda, Y. (2007), New coupling of planar chromatrography with direct analysis in real time mass spectrometry, *J. Chromatogr. A 1143*, 243.

98. Kauppila, T. J.; Talaty, N.; Kuuranne, T.; Kotiaho, T.; Kostiainen, R.; Cooks, R. G. (2007), Rapid analysis of metabolites and drugs of abuse from urine samples by desorption electrospray ionization-mass spectrometry, *Analyst 132*, 868.

99. Zeng, H.; Wu, J. T.; Unger, S. E. (2002), The investigation and the use of high flow column switching LC/MS/MS as a high-throughput approach for direct plasma sample analysis of single and multiple components in pharmacokinetic studies, *J. Pharm. Biomed. Anal. 27*, 967.

100. Shi, Y.; Xiang, R.; Crawford, J.; Colangelo, C.; Horvath, C.; Wilkins, J. (2004), A simple solid phase mass tagging approach for quantitative proteomics, *J. Proteome. Res. 3*, 104.

101. Kassel, D. B. (2001), Combinatorial chemistry and mass spectrometry in the 21st century drug discovery laboratory, *Chem. Rev. 101*, 255.

102. Wang, C.; Vickers, T. J.; Mann, C. K. (1997), Direct assay and shelf-life monitoring of aspirin tablets using Raman spectroscopy, *J. Pharm. Biomed. Anal. 16*, 87.

103. Bell, S. E. J.; Beattie, J. R.; McGarvey, J. J.; Peters, K. L.; Sirimuthu, N. M. S.; Speers, S. J. (2004), Development of sampling methods for Raman analysis of solid dosage forms of therapeutic and illicit drugs, *J. Raman Spectrosc. 35*, 409.

104. Dettmer, K.; Aronov, P. A.; Hammock, B. D. (2007), Mass spectrometry-based metabolomics, *Mass Spectrom. Rev. 26*, 51.

105. Stoeckli, M.; Chaurand, P.; Hallahan, D. E.; Caprioli, R. M. (2001), Imaging mass spectrometry: A new technology for the analysis of protein expression in mammalian tissues, *Nat. Med. 7*, 493.

106. Pacholski, M. L.; Winograd, N. (1999), Imaging with mass spectrometry, *Chem. Rev. 99*, 2977.

107. Wiseman, J. M.; Puolitaival, S. M.; Takats, Z.; Cooks, R. G.; Caprioli, R. M. (2005), Mass spectrometric profiling of intact biological tissue by using desorption electrospray ionization, *Angew. Chem. Int. Ed. 44*, 7094.

108. Ifa, D. R.; Gumaelius, L. M.; Eberlin, L. S.; Manicke, N. E.; Cooks, R. G. (2007), Forensic analysis of inks by imaging desorption electrospray ionization (DESI) mass spectrometry, *Analyst 132*, 461.

109. Ifa, D. R.; Manicke, N. E.; Rusine, A. L.; Cooks, R. G. (2008), Quantitative analysis of small molecules by desorption electrospray ionization (DESI) mass spectrometry from PTFE surfaces, *Rapid Commun. Mass Spectrom. 22*, 503.

110. Ifa, D. R.; Wiseman, J. M.; Song, Q.; Cooks, R. G. (2007), Development of capabilities for imaging mass spectrometry under ambient conditions with desorption electrospray ionization (DESI), *Int. J. Mass Spectrom. 259*, 8.

111. Gardiner, T.; Lewis, J.; Shore, P. (1978), Distribution of clozapine in the rat: Localization in lung, *J. Pharmacol. Exp. Ther. 206*, 151.

112. Dawson, T.; Gehlert, D.; McCabe, R.; Barnett, A.; Wamsley, J. (1986), D-1 dopamine receptors in the rat brain: A quantitative autoradiographic analysis, *J. Neurosci. 6*, 2352.

113. Chou, Y.-H.; Halldin, C.; Farde, L. (2006), Clozapine binds preferentially to cortical D1-like dopamine receptors in the primate brain: A PET study, *Psychopharmacology 185*, 29.

CHAPTER 4

Orbitrap High-Resolution Applications

ROBERT J. STRIFE

Procter & Gamble, 8700 Mason-Montgomery Road, Mason, OH 45040

4.1 HISTORICAL ANECDOTE

In 1999, there was a fairly large community of users of the commercialized Paul quadrupole ion trap (QIT). By that time, the QIT had experienced a rather meteoric rise to prominence over 15 years, as a commercialized and very practical GC-MS^n and LC-MS^n device. Small-molecule nominal mass measurements at unit resolution and at any value of n were routine up to m/q 650. As the limits of operation were pushed by modifying the commercial device, there had been a laboratory demonstration of approximately 100,000 resolution at about m/q 1350 for substance P [1]. Yet the nagging aspect of obtaining at least low ppm mass accuracy, important to the ascertaining of elemental compositions of small-molecule "unknowns," resisted solution [2]. Even so, for those of us who became mass spectrometrists using nontrapping devices, the Paul trap created a fascinating new landscape to design unique experiments in the time domain, which also led to unique solutions to everyday analysis problems [3].

There are unique aspects of trapped-ion experiments that caused our attention to be captured by the curious title of a talk at the American Society for Mass Spectrometry (ASMS) meeting, in Dallas (1999). A paper titled "The Orbitrap: A novel high-performance electrostatic trap" was delivered by Alexander Makarov, late on Monday afternoon. What kind of device could this be?! In short, when I returned home to my job, I went out on a limb and told my managers that "benchtop FT-MS" was not far away. While it was perhaps a bit of bluster at the time, the Orbitrap has experienced an even steeper rise to prominence than the Paul QIT, as it was commercially released a short 5 years later in 2005 and is now utilized in hundreds of laboratories around the world.

Characterization of Impurities and Degradants Using Mass Spectrometry, First Edition.
Edited by Birendra N. Pramanik, Mike S. Lee, and Guodong Chen.

4.2 GENERAL DESCRIPTION OF ORBITRAP OPERATING PRINCIPLES

The ASMS talk was quickly followed by Makarov's formal publication in 2000 [4]. He pointed out that three types of dynamic ion trapping were known and that orbital trapping had first been implemented in a design by Kingdon in 1923 [5]. The Kingdon trap is described as "a wire stretched along the axis of an outer cylinder with flanges enclosing the trapping volume." The wire has a high, fixed DC voltage applied to it, creating a fixed or static electric field about the wire. Only ions that have enough tangential velocity to overcome the attractive force of the field miss the wire and survive in an orbit. Makarov likened their orbits "in some way to orbits of planets or asteroids in the solar system," with the wire being the sun. He also noted that the flanges or endcaps of the device introduce electric field curvature, which constrains the motion of ions along the axis of the wire. We will return to this *critically important axial motion of ions* shortly.

It is interesting to note that Makarov also referred somewhat obliquely to plans to derive *m/q* ratios from the frequency of orbital rotation of ions around the wire in a device utilizing indirect, image current detection. However, drawbacks related to the variance of the initial radius of ion trajectory and ion velocity lead to poor mass resolution and rapid dephasing of the ions in their radial orbits. This lack of orbital coherence translates to rapid loss of image current, among other problems. Thus, Makarov struck out in a different direction for his original design.

Any metal surface to which an electric potential is applied generates an electric field in space [6]. Nearby conductive surfaces, even at ground, will affect the field lines as well. The geometry of the conductive surfaces dictate the geometry of the electric field and its equipotential lines. In 1981, the shape of the outer container of the Kingdon trap was modified by Knight, to add a quadrupole electric field component, along the axis of the device [7], thus trapping ions exhibiting axial harmonic motion. However, there were practical drawbacks to the design, and it was not used to produce mass spectra. Makarov pursued this line of thought about the quadrupole field component and making use of harmonic axial oscillations, by utilizing a unique inner electrode/outer electrode design. A report of this newer design with several very high-resolution mass spectra appeared in 2005 [8].

A 3D view of a typical Orbitrap is shown in Figure 4.1, and Figure 4.2 is a photo of A. Makarov holding the electrode structure in his hand. The device consists of two electrodes, an inner "spindle" and an outer "barrel," split in the middle, with a high DC potential on the spindle halves, while the outer conductive barrel is at ground potential. Thus the device is likened to a cylindrical capacitor with a logarithmic electric field. A quadrupole electric field also results from the electrode shape, and the sum of the two fields is described as "quadro-logarithmic." The fields are mathematically orthoganol and result in two motions: radial orbital motion around the spindle and precise *harmonic oscillations along the axis* due to the quadrupole electric field component. The combined motion induced by the sum of the two field components is likened to a trajectory defined by the perimeter of a rotating ellipse. Makarov shows in his mathematical derivations that only the axial frequency of motion is completely

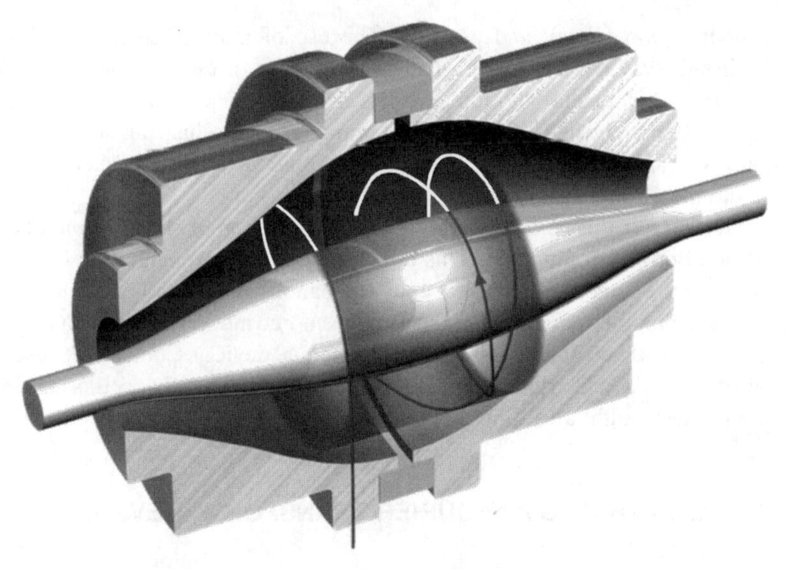

FIGURE 4.1 A 3D, cutaway schematic view of an Orbitrap spindle/barrel electrode assembly (used by permission, Thermo Fisher Scientific Corp.).

FIGURE 4.2 Alexander Makarov, Orbitrap inventor, holding an actual electrode assembly (used by permission, Elsevier Inc.) [12].

independent of ion energy and position. The use of indirect detection of image currents in this way does not suffer the disadvantages of the previously discussed detection of radial orbital motion. Also, coherence of ions in the commercial Orbitrap is effectively helped by a small trapping device just outside the barrel, called the "C-trap." It "squeezes" the ion population into a tighter packet just before injection to the Orbitrap structure.

However, even in this spindle–barrel design, the ions rapidly become out of phase in their radial orbital component. This effect creates the advantage that by using a slight delay in the detection step, frequency sidebands from the radial motion are not observed. The precise frequencies of the axial harmonic motion translate to very high resolution (up to 100,000 at m/q 400 in the commercial device). Cooks et al. described application of an AC electric field to "squeeze" the harmonically oscillating ring of ions to a thinner width, affording a twofold increase in resolution [9].

4.3 THE ORBITRAP IS A "FOURIER TRANSFORM" DEVICE

The popular acronym FT-MS (Fourier transform mass spectrometry) is a somewhat unfortunate acronym, in that it describes a detection methodology, not a mass spectrometer, per sey. In fact, the detection methodology has been demonstrated on other types of mass spectrometers, such as the linear ion trap [10]. However, in commercially available mass spectrometers, indirect detection of image currents and Fourier transform methodology had only been common in the device based on ion cyclotron resonance (ICR). Leaders in this specialty field tend to properly name that technique *Fourier transform ion cyclotron resonance mass spectrometry* (FT-ICR-MS). It seems that a parallel name for Makarov's device would be *Fourier transform axial harmonic orbital trapping mass spectrometry* FT-AHOT-MS. As Makarov stated, "the image current is amplified and processed exactly in the same way as in FT-ICR." While the name "Orbitrap" has become fairly entrenched at this point in time, perhaps the shortened acronym FT-OT is more analogous to FT-ICR.

In spite of the "FT similarity," there are still subtle differences in the operating principles and derived equations for the Orbitrap. For instance, the mass resolution of the device is one-half the frequency resolution. The interested reader is referred to an excellent primer on FT-MS published by the Marshall group [11] as well as Makarov's fundamental paper [4]. The multitude of learnings and concepts (such as the phasing and coherence of ions in the Orbitrap discussed above) already published in the FT-ICR-MS literature, will be readily appreciated by many of those scientists using the Orbitrap for daily solutions of their practical analytical tasks.

Finally, the success of the Orbitrap as a mass spectrometer that is very easy to use (the author's personal experience and opinion) is highlighted by the publication in 2009 of a special issue of the *Journal of the American Society for Mass Spectrometry* (ASMS). The special issue notes Alexander Makarov's Distinguished Contribution in Mass Spectrometry award by the ASMS in 2008, the short history of Orbitrap development, new developments in Orbitrap performance, and several applications covering a broad range of activity [12].

Makarov summarizes that "all trapping mass spectrometers are known to benefit from increasing strength of the trapping field" [13]. In FT-ICR-MS this benefit has been accomplished by the use of stronger magnets. In FT-OT, through fundamental considerations of the equations governing the Orbitrap electric field, Makarov showed one could not only increase the voltage applied to the spindle but also that, by altering the fundamental geometry, namely, the radii of the spindle and the barrel, one should achieve increased performance. Makarov gives a fascinating and detailed description of the physical alterations necessary to actually realize the gains in performance that are theoretically possible. At m/q 196, the observation of fine structure of isotope lines is described at an observed resolution of 600,000. The device is separately recognized as the high-field (HF) Oribtrap. Finally, the current commercially available designs have a m/q range that can extend from 2000 to 4000 Da.

With this short history and abbreviated description of operating principles for the generalist, the discussion below focuses on the analytical figures of merit for the Orbitrap and the solution of trace analytical problems, the focus of this volume.

4.4 PERFORMING EXPERIMENTS IN TRAPPING DEVICES

While there are a wide variety of trapping device designs, most of the work done with commercial devices relies on ionization external to the trapping/mass-measuring device. Before considering the analytical applications of trapping devices, it is a useful to remember that these devices are "gated"; that is, experiments are carried out in discreet steps, using both continuous and pulsed ionization sources (e.g., electrospray and laser desorption). However, the first step is essentially always "open the ion gate, fill the trapping device up with ions, close the entry gate— then decide what you are going to do next!" A simple scheme to acquire an Orbitrap high-resolution mass spectrum at 60,000 resolution is shown in Scheme 4.1.

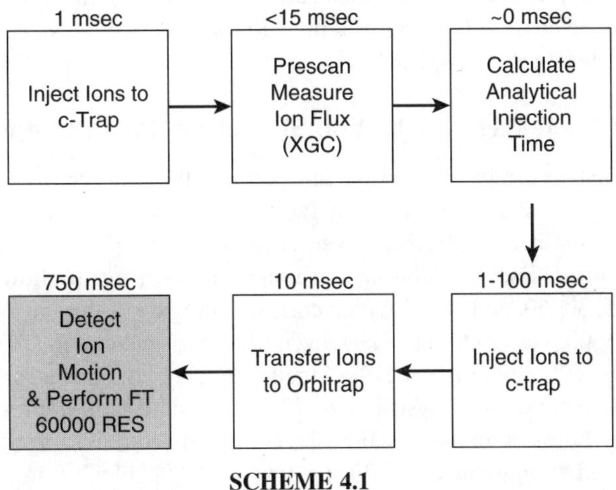

SCHEME 4.1

Note in the scheme that commercial traps of various types generally use *automatic gain control* (AGC). That is, at the beginning of any experiment, the objective is always to optimally fill the trap up with ions. By conducting a very short (10 ms) "collect ions and dump ions" type of procedure (no mass measurement), the onboard computer can make a decision on the length of the injection period, that is, how long it will take to optimally fill up the trap. For example, at the "top" of an HPLC peak, where analyte [C] is relatively high, the ion injection time is relatively short. Near the base of the HPLC peak, where analyte [C] is relatively low, the ion injection time is relatively long. Again, relative to the entire experiment to be performed (such as detection at high resolution or perhaps an MS^n experiment before detection), the AGC function and selected injection time usually occupies a small part of the total scan cycle, such that AGC does not impact the scan-to-scan cycle time too severely.

4.4.1 "Raw" HPLC Data Look Like Infusion Data

In essence, then, the *raw* total-ion current trace (not visible to the operator) as a function of time, in the LC/MS mode, for example, is actually "flat," as for an infusion experiment. The ion current during the HPLC run is then inversely scaled, point by point, to account for the variable injection time used for each scan. The result is a typical chromatogram with discernable chromatographic peaks.

As far as mass measurement and ion statistics are concerned, then, it is likely that the variance of the data at the HPLC peak top, versus that closer to the baseline (say, at 10% of the maximum peak top), does not change, as long as the analyte ion population does not change. Eventually, of course, the ion injection time hits a maximum value and the trap can no longer be optimally filled with analyte ions (i.e., typical background ion current becomes significant in the total ion population). Therefore, rapid *infusion experiments can be used to establish expectations* on mass measurement statistics during HLPC, as long as the number of scans processed and the ion currents examined are similar. Furthermore, a sophisticated microfraction collection device, the Advion Nanomate, performs automated infusion analysis of chromatography runs, collected in a 96 or 384 well plate format, and provides many advantages in structural interrogations (see below).

4.4.2 How Much Mass Resolution Should Be Used During HPLC

Note that the image current observation time period in the scan cycle is critical, for it is directly linked to the achieved resolution. If there is no interference at the *m/q* that one wishes to determine accurately, then lower resolution may suffice, so that higher cycle times are obtained. For the Orbitrap, a simple pulldown menu allows the user to choose 7.5, 15, 30, 60, and 100 K with a corresponding cycle time of approximately 75, 150, 300, 600, and 1000 ms. Thus, even ultra-high-resolution "100 K" experiments are compatible with many HPLC methods using columns as small as 1 mm in diameter. Selectivity of analysis is also affected by choice of *m/q* resolution. This issue has been addressed in the case of residues (e.g., mycotoxins, pesticides) in food and animal feed sources at the 1-30 ppb level [14,15]. Finally, note in FT-ICR, resolution varies *inversely with m/q* [11], and the figures of merit are generally quoted for *m/q* 400. For FT-OT, resolution varies inversely with the square root of *m/q*.

4.5 DETERMINING ELEMENTAL COMPOSITIONS OF "UNKNOWNS" USING AN ORBITRAP

There are generally three key values to obtain in determining the elemental compo sition of an unknown:

1. *Accurate m/q Value.* The position of the monoisotopic peak's centroided top, properly calibrated, is determined with a high degree of accuracy.

2. *Precision of the Measurement.* The *m/q* value is searched against possible elemental compositions. In this step, an error window *must be specified.* For example, search all compositions for *m/q* 614.57109 ± 2 ppm. How does one choose a window? This choice must be tied to the precision of the measurement. It is somewhat sad to say that most experimenters simply assume a precision number on the basis of their experience with a particular instrument. This is a poor analytical practice at best. It is a rather simple matter to perform some replicate measurements of an infused sample containing compounds of known compositions and then to calculate the instrument's precision on the day of the experiment. Again, the optimized filling of the trap by AGC means the ion statistics for infusion experiments will be the very close to those obtained for HPLC peaks, provided (1) scans are chosen where the trap is being optimally filled and (2) the solution concentration [C] chosen for analysis allows AGC to function within its limits. Once a standard deviation of the mean is determined, one can decide, using tables of confidence limits, how often one would like to "be correct"—for example, determination of *m/q* will have a listing of possible elemental compositions that contains the correct answer 99% of the time. The interested reader is referred to a few of the many discussions of these issues in the current literature [16–19].

3. *Heavy-Isotope Pattern Recognition and Fitting.* A measure of how well the "natural" heavy-isotope pattern (commonly referred to as M+1, M+2, M+3) actually fits the calculated patterns for the allowed compositions (from step 2) is examined. Combining this information will usually allow one to narrow the list of candidate compositions sufficiently, such that together with a little orthogonal information (e.g., how this material was synthesized or what was its source), a solution of the elemental composition is often obtained.

Note as the use of ultrahigh resolution becomes more common, it must not be forgotten that the heavy-isotope lines at M+1, M+2, and so on may have resolved, fine structure, that is, a distinct pattern of bumps or shoulders that, much like a "fingerprint," will be useful in eliminating many suspects in the list of candidate compositions. For instance, a compound with many carbon and oxygen atoms and one sulfur atom will have resolved, fine structure in the M+2 line due to $[^{18}O]_1$, $[^{34}S]_1$, and $[^{13}C]_2$. An example at *m/z* 399, calculated at $R = 500{,}000$, is shown in Figure 4.3, while an actual analysis of the compound at $R = 100{,}000$, the current upper limit of the Orbitrap, is shown in Figure 4.4.

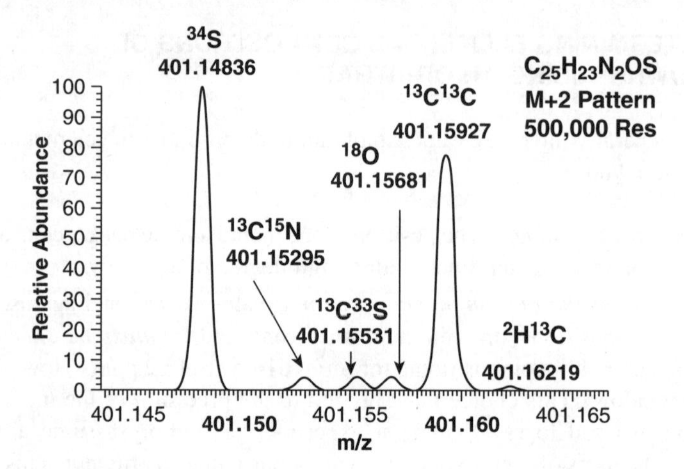

FIGURE 4.3 The theoretical lineshape for the (M + 2) isotope line of the protonated molecule $C_{25}H_{23}N_2OS^+$ at 500,000 resolution, showing the fine structure that may be observed. The isotopic atom (or atom combination) responsible for a line is given at the peak top, with the theoretical total atomic weight.

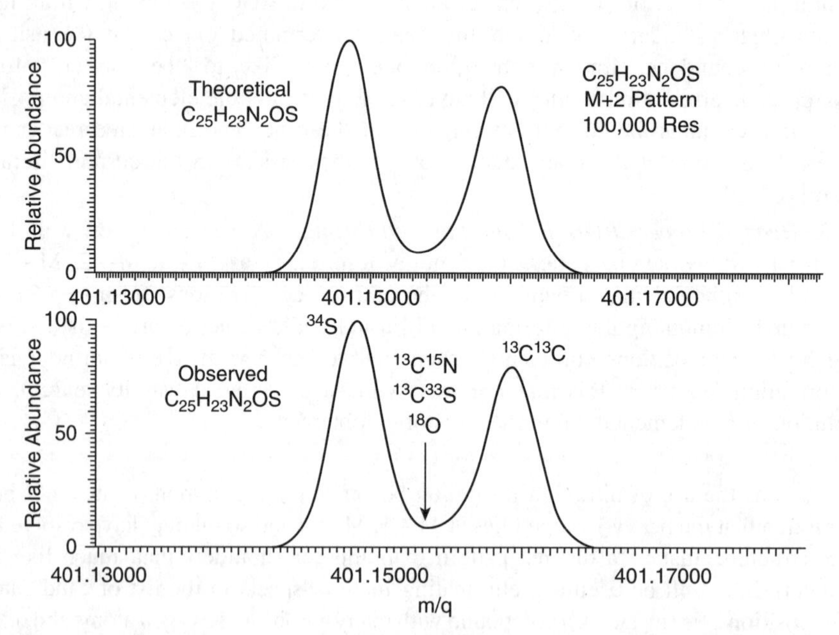

FIGURE 4.4 Actual data at the (M + 2) peak position for the protonated molecular species of Figure 4.3, obtained at 100,000 resolution specification using the Fourier transform orbital trap (FT-OT) or Orbitrap in the electrospray ionization mode. Fine structure is observable and lends to isotopic fingerprinting, to rule out competing candidate compositions.

4.6 ORBITRAP FIGURES OF MERIT IN MASS MEASUREMENT

4.6.1 Accuracy

Consider the electrospray mass spectrum ($R = 100$ K) of a poly(glycerol ester) (PGE) made from a disperse polyglycerol with varying degrees of esterification with coconut oil (a mixture mainly of C12, C14, C16, and C18 saturated fatty acids, along with oleic acid (C18- *cis*-Δ9,10) in Figure 4.5. It is ionized by cationization with sodium ions, ammonium ions, and protons and produces a large mix of "expected" signals that can be checked against a spreadsheet of all theoretically possible values. As industrial chemical manufacturers move into "green" materials that are complex mixtures, the unique capabilities of the Orbitrap (100,000 resolution and isotope fingerprinting) become important. The inset of Figure 4.5 shows some of the PGE data at 100,000 resolution (at m/q 400). Even slight overlap of a particular mass profile by the profile of a separate molecular entity can skew accurate determination of the peak tops. Note how close some of the entities are in m/q space, yet they are cleanly resolved. This is not the case at 30,000 resolution.

Figure 4.6 shows a simple plot of the observed m/q errors versus m/q after averaging of about 30 scans during an infusion of this oligomer mixture of hundreds of compounds. The instrument had been calibrated externally a few hours before acquisition of the data. The errors in accuracy ranged from −1.6 ppm at the low end of the m/q range to as little as 28 ppb (parts per billion) at the high end, or about

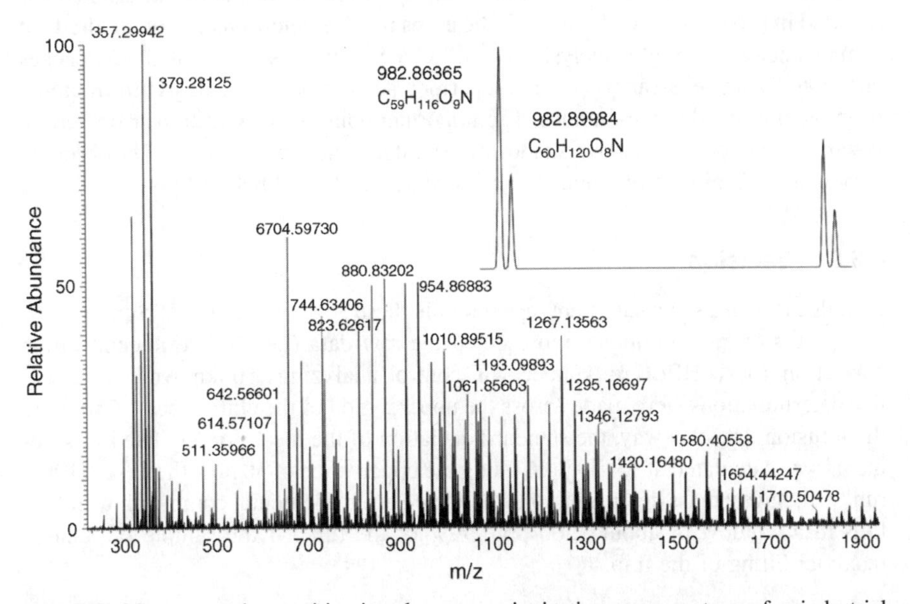

FIGURE 4.5 A complex, positive-ion electrospray ionization mass spectrum of an industrial poly(glycerol ester) material, obtained by infusing the sample to the ESI source of an Orbitrap. Individual molecules can ionize by protonation, ammonium cationization, and sodium cationization, further complicating the spectrum. The inset shows a very short section of the same ESI mass spectrum.

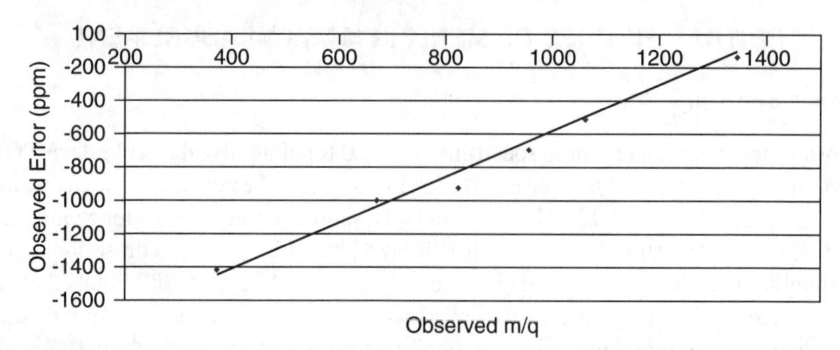

FIGURE 4.6 Actual *m/z* error (*Y*) observed for polyglycerol components as a function of *m/z* (*X*), showing the highly systematic nature of the error.

0.5–0.1 millimass units, respectively. The linear correlation coefficient is extremely high, 0.97. By simply applying this correction line to the other data, it can be shown that *most errors are decreased to less than 200 ppb*. More sophisticated approaches have been described in protein identification [20] and notably extended to metabolomics as well [21]. In the latter application, background ions like copper/acetonitrile clusters (among many others) have been identified and used as "free" lock masses to continuously and internally recalibrate Orbitrap data. The advantage of internal recalibration in FT-MS has been known for some time [16]. Similar to our small-molecule results, a 10-fold reduction of the typically quoted 1–2 ppm errors were reported in proteomics work. Such applications involve determination of hundreds of components in a single analysis by HPLC/MS. These sophisticated approaches undoubtedly are necessary to provide superior results across the analysis time required for proteomics and metabolomics. The important point is that Orbitrap errors have a systematic component that overshadows the random component, and thus the observed error can be significantly reduced to achieve accuracy well below 1 ppm.

4.6.2 Precision

How does the mass measurement repeat within these 30 determinations? Suppose that these infusion measurements represented the raw data (i.e., before injection time correction) for six HPLC runs under AGC control, analyzing an unknown. In each run, five determinations were made across the upper region of the eluting peak. Grouping the infusion data this way, the standard deviation of the mean for any five measurements was calculated at 15–30 ppb. Using the upper value of 30 ppb, the next "HPLC run" will have a mass determination where the correct elemental composition answer falls in a window of about ±500 ppb, 99% of the time, if the sample [C] causes maximal filling of the trap.

4.6.3 Discussion

Note that it is very important to ensure that the conditions for the "known" analysis match those for analysis of the "unknown" when deciding how to set search window

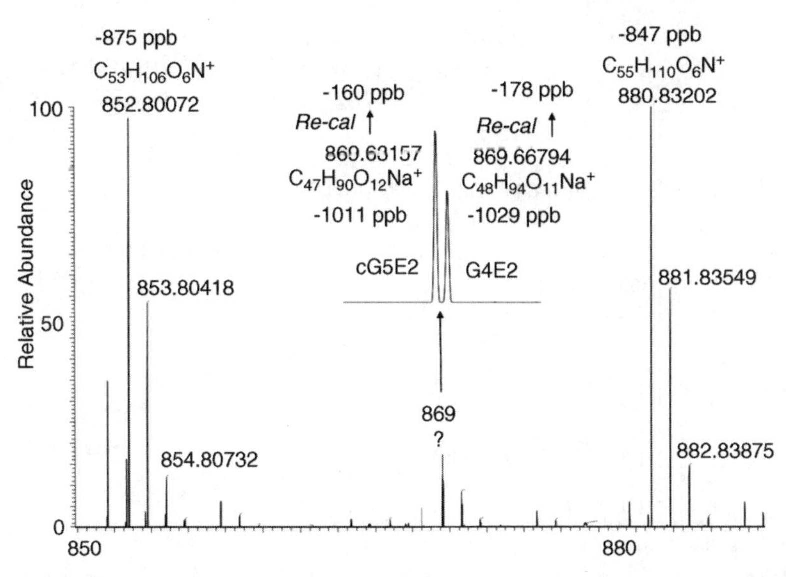

FIGURE 4.7 Improvement in the error observed for trace components at *m/z* 869, after recalibration between *m/z* 852 and *m/z* 880 of known, intense components in the PGE sample.

values. It would be unwise to obtain accuracy and precision for very strong signals and then apply those derived limits to data with very weak signals, where the trap was not optimally filled. Poly(glycerol esters) as a calibrant mixture are ideal, with a wide variety of signals of known compositions over a wide dynamic range, so that appropriate statistical derivations can be performed at any signal intensity (0.1–100% relative at "10×" intervals).

How can an "unknown" peak doublet be solved within a polyglycerol spectrum? Suppose that the small peak at *m/q* 869 is considered as an unknown (Figure 4.7 resolved doublet). A composition search window of 500 ppb is allowed, on the basis of five observations, the standard deviation of the mean, a 99% confidence limit, and derived statistics for other signals of similar intensity. By considering the ratio of M/M + 1 (approximately 2:1), a task that should always be performed, the carbon number lower limit can be reasonably set to 35. By subtracting the mass value (35 × 12) from 869, one obtains the mathematically possible number of atoms for H, N, O, P, S, which are individually allowed as upper atom numbers. Ionization by Na is allowed on the basis of other evidence in the spectrum (atom numbers = 0–1). An "unsaturation number" from −1.5 to 50 is allowed. Then 17 candidate answers are obtained. The true compositions (see Figure 4.7) are weak MNa^+ signals for members of the PGE family. Since the source of the sample was known (orthogonal information), picking the correct answer is simplified. This will not be the case for a true "unknown." Note the importance of the high precision of the Orbitrap. If the search window (based on statistical analysis) had been 1000 ppb, 35 possible answers would be obtained.

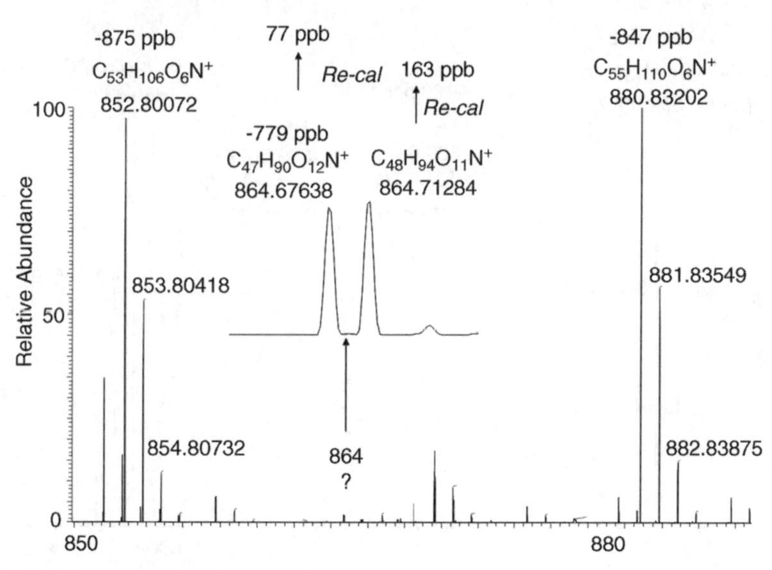

FIGURE 4.8 A recalibration experiment with an extremely weak signal, 10 times lower relative abundance versus *m/z* 869 of Figure 4.7.

To further examine the dynamic range of the Orbitrap, consider *m/q* 864, of 10 times lower intensity (Figure 4.8). These are the corresponding MNH_4^+ forms of *m/q* 869 ions. The mass errors are on the order of 750 ppb but fall by 5–10 times after internal recalibration. In summary, even very weak signals produce the same statistical result, as the Orbitrap has a high dynamic range for detection of "real" mass peaks as opposed to random noise.

How can possible answers in the list be reduced by using heavy-isotope profiles? The list of 17 answers contains the correct answer, $C_{47}H_{90}O_{12}Na$. How would one conclude which answer is correct if this were a true unknown? Sometimes, "orthogonal data" narrows the list, such as how the chemical sample was derived. In materials obtained in chemical commerce, a search of each formula against a worldwide database (e.g., via Scifinder) may show one formula on the list with thousands of associated references, so it becomes a primary suspect. But in a "pure approach" based only on the observed data, the next common step is to compare the observed natural heavy-isotope patterns, so-called M+1 and M+2 lines, against theoretical patterns for each formula in the list, and see which isotope pattern matches best.

A commercially available program called MassWorks has been used to evaluate how faithfully Orbitrap-based analysis captures isotope profiles [22]. This is not a trivial question in devices where ion "clouds" travel in orbits, as they are influenced in their image currents by each other's charges and electric fields. For example, the determination of amphoteracin B ($C_{43}H_{73}NO_{17}$) had 190 possible elemental compositions in a 2-ppm window around its accurate mass, but MassWorks ranked its isotope pattern as the best fitting at most resolutions. Later, 14 other compounds were evaluated, and the results were similar. Although in some cases at 100 K resolving power the isotope pattern-matching performance was actually reduced, it seems that the Orbitrap can perform reasonably well using this commercial offering. No comparisons were made to other commercial software such as iFIT, which runs on time-of-flight mass spectrometers. Another program called Xamine is the Orbitrap vendor's offering.

4.7 HPLC ORBITRAP MS: ACCURATE MASS DEMONSTRATION AND DIFFERENTIATION OF SMALL MOLECULE FORMULAS VERY PROXIMATE IN MASS/CHARGE RATIO SPACE

A separation of six compounds was developed, all with unique elemental compositions and all showing MH^+ 399. The separation is shown in Figure 4.9. Samples were submitted "blind" to an experienced Orbitrap operator, and the task of assigning elemental compositions was undertaken. The data returned consisted of 6–11 *m/q* determinations across the HPLC peak and were evaluated for the standard deviation of the mean. The 99% confidence limit was calculated and used as an initial formula search tolerance window. The allowed atom set was C, H, N, O, P, S, with up to one Na

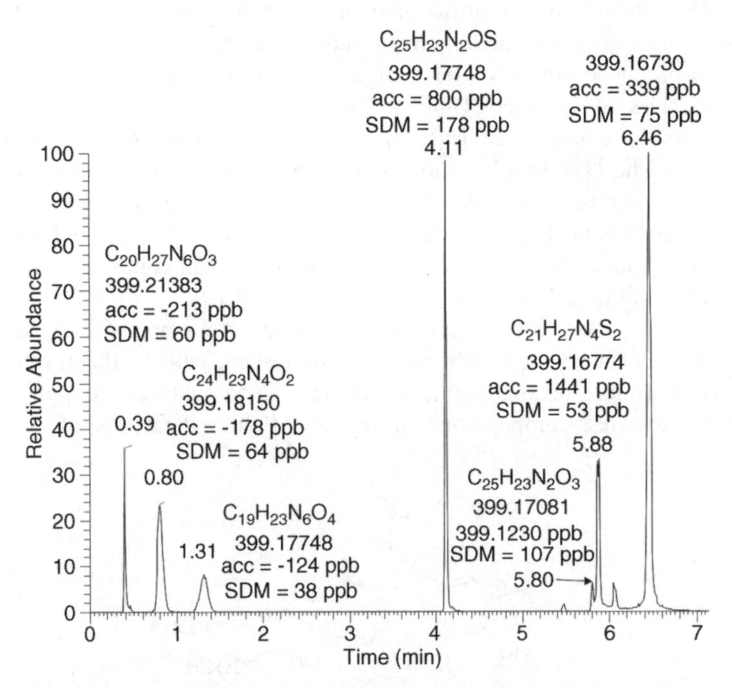

FIGURE 4.9 Mass accuracy and precision obtained in the HPLC/(+) electrospray ionization Orbitrap analysis of a mixture of six compounds, all yielding MH^+ 399 but different decimal masses due to their unique atomic compositions (acc = accuracy; sdm = standard deviation of the mean).

allowed. In half of the cases (peaks 1–3), the standard deviation of the mean was on the order of 40–60 ppb. Only one or two viable formula answers lie within the search window (130–210 ppb range). The correct formula was always returned with the second formula easily rejected by fitting to M+1 and M+2 peak intensities.

In the other four cases (peaks 4–7), where HPLC vertical peak heights were accentuated, the standard deviation of the means covered a range of 200–800 ppb (in the worst-case scenario, only five *m/q* values were obtained) and the correct answer was one of six to eight answers returned, with errors ranging from about 0.4 to 1.0 ppm, slightly outside the range of the initial search parameters. These data had not been corrected by internal recalibration using a known peak, and it is well established that the *m/q* values can be recalibrated if a known *m/q* is present. In fact, an algorithm performing this function is present in the most current, commercial Orbitrap software.

4.8 DETERMINATION OF TRACE CONTAMINANT COMPOSITIONS BY SIMPLE SCREENING HPLC-MS AND INFUSION ORBITRAP MS

A blue dye (structure shown in Figure 4.10) was "screened" by HPLC-UV-MS, and several trace peaks were "tagged" with nominal *m/q* determinations (Figure 4.11). The sample was then simply infused into an ESI source, and accurate Orbitrap *m/q* determinations were undertaken. Some fortunate circumstances were that the parent compound produced not only MH^+ but also two fragments under "normal" ESI conditions (loss of H_2NCH_2COOH twice) that allowed a simple second-order recalibration of the *m/q* range. This improved accuracy on all mass peaks to *better than 100 ppb*. The 99% confidence limits were 350 ppb, based on a standard deviation of the mean ($n = 6$ to mimic HPLC conditions) of *87 ppb*. Applying fairly generic constraints and manually examining theoretical isotope profiles and calculated profiles led to quick solution of several impurity elemental compositions with few other formulas falling in the search window (Figure 4.12).

In closing this section on determination of elemental compositions, it is noteworthy that a heuristic approach has been described, called "the seven golden rules." Here a more detailed approach is taken to the automated application of chemical rules that compositions must obey. Public databases of compound

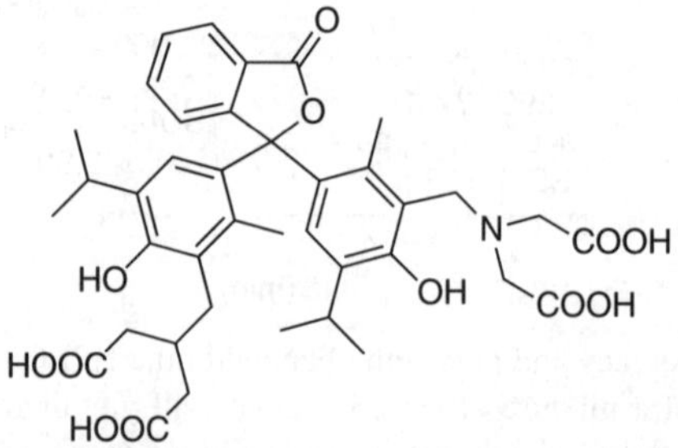

FIGURE 4.10 Structure of a dye subjected to impurity analysis using FT-OT.

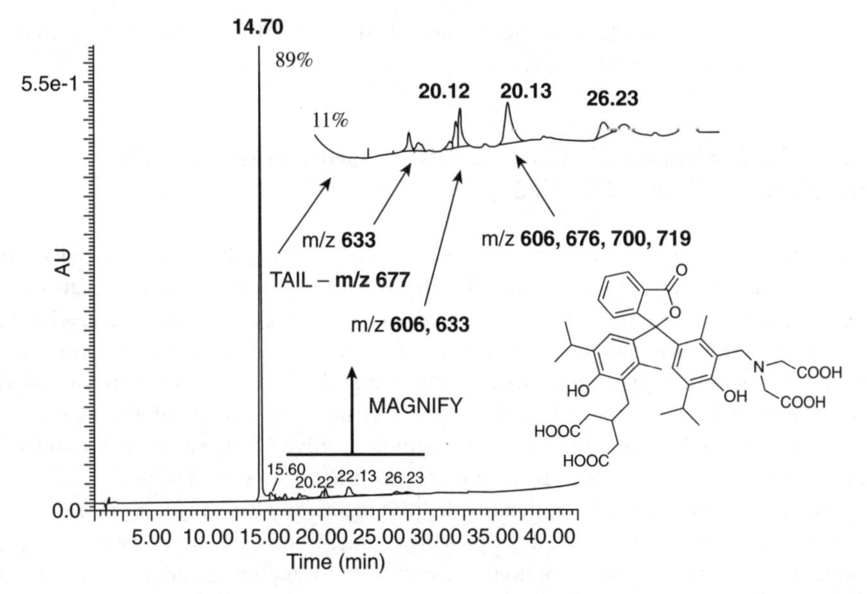

FIGURE 4.11 HPLC/UV/(+) ion ESI—quadrupole MS analysis of a dye, revealing impurities on the main component's tail that were tagged with nominal *m/q* assignments.

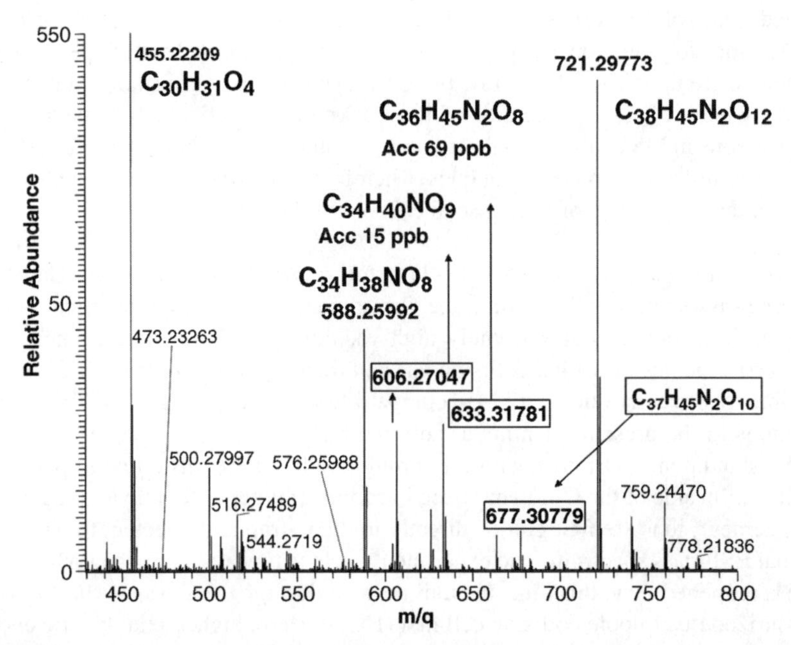

FIGURE 4.12 Infusion/(+) ESI Orbitrap MS analysis of dye impurity formulas by using internal recalibration based on parent material protonated molecule and fragment ions; <100 ppb errors are obtained.

formulae identities are automatically searched. The software program is freely available via the Internet [23].

4.9 DETERMINING SUBSTRUCTURES: ORBITRAP TANDEM MASS SPECTROMETRY (MS^n)

Many mass spectrometrists have had the experience where a chromatographer visits the lab and requests, "please tell me what that tiny peak is in my chromatogram—it's not supposed to be there." Mass spectrometry can often accomplish the task without resorting to extensive isolation or preconcentration/purification schemes—and in the best outcomes, the suggested impurity might be readily synthesized or even purchased for a simple head-to-head LC-MS comparison and verification of the elemental composition. In summary, this would be "structure solved," unless one has stumbled on a coeluting isomer in the HPLC part of the structural proof design.

So, while elemental compositions are helpful in actually "solving a structure," they are only a compilation of the atoms present. Tandem mass spectrometry or MS^n (in time, as with trapping devices, or in space, as with non-trapping devices) is necessary to fragment ions and build substructural hypotheses [24–26]. Further, in a non-optimum situation, solving the structure of a trace impurity entails suggesting detailed bond-to-bond connectivity, which is not really an inherent capability of mass spectrometry.

In this scenario, a knowledge base of the tandem mass spectral behavior of related available materials is established. Specifically, closely related materials, whose atom-to-atom connectivity was determined by using a host of spectroscopic techniques (e.g., NMR, MS, X ray, IR, UV), is created by performing MS^n experiments, where $n \geq 2$. The identity of the "unknown" is obtained by inference or comparisons in MS analyses, as the amount of material available is seldom sufficient for successful use of other techniques. Therefore, the structural proof endeavor is really a team approach of spectroscopies most of the time.

How does one perform MS^n ($n > 1$) using a standalone Orbitrap? Collision-induced dissociation (CID) within the Orbitrap structure is not a practical option. The Orbitrap operates at extremely high vacuum, and the cycle time needed to pressurize the device with a collision gas and then depressurize the device is not a feasible way to carry out "fast" MS^n compatible with chromatography. Two current solutions to the present dilemma are offered here.

A "standalone" Orbitrap with a trapping quadrupole collision cell opposite to the injection side of the C-trap has been introduced (Figure 4.13). In a "ping-pong" arrangement, ions are first guided directly to the C-trap and injected, to obtain the normal ESI mass spectrum. The experiment cycle is repeated (i.e., the ion injection gate is reopened), and this time, the ions are sent through the C-trap to the *trapping*, pressurized quadrupole collision cell first (HCD). Here, higher axial kinetic energy of the ions causes energetic collisions, and new fragment ions are formed. The fragment ions are then shunted directly back to the C-trap, where the experiment continues, as normal, with accurate mass measurement.

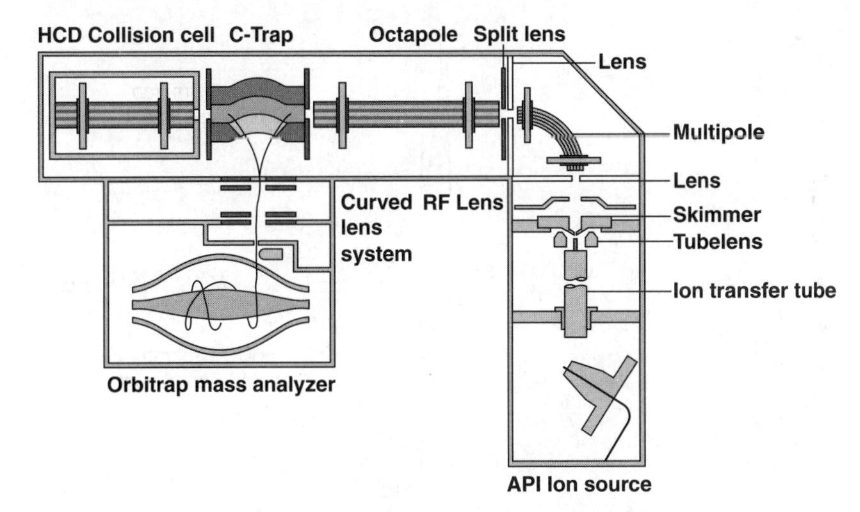

FIGURE 4.13 Schematic representation of an Orbitrap fitted with a collision cell (called the "HCD option"), allowing MS/MS-like data to be obtained for pure or cleanly separated (e.g., inline HPLC) components (used by permission, Thermo Fisher Scientific Corp., 2008).

In Figure 4.14, the fast HPLC analysis of a simple, brewed-coffee filtrate is shown under normal (upper) and HCD conditions. First, the ratio of signal to chemical noise is improved significantly for the peaks at retention time 3.59 and 3.76 min using HCD. This is quite possibly due to the reduction of solvent cluster ions in the HCD experiment—they are lost because they fragment to small, nontrapped entities. This is an advantage of HCD: enhancement of *S/N* of trace components in the total-ion current trace. A similar effect is achieved in the use of field asymmetric waveforms on the "backend" of some electrospray ionization devices.

The mass spectra are shown in Figure 4.15. The suspected formula of the parent was derived by the procedures described earlier in this chapter. From the HCD spectrum, it can be reasonably ascertained that the molecule bears a caffeic acid residue.

Is the HCD experiment a tandem mass spectrometry (MS/MS) experiment? The term MS/MS or MS^2 is very ambiguous—it can imply any number of experiments [27]. The ambiguity only grows with MS^3, MS^4, and son on, along with considerations of tandem-in-space and tandem-in-time instruments (trapping devices). For the purpose of this discussion, the context will be restricted to the most common experiment in trapping devices: sequential product ion analysis. In the first stage, "purification" is effected by selecting only the *m/z* of interest, prior to subjecting the selected ions to "activation" (most commonly using a collision gas) to produce ionic fragments and neutral fragments; the product ions then undergo a second stage of *m/z* analysis. Typically a full "scan" of the product ion collection, whether by direct or indirect detection approach, is obtained.

In this purest sense, HCD is not tandem mass spectrometry. For example, all components coeluting from an HPLC column will simultaneously enter the HCD

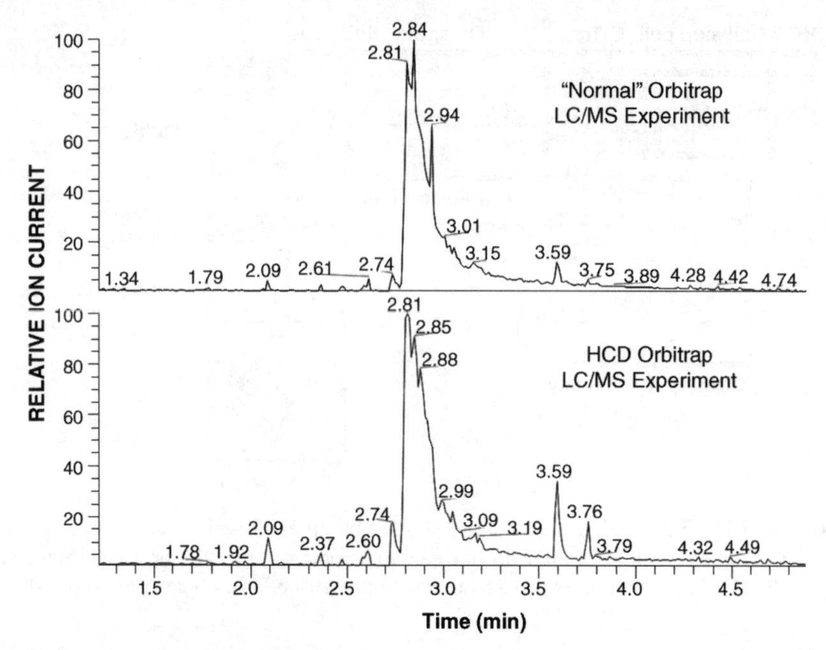

FIGURE 4.14 Comparison of an HPLC/(+) ESI Orbitrap MS analysis of a coffee filtrate, in the "normal mode" (upper trace) versus the "HCD" mode (lower trace).

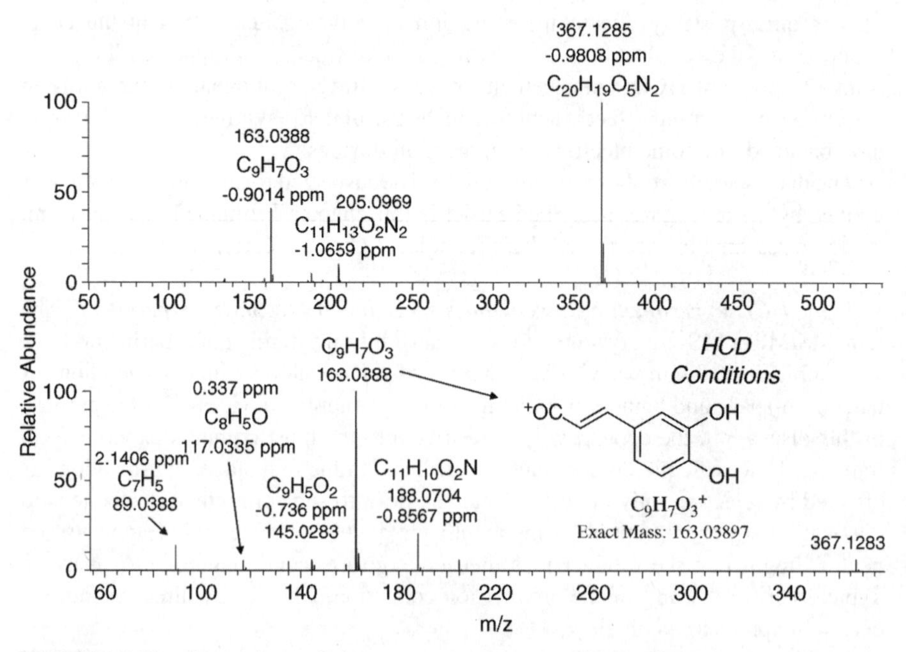

FIGURE 4.15 Comparison of the spectra associated with the HPLC peak at 3.59 min obtained in the "normal" mode of Orbitrap operation versus the "HCD" mode.

collision cell and produce a fragment–ion–mixture spectrum. Because the mass-measuring device is an Orbitrap, it may be possible to sort some fragments into groups, based on considerations of derived molecular formulas. For instance, a C_9 precursor cannot be related to a C_{10} fragment; and if the components are cleanly separated in the HPLC domain, the only essential difference is that HCD spectra will always bear isotope peaks in the collisionally derived spectrum. Thus, HCD may be a useful acronym for clearly differentiating the experiment from tandem mass spectrometry.

4.10 MULTIANALYZER (HYBRIDIZED) SYSTEM: THE LINEAR ION TRAP/ORBITRAP FOR MS/MS AND HIGHER-ORDER MS^n, $n > 2$

Curiously, the first commercially released Orbitrap instrument was actually a hybridized device, and was marketed commercially as the "LTQ-FT" (linear trap quadrupole–Fourier transform). In this hybrid design, the first mass analysis device is a linear ion trap. The second mass analysis device is the Orbitrap. Thus, tandem mass spectrometry may be performed on this device.

It is not the purpose of this chapter to delineate linear ion trap operation, but a few brief points are salient to the discussion:

1. The linear ion trap has a faster cycle time than does the Orbitrap and uses mass-selective instability to eject ions orthogonally to a conversion dynode (direct detection) for *m/q* analysis. Alternatively, ions may be passed "in line" through the device, to the Orbitrap.
2. In the LTQ-Orbitrap configuration, all MS-MS experiments are carried out using the LTQ, generally with unit mass isolation of precursor ions in small-molecule work.
3. Because the device is a trap, multi-stage mass spectrometry (or so-called MS^n) may be performed. MS^n is defined here as sequential product ion scans in time.
4. At any stage of n in the linear ion trap, the product ions may be transferred to the Orbitrap for accurate *m/z* determination, *instead of being ejected* and detected directly by the LTQ.
5. Because the detection of image currents is relatively slow, many LTQ MS^n experiments may be carried out with direct detection, *while an FT-based mass measurement is in progress*; thus, multiplexed experiments are possible with this hybridized device.
6. Because of the different cycle times of the linear trap and the Orbitrap, typically only some of the LTQ MS^n experiments have associated Orbitrap data at higher resolution and accuarcy.
7. It is possible to perform all MS^n experiments in a data-dependent fashion.
8. If one has sufficient time (infusion experiments), the entire fragmentation genealogy may be examined. This capability is greatly enhanced by modern "infusion" equipment such as the Advion Triversa Nanomate (see discussion below).

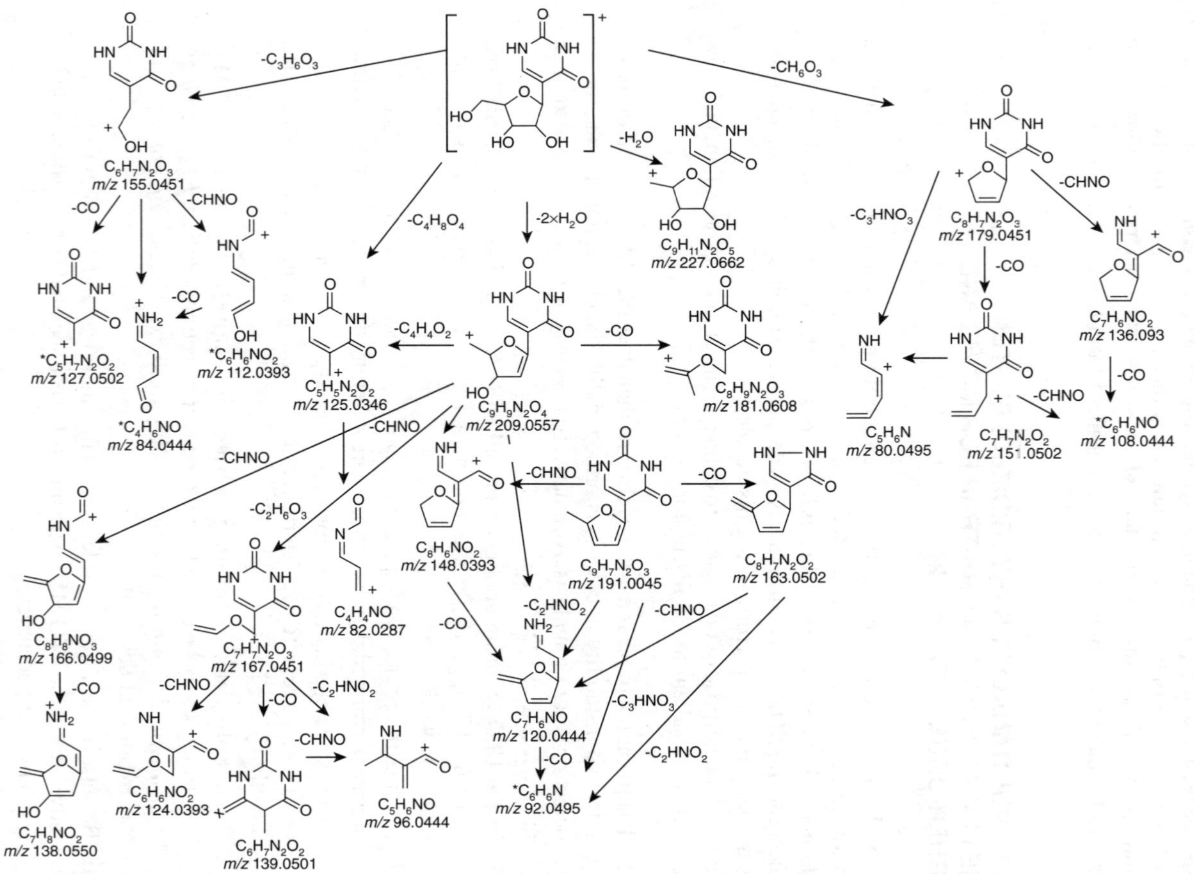

FIGURE 4.16 An accurate mass annotated fragmentation geneaology (family tree) of pseudouridine, used to discover other novel structural metabolites; the map is based on Orbitrap data (used by permission, Wiley Interscience) [28].

In the previously cited example of the blue dye, a fragment ion geneaology of the parent material was constructed using infusion–MS^n. By selectively storing ions of the impurity *m/z* values in the LTQ, subjecting them to MS^n with accurate *m/q* determination, and examining the various impurity peaks, one could suggest several secure identities for the various impurities. The approach has been applied in the pursuit of identification of uremic metabolites in hemodialysate. A detailed fragmentation map of the known compound pseudouridine was obtained (Figure 4.16), and the fragments were annotated with elemental compositions. Subsequently, using MS^n on an "unknown," a novel, previously unreported uremic metabolite was proposed [28].

Is there a case for infusion methodology? Obviously, one can observe a sample for a much longer time period during infusion versus HPLC. A large leap forward in sophisticated fraction collection and infusion analysis is the Advion Triversa Nanomate, for the front end of the mass spectrometer. Fractions from even low-flow columns are captured in a 96 or 384-well format in receivers with their own electrospray nozzle. The collection plates are "chip-based" technology and manufactured at relatively low cost. As a nozzle comes into position for *m/q* analysis, the individual cell contents can be electrosprayed for several minutes easily, due to low consumption rates of nanoliters per minute. In the data-dependent scan mode, MS^n genealogical trees can be readily generated, and annotated with elemental compositions and accurate masses. This is the "ultimate" experiment, for it preserves the HPLC chromatogram at sufficient HPLC peak definition and allows a non-time-restricted, non-volume-restricted interrogation of the fractions (i.e., there is more than enough sample in the fraction to allow detailed map generation). Of particular note is an investigation of oligosaccharide sequences, by infusion using an ion trapping device [29,30] for analysis. Extension to use of an Orbitrap would be straightforward.

4.11 MASS MAPPING TO DISCOVER IMPURITIES

An application heretofore exclusive to FTICRMS is Kendrick mapping [31], a compact, two dimensional "self-organizing" map of materials in complex mixtures. For instance, it has been shown that in diluted crude oil, it is possible to observe signals for over 10,000 compounds in a 1000 Da range, when very high resolution is applied (as in FT-ICR-MS, with 1,000,000 resolution at *m/q* 400). Interpreting such a massive quantity of mass peaks—indeed, to even visualize it—would require such lateral spreading of the *m/q* axis that the spectrum might be several meters wide.

Yet, the same information can be compactly organized in a "one page" 2D mass map. In this technique, the mass axis is normalized to $CH_2 = 14.0000$, instead of $C = 12.00000$. The result is that a particular family of compounds, differing only in the number of (CH_2) within their respective formulas, would all have the same decimal mass (mass defect). This transformation is essentially achieved by multiplying the conventional *m/q* values by 0.998883 or 14.0/14.01565, where the latter number is the IUPAC (International Union of Pure and Applied Chemistry) mass of the CH_2 unit. The resulting mass numbers are called *Kendrick masses*. For example, in this approach the Kendrick mass of indole is 116.9271, and that of ethyl-substituted indole is 144.9271. On the other hand, alkyl-substituted indole-*n*-oxides have Kendrick masses with a decimal value of 0.9041. Thus, it is easy to understand that by plotting the derived whole-number part (X) versus the decimal part (Y), one may observe that compounds that differ only by methylene units will line up in horizontal rows, separated vertically by their different classes (e.g., alkylindoles, alkylindole *N*-oxides).

As we pointed out, the Orbitrap is an FT-MS device. Can it be used to map complex mixtures? The answer is yes; our laboratory has been mapping industrial raw materials since 2008. While doing this at a maximum of only 100,000 resolution, it works well for our applications [32]. There may, indeed, be mapping exercises that do not work so well using the Orbitrap, if higher resolution is required to separate the components. Nevertheless, higher Orbitrap operating resolution is clearly being pursued.

In our work, any repeating oligomer unit may be compelled to self-organize in these maps. In the case of PGEs, where the length of glycerol chains varies by $[C_3H_6O_2]$ repeating units, using a mass normalizing factor of 74/74.03677 or 0.999503 and applying it to the *m*/*q* values in Figure 4.5, one obtains the map shown in Figure 4.17. The esterified materials are neatly grouped into classes (monoester, diester, triester, etc.). Cyclic and acyclic glycerols occupy distinct positions, and glycerol repeat numbers are color-coded in horizontal rows. Moreover, variance by ester chain length is observed in a straight line, running at a distinct angle across the map.

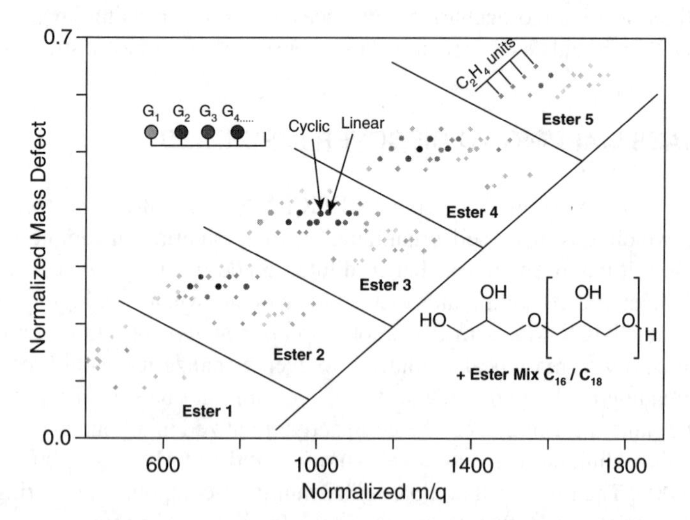

FIGURE 4.17 A self-organizing normalized mass map of the complex polyglycerol spectrum shown in Figure 4.5. Compound classes align linearly depending on their formula relationships and are readily speciated and classified. Data were obtained at 100,000 resolution (at *m*/*q* 4000) using an Orbitrap [32].

It is important to note that perhaps the overriding important factor in generating such maps is the *m/q* precision, because this determines the extent to which a dot position will "wiggle" or move about. The "wiggle" will determine whether a straight-line relationship can be recognized. The rows in the Orbitrap polyglycerol map are very neatly aligned. Mass accuracy and precision are also very important for *simplifying the map*. The combining of *m/q* signals from a single molecular entity (MH^+, MNa^+, MNH_4^+ in this example) into one representative "dot" position reduces clutter in the map. The same is true for recognizing [^{13}C] isotope peaks and collapsing their intensities into the position for the monoistopic line in the spectrum. By accurately identifying such signals by the accurate *m/q* differences and combining them at one position, one obtains a single representative "dot" for each formula.

We have used these maps to classify raw materials by derived values such as the average ester chain length, the average number of glycerols, and even the degree of unsaturation or iodine value (a wet-chemical procedure involving the consumption of iodine by C=C in fat samples). By comparison to traditional wet-chemical approaches we have obtained an agreement that is reasonable enough for manufacturing purposes to allow replacement of the wet-chemical methods. Further, pertinent to this volume on trace analysis, we have been able to single out unrelated impurities down to 0.1%. They appear as unaligned dots on the map; that is, they are not in the same class and appear out of position. Further, if the impurity was reactive, as in an ethoxylation reaction, an out-of-place "impurity line" appears. A 1% level of polyethoxylated "contaminant" is readily detected. Finally, other data mapping strategies reported for FT-ICR [33] should work for FT-OT, as long as compounds are resolved in the *m/q* domain at 100,000.

4.12 THE CURRENT PRACTICE OF ORBITRAP MASS SPECTROMETRY

The principle aim of this chapter has been to provide a somewhat instructive overview and has intentionally focused on the narrow issues of resolution, mass accuracy, mass precision, isotope patterns, and accurate mass MS^n (hybrid instruments) for trace impurity identification. A review of the current literature shows the Orbitrap being used in widespread laboratory applications. Over 300 references to the Orbitrap between the years 2005 and 2009 were found, with roughly a third of the applications devoted to small-molecule analyses and about two-thirds devoted to proteomics. From 2007 to 2008, the number of citations approximately quadrupled, a testament to the rapid growth of Orbitrap mass spectrometry.

It suffices to say that many talented investigators are at work utilizing the novel and lower-cost (vs. those of FT-ICR-MS) capabilities of the Orbitrap. A few application areas are highlighted here to provide a view of the horizon covered in Orbitrap applications. Regulation of residues in the food stream is an important issue. In regard to *m/q* resolution, it was shown that the anabolic steroid trenbolone was inadequately resolved in time-of-flight MS-based analysis (15,000 resolution) from background, resulting in poor mass accuracy, whereas the Orbitrap could cleanly resolve the analyte peak's *m/q* value. A similar trend in hair analysis for dopants (steroids) was cited [34]. Residual pharmaceuticals and their transformation products in drinking-water supplies has also become an environmental problem and was investigated using an Orbitrap [35]. Applications in drug metabolism abound [36–38]. Direct analysis of natural products has been performed [39]. Quantitation has been addressed in several publications, since reproducibly determining the *m/q* of a mass accurately at high precision and resolution naturally increases the selectivity of analysis [40].

4.13 CONCLUSION

It has been fairly well established at this point in time that, while the Orbitrap is often referred to as a "1–2 ppm" accuracy device, this is really an understatement of the capability of the device. Its capabilities clearly can be made to fall in the 100 ppb range in precision and accuracy with the right experimental approach. As to fitting of isotope profiles, for structure elucidation, TOF does a better job currently, but the Orbitrap offers "isotope fingerprinting" at 100,000 resolution, for some of the heavy-isotope lines that might typically be present. In the near future, it will be interesting to see where the limits of electronics place high-performance Orbitraps relative to ICR devices with their increasingly large magnets. At the very least, choosing a structure elucidation tool in the FT-MS arena will require careful evaluation of both technologies.

REFERENCES

1. Williams, J. D.; Cox, K. A.; Cooks, R. G.; Kaiser, R. E., Jr.; Schwartz, J. C. (1991), High mass- resolution using a quadrupole ion-trap mass spectrometer, *Rapid Commun. Mass Spectrom. 5*(7), 327–329.
2. Londry, F. A.; March, R. E. (1995), Systematic factors affecting high mass-resolution and accurate mass assignment in a quadrupole ion trap, *Int. J. Mass Spectrom. Ion Process. 144*(1/2) 87–103.
3. March, R. E., Todd, J. F. J. (1995), *Practical Aspects of Ion Trap Mass Spectrometry*, CRC Press, New York, Vols. 1–3.
4. Makarov, A. (2000), Electrostatic axially harmonic orbital trapping: A high performance technique of mass analysis, *Anal. Chem. 72*, 1156–1162.
5. Kingdon, K. H. (1923), *Phys. Rev. 21*, 408–418.
6. Haliday, D.; Resnick, R. (1962), *Physics*, Wiley, New York, pp. 647–720.
7. Knight, R. D. (1981), Storage of ions from laser-produced plasmas, *Appl. Phys. Lett. 38*, 221–222.
8. Hu, Q.; Noll, R. J.; Li, H.; Makarov, A.; Hardman, M.; Cooks, R. G. (2005), The Orbitrap: A new mass spectrometer, *J. Mass Spectrom. 40*, 430–443.
9. Perry, R. H.; Hu, Q.; Salazar, G. A.; Cooks, R. G. (2009), Rephasing ion packets in the Orbitrap mass analyzer to improve resolution and peak shape, *J. Am. Soc. Mass Spectrom. 20*, 1397–1404.

10. Syka, J. E. P.; Bai, D. L.;et al. (2001), A linear quadrupole ion trap Fourier transform mass spectrometer, a new tool for proteomics, *Proc. 49th ASMS Conf. Mass Spectrometry and Allied Topics*, Chicago.
11. Marshall, A. G.; Hendrickson, C. L.; Jackson, G. S. (1998), Fourier transform ion cyclotron resonance mass spectrometry: A primer, *Mass Spectrom. Rev. 17*, 1–35.
12. Brodbelt, J. (2009), Focus in honor of Alexander Makarov, recipient of the 2008 award for a distinguished contribution in mass spectrometry, *J. Am. Soc. Mass Spectrom. 20*, i–iii (editorial focus; see articles within this issue).
13. Makarov, A.; Denisov, E.; Lange, O. (2009), Performance evaluation of a high field Orbitrap mass analyzer, *J. Am. Soc. Mass Spectrom. 20*, 1391–1396.
14. Kellman, M.; Weighaus, A.; Muenster, H. (2009), *High Resolution and Precise Mass Accuracy: A Perfect Combination for Food and Feed Analysis in Complex Matrices*, Application Note 30163, Thermo Fisher Scientific, San Jose, CA.
15. Kellmann, M.; Muenster, H.; Zomer, P.; Mol, H. (2009), Full scan MS in comprehensive qualitative and quantitative residue analysis in food and feed matrices: How much resolving power is needed? *J. Am. Soc. Mass Spectrom. 20*, 1464–1476.
16. Zhang, L. K.; Rempel, D.; Pramanik, B. N.; Gross, M. L. (2005), Accurate mass measurements by Fourier transform mass spectrometry, *Mass Spectrom. Rev. 24*, 286–309.
17. Tyler, A. N.; Clayton, E.; Green, B. N. (1996), Exact mass measurement of polar organic molecules at low resolution using electrospray ionization and a quadrupole mass spectrometer, *Anal. Chem. 68*, 3561–3569.
18. Blom, K. F. (2001), Estimating the precision of exact mass measurements on an orthogonal time-of-flight mass spectrometer, *Anal. Chem. 73*, 715–719.
19. Gross, M. L. (1994), Accurate masses for structure confirmation, *J. Am. Soc. Mass Spectrom. 5*, 57.
20. Cox, J.; Mann, M. (2009), Computational principles of determining and improving mass precision and accuracy for proteome measurements in an Orbitrap, *J. Am. Soc. Mass Spectrom. 20*, 1477–1485.
21. Scheltema, R. A.; Kamleh, A.; Wildridge, D.; Ebikeme, C.; Watson, D. G.; Barrett, M. P.; Jansen, R. C.; Breitling, R. (2008), Increasing the mass accuracy of high-resolution LC-MS data using background ions—a case study on the LTQ-Orbitrap, *Proteomics 8*, 4647–4656.
22. Erve, J. C.; Gu, M.; Wang, Y.; DeMaio, W.; Talaat, R. E. (in press), Spectral accuracy of molecular ions in an LTQ/Orbitrap mass spectrometer and implications for elemental composition determination, *J. Am. Soc. Mass Spectrom.*
23. Kind, T.; Fiehn, O. (2007), Seven golden rules for heuristic filtering of molecular formulas obtained by accurate mass spectrometry, *BMC Bioinform. 8*, 105.
24. Strife, R. J.; Robosky, L. C.; Garrett, G.; Ketcha, M. M.; Shaffer, J. D.; Zhang, N. (2000), Ion trap MS^n genealogical mapping—approaches for structure elucidation of novel products of consecutive fragmentations of morphinans, *Rapid Commun. Mass Spectrom. 14*, 250–260.
25. Strife, R. J. (1999), Structure elucidation by ion trap sequential mass spectrometry of radical cations formed in low-energy charge exchange reactions, *Rapid Commun. Mass Spectrom. 13*, 759–763.
26. Strife, R. J.; Ketcha, M. M.; Schwartz, J. (1997), Multi-stage mass spectrometry for the isolation and structure elucidation of components in a crude extract, *J. Mass Spectrom. 32*, 1226–1235.

27. Schwartz, J. C.; Wade, A. P.; Enke, C. G.; Cooks, R. G. (1990), Systematic delineation of scan modes in multidimensional mass spectrometry, *Anal. Chem. 62*, 1809–1818.
28. Godfrey, A. R.; Williams, C. M.; Dudley, E.; Newton, R. P.; Willshaw, P.; Mikhail, A.; Bastin, L.; Brenton, A. G. (2009), Investigation of uremic analytes in hemodialysate and their structural elucidation from accurate mass maps generated by a multi-dimensional chromatography/mass spectrometry approach, *Rapid Commun. Mass Spectrom. 23*, 3194–3204.
29. Zamfir, A.; Vakhrushev, S.; Sterling, A.; Niebel, H. J.; Allen, M.; Peter-Katalinic, J. (2004), Fully automated chip-based mass spectrometry for complex carbohydrate system analysis, *Anal. Chem. 76*, 2046–2054.
30. Almeida, R.; Mosoarca, C.; Chirita, M.; Udrescu, V.; Dinca, N.; Vukelic, Z.; Allen, M.; Zamfir, A. (2008), Coupling of fully automated chip-based electrospray ionization to high-capacity ion trap mass spectrometer for ganglioside analysis, *Anal. Biochem. 378*, 43–52.
31. Hughey, C.; Hendrickson, C.; Rodgers R.; Marshall, A. (2001), Kendrick mass defect spectrum: A compact visual analysis for ultra-high-resolution broadband mass spectra, *Anal. Chem. 73*, 4676–4681.
32. Strife, R.; Mangels, M. (2010), Normalized mass mapping of Orbitrap data to define complex, polyglycerol-based raw material compositions, *Rapid Commun. Mass Spectrom. 24*, 1497–1501
33. Artemenko, K.; Zubarev, A.; Samgina, T.; Lebedev, A.; Savitski, M.; Zubarev, A. (2009), Two dimensional mass mapping as a general method of data representation in comprehensive analysis of complex molecular mixtures, *Anal. Chem. 81*, 3738–3745.
34. Vanhaecke, L.; Verheyden, K.; Vanden-Bussche, J.; Schoutsen, F.; DeBrabander, H. (2009), UHPLC coupled with Fourier transform Orbitrap for residue analysis, *LC-GC Eur. 22*, 364–370.
35. Radjenovic, J.; Petrovic, M.; Barcelo, D. (2009), Complementary mass spectrometry and bioassays for evaluating pharmaceutical-transformation products in treatment of drinking water and wastewater, *Trends Anal. Chem. 28*, 562–580.
36. Chen, G.; Daaro, I.; Pramanik, B. N.; Piwinski, J. J. (2009), Structural characterization of in-vitro rat liver microsomal metabolites of antihistamine desloratidine using LTQ-Orbitrap hybrid mass spectrometer with online hydrogen/deuterium exchange HR-LC/MS, *J. Mass Spectrom. 44*, 203–213.
37. Semenistaya, E. N.; Virus, E. D.; Rodchenkov, G. M. (2009), Determination of sulfates and glucuronides of endogenic steroids in biofluids by high-performance liquid chromatography Orbitrap mass spectrometry, *Russ. J. Phys. Chem. A 83*, 530–536.
38. Erve, J. C. L.; DeMaio, W.; Talaat, R. E. (2008), Rapid metabolite identification with sub-parts-per-million mass accuracy from biological matrices by direct infusion nanoelectrospray ionization after clean-up on a ZipTip and LTQ/Orbitrap mass spectrometry, *Rapid Commun. Mass Spectrom. 22*, 3015–3026.
39. Jackson, A. U.; Tata, A.; Wu, C.; Perry, R. H.; Haas, G.; West, L.; Cooks, R. G. (2009), Direct analysis of stevia leaves for diterpene glycosides by desorption electrospray ionization mass spectrometry, *Analyst 134*, 867–874.
40. Zhang, N. R.; Yu, S.; Tiller, P.; Yeh, S.; Mahan, E.; Emary, W. B. (2009), Quantitation of small molecules using high resolution accurate mass spectrometers—a different approach for biological samples, *Rapid Commun. Mass Spectrom. 23*, 1085–1094.

CHAPTER 5

Structural Characterization of Impurities and Degradation Products in Pharmaceuticals Using High-Resolution LC-MS and Online Hydrogen/Deuterium Exchange Mass Spectrometry

GUODONG CHEN

Bristol-Myers Squibb, Route 206 & Province Line Road, Princeton, NJ 08543

BIRENDRA N. PRAMANIK

Merck and Co., 2015 Galloping Hill Road, Kenilworth, NJ 07033

5.1 INTRODUCTION

The typical drug discovery process involves disease target identification, lead compound generation and optimization, preclinical testing, a scaleup process, formulation studies, and clinical trials [1]. One important consideration in the drug discovery process is drug safety, including impurities and degradation products in pharmaceuticals. Impurities present in active pharmaceutical ingredients (APIs) need be identified to ensure that no mutagenic or toxic substances will be administered to patients. A drug product degradation profile should be established to guide stable formulation and provide suitable drug shelf-life assignment. Drug regulatory agencies also have requirements for identification of the purity of a pharmaceutical. For example, Food and Drug Administration (FDA) guidelines

Characterization of Impurities and Degradants Using Mass Spectrometry, First Edition.
Edited by Birendra N. Pramanik, Mike S. Lee, and Guodong Chen.

state that impurities present at a level of 0.1% or above relative to API must be identified. These activities are designed to ensure that toxicological and pharmacological effects are only from drug substances and not due to impurities and degradants. Structural characterization of impurities and degradation products in bulk drug substances is an integral part of pharmaceutical product development. Analysis of these low-level unknown impurities and degradants can be a significant analytical challenge [2].

As one of the prime analytical techniques, mass spectrometry (MS) has become the method of choice for analysis of impurities and degradants because of its distinct analytical features [3]. A general MS-based strategy can be developed to analyze impurities and degradants [4]. The first step is to determine the molecular weight (MW) of the unknown, carried out by either electrospray ionization (ESI) [5,6] for polar compounds and atmospheric-pressure chemical ionization (APCI) [7] for nonpolar compounds or any other ionization methods whenever suitable. A second step is to obtain elemental composition of molecular ions for an unknown by high-resolution (HR) MS experiments. This can be performed using magnetic sector instruments, time-of-flight (TOF) [8,9], Fourier transform ion cyclotron resonance (FT-ICR) [10,11] and Orbitrap MS [12–15]. Tandem MS (MS/MS) experiments are usually followed to yield structural information by fragmentation studies of the unknown and parent molecule [16,17]. In case of complex mixtures, hyphenated analytical techniques such as high-performance liquid chromatography (HPLC)-MS (LC-MS) and gas chromatography-MS (GC-MS) can be used to obtain MW information of individual components, including their elemental compositions in HR-MS mode, as well as fragmentation information from LC-MS/MS experiments. Once a tentative structure is proposed, one can perform chemical derivatization, including hydrogen/deuterium (H/D) exchange experiments [18–27] to provide additional structural evidence. The final structural assignment is confirmed by nuclear magnetic resonance (NMR) experiments on isolated sample fraction or a synthetic standard based on the proposed structure. It is important to note that LC-MS or LC-MS/MS combines the high-resolution separation capability of HPLC with superior MS detection and characterization ability, playing important roles in structural characterization of impurities and degradants. The use of HR-LC/MS and HR-LC-MS/MS is critical in providing accurate mass data for both molecular ions and fragment ions. Another approach using online H/D exchange LC-MS is also very effective in structural elucidation studies. This method measures the difference in molecular weight of a compound before and after the deuterium exchange to determine the exchangeable hydrogen atoms in a molecule. The exchangeable hydrogen atoms are usually bound to N, O, or S atoms in functional groups such as OH–, NH–, NH_2–, COOH–. Deuterated mobile phases such as D_2O or CH_3OD can be deployed for online LC-MS analysis of mixtures. The change of chromatographic retention time due to the use of deuterated mobile phases is not an issue because of the use of mass identifications. This approach provides accurate measurements of exchangeable hydrogen atoms in a molecule to assist structural elucidation.

5.2 CHARACTERIZATION OF IMPURITIES

Impurities in pharmaceuticals are generated largely during the synthetic process from starting materials, intermediates, and byproducts. Generally, impurities in starting materials and intermediates do not have to be analyzed as part of requirements by the regulatory agencies. However, these impurities may contain components that could affect the purity of the final manufactured pharmaceutical. Byproducts are often produced during synthesis, and are one of the major sources of pharmaceutical impurities. The identification of by-products can be used to optimize manufacturing process in minimizing impurities and maximizing yield. This will also help analytical testing optimization.

Traditional approach in impurity identification includes isolation and purification by offline HPLC, followed by characterization using Fourier transform (FT)-IR, NMR, MS, and X-ray crystallography, and other methods. A relatively large amount of sample is needed for analysis, and the process can be very labor-intensive. In contrast, LC-MS and LC-MS/MS are highly sensitive techniques requiring a small amount of material for analysis (less than 1 μg). In some cases, if impurities are found to be at very low levels in the drug substance, extraction procedures might be used to preconcentrate them to detectable levels. Two examples are illustrated below, including impurity identification in mometasone furoate drug substance [1,2,28] and structural identification of enol tautomer impurity in a potent hepatitis C virus (HCV) protease inhibitor [26].

5.2.1 Mometasone Furoate

Mometasone furoate is a highly potent chlorinated corticosteroid. It has been used in the treatment of glucocorticoid responsive dermatologic disorders as topical formulations of ointments, in seasonal and perennial allergic rhinitis as an aqueous intranasal spray, and in asthma as a dry powder.

In the course of large-scale production of mometasone furoate, several very low-level impurities were detected, as shown in total-ion chromatogram (Figure 5.1). LC-MS experiments were performed using a C18 column under isocratic conditions with methanol and 2 mM ammonium acetate in water (75/25) at a flow rate of 1 mL/min. Peak B shows two coeluting components with protonated molecular ions $[M + H]^+$ at *m/z* 535 and 581, as supported by formation of ammoniated adducts (*m/z* 552, 598) and sodiated adducts (*m/z* 557, 603) in the low-resolution ESI mass spectrum (Figure 5.2). The data suggest molecular weights of these two components to be 534 and 580 Da, respectively.

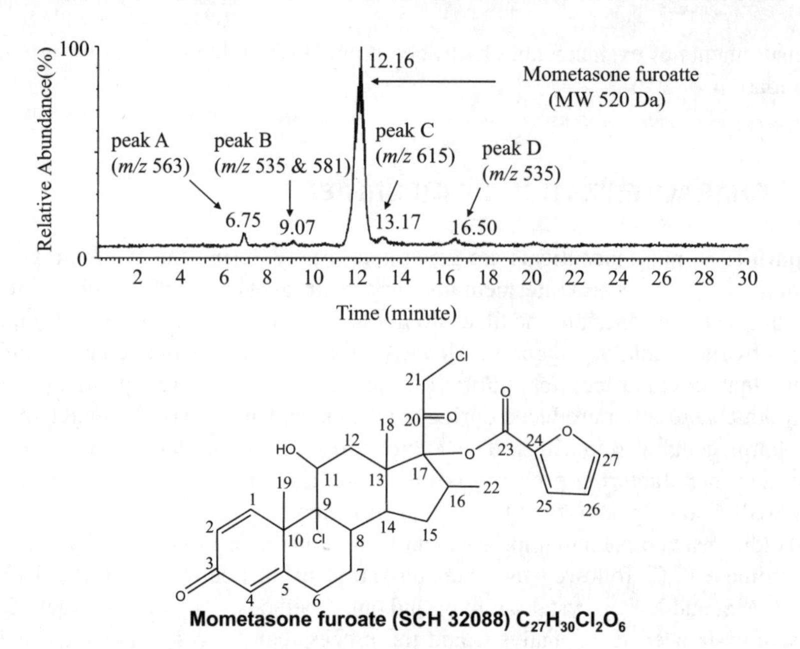

FIGURE 5.1 LC-MS total-ion chromatogram of mometasone furoate sample. Detected impurity molecular ions (*m/z*) are indicated for impurity peaks, and the structure of mometasone furoate is shown in the inset. (Reprinted from Ref. 2, with permission of John Wiley & Sons, Ltd.)

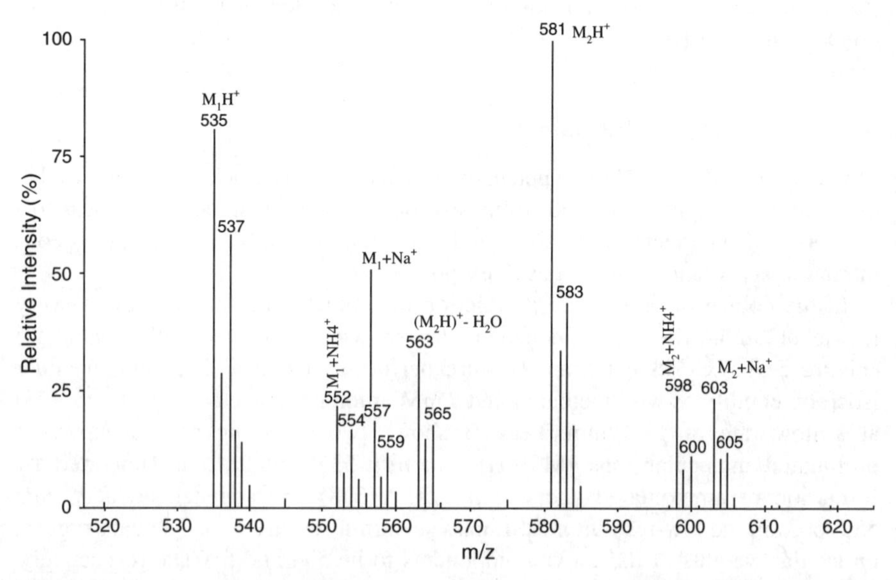

FIGURE 5.2 ESI mass spectrum from peak B, indicating two coeluting components with molecular ions at *m/z* 535 and 581, respectively.

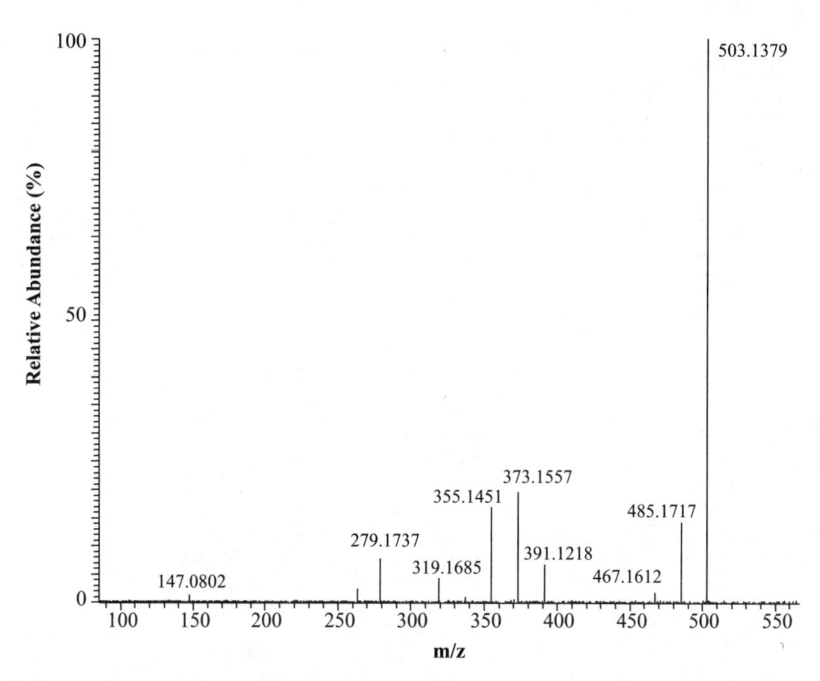

FIGURE 5.3 Product ion mass spectrum of mometasone furoate molecular ion at *m/z* 521 obtained from a LTQ-Orbitrap mass spectrometer.

Their isotopic patterns also indicate two chlorine patterns for *m/z* 535 and one chlorine pattern for *m/z* 581.

To obtain structural information on the unknown impurities, high-resolution tandem MS experiments were carried out on mometasone furoate molecular ion at *m/z* 521 ($C_{27}H_{31}O_6Cl_2$, $[M + H]^+$)using LTQ-Orbitrap mass spectrometer at a resolution of 30,000 in the FT-MS mode (Figure 5.3). As one of the latest HR-MS instrumentation, LTQ-Orbitrap has the capability of performing HR-LC-MS and LC-MS^n experiments with high resolving power (< 100,000), excellent mass accuracy (<3 ppm with external calibration), and wide dynamic range (> 10^3) [14,15]. Common fragmentations observed for mometasone furoate include the loss of water, furoate ring, HCl, and the cleavages of the steroid rings. The base peak in the mass spectrum is the product ion at *m/z* 503 due to the loss of water from the molecular ion at *m/z* 521. Further loss of furoate ring from *m/z* 503 leads to the fragment ion at *m/z* 391. Further loss of HCl group from *m/z* 391 results in the formation of *m/z* 355. Two product ions at *m/z* 279 and 319 are likely generated by the loss of ClCHCO moiety and HCl from *m/z* 355, respectively. Another fragmentation pathway involves *m/z* 521 → 485 (loss of HCl from *m/z* 521) → 373 (loss of furoate ring from *m/z* 485) → 355 (loss of water from *m/z* 373). The product ion at *m/z* 485 could also lose water to yield *m/z* 467, which subsequently loses furoate ring to form *m/z* 355. A low-abundance fragment ion at *m/z* 147 is likely due to cleavage of the steroid ring. Scheme 5.1 summarizes

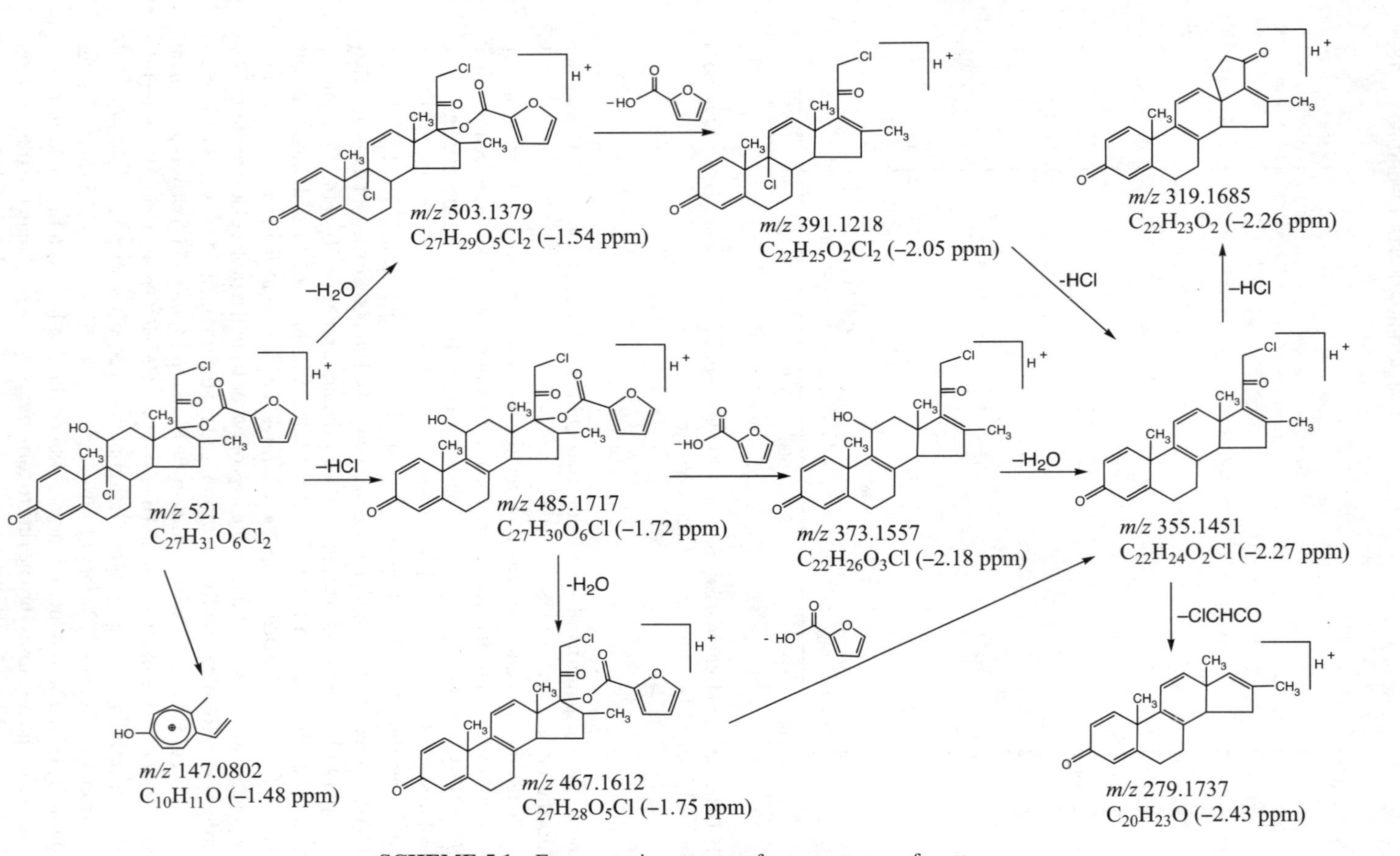

SCHEME 5.1 Fragmentation patterns for mometasone furoate.

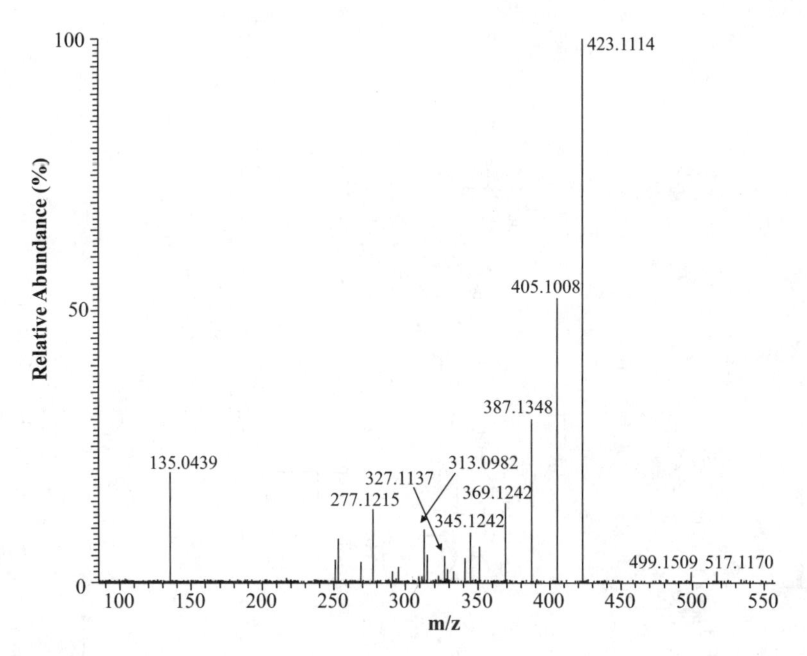

FIGURE 5.4 Product ion mass spectrum of impurity molecular ion at *m/z* 535 (peak B) obtained from a LTQ-Orbitrap mass spectrometer.

fragmentation patterns for mometasone furoate, as supported by accurate mass data with excellent accuracy (< 3 ppm). For impurity ion at *m/z* 535 (peak B), its accurate mass data give the elemental composition of $C_{27}H_{29}O_7Cl_2$ ($[M + H]^+$, −0.41 ppm), suggesting the addition of one oxygen atom with the reduction of two hydrogen atoms to mometasone furoate molecule. The most abundant fragment ion at *m/z* 423 in the product ion mass spectrum of *m/z* 535 corresponds to the loss of furoate ring from *m/z* 535 (Figure 5.4). The loss of water, HCl, and $ClCH_2COH$ moiety from *m/z* 423 produces the fragment ions at *m/z* 405, 387, and 345. Two very low-abundance fragment ions at *m/z* 517 and 499 are results of loss of water and HCl from molecular ion at *m/z* 535. The consecutive loss of water, HCl, and furoate ring from *m/z* 535 leads to the product ion at *m/z* 369. A highly abundant fragment ion at *m/z* 135 (absent in the product ion mass spectrum of *m/z* 521 under similar activation conditions) indicates a stabilized product ion, suggesting a 6-keto structure for this impurity (Figure 5.4). Detailed fragmentation patterns along with accurate mass data are illustrated in Scheme 5.2. The 6-keto structure was further confirmed by a synthetic standard [1].

Initial characterization on impurity ion at *m/z* 581 was performed using a low-resolution triple-quadrupole mass spectrometer. Its low-resolution product ion mass spectrum appeared to be fit, with the original proposed tentative structure

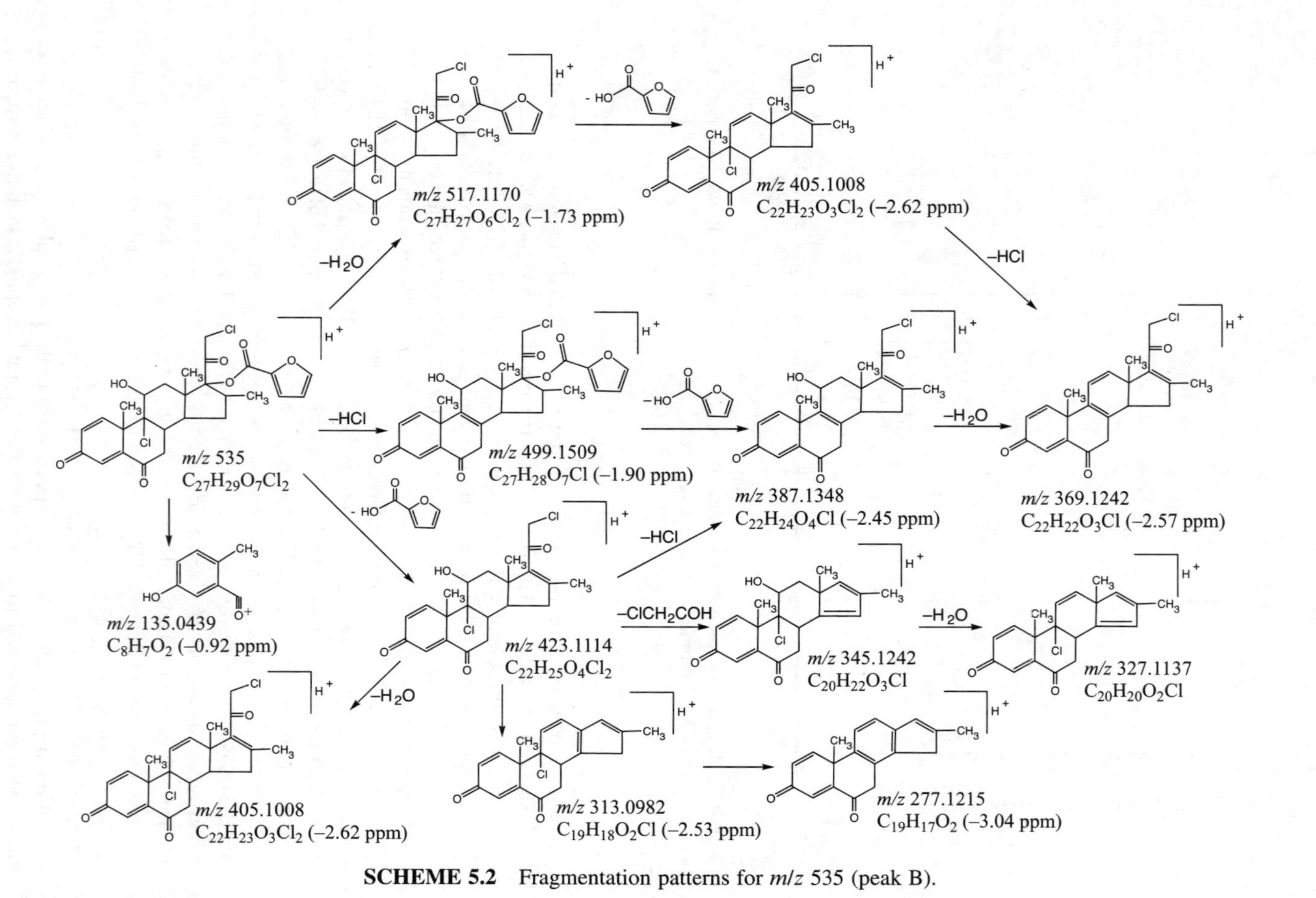

SCHEME 5.2 Fragmentation patterns for *m/z* 535 (peak B).

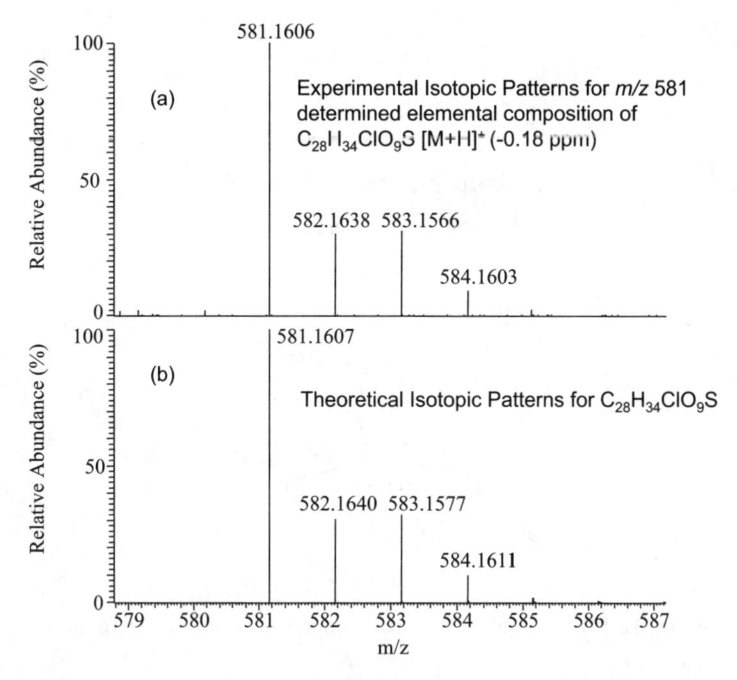

FIGURE 5.5 (a) Experimental isotopic distribution patterns for impurity molecular ion at *m/z* 581 (peak B), (b) theoretical isotopic distribution patterns based on elemental composition of $C_{28}H_{34}O_9ClS$ for *m/z* 581.

being 21-furoate (where 21-Cl was replaced by a furoate ring). However, accurate mass measurement on *m/z* 581 has a mass accuracy error of > 10 ppm on the tentative structure of 21-furoate mometasone furoate. Clearly, the initially proposed 21-furoate structure is not supported by accurate mass data. On the basis of detected isotopic patterns and likely, possible element combinations, the best possible elemental composition for *m/z* 581 was determined to be $C_{28}H_{34}O_9ClS$, with a mass accuracy of 0.18 ppm (Figure 5.5). Compared with the elemental composition of mometasone furoate ($C_{27}H_{31}O_6Cl_2$, $[M + H]^+$), the net addition for *m/z* 581 is the moiety of CH_3O_3S, with reduction of one chlorine atom. Two possible structures include open structure involving replacement of 21-Cl and closed structure involving 20-keto and replacement of 21-Cl (Scheme 5.3). The number of exchangeable hydrogen atoms is varied for these two structures: one exchangeable hydrogen atom for open structure and two exchangeable hydrogen atoms for closed structure. On-line H/D exchange experiments were performed, and molecular ions were found to be shifted to *m/z* 583.1731 ($C_{28}H_{32}D_2O_9ClS$, $[M + D]^+$, -0.27 ppm), indicating one exchangeable hydrogen atom in the molecule. The result is consistent with this impurity being an open structure. One possible mechanism for formation of this structure for impurity ion at *m/z*

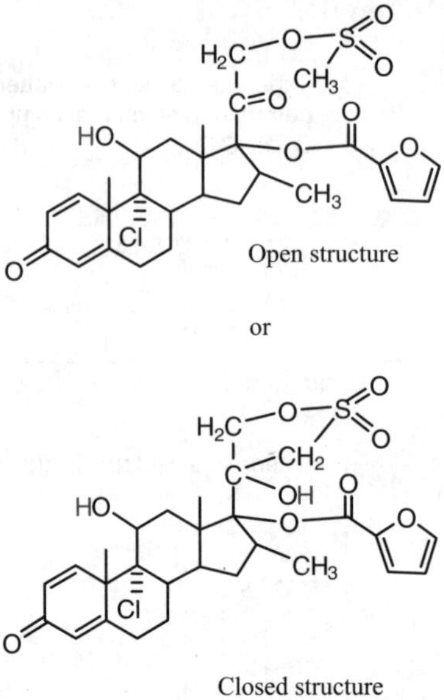

Open structure

or

Closed structure

SCHEME 5.3 Possible structures of impurity ion at *m/z* 581 (peak B).

581 involves reactions with CH_3SO_2Cl (reagent) (Scheme 5.4). Further HR-LC-MS/MS experiments on *m/z* 581 support its structural assignment as the replacement of 21-Cl with the sulfur moiety, as evidenced by accurate mass measurements of product ions (Figure 5.6). The ready loss of water from *m/z* 581 generates the most abundant product ion at *m/z* 563. Further loss of furoate ring

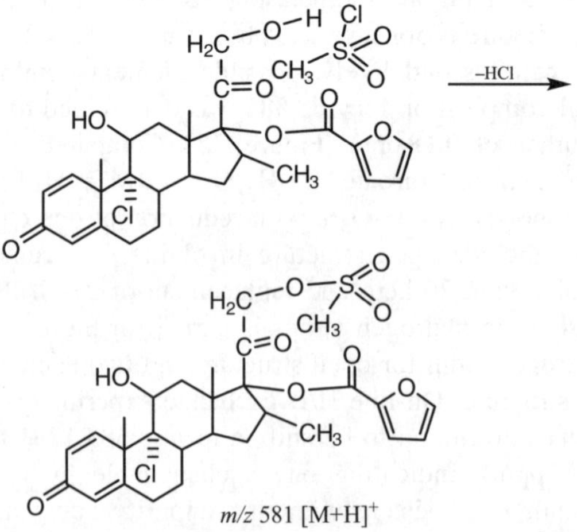

–HCl

m/z 581 $[M+H]^+$

SCHEME 5.4 Possible mechanism for formation of *m/z* 581 (peak B).

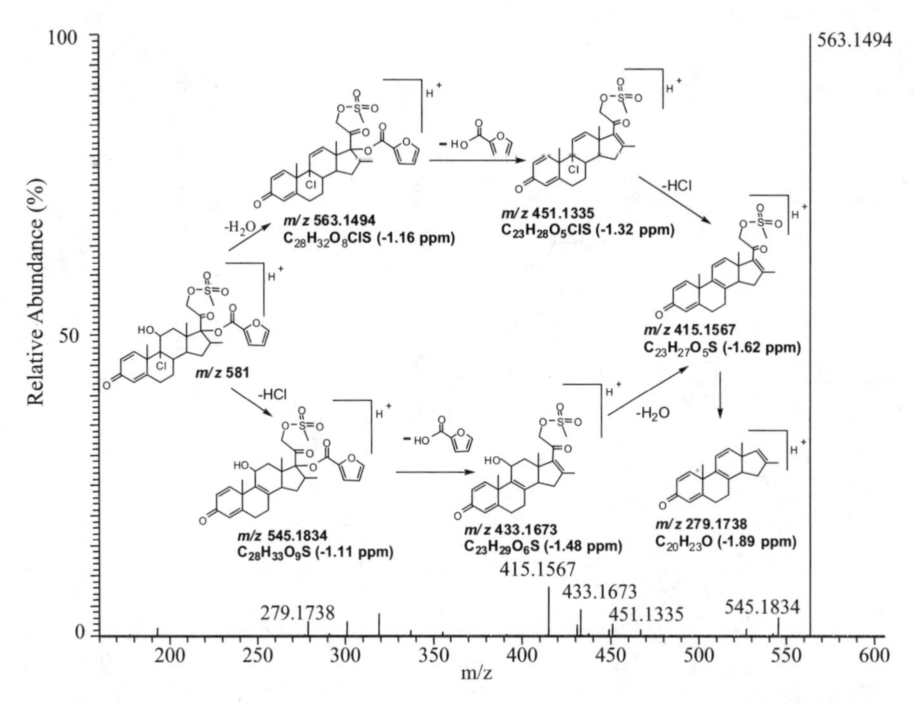

FIGURE 5.6 Product ion mass spectrum of impurity molecular ion at *m/z* 581 (peak B) obtained from a LTQ-Orbitrap mass spectrometer.

from *m/z* 563 produces the fragment ion at *m/z* 451. The molecular ion at *m/z* 581 can also lose HCl to yield the product ion at *m/z* 545. Additional loss of furoate ring from *m/z* 545 generates the fragment ion at *m/z* 433. A low-abundance product ion at *m/z* 415 is a result of loss of water, HCl, and furoate ring from *m/z* 581.

Another sulfur-containing impurity ion at *m/z* 563 (peak A) is closely related to *m/z* 581. In fact, further loss of water from impurity ion at *m/z* 581 might lead to the formation of *m/z* 563. The HR-LC-MS/MS product ion mass spectrum of *m/z* 563 exhibits distinct fragmentation patterns (Figure 5.7). The dominant fragment ions are *m/z* 433 (loss of furoate ring from *m/z* 545), 415 (loss of furoate ring from *m/z* 527), and 397 (loss of water, HCl, and furoate ring from *m/z* 563). The molecular ion at *m/z* 563 can lose water and HCl to generate abundant fragment ions at *m/z* 545 and 527, respectively. Detailed fragmentation patterns for *m/z* 563 along with accurate mass data are shown in Scheme 5.5. Impurity ion at *m/z* 615 (peak C) eluted closely with mometasone furoate and was characterized as a 11-furoate structure by HR-LC-MS/MS experiments (Figure 5.8, Scheme 5.6). As the 11-OH group is replaced by a 11-furoate group, impurity ion at *m/z* 615 does not generate direct water-loss fragment ions in the product ion mass spectrum. The ready loss of furoate ring from *m/z* 615 gives the most abundant fragment ion at

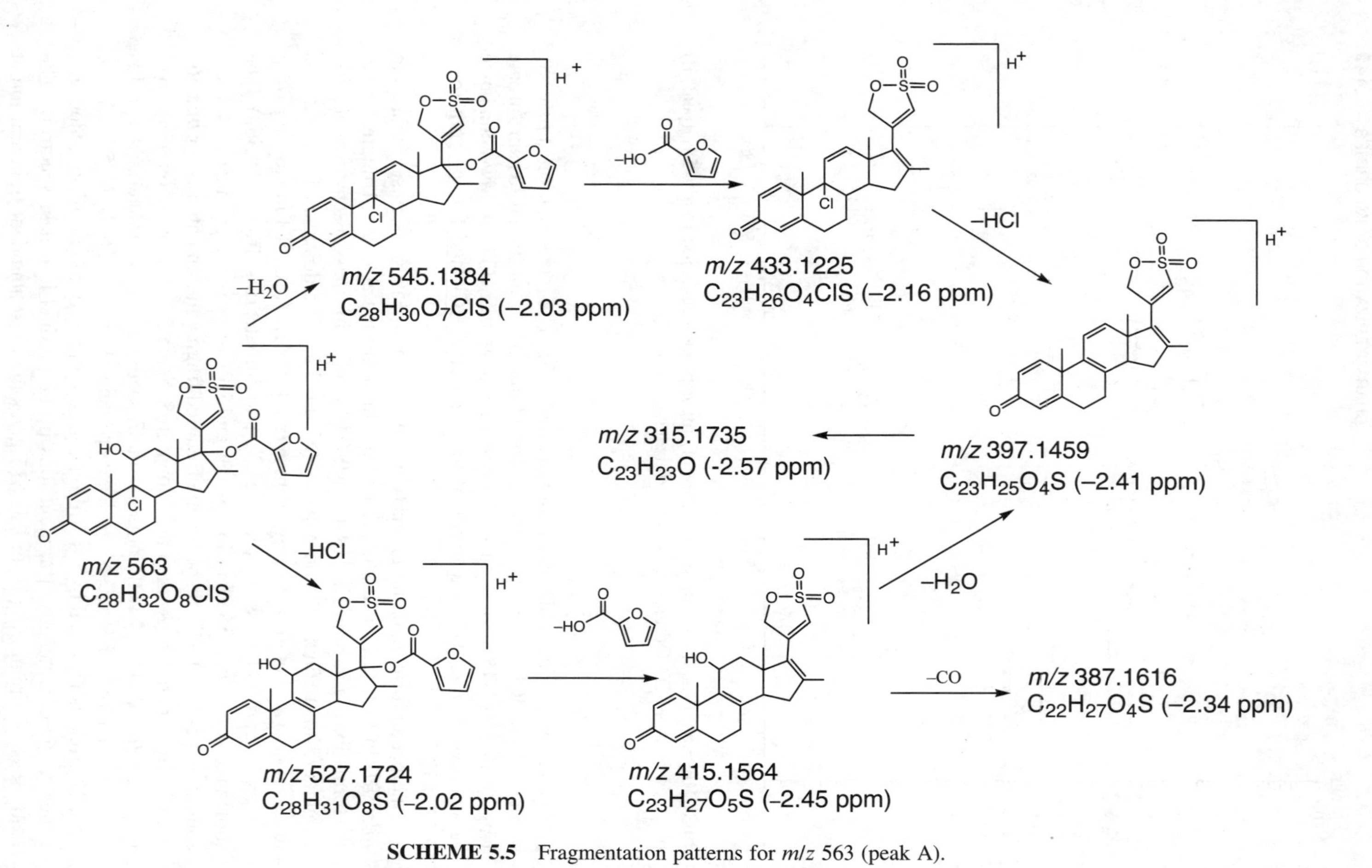

SCHEME 5.5 Fragmentation patterns for *m/z* 563 (peak A).

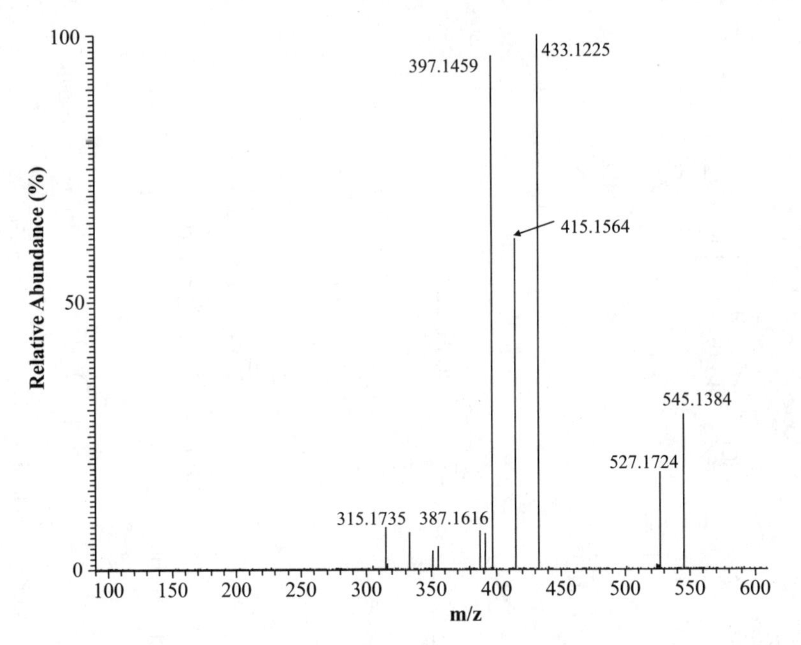

FIGURE 5.7 Product ion mass spectrum of impurity molecular ion at *m/z* 563 (peak A) obtained from a LTQ-Orbitrap mass spectrometer.

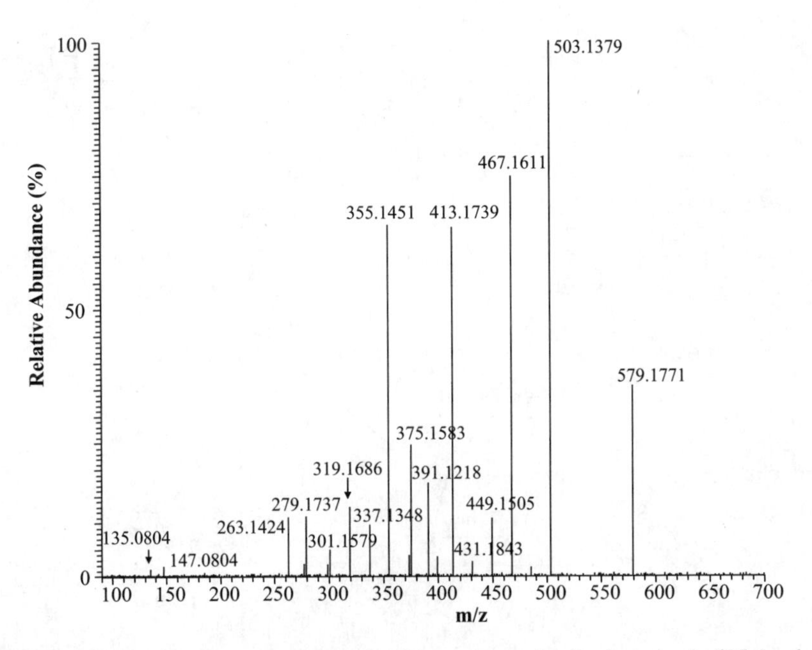

FIGURE 5.8 Product ion mass spectrum of impurity molecular ion at *m/z* 615 (peak C) obtained from a LTQ-Orbitrap mass spectrometer.

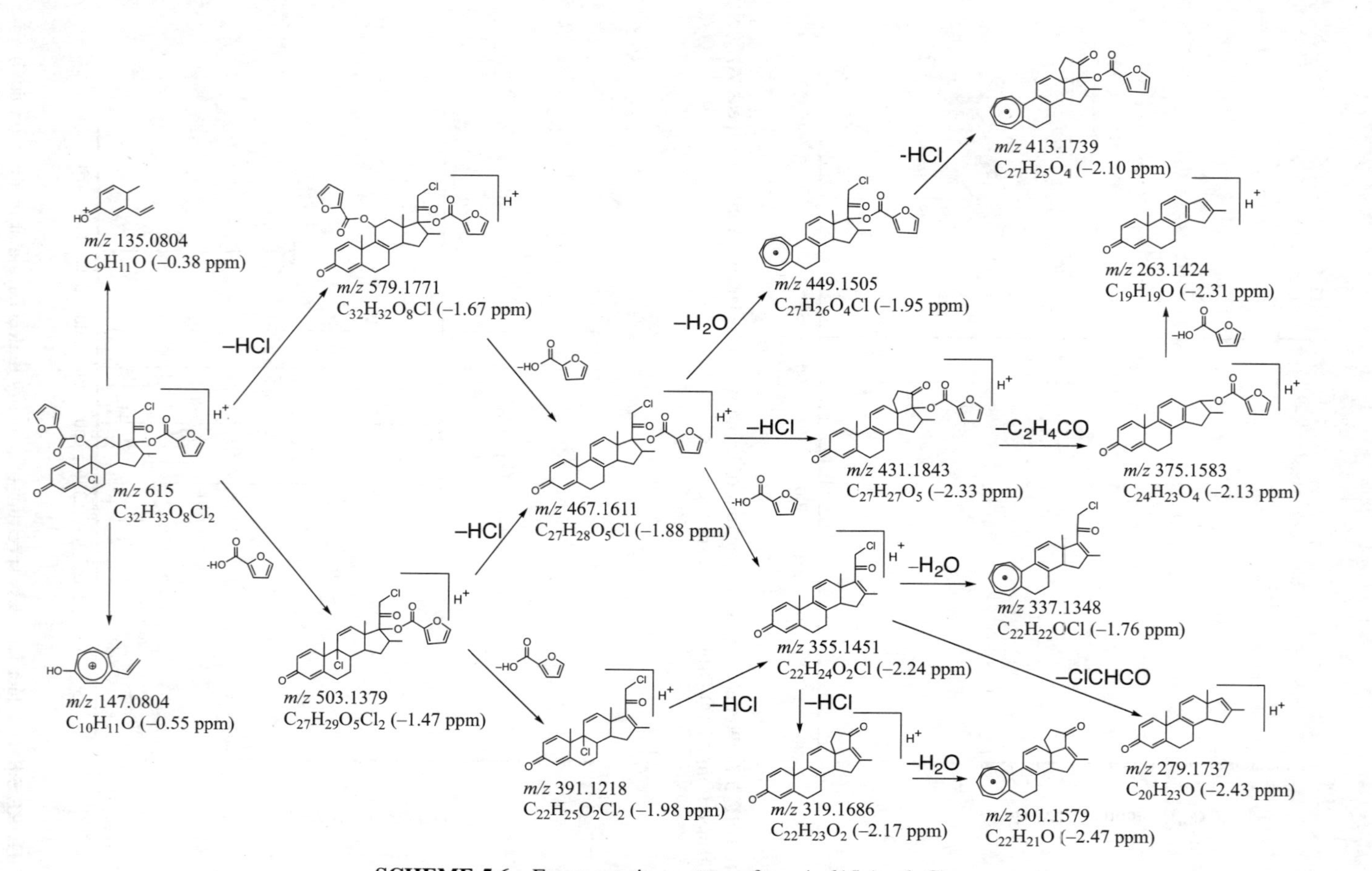

SCHEME 5.6 Fragmentation patterns for *m/z* 615 (peak C).

m/z 503. Another abundant fragment ion at *m/z* 579 is a result of the loss of HCl from the molecular ion at *m/z* 615. These two fragment ions at *m/z* 503 and 579 can be further activated to produce a number of other fragment ions, as shown in Scheme 5.6. Accurate mass measurements of these fragment ions support their structural assignments.

Impurity ion at *m/z* 535 (peak D) has the same nominal MWs of 534 Da as the 6-keto impurity in peak B. However, their accurate mass data give different elemental compositions. The impurity ion at *m/z* 535 (peak D) has a measured elemental composition of $C_{28}H_{33}O_6Cl_2$ ($[M + H]^+$, −0.41 ppm), a net addition of CH_2 moiety to the mometasone furoate molecule. HR-LC-MS/MS experiments on this impurity ion at *m/z* 535 also exhibit fragmentation behaviors (Figure 5.9) very different from those of impurity ion at *m/z* 535 (peak B) (Figure 5.4). The dominant fragment ion at *m/z* 517 in the product ion mass spectrum of *m/z* 535 (peak D) is due to the loss of water from the molecular ion at *m/z* 535. The loss of HCl from *m/z* 535 gives the fragment ion at *m/z* 499. These abundant fragment ions at *m/z* 517 and 499 dissociate to generate additional product ions (Scheme 5.7). There is also a very low-abundance fragment ion at *m/z* 161 corresponding to $C_{11}H_{13}O$, suggesting a 6-methyl structure for this impurity. Complete analysis of fragmentation patterns of the impurity ion at *m/z* 535 (peak D) supports this 6-methyl structure.

The structures of all five impurities in the sample are shown in Scheme 5.8. Note that mass accuracy data for measured elemental compositions of impurities

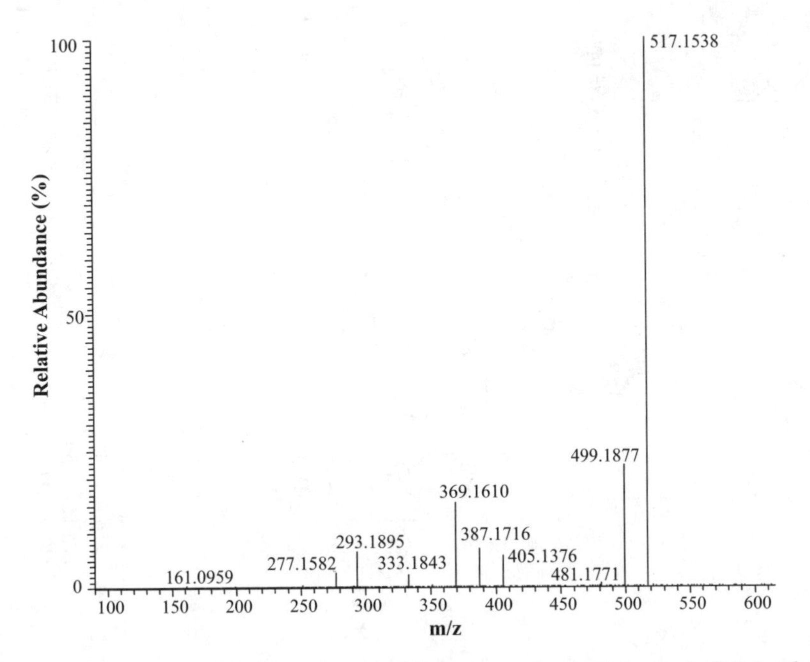

FIGURE 5.9 Product ion mass spectrum of impurity molecular ion at *m/z* 535 (peak D) obtained from a LTQ-Orbitrap mass spectrometer.

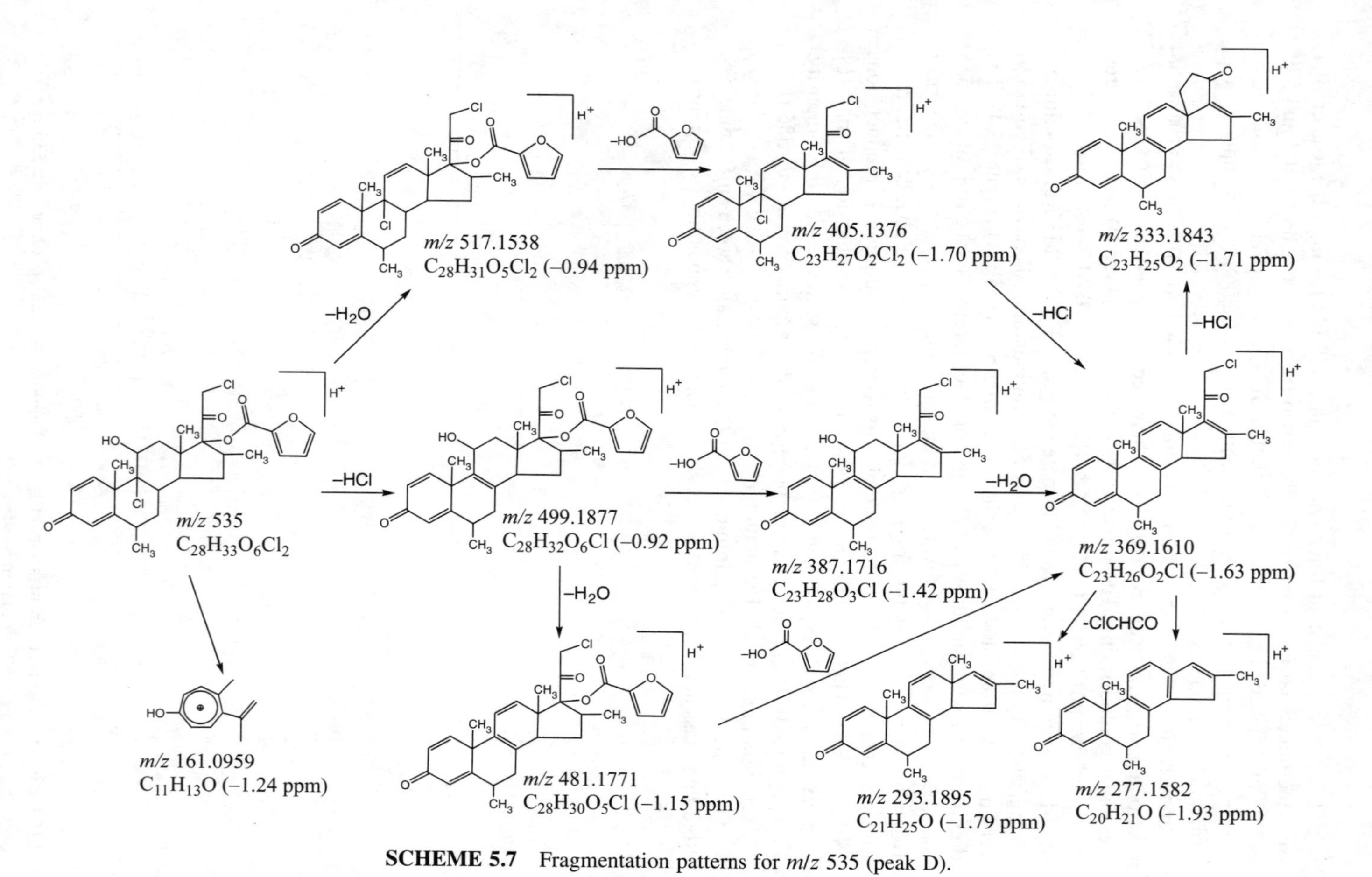

SCHEME 5.7 Fragmentation patterns for *m/z* 535 (peak D).

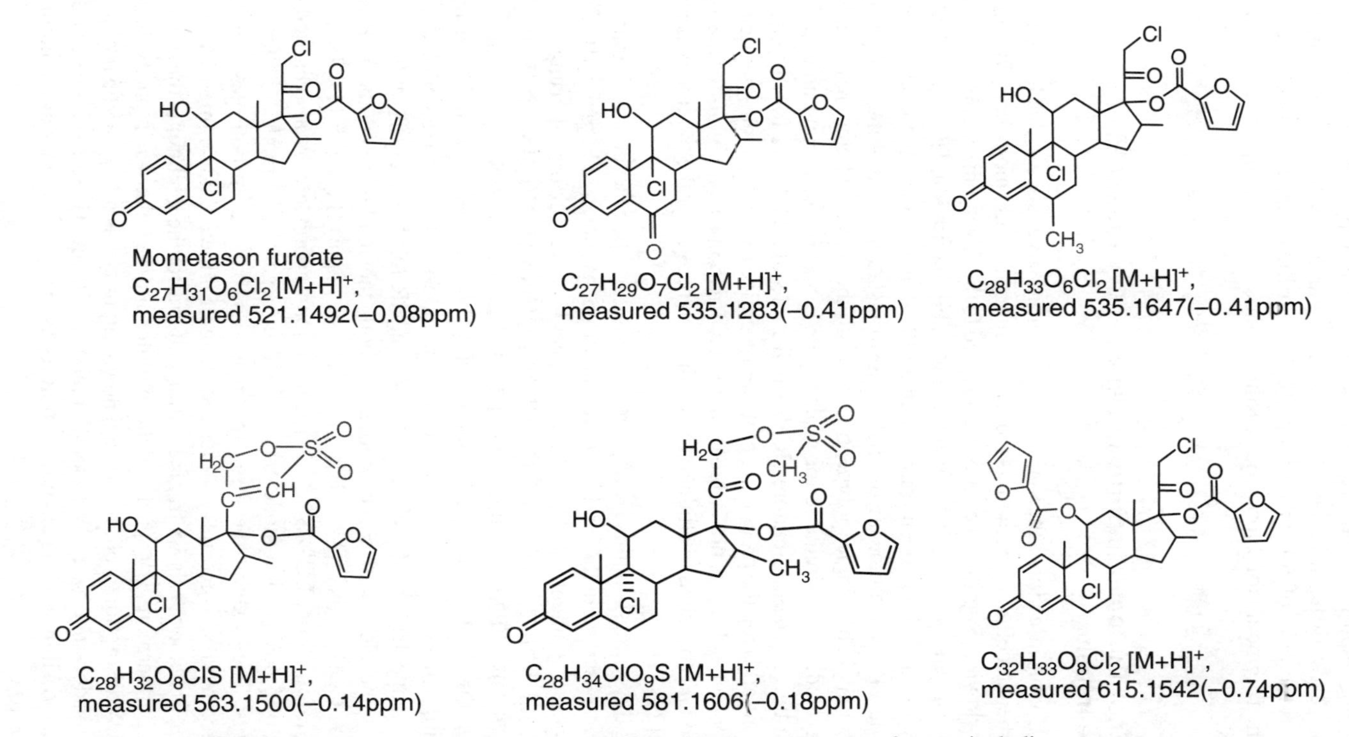

SCHEME 5.8 Proposed structures of impurities related to mometasone furoate, including accurate mass data on elemental compositions (reprinted from Ref. 2, with permission of John Wiley & Sons, Ltd.).

and drug substance are within 1 ppm even with large differences in ion abundance among impurities and drug substance. The low-ppm mass accuracy data allow the ready differentiation of isomeric impurities with the same nominal mass, as in the case of 6-keto and 6-methyl impurities. Accurate mass measurements on fragment ions provide further evidence in support of proposed structures.

5.2.2 Enol Tautomer Impurity in Hepatitis C Virus (HCV) Protease Inhibitor

Hepatitis C virus infection is the major cause of chronic liver disease that leads to liver cirrhosis, hepatocellular carcinoma, and liver failure, affecting more than 170 million people worldwide. Significant efforts are directed to the development of therapies that target key enzymes critical to HCV replication and maturation. HCV is a positive-strand RNA with a single open frame of about 9600 nucleotides. It encodes a single polypeptide of 3000 amino acids that is posttranslationally modified to produce mature virions. The single polypeptide has all the structural and nonstructural proteins, including trypsinlike serine protease NS3. NS3 protease plays an important role in the development of mature HCV. Small-molecule inhibitors for this enzyme could potentially stop the processing of the polypeptide required for viral replication. A novel, potent, selective, orally bioavailable HCV NS3 protease inhibitor has been discovered and advanced to clinical trials in human beings for the treatment of HCV infection [29–31].

During the synthesis of this HCV protease inhibitor, a trace level of an impurity of interest was observed in the total ion chromatogram, as shown in Figure 5.10. These reversed-phase gradient LC-ESI/MS experiments were performed using a C18 column with mobile phases of $CH_3CN/CH_3OH/5\,mM\ NH_4OAc$ (5/5/90) and CH_3CN/CH_3OH (80/20) at a flow rate of 1 mL/min. HR-LC/MS experiments were operated in the FTMS mode at a resolution of 30,000 (external calibration) using a LTQ-Orbitrap mass spectrometer. The unknown impurity elutes at a retention time of 21.54 min, while the inhibitor elutes at 19.84 min. HR-LC/MS analysis of the sample gives the same elemental compositions for both the inhibitor and the unknown impurity as $C_{27}H_{46}O_5N_5$ ($[M + H]^+$, m/z 520, < 1 ppm) (Figure 5.11). To obtain structural information, LC/MS^n experiments in ion trap mode were also carried out on m/z 520 in positive-ion mode and m/z 518 in negative-ion mode for these two components. They exhibit the same fragmentation patterns in both positive- and negative-ion modes, suggesting isomeric structure for the unknown impurity. As an illustration, Scheme 5.9 shows fragmentation patterns for the inhibitor in the positive-ion mode. Furthermore, UV spectra of the inhibitor and the unknown impurity display different UV absorption profiles. The unknown impurity has a λ_{max} at 265 nm, while the inhibitor does not have UV absorption above 220 nm. These experimental data suggest the possibility of the unknown impurity as the enol tautomer of the inhibitor (Scheme 5.10).

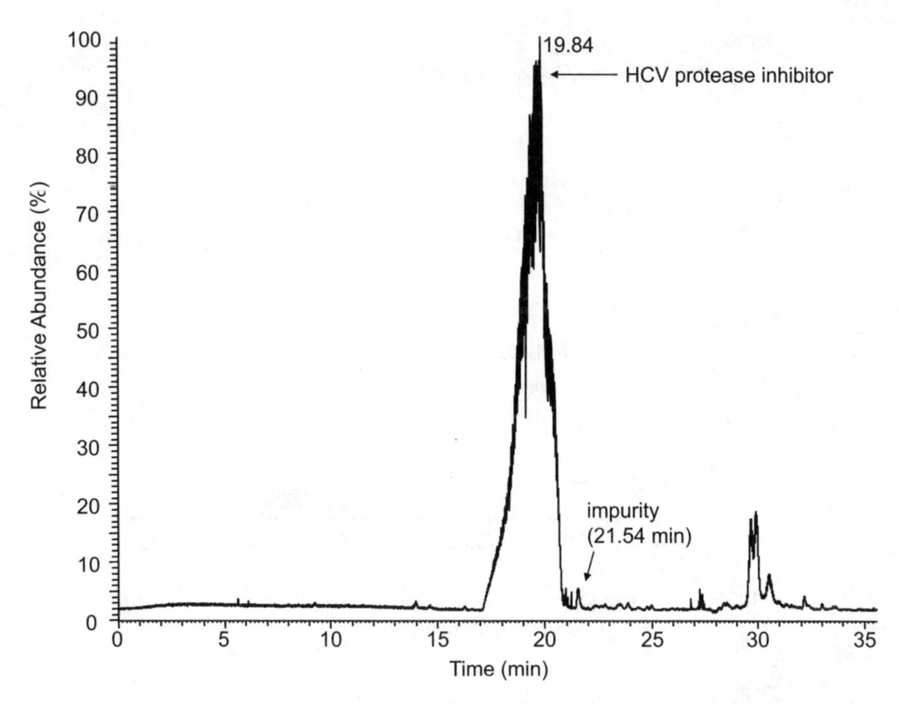

FIGURE 5.10 High-resolution LC-MS total-ion chromatogram of a drug substance HCV protease inhibitor sample obtained from a LTQ-Orbitrap mass spectrometer (reprinted from Ref. 26, with permission of John Wiley & Sons, Ltd.).

To establish the proposed tautomer structure, online H/D exchange HR-LC/MS experiments using deuterated solvents including deuterium oxide and methanol-D_4 were performed. As expected, molecular ions for the inhibitor were shifted to *m/z* 526 ($C_{27}H_{40}D_6O_5N_5$ $[M + D]^+$, 0.02 ppm), while molecular ions for the unknown impurity were changed to *m/z* 527 ($C_{27}H_{39}D_7O_5N_5$ $[M + D]^+$, 0.14 ppm) (Figure 5.12). Clearly, there are five exchangeable hydrogen atoms for the inhibitor and six exchangeable hydrogen atoms for the impurity, consistent with the proposed tautomer structure (Scheme 5.10). As a 1,2-dicarbonyl compound (keto amide), the inhibitor may exist as the enol form, as reported in the literature on the enol form of a 1,2-ketocarbonyl compound. It is possible that the tautomer structure is shifted to the original keto amide structure under current experimental conditions (high temperature at ESI source), and this can lead to the same fragmentation patterns as observed in multistage MS experiments. Further structural confirmation was performed by NMR analysis on an isolated impurity fraction collected at a very low temperature. The impurity sample was also derivatized to form a more stable chemical derivative structure of "enol ether" by reaction of the impurity with diazomethane (CH_2N_2), as supported by NMR experiments. This study

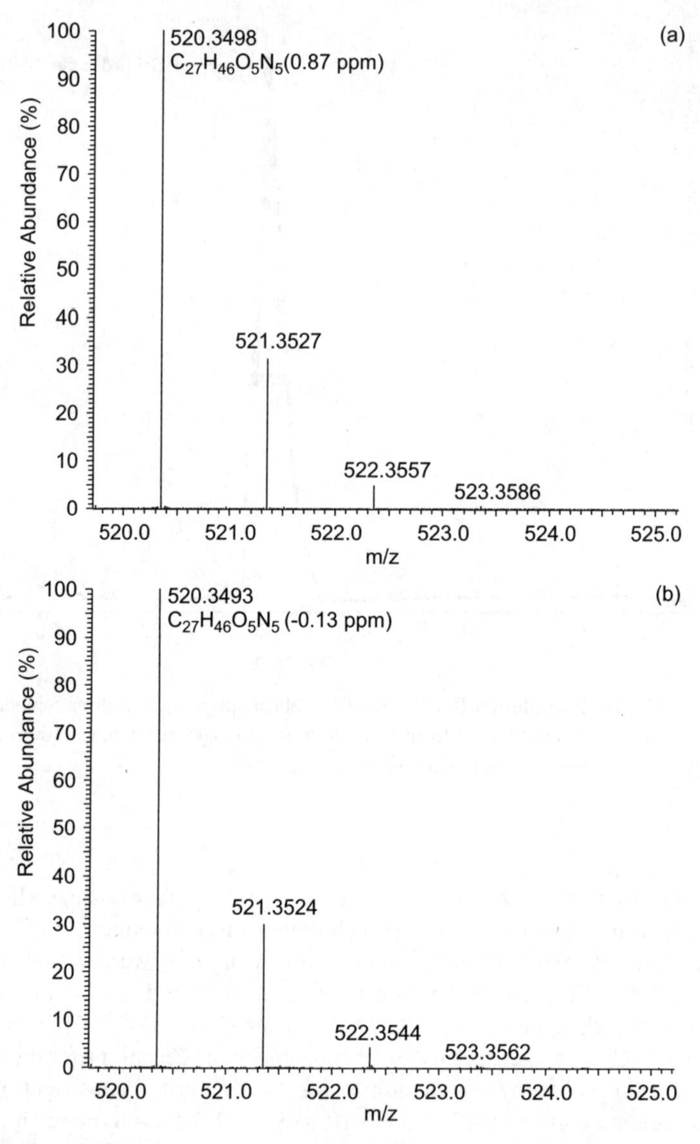

FIGURE 5.11 (a) HR-LC/ESI-MS mass spectrum of HCV protease inhibitor obtained from a LTQ-Orbitrap mass spectrometer; (b) HR-LC/ESI-MS mass spectrum of the unknown impurity obtained from a LTQ-Orbitrap mass spectrometer (reprinted from Ref. 26, with permission of John Wiley & Sons, Ltd.).

demonstrates the practical utility of online H/D exchange HR-LC/MS for differentiation of enol and keto tautomers, representing a unique and simplified approach to rapid structural identifications of similar types of other unknown impurities in drug substances.

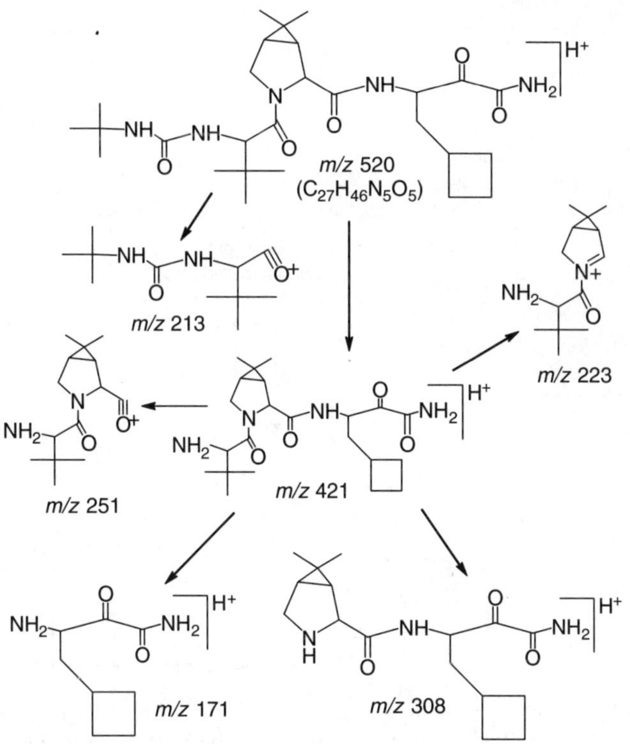

SCHEME 5.9 Positive-ion fragmentation patterns for HCV protease inhibitor (reprinted from Ref. 26, with permission of John Wiley & Sons, Ltd.).

5.3 CHARACTERIZATION OF DEGRADATION PRODUCTS

Degradation profiles of pharmaceuticals are integral parts of safety and potency assessments of drug candidates for clinical trials. Drug degradation in formulated pharmaceutical products is often unpredictable and highly complex. Degradation products can arise from ingredients used in dosage formulation and/or in the process of formulation where a number of factors such as temperature, humidity, and light may all play a role. Degradation products can be generated from various processes such as hydrolysis, oxidation, adduct formation, dimerization, and rearrangement.

In order to accelerate the drug development process, various stress-testing methods had been designed to simulate stresses that the compound might experience during production processes and storage. These methods exposed drug candidates to forced-degradation conditions such as acid, base, heat, oxidation, and exposure to light. A rapid and successful identification of degradation products can enhance our understanding of the degradation mechanism of

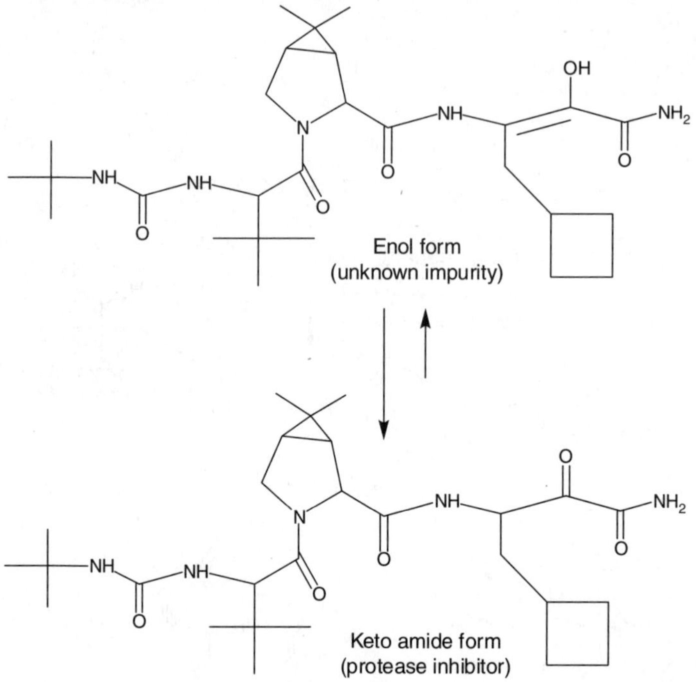

Enol form
(unknown impurity)

Keto amide form
(protease inhibitor)

SCHEME 5.10 Equilibrium between keto amide form and enol form of HCV protease inhibitor (reprinted from Ref. 26, with permission of John Wiley & Sons, Ltd.).

drug candidates and improve the formulation development for clinical studies. Two case studies are described below, including characterization of degradants of the complex oligosaccharide antibiotics everninomicins and the antifungal agent posaconazole.

5.3.1 Everninomicin

Everninomicins are an important class of oligosaccharide antibiotics produced by *Micromonospora carbonacea* [32]. These compounds are highly active against gram-positive bacteria, including methicillin-resistant *Staphylococcus aureus* and vancomycin-resistant *enterococci* [33–35]. Structures of these antibiotics are very complex with an eight-sugar backbone and two *ortho*-ester functionalities. Other uncommon structural features include a nitrosugar, a completely substituted aromatic ester containing two chlorines, and a methylene dioxy group. The early structural characterization work on everninomicins employed a Ganguly–Sarre chemical degradation procedure [36–38]. Subsequent studies were carried out primarily by fast-atom bombardment (FAB) MS with the ready formation of cationized molecules when NaCl or KCl was added to the matrix [1,39–44]. The use of FAB-MS on sector instruments is limited to relatively poor sensitivity and

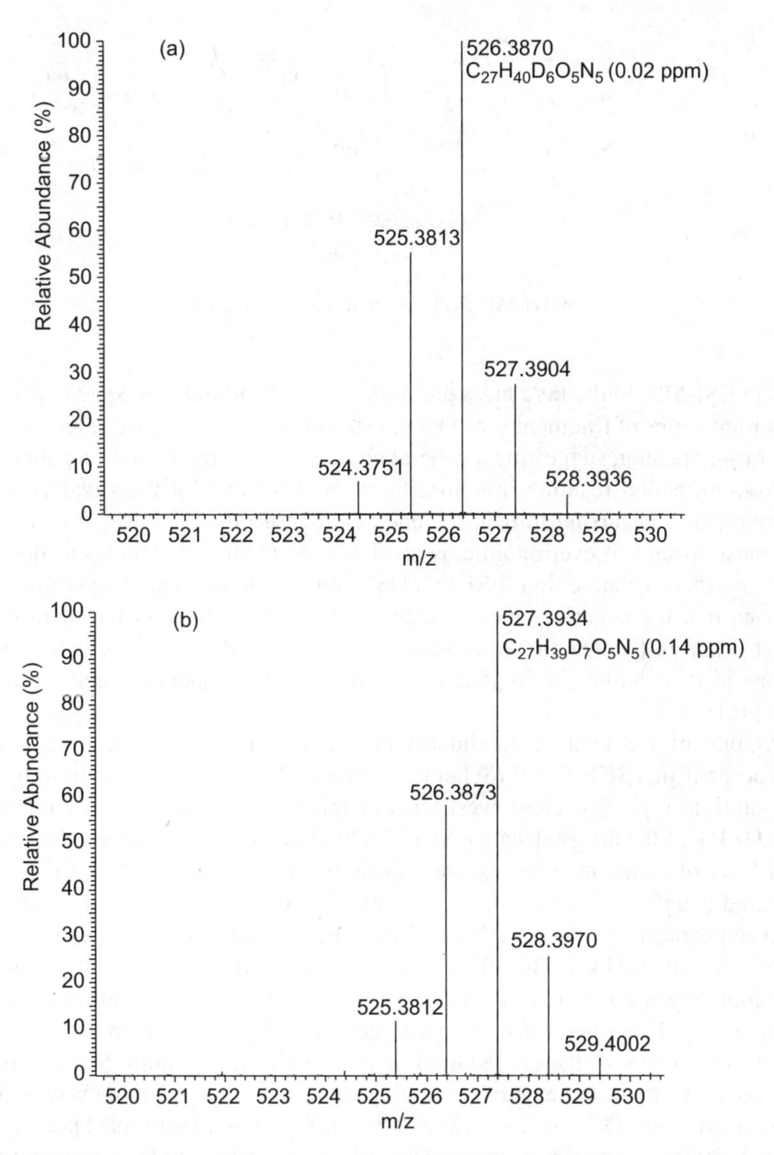

FIGURE 5.12 (a) Online H/D exchange HR-LC/ESI-MS mass spectrum of HCV protease inhibitor obtained from a LTQ-Orbitrap mass spectrometer; (b) online H/D exchange HR-LC/ESI-MS mass spectrum of the unknown impurity obtained from a LTQ-Orbitrap mass spectrometer (reprinted from Ref. 26, with permission of John Wiley & Sons, Ltd.).

restricted MS/MS capability for detailed fragmentation studies. Coupling ESI with multistage quadrupole iontrap MS provides the capability of studying fragmentation patterns of this class of compounds [45,46]. It was found that the addition of sodium chloride (~1 μg/mL) facilitates the formation of abundant metal complex

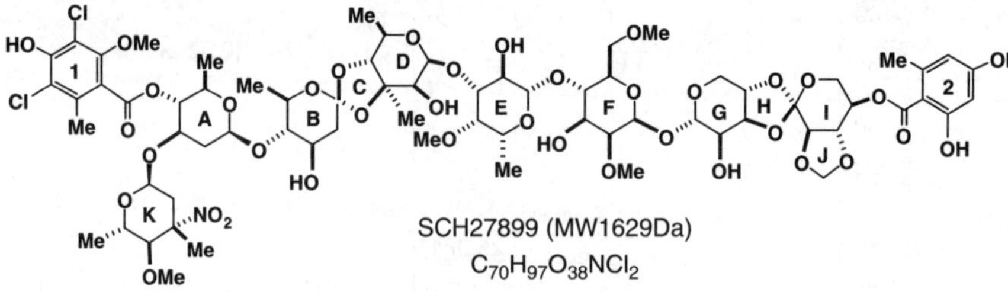

SCH27899 (MW1629Da)

$C_{70}H_{97}O_{38}NCl_2$

SCHEME 5.11 Structure of SCH27899.

ions in ESI-MS. Multistage mass analysis (MS^n) of the sodiated species yields an important series of fragment ions that are specific for sugar sequence and for some sugar ring opening, suggesting a general charge-remote fragmentation pattern with the sodium cation residing in a specific, central location of the sugar chain and fragmentation occurring to trim the end of the molecule [45]. Negative ion ESI-MS mass spectra of everninomicins display deprotonated molecules as dominant ions. Further negative ion ESI-MS/MS data on deprotonated everninomicins indicate that the negative charge likely resides in the deprotonated dichlorophenoxyl group in the substituted aromatic ester ring, and the fragmentation occurs remote to this charge site in generating simple sugar sequence-specific fragment ions [46].

As one of the leading candidates in everninomicins to treat drug-resistant microorganisms, SCH27899 (Scheme 5.11) was degraded in ammonium hydroxide solution in pH 10. The lowest level of minor components in the sample was about 0.1% of the drug substance SCH27899, based on UV absorbance at 268 nm. This level of component represented about 100 ng from a 20-µL injection of the prepared 5 mg/mL degraded mixtures. The degradation solution in methanol was analyzed in negative-ion HR-LC-ESI/MS using a double-focusing magnetic sector mass spectrometer [47]. The HPLC gradient was run from 65% mobile phase B to 80% mobile phase B in 15 min on a C18 column at a flow rate of 1 mL/min (mobile phase A: 20 mM ammonium acetate in water, mobile phase B: methanol). The total ion chromatogram of the degradation mixture is shown in Figure 5.13. A total of nine components were detected, including peaks eluting at 6.1 min (MW 713 Da), 7.5 min (MW 952 Da), 10.2 min (MW 1497 Da), 12.1 min (MW 695 Da), 13.3 min (MW 1479 Da), 13.5 min (MW 855 Da), 14.8 min (MW 1617 Da), 15.7 min (MW 1647 Da), and 27.4 min (MW 1661 Da). The peak broadening in the chromatogram is likely caused by overloaded column and ESI/magnetic sector interface. A mass measurement accuracy of 0.4–1.9 ppm was obtained for all detected components, using PEG sulfates as internal calibration compound covering peaks between 100 and 2000 Da (singly charged monosulfates and doubly charged disulfates) at a resolving power of 5000 with electric field linear scan mode (Table 5.1). The PEG sulfate solution in methanol (10 µg/mL) was introduced to the LC flow at 50 µL/min in postcolumn and prior to splitting at the ESI source. The mass scan ranges were determined by bracketing the analytes by the references ions. The

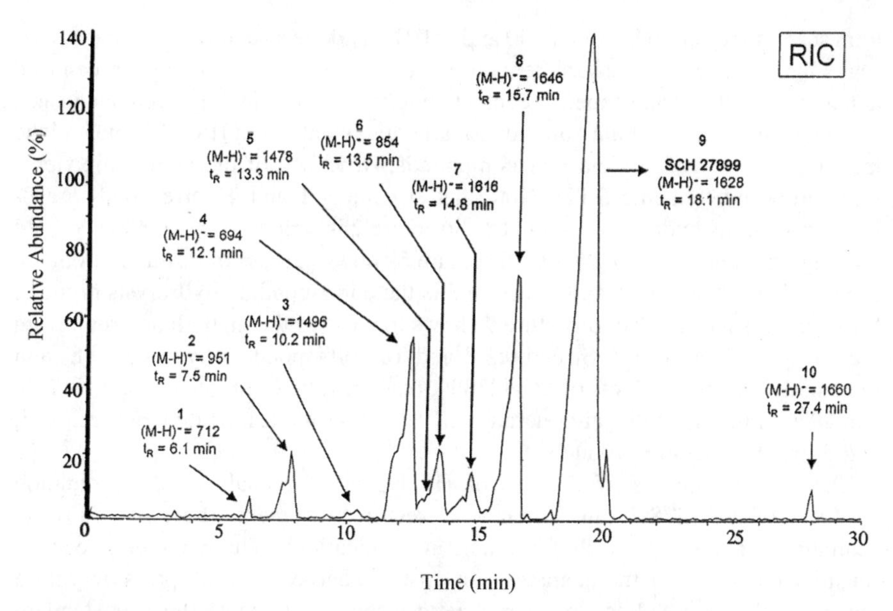

FIGURE 5.13 HR-LC/MS reconstructed ion chromatogram (RIC) of SCH27899 sample degraded in ammonium hydroxide solution at pH 10, displaying all the identified components 1–10 from a magnetic sector mass spectrometer (reprinted from Ref. 47, with permission of John Wiley & Sons, Ltd.).

TABLE 5.1 Exact Mass Measurements and Elemental Compositions for Components 1–10

Number	t_R (min)	$(M–H)^-$	Elemental Composition	Theoretical Value	Experimental Value	Δ (ppm)
1	6.1	712[a]	$C_{29}H_{40}O_{15}NCl_2$	714.1733(^{37}Cl)	714.1746	1.8
2	7.5	951	$C_{41}H_{59}O_{25}$	951.3345	951.3353	1.1
3	10.2	1496	$C_{62}H_{92}O_{36}NCl_2$	1496.4776	1496.4781	0.4
4	12.1	694	$C_{29}H_{38}O_{14}NCl_2$	694.1669	694.1678	1.3
5	13.3	1478	$C_{62}H_{90}O_{35}NCl_2$	1478.4670	1478.4689	1.2
6	13.5	854	$C_{36}H_{50}O_{18}NCl_2$	854.2405	854.2414	1.1
7	14.3	1616	$C_{69}H_{96}O_{38}NCl_2$	1616.4957	1616.4987	1.9
8	15.7	1646	$C_{70}H_{98}O_{39}NCl_2$	1646.5093	1646.5082	0.7
9	18.1	1628	$C_{70}H_{96}O_{38}NCl_2$	1628.4987	1628.4999	0.7
10	27.4	1660	$C_{71}H_{100}O_{39}NCl_2$	1660.5250	1660.5262	0.8

[a]The most abundant ion was observed at *m/z* 712 (^{35}Cl), however, because of reference interference at *m/z* 712, the exact mass measurement was performed on the peak at *m/z* 714 (^{37}Cl).

Source: Reprinted from Ref. 47, with permission of John Wiley & Sons, Ltd.

total scan range varied between 100 and 150 Da, and the scan rate was set at 8 s per scan. Accurate mass measurements were obtained using peak top or centroid depending on the peak shape for each analyte and corresponding reference peaks. From the empirical formula obtained for all components from HR-MS data and the known chemistry of this class of compounds, structures of the nine components were proposed (Scheme 5.12). Note that structures **1** and **4** correspond to a ß-lactone 4 and its hydrolysis product 1. Structure **2** is a right-side fragment formed via ring opening of *ortho*-ester C. Structure **5** shows the loss of terminal aromatic group 2 from SCH27899, and structure **3** is the corresponding hydrolysis product. Structure **6** is a d-lactone. Structure **7** shows the loss of the methylene group from the J-ring to form a diol. Structures **8** and **10** correspond to hydrolysis (**8**) and methanolysis (**10**) products of SCH27899, respectively. Components **2**, **4**, **6**, **7**, **8**, and **10** are impurities that are related to SCH27899 samples, while structures **1**, **3**, and **5** are degradation products.

Despite the success of the use of HR-LC/MS for analysis of degradation products of SCH27899 on a magnetic sector instrument, there are intrinsic limitations to this approach for practical applications. These include reduced sensitivity because of the adaptation of an ESI source at a high pressure with a lower acceleration voltage on a magnetic sector instrument. Internal standard must be used in HR-MS measurements in order to obtain accurate mass data on a magnetic sector instrument. The use of internal standards may introduce matrix effects on the ionization of analytes. Expertise is also required for the operation of such instrumentation. More recent introduction of HR-LC/MS instrumentation such as the LTQ-Orbitrap mass spectrometer greatly enhances HR-LC/MS capability with ease of operation. In the case of everninomicins, SCH27899 sample from a different batch was degraded under base conditions. Negative-ion HR-LC/MS experiments were carried out using FT-MS mode at a resolution of 7500 (external calibration) on a LTQ-Orbitrap mass spectrometer. The reversed-phase LC was run on a C18 column with mobile phase A of 10 mM ammonium acetate in water and mobile phase B of methanol. The total-ion chromatogram obtained for the sample solution is displayed in Figure 5.14. On the basis of negative-ion HR-LC/MS data, two major degradants (A and C) and one impurity (B) are identified with excellent mass accuracy within 3 ppm, and proposed structures are illustrated in Scheme 5.13. To further verify their structures, on-line H/D exchange HR-LC/MS experiments were performed on the sample. Deprotonated SCH27899 molecular ion at *m/z* 1628 is shifted to *m/z* 1635, indicating eight exchangeable hydrogen atoms in the molecule (Scheme 5.14). Accurate mass measurements confirm this finding with the elemental composition of $C_{70}H_{89}D_7O_{38}NCl_2$ for SCH27899 after H/D exchange experiments ($[M-D]^-$, 1635.5430, 0.54 ppm). Degradants A and C and impurity B all show corresponding shifts of deprotonated molecular ions, consistent with the number of exchangeable hydrogen atoms in the molecules (Scheme 5.15). In addition, accurate mass data can be readily obtained for fragment ions during collision-induced activation of selected ions, providing informative structural information.

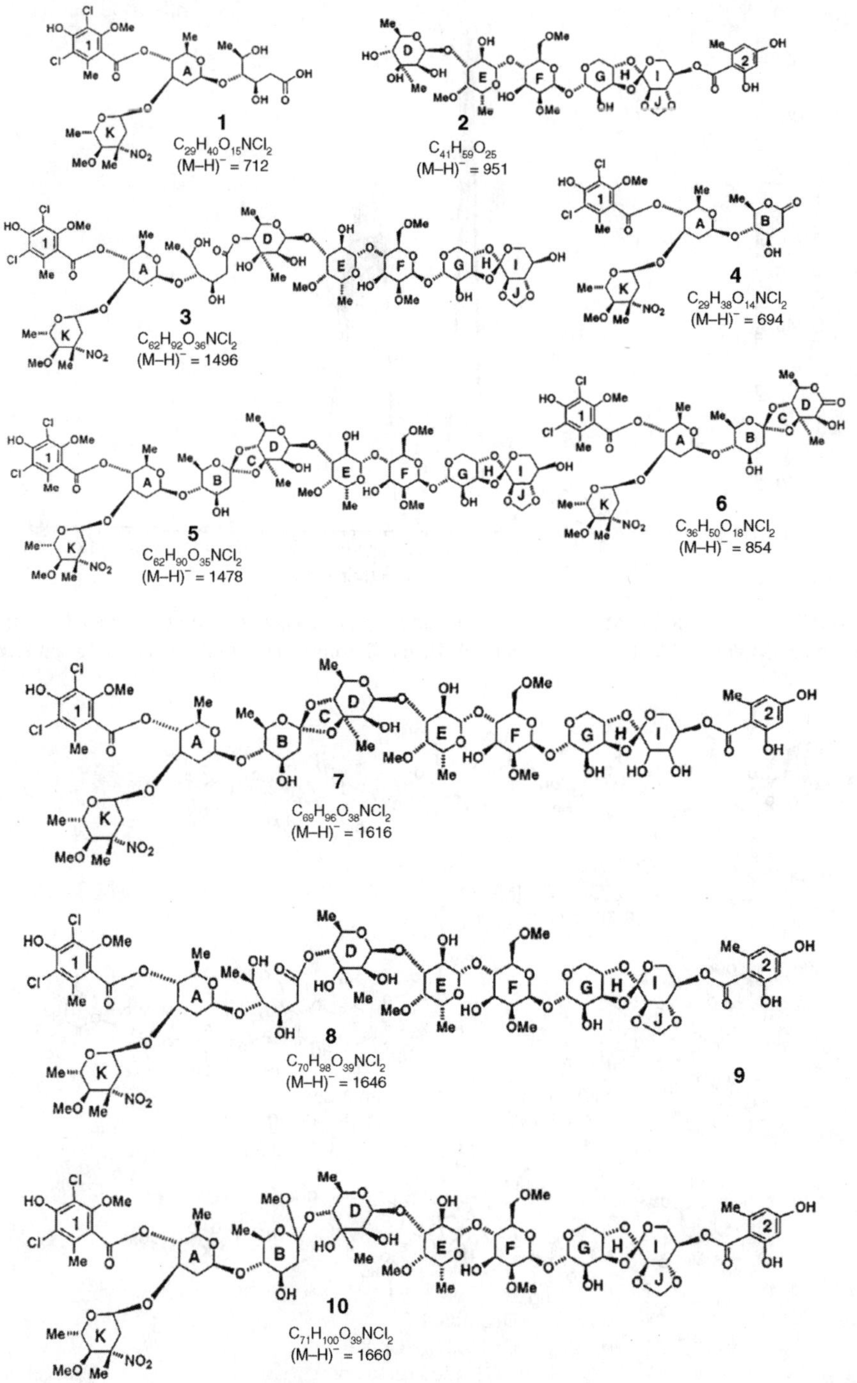

1 $C_{29}H_{40}O_{15}NCl_2$ $(M–H)^- = 712$

2 $C_{41}H_{59}O_{25}$ $(M–H)^- = 951$

3 $C_{62}H_{92}O_{36}NCl_2$ $(M–H)^- = 1496$

4 $C_{29}H_{38}O_{14}NCl_2$ $(M–H)^- = 694$

5 $C_{62}H_{90}O_{35}NCl_2$ $(M–H)^- = 1478$

6 $C_{36}H_{50}O_{18}NCl_2$ $(M–H)^- = 854$

7 $C_{69}H_{96}O_{38}NCl_2$ $(M–H)^- = 1616$

8 $C_{70}H_{98}O_{39}NCl_2$ $(M–H)^- = 1646$

9

10 $C_{71}H_{100}O_{39}NCl_2$ $(M–H)^- = 1660$

SCHEME 5.12 Proposed structures for the detected mixture components 1–10 (reprinted from Ref. 47, with permission of John Wiley & Sons, Ltd.).

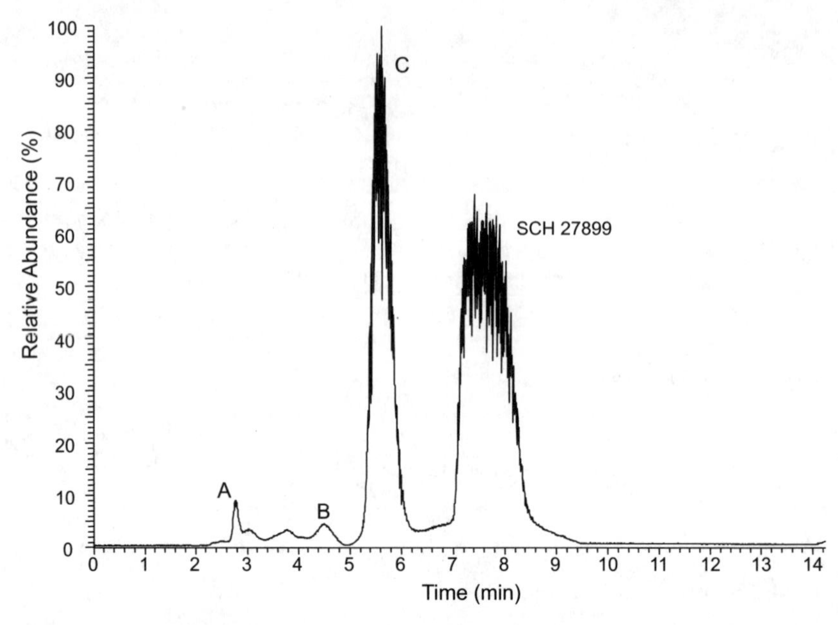

FIGURE 5.14 HR-LC/MS total-ion chromatogram of SCH27899 sample degraded in base solution, displaying identified components A, B, and C from a LTQ-Orbitrap mass spectrometer.

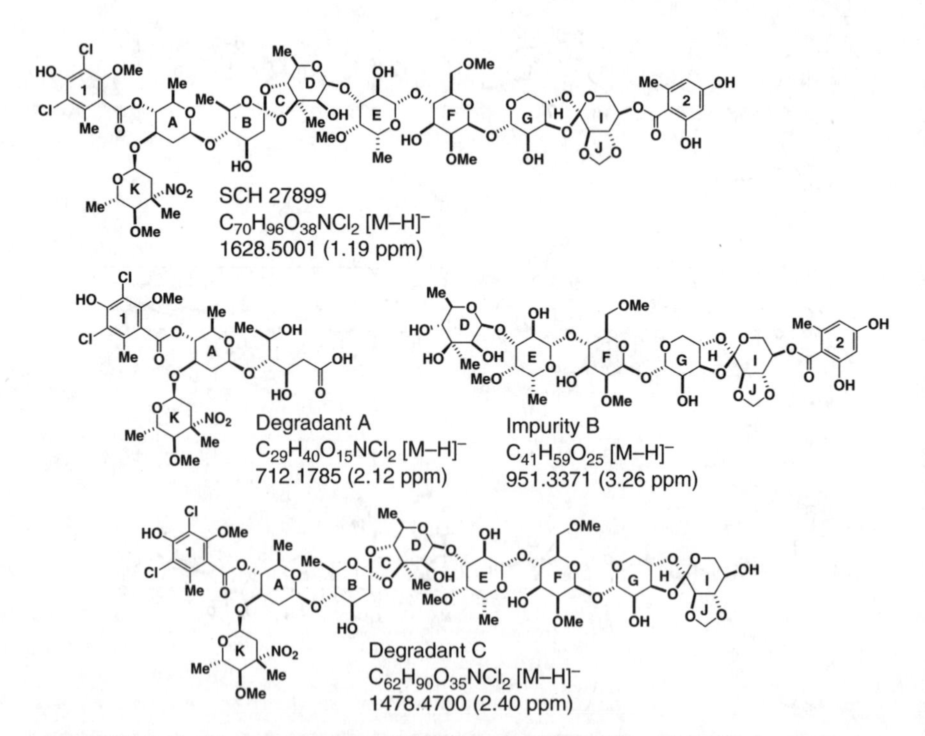

SCHEME 5.13 Negative-ion HR-LC/MS data for SCH27899 and detected components.

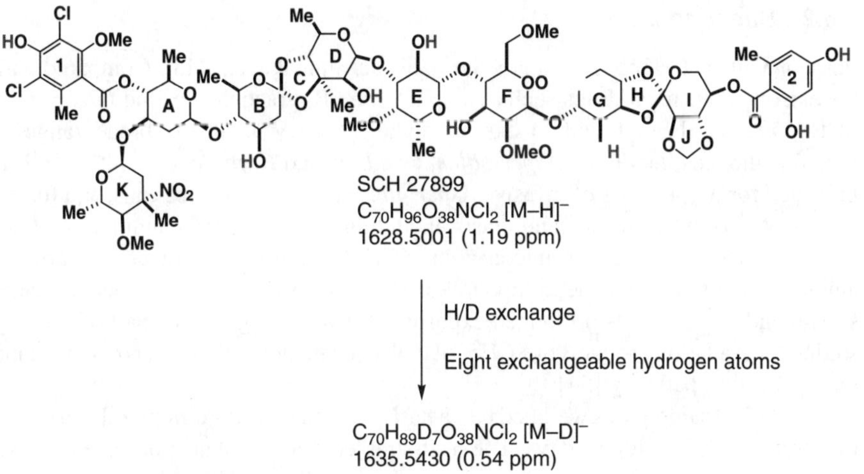

SCH 27899
$C_{70}H_{96}O_{38}NCl_2$ $[M-H]^-$
1628.5001 (1.19 ppm)

H/D exchange
Eight exchangeable hydrogen atoms

$C_{70}H_{89}D_7O_{38}NCl_2$ $[M-D]^-$
1635.5430 (0.54 ppm)

SCHEME 5.14 H/D exchange HR-LC/MS data for SCH27899.

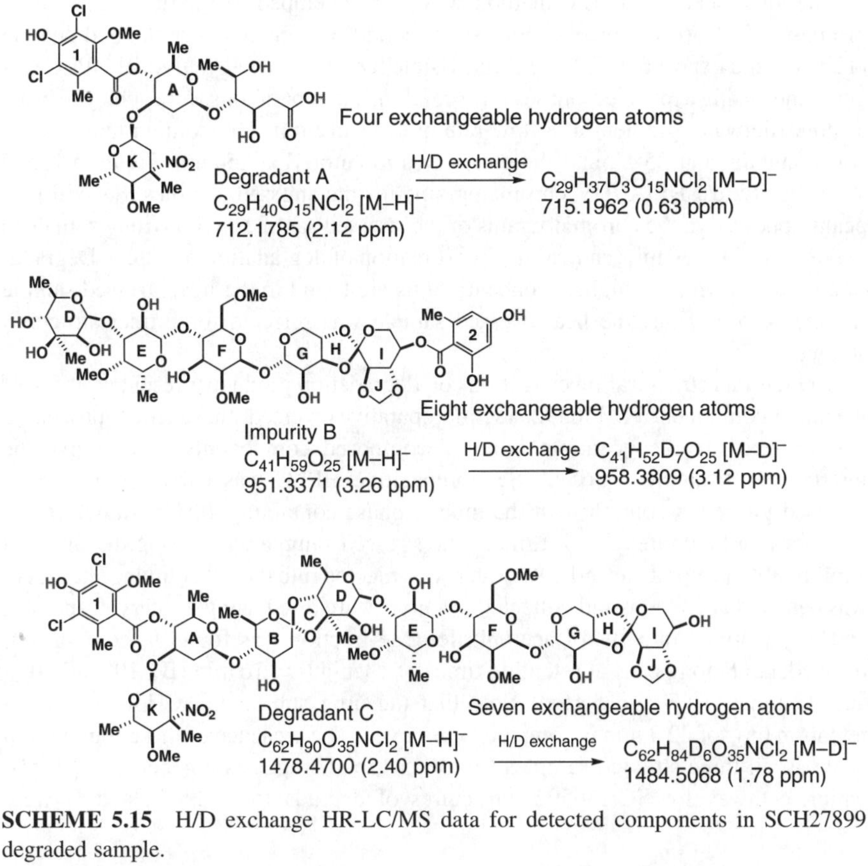

Four exchangeable hydrogen atoms

Degradant A
$C_{29}H_{40}O_{15}NCl_2$ $[M-H]^-$
712.1785 (2.12 ppm)

H/D exchange

$C_{29}H_{37}D_3O_{15}NCl_2$ $[M-D]^-$
715.1962 (0.63 ppm)

Eight exchangeable hydrogen atoms

Impurity B
$C_{41}H_{59}O_{25}$ $[M-H]^-$
951.3371 (3.26 ppm)

H/D exchange

$C_{41}H_{52}D_7O_{25}$ $[M-D]^-$
958.3809 (3.12 ppm)

Seven exchangeable hydrogen atoms

Degradant C
$C_{62}H_{90}O_{35}NCl_2$ $[M-H]^-$
1478.4700 (2.40 ppm)

H/D exchange

$C_{62}H_{84}D_6O_{35}NCl_2$ $[M-D]^-$
1484.5068 (1.78 ppm)

SCHEME 5.15 H/D exchange HR-LC/MS data for detected components in SCH27899 degraded sample.

5.3.2 Posaconazole

Posaconazole (SCH56592) is a novel triazole antifungal agent [48]. Compared with the existing antifungal drugs such as amphotericin B, itraconazole, and fluconazole, SCH56592 has been found to exhibit higher potency against a broad range of fungal pathogens, inclduing *Asperigillus*, *Candida,* and *Cryptococcus* [49–53]. It is indicated for prophylaxis of invasive *Aspergillus* and *Candida* infections, and for the treatment of oropharyngeal candidiasis, including oropharyngeal candidiasis refractory to itraconazole and/or fluconazole. SCH56592 drug substance is stable at ambient conditions, but the compound starts to form degradation products under stress conditions such as prolonged exposure to heat and light. As a part of stability studies, degradation products of SCH56592 were characterized in order to understand its degradation pathway [54].

SCH56592 sample was weighed into a flask covered with aluminum foil and stored in an oven at 150°C for 12 days. The heat-stressed drug substance became a brown chunky mass and was ground to powder. Similarly, SCH56592 samples were separately subjected to near-UV (UVA) fluorescent light [1×10^4 (W·h)/m^2] and white (visible) fluorescent light (7.6×10^6 lux·h) at room temperature.

The analytical HPLC/UV method was first developed to monitor degradation profiles of SCH56592 under various stress conditions, including prolonged heating (12 days) and exposure to UV light and visible light (30 days). Analytical HPLC was performed using a reversed-phase chiral column, and separations were achieved by an acetonitrile/water gradient at a flow rate of 0.75 mL/min. The column temperature was maintained at 35°C, and the eluent was monitored at 254 nm. Figure 5.15b–d shows analytical HPLC/UV chromatograms for the stressed samples. Several new peaks appeared in the chromatograms of the stressed samples, with strong signals in the region of 11–30 min, indicating the formation of degradation products. Degradation products with the highest concentrations are found in the heat-stressed sample (Figure 5.15d). Thus, the heat-stressed sample was selected for further structural studies.

To facilitate structural investigations on degradation products present at $< 5\%$ of the major component SCH56592, a semipreparative reversed-phase HPLC procedure was utilized to preconcentrate minor decomposed components and remove the interference from SCH 56592. Semipreparative HPLC was carried out using a reversed-phase C8 column with the mobile phase containing 40% tetrahydrofuran in water at a flow rate of 10 mL/min. The stressed sample with 50 mg dissolved in 1 mL of 80% acetonitrile and 20% water was injected into the C8 column. The eluent was detected at 254 nm, and collected fractions were dried using a rotary evaporator and reconstituted in 3 mL of acetonitrile. One fraction was found to contain four major degradation peaks at retention times of 13 min (D), 16 min (B), 19.5 min (C), and 25 min (A) (Figure 5.15e). Note that the drug substance SCH56592 with a retention time of 22.3 min is completely removed. The presence of these four elution peaks in all three stressed samples (Figure 5.15b–d) suggests a generalized degradation pathway for SCH56592. Structures of degradants A, B, C, and D were completely elucidated by LC-NMR, LC/MS, and LC/MS/MS, as illustrated in Scheme 5.16.

The LC-NMR method combines high-field NMR instrumentation with an HPLC separation system, enabling online collection of NMR spectra on individual components of mixtures [55,56]. Compared with the traditional NMR method, LC-NMR has a lower detection limit (to micrograms of materials) and allows NMR experiments to

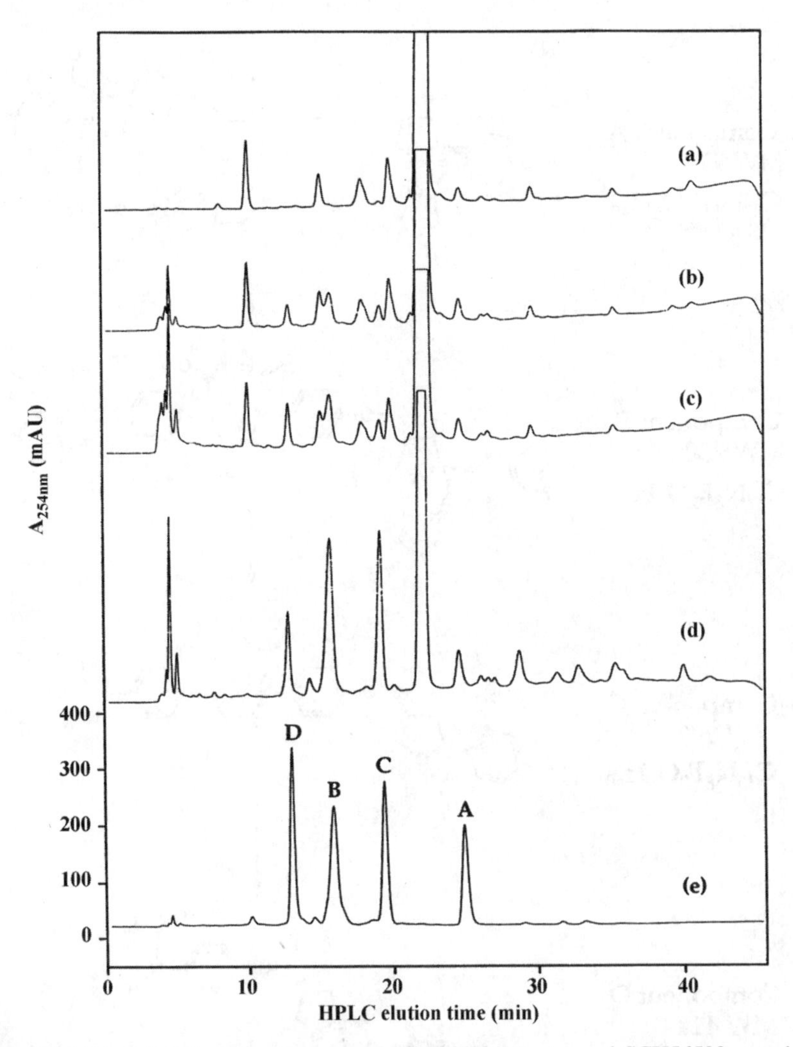

FIGURE 5.15 LC-UV chromatograms of SCH56592 (a), stressed SCH56592 sample by visible light (b), UV light (c), heat (d) as well as the partially purified mixture containing the major heat-stressed degradation products of SCH56592 (e) (reprinted from Ref. 54, with permission of Elsevier Science B.V.).

(a) SCH 56592
MW 700
$C_{37}N_8F_2O_4H_{42}$

(b) Component A
MW 714
$C_{37}N_8F_2O_5H_{40}$

(c) Component B
MW 730
$C_{37}N_8F_2O_6H_{40}$

(d) Component C
MW 702
$C_{36}N_8F_2O_5H_{40}$

(e) Component D
MW 414
$C_{21}N_4F_2O_3H_{20}$

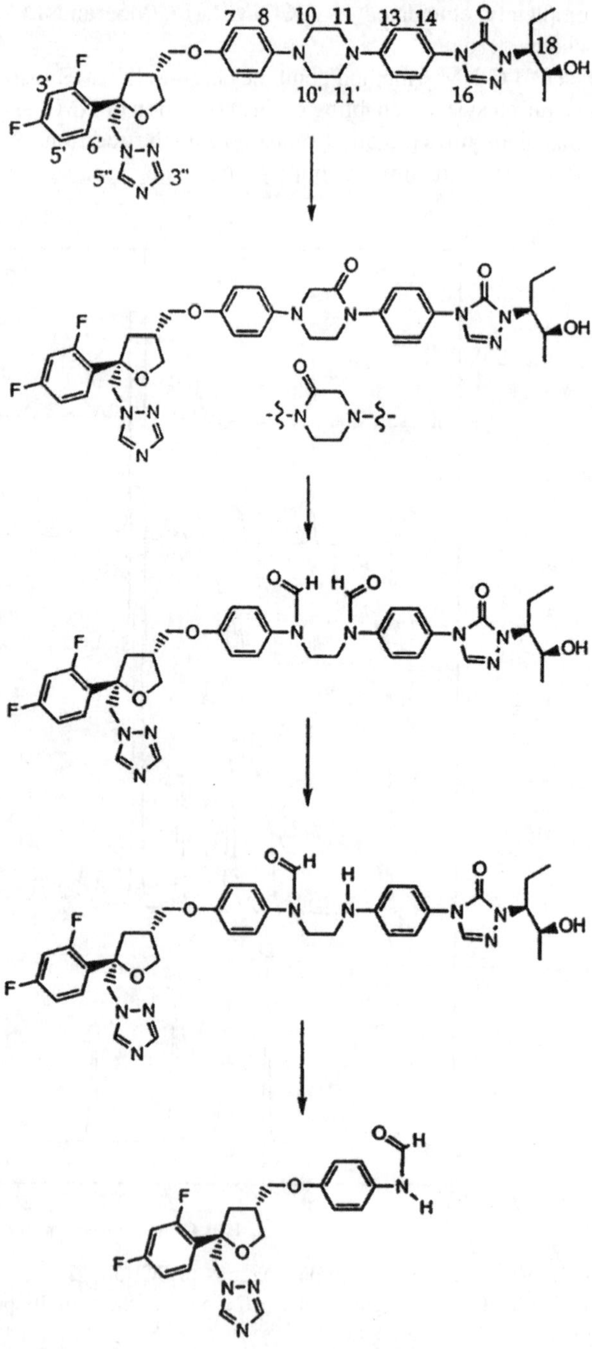

SCHEME 5.16 Structures of SCH56592 (a) and four major degradants A–D (reprinted from Ref. 54, with permission of Elsevier Science B.V.).

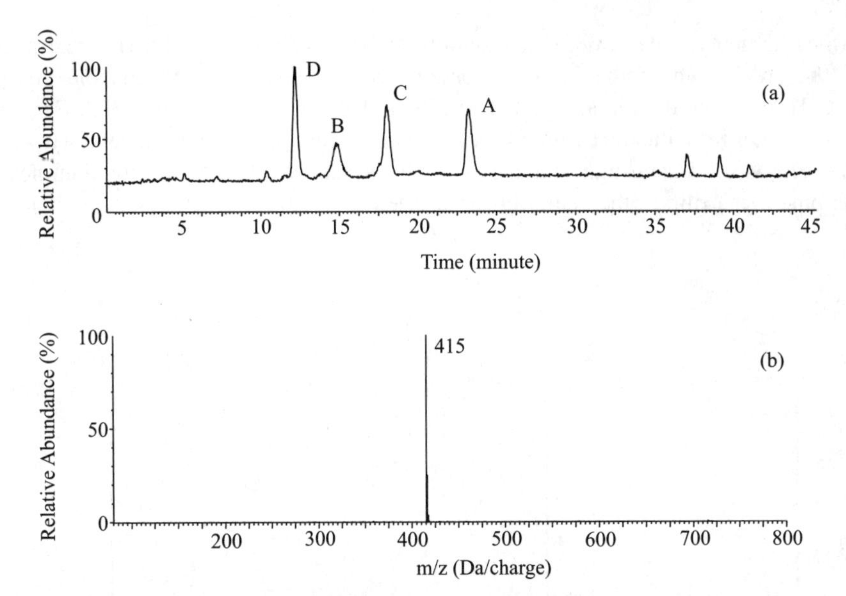

FIGURE 5.16 LC/MS total-ion chromatogram of the partially purified mixture containing the major heat-stressed degradation products of SCH56592 (a), and the ESI mass spectrum of degradant D (b) (reprinted from Ref. 54, with permission of Elsevier Science B.V.).

be conducted directly on the chromatographic resolved elution peaks. LC-NMR has been successfully applied in structural determinations of drug metabolites, chemical impurities, degradants, and components of natural product crude extracts. In this case, 1D and 2D proton NMR spectra for all degradants were obtained in "stop–flow" mode at 20°C. Analysis of the relevant resonances in all spectral regions led to the structural elucidation of degradation products.

Liquid Chromatography–Mass Spectrometric data provide MW information for individual components under optimized ESI conditions using a low-resolution triple-quadrupole mass spectrometer. Figure 5.16a shows a total-ion chromatogram for the degradation mixture of SCH56592, detecting four major degradants—A, B, C, and D—along with some other minor components. The corresponding mass spectrum for peak D is illustrated in Figure 5.16b, displaying the protonated molecular ion at *m/z* 415. Thus, the MW of degradant D is 414 Da. Similarly, the MWs of degradants A, B, and C are determined to be 714, 730, and 702 Da, respectively. Further LC-MS/MS experiments on a triple-quadrupole mass spectrometer (collision gas Ar, collision energy 60 V) were then performed on these molecular ions in order to obtain structural information by fragmentation studies.

Compared with the MW (700 Da) of SCH56592, degradant B (MW 730 Da) has an additional 30 mass units and might be the oxidized product of SCH56592 via the addition of two oxygen atoms and the reduction of two hydrogen atoms in the molecule. The major product ions from MS/MS experiments include *m/z* 685, 441, 372, 317, and 299 (Figure 5.17a). The loss of a CH_3CHOH- moiety from *m/z* 731 leads to the fragment ion at *m*/*z* 685. Two major fragment ions at *m*/*z* 441 and 317 are results of the breakdown of the center piperazine ring. The loss of the triazole group ($-C_2H_2N_3$) from the ion at *m*/*z* 441 results in the fragment ion at *m*/*z* 372. Further loss of water from the ion at *m*/*z* 317 leads to the formation of the fragment ion at *m*/*z* 299. The MS/MS data suggest that SCH56592 is cleaved into *N*,*N*′-formyldiamine at the piperazine ring in the center of the molecule.

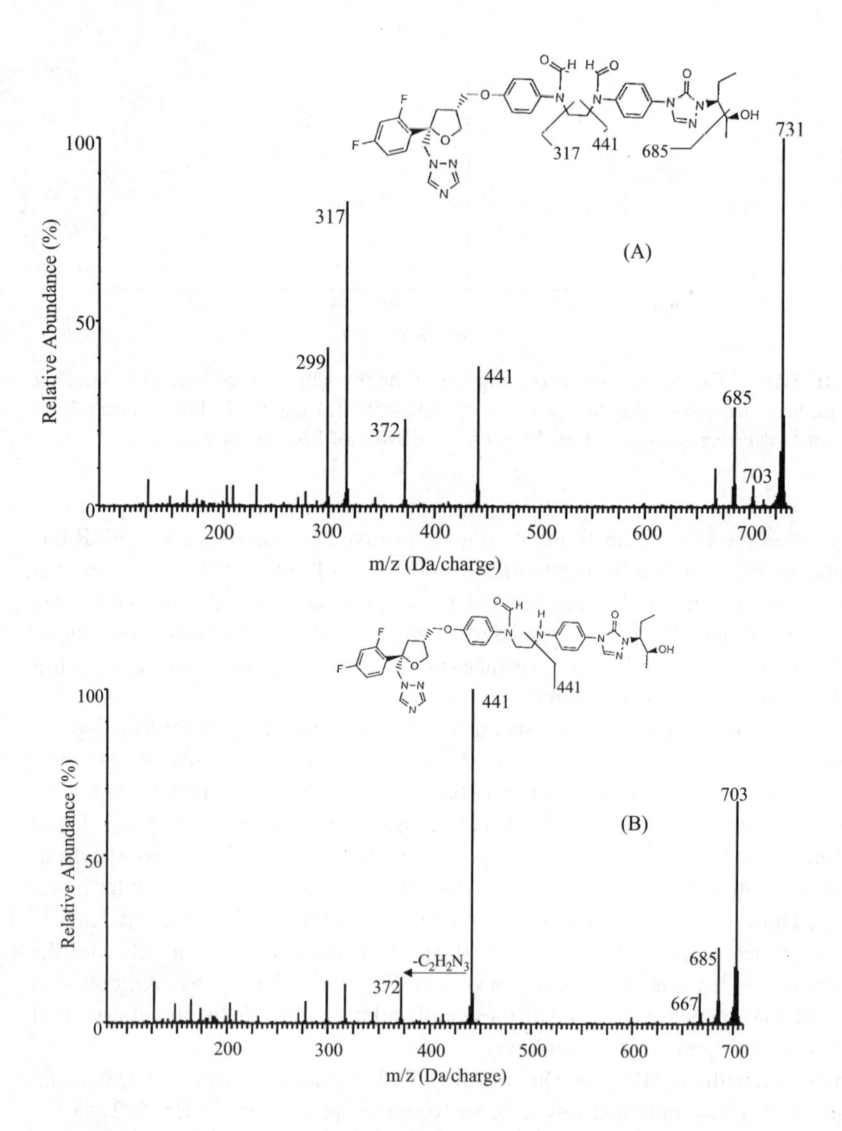

FIGURE 5.17 Product ion mass spectra of protonated molecular ions at *m*/*z* 731 (degradant B) (a), 703 (degradant C) (b), and 415 (degradant D) (c) (reprinted from Ref. 54, with permission of Elsevier Science B.V.).

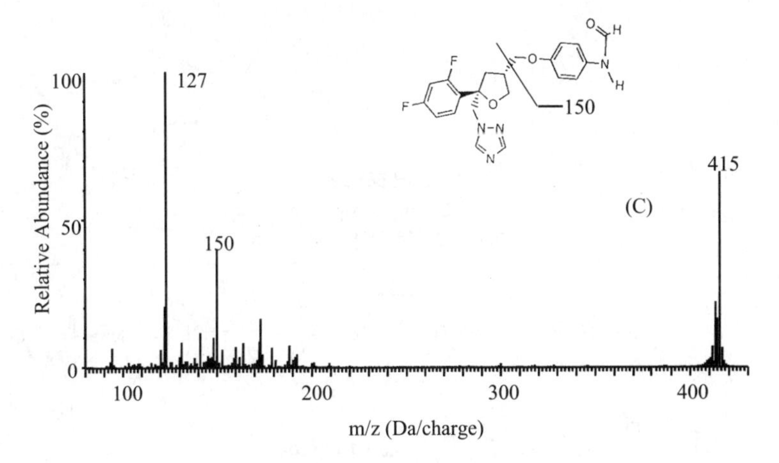

FIGURE 5.17 (*Continued*)

Degradant C (MW 702 Da) represents a possible loss of a carbonyl group (−CO) from degradant B (MW 731 Da). The most abundant fragment ion at *m/z* 441 corresponds to the left portion of the molecule via cleavage of the C-N bond in the open form of the center piperazine ring (Figure 5.17b). The further loss of the triazole group (-$C_2H_2N_3$) from *m/z* 441 yields the fragment ion at *m/z* 372. The consecutive loss of water from the molecular ion at *m/z* 703 leads to the fragment ions at *m/z* 685 and 667. Degradant D has a MW of 414 Da, indicating a possible breakdown of SCH56592 molecule. Its product ion mass spectrum displays two abundant fragment ions at *m/z* 127 and 150 (Figure 5.17c). The most intense fragment ion at *m/z* 127 is likely a stable difluorobenzonium ion. Degradant A has a MW of 714 Da, suggesting a possible oxidative product of SCH56592 (MW 700 Da) with the reduction of two hydrogen atoms in the molecule. However, its MS/MS data exhibit very limited fragment ions for structural information. Extensive LC-NMR experiments were carried out to obtain its structural information, indicating two oxidative products at C10 and C11 positions for degradant A.

It is important to note that low-resolution LC-MS and LC-MS/MS data provide useful information on structural characterization of degradation products such as in the case of SCH56592. However, no accurate mass data are available on both molecular ions and fragment ions using low-resolution quadrupole-based instrumentation. To gain further confidence in structural assignments, HR-LC/MS and HR-LC-MS/MS experiments were performed on these four degradants using a LTQ-Orbitrap mass spectrometer (resolution 15,000). Accurate mass measurements are readily obtained on all degradants with excellent mass accuracy ($<$ 2 ppm) (Scheme 5.17). Clearly, the results eliminate the ambiguities for the elemental compositions of degradants with low-resolution mass data. Another advantage is confident structural assignment for fragment ions using accurate mass data in HR-LC-MS/MS mode. For

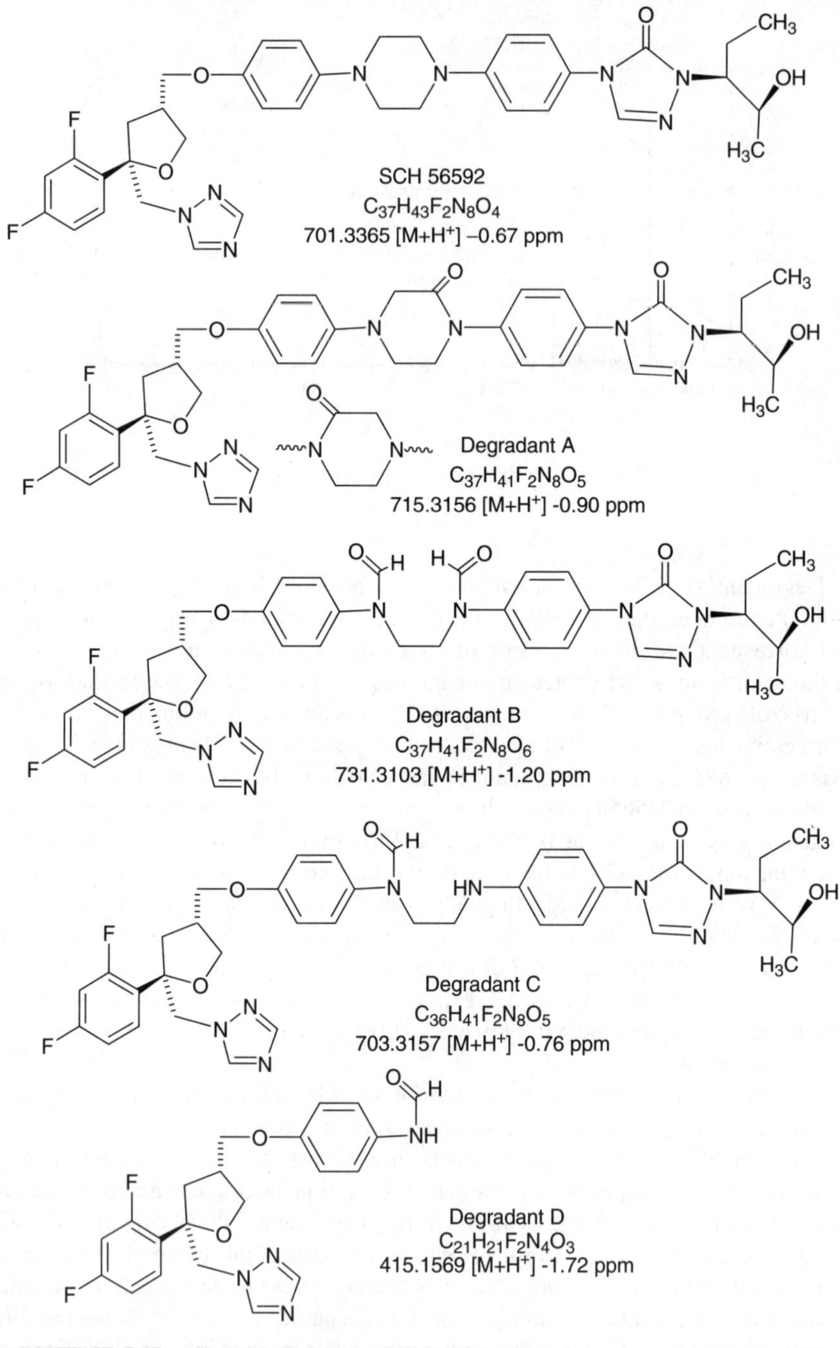

SCH 56592
$C_{37}H_{43}F_2N_8O_4$
701.3365 [M+H$^+$] –0.67 ppm

Degradant A
$C_{37}H_{41}F_2N_8O_5$
715.3156 [M+H$^+$] -0.90 ppm

Degradant B
$C_{37}H_{41}F_2N_8O_6$
731.3103 [M+H$^+$] -1.20 ppm

Degradant C
$C_{36}H_{41}F_2N_8O_5$
703.3157 [M+H$^+$] -0.76 ppm

Degradant D
$C_{21}H_{21}F_2N_4O_3$
415.1569 [M+H$^+$] -1.72 ppm

SCHEME 5.17 HR-LC/MS data for SCH56592 and its four major degradants A–D.

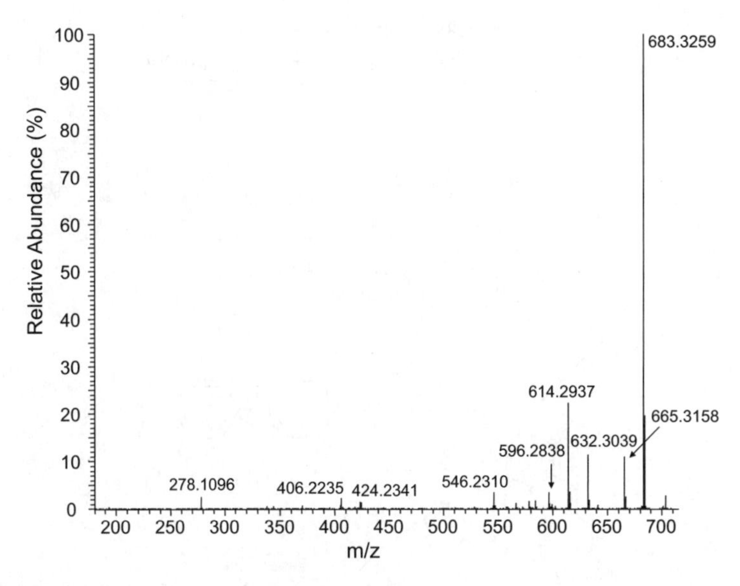

FIGURE 5.18 Product ion mass spectrum of SCH56592 molecular ion at *m/z* 701 obtained from a LTQ-Orbitrap mass spectrometer.

example, SCH56592 produces a number of fragment ions in CID experiments (Figure 5.18), and its fragmentation patterns can be described with the support of accurate mass measurements on fragment ions (Scheme 5.18). The high-resolution product ion mass spectra of these four degradants are illustrated in Figures 5.19–5.22.

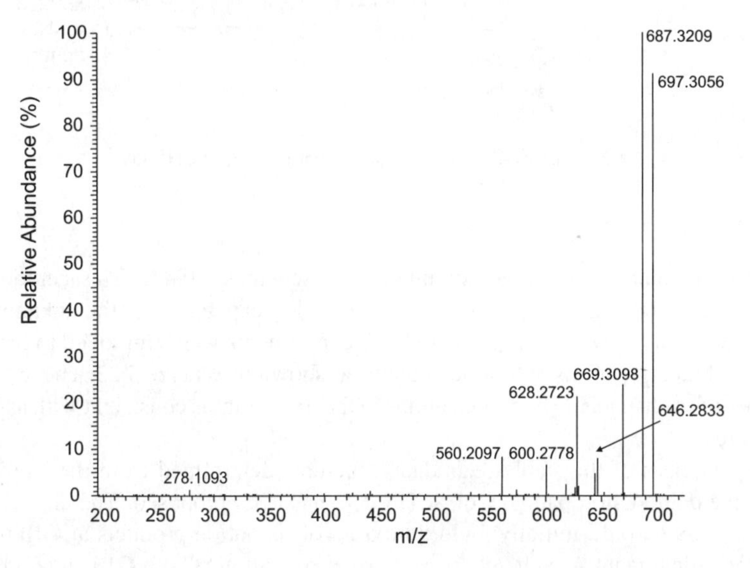

FIGURE 5.19 Product ion mass spectrum of degradant A of SCH56592 degradation products at *m/z* 715 obtained from a LTQ-Orbitrap mass spectrometer.

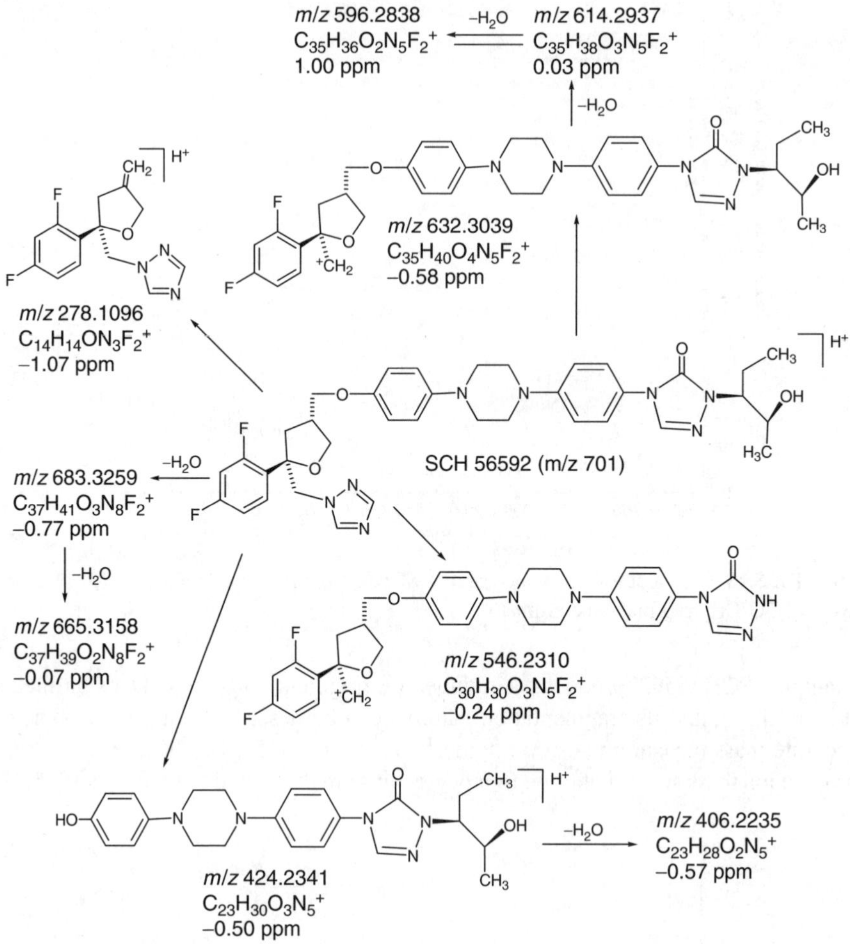

m/z 596.2838
$C_{35}H_{36}O_2N_5F_2^+$
1.00 ppm

m/z 614.2937
$C_{35}H_{38}O_3N_5F_2^+$
0.03 ppm

m/z 632.3039
$C_{35}H_{40}O_4N_5F_2^+$
−0.58 ppm

m/z 278.1096
$C_{14}H_{14}ON_3F_2^+$
−1.07 ppm

SCH 56592 (m/z 701)

m/z 683.3259
$C_{37}H_{41}O_3N_8F_2^+$
−0.77 ppm

m/z 665.3158
$C_{37}H_{39}O_2N_8F_2^+$
−0.07 ppm

m/z 546.2310
$C_{30}H_{30}O_3N_5F_2^+$
−0.24 ppm

m/z 424.2341
$C_{23}H_{30}O_3N_5^+$
−0.50 ppm

m/z 406.2235
$C_{23}H_{28}O_2N_5^+$
−0.57 ppm

SCHEME 5.18 Fragmentation patterns for SCH56592.

Their fragmentation patterns are summarized in Schemes 5.19–22. The accurate mass data can greatly enhance structural characterization capability for the unknowns. In addition, online H/D exchange HR-LC/MS experiments were employed to facilitate structural identifications of four degradants, as shown in Scheme 5.23. The measured number of exchangeable hydrogen atoms in the molecules is consistent with assigned structures.

On the basis of the stable degradant structures determined from the studies, an oxidative degradation pathway of SCH56592 has been proposed. The air oxidation of SCH56592 would initially yield a mixture of oxidation products at C10 or C11 positions (degradant A, Scheme 5.16). Further oxidation at both C10 and C11 leads to the formation of *N, N′*-diformyl structure in degradant B. Deformylation of

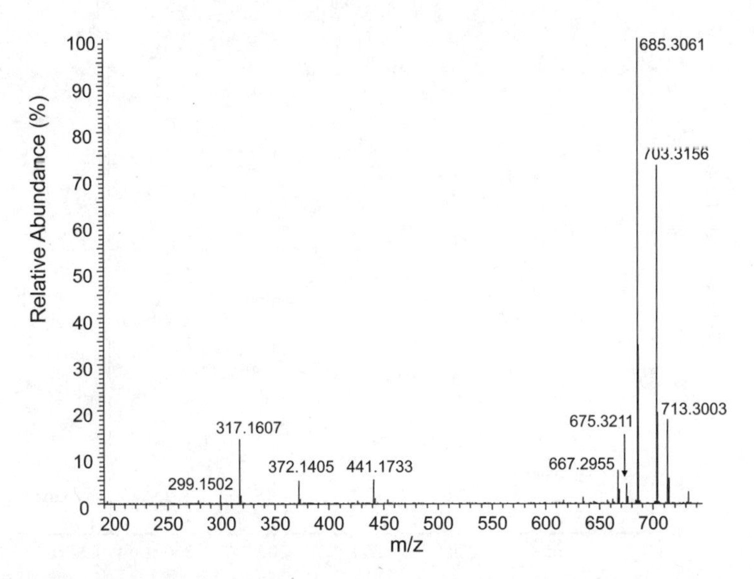

FIGURE 5.20 Product ion mass spectrum of degradant B of SCH56592 degradation products at *m/z* 731 obtained from a LTQ-Orbitrap mass spectrometer.

degradant B forms degradant C with one *N*-formyl group remaining and one secondary amine. Subsequent oxidative cleavage in degradant C causes breakdown of the structure into degradant D. The proposed oxidative degradation pathway is believed to be a dominant one under the stress conditions in the studies.

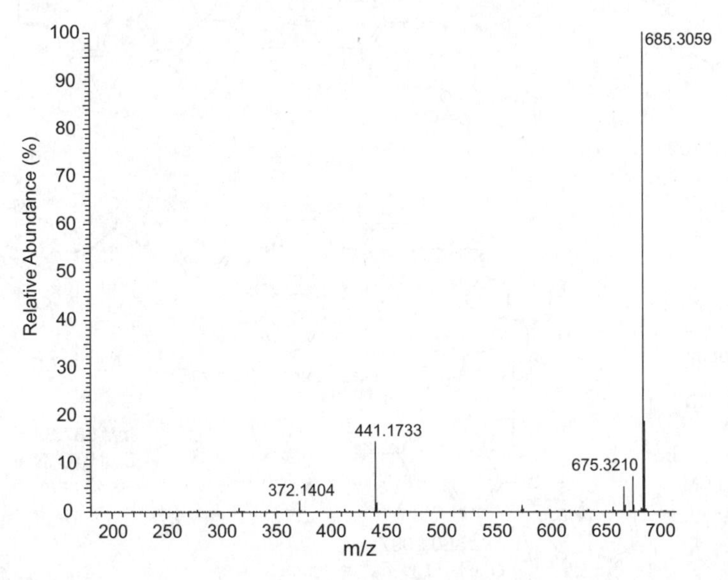

FIGURE 5.21 Product ion mass spectrum of degradant C of SCH56592 degradation products at *m/z* 703 obtained from a LTQ-Orbitrap mass spectrometer.

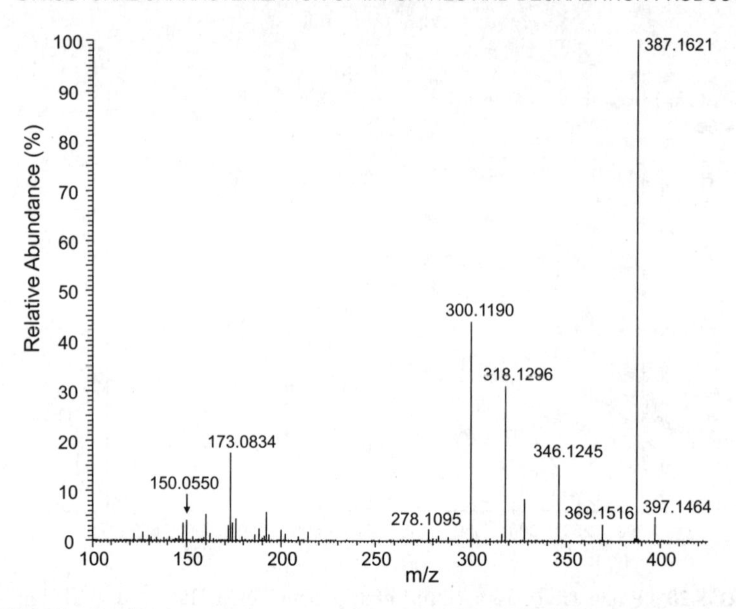

FIGURE 5.22 Product ion mass spectrum of degradant D of SCH56592 degradation products at *m/z* 415 obtained from a LTQ-Orbitrap mass spectrometer.

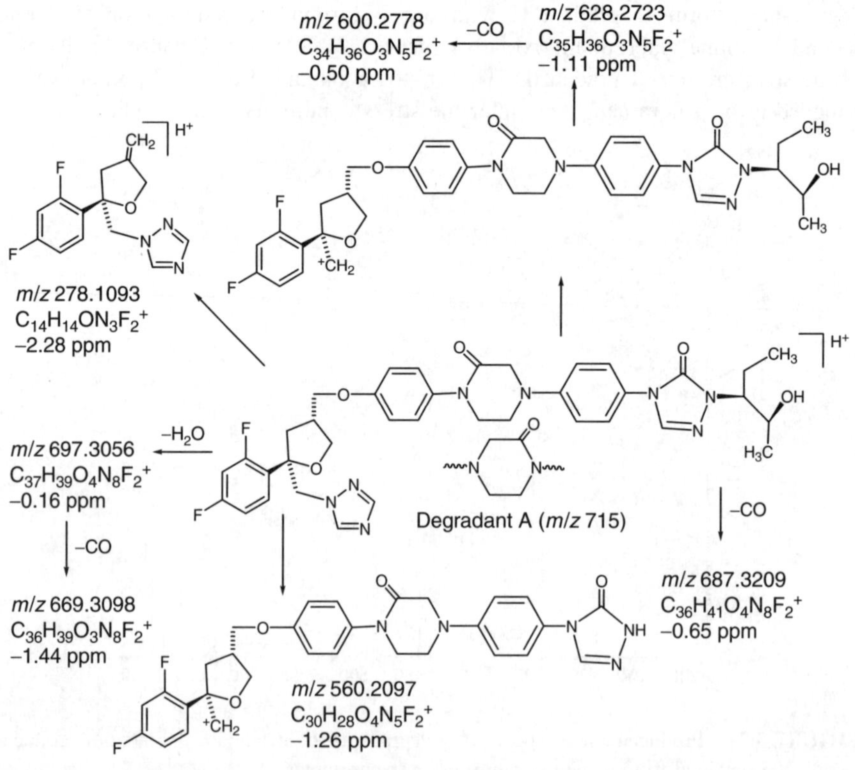

m/z 600.2778
$C_{34}H_{36}O_3N_5F_2^+$
−0.50 ppm

−CO

m/z 628.2723
$C_{35}H_{36}O_3N_5F_2^+$
−1.11 ppm

m/z 278.1093
$C_{14}H_{14}ON_3F_2^+$
−2.28 ppm

−H_2O

m/z 697.3056
$C_{37}H_{39}O_4N_8F_2^+$
−0.16 ppm

−CO

m/z 669.3098
$C_{36}H_{39}O_3N_8F_2^+$
−1.44 ppm

Degradant A (*m/z* 715)

−CO

m/z 687.3209
$C_{36}H_{41}O_4N_8F_2^+$
−0.65 ppm

m/z 560.2097
$C_{30}H_{28}O_4N_5F_2^+$
−1.26 ppm

SCHEME 5.19 Fragmentation patterns for degradant A (*m/z* 715) of SCH 56592 sample.

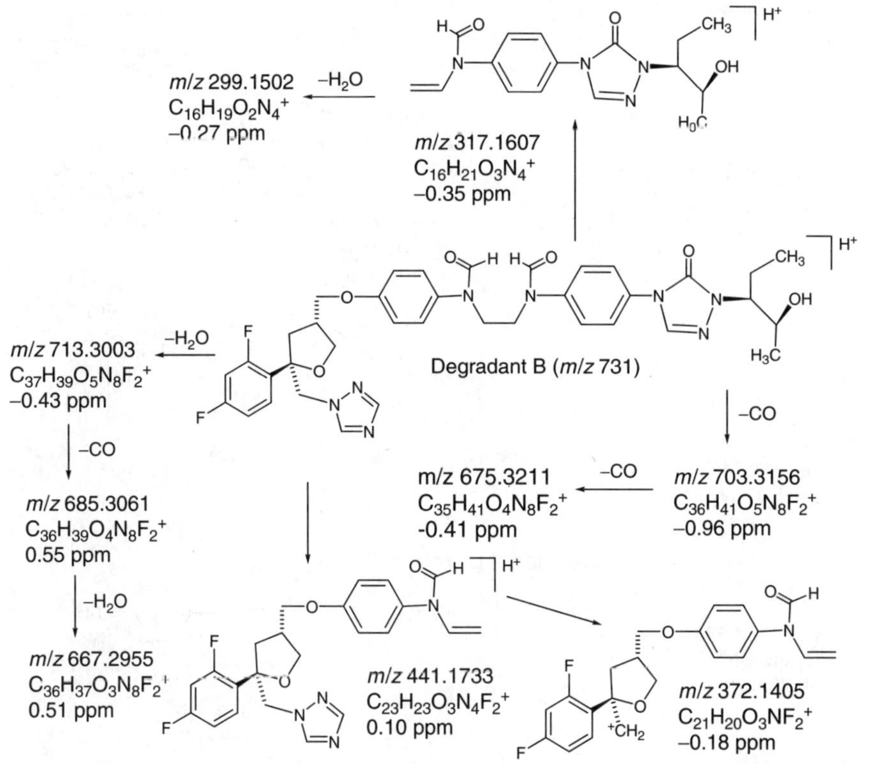

m/z 299.1502
$C_{16}H_{19}O_2N_4^+$
−0.27 ppm

m/z 317.1607
$C_{16}H_{21}O_3N_4^+$
−0.35 ppm

Degradant B (*m/z* 731)

m/z 713.3003
$C_{37}H_{39}O_5N_8F_2^+$
−0.43 ppm

m/z 685.3061
$C_{36}H_{39}O_4N_8F_2^+$
0.55 ppm

m/z 667.2955
$C_{36}H_{37}O_3N_8F_2^+$
0.51 ppm

m/z 675.3211
$C_{35}H_{41}O_4N_8F_2^+$
-0.41 ppm

m/z 703.3156
$C_{36}H_{41}O_5N_8F_2^+$
−0.96 ppm

m/z 441.1733
$C_{23}H_{23}O_3N_4F_2^+$
0.10 ppm

m/z 372.1405
$C_{21}H_{20}O_3NF_2^+$
−0.18 ppm

SCHEME 5.20 Fragmentation patterns for degradant B (*m/z* 731) of SCH56592 sample.

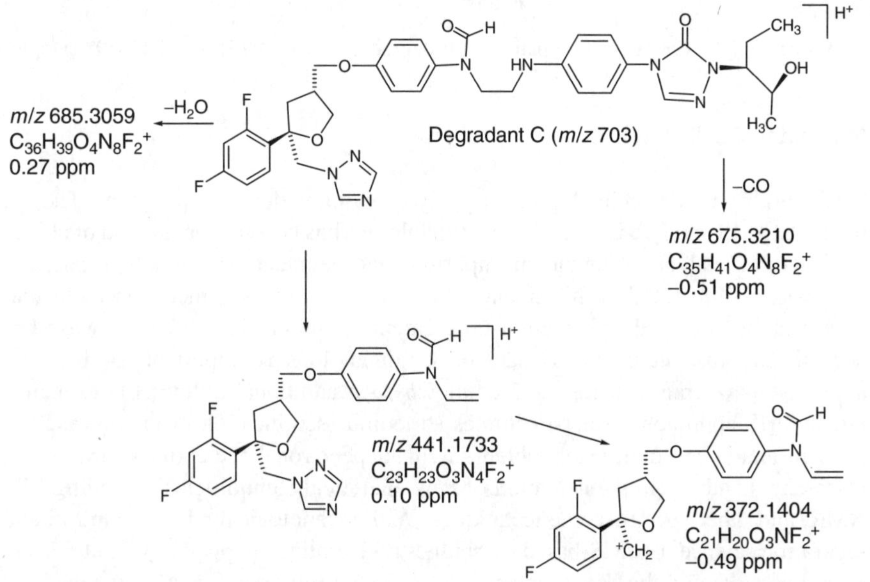

Degradant C (*m/z* 703)

m/z 685.3059
$C_{36}H_{39}O_4N_8F_2^+$
0.27 ppm

m/z 675.3210
$C_{35}H_{41}O_4N_8F_2^+$
−0.51 ppm

m/z 441.1733
$C_{23}H_{23}O_3N_4F_2^+$
0.10 ppm

m/z 372.1404
$C_{21}H_{20}O_3NF_2^+$
−0.49 ppm

SCHEME 5.21 Fragmentation patterns for degradant C (*m/z* 703) of SCH56592 sample.

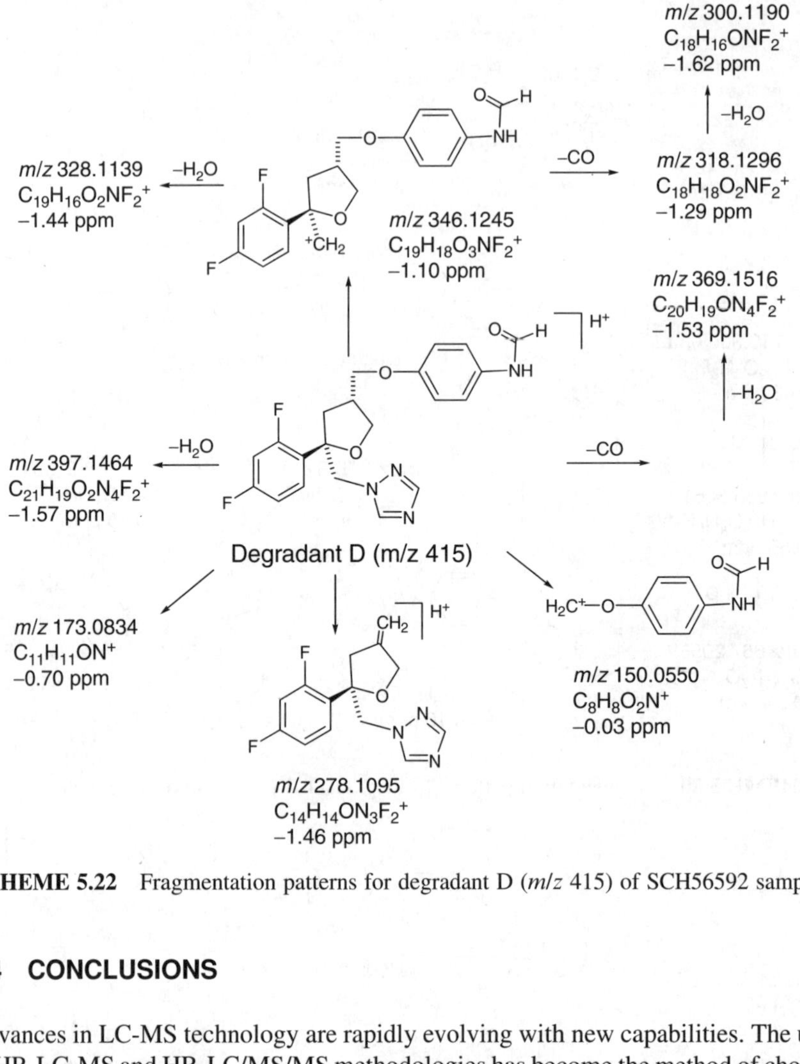

SCHEME 5.22 Fragmentation patterns for degradant D (*m/z* 415) of SCH56592 sample.

5.4 CONCLUSIONS

Advances in LC-MS technology are rapidly evolving with new capabilities. The use of HR-LC-MS and HR-LC/MS/MS methodologies has become the method of choice in structural analysis of unknown impurities and degradation products in pharmaceuticals, as illustrated in this chapter. Accurate mass measurements on molecular ions can provide vital information on elemental compositions of unknowns. The accessibility to accurate mass data on fragment ions is important to structural assignments of fragment ions and unknowns. An additional online H/D exchange HR-LC/MS approach further facilitates structural identification of unknowns. The ultimate analytical solution to problem solving depends on the scientific knowledge of researchers and the implementation of new analytical technologies, including MS, NMR, and other spectroscopic techniques. With technological advances in MS and separation science, LC/MS-based techniques will continue to play important roles in drug discovery and the development process in the future.

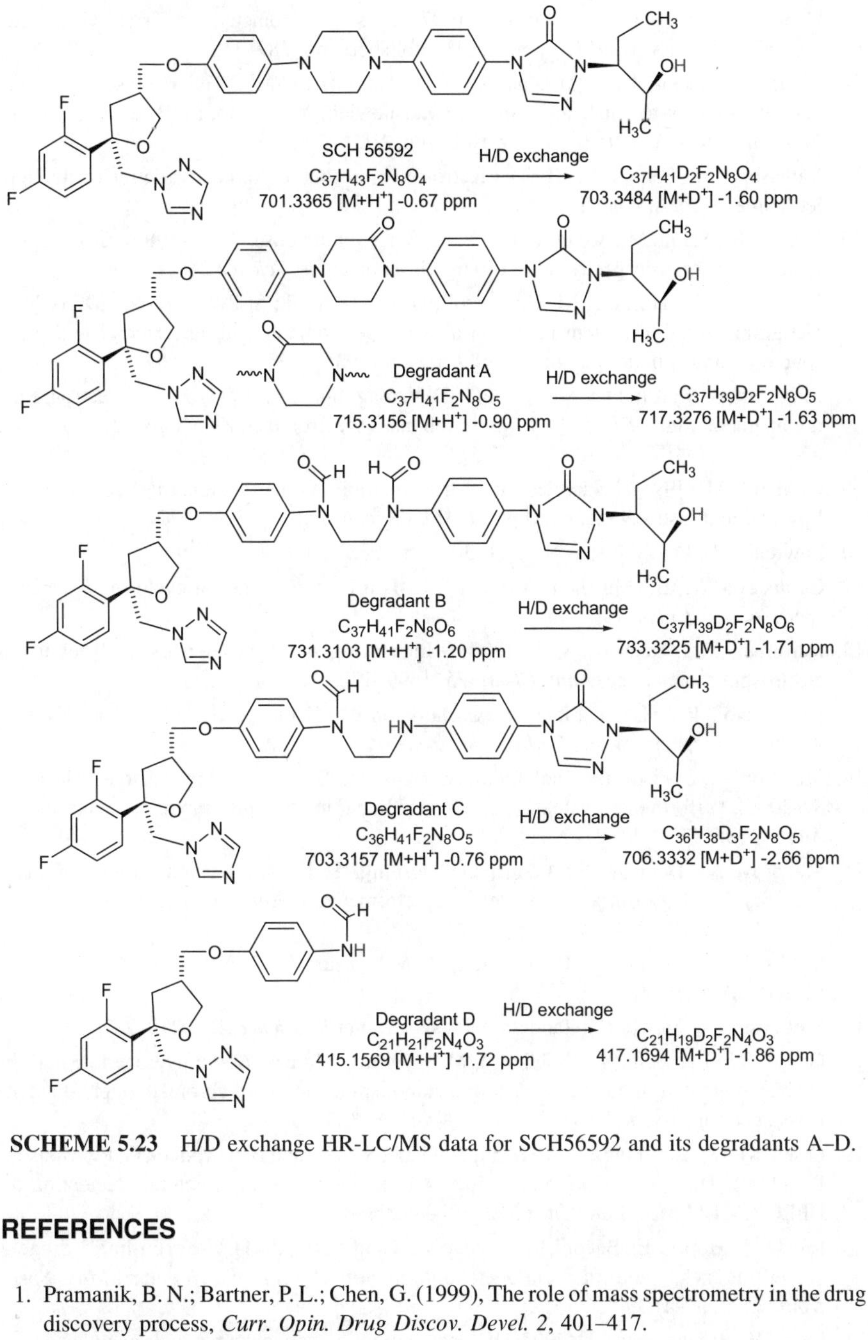

SCH 56592
$C_{37}H_{43}F_2N_8O_4$
701.3365 [M+H$^+$] -0.67 ppm
H/D exchange
$C_{37}H_{41}D_2F_2N_8O_4$
703.3484 [M+D$^+$] -1.60 ppm

Degradant A
$C_{37}H_{41}F_2N_8O_5$
715.3156 [M+H$^+$] -0.90 ppm
H/D exchange
$C_{37}H_{39}D_2F_2N_8O_5$
717.3276 [M+D$^+$] -1.63 ppm

Degradant B
$C_{37}H_{41}F_2N_8O_6$
731.3103 [M+H$^+$] -1.20 ppm
H/D exchange
$C_{37}H_{39}D_2F_2N_8O_6$
733.3225 [M+D$^+$] -1.71 ppm

Degradant C
$C_{36}H_{41}F_2N_8O_5$
703.3157 [M+H$^+$] -0.76 ppm
H/D exchange
$C_{36}H_{38}D_3F_2N_8O_5$
706.3332 [M+D$^+$] -2.66 ppm

Degradant D
$C_{21}H_{21}F_2N_4O_3$
415.1569 [M+H$^+$] -1.72 ppm
H/D exchange
$C_{21}H_{19}D_2F_2N_4O_3$
417.1694 [M+D$^+$] -1.86 ppm

SCHEME 5.23 H/D exchange HR-LC/MS data for SCH56592 and its degradants A–D.

REFERENCES

1. Pramanik, B. N.; Bartner, P. L.; Chen, G. (1999), The role of mass spectrometry in the drug discovery process, *Curr. Opin. Drug Discov. Devel. 2*, 401–417.
2. Chen, G.; Pramanik, B. N.; Liu, Y. H.; Mirza, U. A. (2007), Applications of LC/MS in structural identifications of small molecules and proteins in drug discovery, *J. Mass Spectrom. 42*, 279–287.

3. Cooks, R. G.; Chen, G.; Wong, P. (1997), Mass spectrometers, in Trigg, G. L., ed., *Encyclopedia of Applied Physics*, VCH Publishers, pp. 289–330.
4. Chen, G.; Zhang, L. K.; Pramanik, B. N. (2007), LC/MS: Theory, instrumentation and applications to small molecules, in Kazakevich, Y., LoBrutto, R. eds., *HPLC for Pharmaceutical Scientists*, Wiley, Hoboken, NJ, pp. 281–346.
5. Yamashita, M.; Fenn, J. B. (1984), Electrospray ion source. Another variation on the free-jet theme, *J. Phys. Chem. 88*, 4451–4459.
6. Fenn, J. B.; Mann, M.; Meng, C. K.; Wong, S. F.; Whitehouse, C. M. (1989), Electrospray ionization for mass spectrometry of large biomolecules, *Science 246*, 64–71.
7. Horning, E. C.; Horning, M. G.; Carroll, D. I.; Dzidic, I.; Stillwell, R. N. (1973), New picogram detection system based on a mass spectrometer with an external ionization source at atmospheric pressure, *Anal. Chem. 45*, 936–943.
8. Mamyrin, B. A.; Karataev, V. I.; Shmikk, D. V.; Zagulin, V. A. (1973), Mass reflectron. New nonmagnetic time-of-flight high-resolution mass spectrometer, *Zh. Eksper. Teor. Fiz. 64*, 82–89.
9. Guilhaus, M. (1995), Principles and instrumentation in time-of-flight mass spectrometry. Physical and instrumental concepts, *J. Mass Spectrom. 30*, 1519–1532.
10. Lawrence, E. O.; Edlefsen, N. E. (1930), *Science 72*, 376.
11. Comisarow, M. B.; Marshall, A. G. (1974), Fourier transform ion cyclotron resonance spectroscopy, *Chem. Phys. Lett. 25*, 282–283.
12. Hardman, M.; Makarov, A. A. (2003), Interfacing the orbitrap mass analyzer to an electrospray ion source, *Anal. Chem. 75*, 1699–1705.
13. Hu, Q.; Noll, R. J.; Li, H.; Makarov, A.; Hardman, M.; Cooks, R. G. (2005), The Orbitrap: A new mass spectrometer, *J. Mass Spectrom. 40*, 430–443.
14. Makarov, A.; Denisov, E.; Kholomeev, A.; Balschun, W.; Lange, O.; Strupat, K.; Horning, S. (2006), Performance evaluation of a hybrid linear ion trap/orbitrap mass spectrometer, *Anal. Chem. 78*, 2113–2120.
15. Makarov, A.; Denisov, E.; Lange, O.; Horning, S. (2006), Dynamic range of mass accuracy in LTQ orbitrap hybrid mass spectrometer, *J. Am. Soc. Mass Spectrom. 17*, 977–982.
16. Cooks, R. G,; Beynon, J. H.; Caprioli, R. M.; Lester, G. R. (1973), *Metastable Ions*, Elsevier, Amsterdam.
17. McLafferty, F. W. (1981), Tandem mass spectrometry, *Science 214*, 280–287.
18. Ohashi, N.; Furuuchi, S.; Yoshikawa, M. (1998), Usefulness of the hydrogen-deuterium exchange method in the study of drug metabolism using liquid chromatography-tandem mass spectrometry, *J. Pharm. Biomed. Anal. 18*, 325–334.
19. Olsen, M. A.; Cummings, P. G.; Kennedy-Gabb, S.; Wagner, B. M.; Nicol, G. R.; Munson, B. (2000), The use of deuterium oxide as a mobile phase for structural elucidation by HPLC/UV/ESI/MS, *Anal. Chem. 72*, 5070–5078.
20. Liu, D. Q.; Hop, C. E.; Beconi, M. G.; Mao, A.; Chiu, S. H. (2001), Use of on-line hydrogen/deuterium exchange to facilitate metabolite identification, *Rapid Commun. Mass Spectrom. 15*, 1832–1839.
21. Lam, W.; Ramanathan, R. (2002), In electrospray ionization source hydrogen/deuterium exchange LC-MS and LC-MS/MS for characterization of metabolites, *J. Am. Soc. Mass Spectrom. 13*, 345–353.

22. Tolonen, A.; Turpeinen, M.; Uusitalo, J.; Pelkonen, O. (2005), A simple method for differentiation of monoisotopic drug metabolites with hydrogen-deuterium exchange liquid chromatography/electrospray mass spectrometry, *Eur. J. Pharm. Sci. 25*, 155–162.

23. Wolff, J. C.; Laures, A. M. (2006), "On the fly" hydrogen/deuterium exchange liquid chromatography/mass spectrometry using a dual-sprayer atmospheric pressure ionisation source, *Rapid Commun. Mass Spectrom. 20*, 3769–3779.

24. Novak, T. J.; Helmy, R.; Santos, I. (2005), Liquid chromatography-mass spectrometry using the hydrogen/deuterium exchange reactions as a tool for impurity identification in pharmaceutical process development, *J. Chromatogr. B 825*, 161–168.

25. Liu, D. Q.; Hop, C. E. C. A. (2005), Strategies for characterization of drug metabolites using liquid chromatography-tandem mass spectrometry in conjunction with chemical derivatization and on-Line H/D exchange approaches, *J. Pharm. Biomed. Anal. 37*, 1–18.

26. Chen, G.; Khusid, A.; Daaro, I.; Irish, P.; Pramanik, B. N. (2007), Structural identification of trace level enol tautomer impurity by on-line hydrogen/deuterium exchange HR-LC/MS in a LTQ-Orbitrap hybrid mass spectrometer, *J. Mass Spectrom. 42*, 967–970.

27. Chen, G.; Daaro, I.; Pramanik, B. N.; Piwinski, J. J. (2009), Structural characterization of in vitro rat liver microsomal metabolites of antihistamine desloratadine using LTQ-Orbitrap hybrid mass spectrometer in combination with online hydrogen/deuterium exchange HR-LC/MS, *J. Mass Spectrom. 44*, 203–213.

28. Chen, G.; Daaro, I.; Pramanik, B. N. (2007), Structural identifications of mometasone furoate steroid related impurities in a LTQ-Orbitrap hybrid mass spectrometer, *Proc. 55th ASMS Conf. Mass Spectrometry and Allied Topics*, Abstract A074176, Indianapolis, June 3 – 7.

29. Venkatraman, S.; Bogen, S. L.; Arasappan, A.; Bennett, F.; Chen, K.; Jao, E.; Liu, Y. T.; Lovey, R.; Hendrata, S.; Huang, Y.; Pan, W.; Parekh, T.; Pinto, P.; Popov, V.; Pike, R.; et al. (2006), Discovery of (1R,5S) N-[3-amino-1-(cyclobutylmethyl)-2,3-dioxo-propyl]- 3-[2(S)-[[[(1,1-dimethylethyl)amino]carbonyl]amino]-3,3-dimethyl-1-oxo-butyl]- 6,6-dimethyl-3-azabicyclo[3.1.0]hexan-2(S)-carboxamide (SCH 503034), a selective, potent, orally bioavailable hepatitis C virus NS3 protease inhibitor: A potential therapeutic agent for the treatment of hepatitis C infection, *J. Med. Chem. 49*, 6074–6086.

30. Malcolm, B. A.; Liu, R.; Lahser, F.; Agrawal, S.; Belanger, B.; Butkiewicz, N.; Chase, R.; Gheyas, F.; Hart, A.; Hesk, D.; Ingravallo, P.; Jiang, C.; Kong, R.; Lu, J.; Pichardo, J.;et al. (2006), SCH 503034, a mechanism-based inhibitor of hepatitis C virus NS3 protease, suppresses polyprotein maturation and enhances the antiviral activity of alpha interferon in replicon cells, *Antimicrob. Agents Chemother. 50*, 1013–1020.

31. Tong, X.; Chase, R.; Skelton, A.; Chen, T.; Wright-Minogue, J.; Malcolm, B. A. (2006), Identification and analysis of fitness of resistance mutations against the HCV protease inhibitor SCH 503034, *Antiviral Res. 70*, 28–38.

32. Weinstein, M. J.; Wagman, G. H.; Oden, E. M.; Luedemann, G. M.; Sloane, P.; Murawski, A.; Marquez, J. (1965), Purification and biological studies of everninomicin B, *Antimicrob. Agents Chemother. 5*, 821–827.

33. Girijavallabhan, V. M.; Ganguly, A. K. (1992), *Kirk-Othmer Encyclopedia of Chemical Technology*, 4th ed., Wiley, New York, Vol. *3*, p. 259.

34. Ganguly, A. K.; McCormick, J. L.; Saksena, A. K.; Das, P. R.; Chan, T. M. (1999), Chemical modifications and structure activity studies of ziracin and related everninomicin antibiotics, *Bioorg. Med. Chem. Lett. 9*, 1209–1214.
35. Foster, D. R.; Rybak, M. J. (1999), Pharmacologic and bacteriologic properties of SCH 27899 (Ziracin), an investigational antibiotic from the everninomicin family, *Pharmacotherapy 19*, 1111–1117.
36. Ganguly, A. K.; Sarre, O. Z.; Greeves, D.; Morton, J. (1975), Structure of everninomicin D-1, *J. Am. Chem. Soc. 97*, 1982–1985.
37. Ganguly, A. K.; Szmulewicz, S. (1975), Structure of everninomicin C, *J. Antibiot. 28*, 710–712.
38. Ganguly, A. K.; Saksena, A. K. (1975), Structure of everninomicin B, *J. Antibiot. 28*, 707–709.
39. Ganguly, A. K.; Pramanik, B. N.; Girijavallabhan, V. M.; Sarre, O.; Bartner, P. L. (1985), The use of fast atom bombardment mass spectrometry for the determination of structures of everninomicins, *J. Antibiot. 38*, 808–812.
40. Pramanik, B. N.; Ganguly, A. K. (1986), Fast atom bombardment mass spectrometry: A powerful technique for study of oligosaccharide antibiotics, *Indian J. Chem. 25B*, 1105–1111.
41. Pramanik, B. N.; Das, P. R. (1989), Molecular ion enhancement using salts in FAB matrices for studies on complex natural products, *J. Nat. Products 52*, 534–546.
42. Ganguly, A. K.; Pramanik, B. N.; Chan, T. M.; Liu, Y. H.; Morton, J.; Girijavallabhan, V. M. (1989), The structure of new oligosaccharide antibiotics, 13–384 components 1 and 5 *Heterocycles 28*, 83–88.
43. Ganguly, A. K.; McCormick, J. L.; Chan, T. M.; Saksena, A. K.; Das, P. R. (1997), Determination of the absolute stereochemistry at the C16 orthoester of everninomicin antibiotics;a novel acid-catalyzed isomerization of orthoesters, *Tetrahedron Lett. 38*, 7989–7992.
44. Bartner, P. L.; Pramanik, B. N.; Saksena, A. K.; Liu. Y. H.; Das, P. R.; Sarre, O.; Ganguly, A. K. (1997), Structural elucidation of everninomicin-6, a new oligosaccharide antibiotic, by chemical degradation and FAB-MS methods, *J. Am. Soc. Mass Spectrom. 8*, 1134–1140.
45. Chen, G.; Pramanik, B. N.; Bartner, P. L.; Saksena, A. K.; Gross, M. L. (2002), Multiple-stage mass spectrometric analysis of complex oligosaccharide antibiotics (everninomicins) in a quadrupole ion trap, *J. Am. Soc. Mass Spectrom. 13*, 1313–1321.
46. Ganguly, A. K.; Chen, G.; Pramanik, B. N.; Daaro, I.; Luk, E.; Bartner, P. L.; Saksena, A. K.; Girijavallabhan. V. M. (2003), Negative ion multiple-stage mass spectrometric analysis of complex oligosaccharides (everninomicins) in a quadrupole ion trap: implications for charge-remote fragmentation, *Arch. Org. Chem. 2003*(iii), 31–44.
47. Shipkova, P. A.; Heimark, L.; Bartner, P. L.; Chen, G.; Pramanik, B. N.; Ganguly, A. K.; Cody, R. B.; Kusai, A. (2000), High-resolution LC/MS for analysis of minor components in complex mixtures: negative ion ESI for identification of impurities and degradation products of a novel oligosaccharide antibiotic, *J. Mass Spectrom. 35*, 1252–1258.
48. Nomeir, A. A.; Pramanik, B. N.; Heimark, L.; Bennett, F.; Veals, J.; Bartner, P.; Hilbert, M.; Saksena, A.; McNamara, P.; Girijavallabhan, V.; Ganguly, A. K.; Lovey, R.; Pike, R.; et al. (2008), Posaconazole (Noxafil, SCH 56592), a new azole antifungal drug, was a discovery based on the isolation and mass spectral characterization of a circulating metabolite of an earlier lead (SCH 51048), *J. Mass Spectrom. 43*, 509–517.

49. Perfect, J. R.; Cox, G. M.; Dodge, R. K.; Schell, W. A. (1996), In vitro and in vivo efficacies of the azole SCH56592 against Cryptococcus neoformans, *Antimicrob. Agents Chemother. 40*, 1910–1913.
50. Galgiani, J. N.; Lewis, M. L. (1997), In vitro studies of activities of the antifungal triazoles SCH56592 and itraconazole against Candida albicans, Cryptococcus neoformans, and other pathogenic yeasts, *Antimicrob. Agents Chemother. 41*, 180–183.
51. Graybill, J. R.; Bocanegra, R.; Najvar, L. K.; Luther, M. F.; Loebenberg, D. (1998), SCH56592 treatment of murine invasive aspergillosis, *J. Antimicrob. Chemother. 42*, 539–542.
52. Uchida, K.; Yokota, N.; Yamaguchi, H. (2001), In vitro antifungal activity of posaconazole against various pathogenic fungi, *Int. J. Antimicrob. Agents 18*, 167–172.
53. Frampton, J. E.; Scott, L. J. (2008), Posaconazole: A review of its use in the prophylaxis of invasive fungal infections, *Drugs 68*, 993–1016.
54. Feng, W.; Liu, H.; Chen, G.; Malchow, R.; Bennett, F.; Lin, E.; Pramanik, B.; Chan, T. M. (2001), Structural characterization of the oxidative degradation products of an antifungal agent SCH 56592 by LC-NMR and LC-MS, *J. Pharm. Biomed. Anal. 25*, 545–557.
55. Corcoran, O.; Spraul, M. (2003), LC-NMR-MS in drug discovery, *Drug Discov. Today 8*, 624–631.
56. Exarchou, V.; Krucker, M.; van Beek, T. A.; Vervoort, J.; Gerothanassis, I. P.; Albert, K. (2005), LC-NMR coupling technology: Recent advancements and applications in natural products analysis, *Magn. Reson. Chem. 43*, 681–687.

CHAPTER 6

Isotope Patten Recognition on Molecular Formula Determination for Structural Identification of Impurities

MING GU

Cerno Bioscience, 14 Commerce Drive, Danbury, CT 06810

6.1 INTRODUCTION

Pharmaceutical impurities are unwanted components that may come from the processes of organic synthesis, formulation, and storage of pharmaceuticals. Identification of these impurities is of great concern for both pharmaceutical industries and regulatory agencies because even small amounts of impurities in the pharmaceuticals can significantly compromise the efficacy of drug products or cause adverse drug reactions. To identify the molecular structures of these impurities, it is critical to determine their elemental compositions. For drug-related impurities such as those found from organic synthesis or long-term stability studies of drug products, they are quite predictable, resulting from common modifications such as oxidation or reduction. The determination of their elemental compositions often serves as independent confirmation for their identities. On the other hand, many impurities may have nothing to do with drug molecules and are considered as the true unknowns. They are often detected in the processes such as formulation or storage as so-called extractables and leachables. Identification of this type of true unknown impurity can be as difficult as finding a needle in a haystack. To obtain their empirical formula is the crucial first step to further elucidate their molecular structures.

The conventional mass spectrometric (MS) approach to obtaining a formula of unknowns is to perform accurate mass measurements for the unknown ions and calculate possible formulas based on a set of parameters, including possible chemical elements and their lower and upper limits, mass tolerance, the electron states

Characterization of Impurities and Degradants Using Mass Spectrometry, First Edition.
Edited by Birendra N. Pramanik, Mike S. Lee, and Guodong Chen.

(odd-electron ions or even-electron ions), the number of charges, and the range of double-bond equivalents (DBE). The candidate formulas are typically ranked according to mass accuracy. Whether correct unknown molecules are positively identified essentially depends on the search parameter of mass tolerance, which is ultimately dictated by accurate mass measurement performance of the instruments used. With the key focus on high mass accuracy, this approach greatly benefits from the state of the art of high-resolution mass spectrometers such as Q-TOF, Orbitrap, and FT-ICR, which routinely deliver the high mass accuracy of 5 ppm or better. As a commonly accepted and practically useful benchmark for accurate mass measurements, the mass accuracy at 5 ppm is sufficient to unique identify the formula for unknowns with molecular weights of <150 Da [1,2]. However, as the molecular weight of the unknown increases, the number of theoretically calculated formulas grows exponentially. For example, formula search with the most common elements of C, H, N, O, S, and P and the mass accuracy level of 5 ppm will result in as many 257 formulas for an unknown at *m/z* of 600. Even at the high mass accuracy of 0.1 ppm, there are still five possible formulas. This example clearly shows that the elemental composition determination for an unknown exclusively relying on mass accuracy could be economically prohibitive or technically infeasible.

Evidently, alternative methods taking advantage of mass spectral isotope patterns for formula identification have been developed [3–6] and are becoming increasingly popular. On the same topic in two separate articles, Kind and Fiehn made it clear that mass accuracy of 1 ppm is not sufficient to determine elemental composition of unknown metabolites in metabonomics applications. They also identified isotope pattern recognition as one of seven golden rules for filtering molecular formulas by mass spectrometry that can effectively remove more than 95% false positives from a given formula search. Mass spectrometer manufacturers continue to add the feature of isotope pattern recognition [7] in, or to promote its power with, their data systems [8]. Independent software developers [9,10] are making innovative products available for end users. The objectives of this chapter are to review the latest developments in commercial software designed to facilitate formula identification using isotope pattern recognition and to demonstrate its powerful filtering effects to drastically remove false positives through many examples, including identification of impurities in pharmaceuticals.

6.2 THREE BASIC APPROACHES TO ISOTOPE PATTERN RECOGNITION

It is well known that the isotope pattern of molecules is a fingerprint of the molecules. Formula identification employing isotope pattern recognition is simple quantitative measurements of the similarities between experimentally measured isotope patterns and theoretically calculated ones. This method has been used to identify unknown compounds for a long time. For example, $^{13}C/^{12}C$ ratios are used for determining the number of carbons in small organic molecules, and the $(M + 2)/M$

peak ratio of 0.33 characterizes a compound containing a single element of chlorine [11]. Instead of calculating isotope ratios using pencil, paper, and ruler (straightedge) as in the past, modern technologies not only effectively automate the process but also provide highly accurate quantitative results through sophisticated mathematics and statistics. Erve et al. [12] conducted an extensive review on many published methods for utilizing isotope information in the context of elemental composition determination. Based on methodology by which isotopic information is processed, they classified the methods into three categories; described in sections 6.2.1–6.2.3.

6.2.1 With Centriod Data

Although original or raw mass spectral data are always acquired in profile or continuum mode by any mass spectrometers, the profile data are often reduced either at a hardware level or by postacquisition data processing to centriod spectra largely for historical reasons of limited computer space. The centriod spectra typically have a few discrete integer numbers representing the intensities of monoisotope peak (M), M + 1, M + 2, and M + 3 peaks for a given ion. Isotope pattern recognition based on the centriod spectra obviously is straightforward and convenient to implement.

This class of methods includes early work by Grange and coworkers [13,14]. In a double-focusing high-resolution sector instrument, they developed data acquisition procedures with optimized scan speed, sensitivity, and resolution, termed *mass peak profiling from selected ion recording data* (MPPSIRD), to achieve high mass accuracy and high-resolution spectra for monoisotope (M), M + 1, and M + 2 peaks. These peaks were centered to compare with calculated exact masses and intensities of possible formulas generated by in-house-developed software, a profile generation model (PGM). Their approach was successfully demonstrated by positive determination of the elemental composition of an environmental contaminant that was not found through library matches. Amirav [15] described a patented method with an approach similar to that using isotope abundance analysis (IAA) mostly for GC-MS analysis either with supersonic molecular beam (SMB) or electron ionization [16] for compound identification or confirmation. With mass accuracy of ~0.1–0.5 Da from a typical unit mass resolution MS, he achieved a reasonably good result of hit 5 of formula identification of (i.e., identified five formulas for) dimethoate by IAA. The performance of this method in the determination of elemental composition (DEC) is less than perfect and was attributed to the high mass tolerance leading to an excessive number of possible candidate formulas. Another patented approach was reported by Zweigenbaum [17], who utilized isotope abundance and mass defects of the prominent isotope peaks. Instead of employing the accurate mass of only monoisotope peak, this method essentially uses accurate masses of all the isotope peaks and their abundance for DEC. In the case of a dimer of nonylphenol with elemental composition of $C_{30}H_{45}O_2$ in which the M + 1 peak is dominated by a ^{13}C peak, he calculated both the intensities and accurate mass derived from mass defect of ^{13}C and obtained a great match between theoretical and measured spectra. This method depends on highly accurate centroiding processing to obtain accurate mass measurements for M+1, M+2, and so on. Unlike the monoisotope peak (M) as a singlet, M+1, M+2, and M+3 peaks consisted of multiple isotopic components. For an example of molecules containing elements C, H, N, O, and S, the major contribution to M+1 peak comes from ^{15}N and ^{13}C, while the M+2 peak is dominated by overlapped ^{34}S and $^{13}C^{13}C$ peaks. As more elements are included in molecules and molecular weight increases, the overlap becomes much more complex. It will be very challenging to obtain accurate mass measurements for these isotope peaks.

Commercial software products in this class include i-FIT through MassLynx software (Waters, Billerica, MA) and a more recent addition to this type of methodology, the FuzzyFit (Kisotopic Solutions, Manchester, UK), described by Hobby and coworkers [10] (see more detailed discription in next paragraph below). Although a detailed algorithm of i-FIT was not reported, a brief methodological mention was made by Mather and coworkers [18]. They described that i-FIT is a measure of the likelihood of a collection of peaks in the spectrum matching a theoretical isotope model. Calculated as the loglikelihood based on an χ^2 distribution, the i-FIT scores zero for a perfect isotope match. In the i-FIT analysis of two groups of compounds of pharmaceuticals and pesticides, Hobby [19] showed that i-FIT significantly reduced possible false-positive hits, and ECD performance by i-FIT was dependent on possible elements in the molecules under investigation. For the pharmaceutical molecules, 9 of 12 compounds were ranked as the top three by i-FIT with search elements including C, H, N, O, S, Cl, and Br, while only 4 out of 10 pesticides made it to the top three when additional elements of P, F, or Na were included in the i-FIT search. As he pointed out, phosphorus and fluorine do not have isotopes, and their presence in a molecule will not significantly alter the observed isotope distribution. The addition of phosphorus or fluorine to a calculation of elemental compositions will result in a significantly higher number of proposed formulas. He estimated that there is an approximate one order of magnitude increase in the number of elemental compositions as a result of inclusion of phosphorus and fluorine.

Spectral Simplicity is another software package offered by Kistope Solutions [10]. This software design was based on a spectral correlation algorithm called "FuzzyFit", which is a probability-based match between experimentally measured spectra and the proposed formulas at a given level of statistical significance. The authors described the algorithm as a dynamic self-optimizing methodology that requires mass spectral calibration of a set of known standards under experimental conditions similar to those of the unknowns in the analysis. In addition to the calculation of accurate mass for all the isotopic peaks, which is similar to Zweigenbaum's approach, this method utilizes heuristic atom ratio filtering to simplify the list of possible formulas. Since the addition of sulfur or phosphorus requires two oxygen atoms and the addition of nitrogen is always associated with the addition of a carbon atom, any molecular formula with the ratios of $C:N < 1.5$, $O:S < 2$, and $O:P < 2$ are excluded from candidate formulas in this method. Shown in an application of unknown metabolite identification, the correct metabolite of $C_{15}H_{15}N_2O_8$ was found as the best fit only after the rules of heuristic atom ratios were applied to filter out three out of four likely compounds, $C_{19}H_{17}N_2OP_2$, $C_{12}H_7N_{12}O_2$, $C_{15}H_{15}N_2O_8$, and $C_{11}H_{23}N_4OP_4$, resulting from FuzzyFit analysis. As can be seen, the formula of both $C_{19}H_{17}N_2OP_2$ and $C_{11}H_{23}N_4OP_4$ has a O : P ratio of <2 and the formula $C_{12}H_7N_{12}O_2$ has a C : N ratio of <1.5.

Isotope pattern recognition with centriod spectra has proved to be more effective than the approaches by mass accuracy only for formula determination. However, centroid isotope data are only marginally useful because of the gross approximations of these overlapped fine structures in the mass spectral centroiding process, as pointed out by Wang [20] since the fine and ultrafine structures of heavier isotopes typically overlap with each other because of the very close proximity of their *m/z* values, even on FT-ICR-MS instruments with R~1,000,000 It is in these overlapped fine structures that the information most relevant to elemental composition is contained.

6.2.2 With Profile Data without Peak Shape Calibration

This approach employs profile mass spectral data without centroiding for DEC, assuming the measured spectra to have a peak shape such as a Gaussian distribution [13,21]. Realizing the rich information contained in the profile spectra, Grange [13] demonstrated that even a partial profile spectrum of M + 2 played a critical role in the identification of a compound. On the basis of the accurate masses of M, M + 1, and M + 2 and their intensities of centriod spectra, a total of 12 candidate formulas were filtered out to the final two formulas as $C_{24}H_{44}N_2PS$ and $C_{28}H_{41}NS$. The correct elemental composition of the unknown, $C_{28}H_{41}NS$, was identified by comparing the M + 2 portion of profile spectra from experimental data with theoretically calculated spectra as shown in Figure 6.1. A similar approach was independently proposed by Fernandez-de-Cossio et al. [21] for proteomics applications.

The SigmaFit™ approach available through Bruker Daltonics' software is the only commercial product in this class. The exact details of its algorithms are unknown because of their proprietary nature, although a reference to a patent in the context of precise mass positioning is available [22]. Bristow and coworkers [23] conducted an extensive evaluation of the performance of the SigmaFit approach. With a composition of various elements including C, H, N, O, S, F, P, Cl, and Br and a mass range of 218–587, 11 compounds were measured at various abundance levels and their SigmaFit scores were calculated. As shown in Figure 6.2, the best rankings of (>10) were achieved when the spectra were acquired at abundance levels from 1×10^4 to 1×10^5, which were optimized for the best mass accuracy, while the rankings deteriorated as lower or higher levels of signals caused either poor signal-to-noise ratio or signal saturation, respectively. They also found SigmaFit performance to be influenced by isotope complexity. The compounds consisting of elements C, H, N, O, and F with simple isotope patterns consistently ranked higher, between the top first and fourth, than did those containing additional Br and Cl, resulting in complex isotope patterns.

Although superior to the first class of methods with avoidance of centroiding, this approach also entails significant errors, for several reasons, including the fact that all

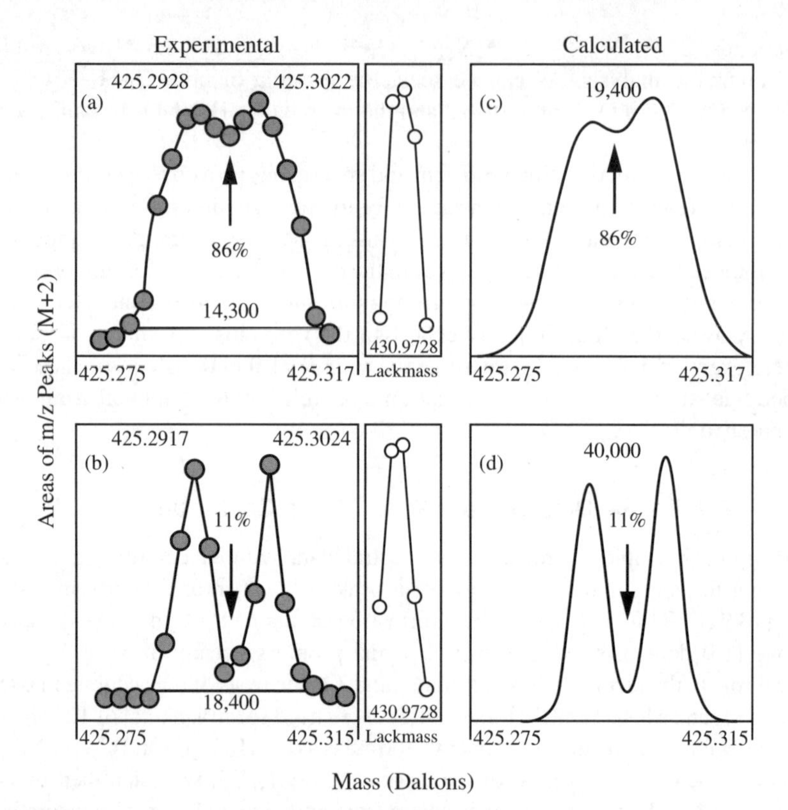

FIGURE 6.1 (a,b) Experimental M + 2 profiles obtained with 20,400 and 37,600 resolution; (c,d) calculated M + 2 profiles for $C_{28}H_{41}NS$. Resolutions 14,300 and 18,400 were apparent resolutions calculated from the plotted profile widths at 5% of the maxima; 19,400 and 40,000 were the resolutions entered into the model to provide calculated profiles. (Reprinted from Ref. [13], with permission from the American Chemical Society.)

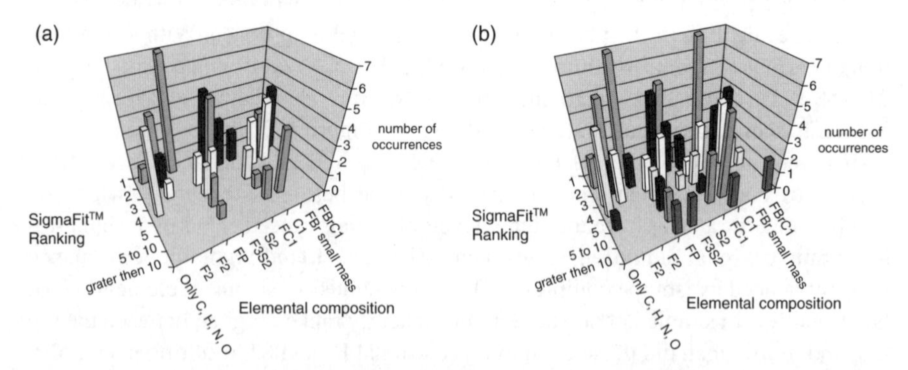

FIGURE 6.2 The relationship between SigmaFit ranking and elemental composition for (a) ion abundances in the range from 1×10^4 to 1×10^5 and (b) all ion abundances in the study (reprinted from Ref. [23], with permission from John Wiley & Sons, Inc.).

mass spectral profile mode data come with a given but not accurately known or defined mass spectral peak shape function. This function describes the statistical distribution of a population of ions of a given isotope along the *m/z* axis, and the full width at half-maximum (FWHM) of this distribution approximates instrument resolution. Since the theoretical isotope distribution is a discrete distribution representing mass spectral data measured on a mass spectrometer of both infinite resolving power and linear dynamic range, an assumed peak shape function is required to convert the theoretical isotope distribution into profile-mode mass spectral data before these data can be compared with experimentally observed mass spectral data. In this conversion process, the assumed peak shape function is superimposed onto a theoretical isotope abundance distribution for comparison. Grange observed that such an assumption could lead to as much as 2.5% error [13], which is too large to allow differentiation of closely related elemental compositions, thereby limiting the power of this methodology. It should be noted here that although some mass spectral systems are known to possess certain types of peak shape functions, these functions, as measured from an actual system, typically vary from one instrument to another or even from one tune or data acquisition mode to another, giving rise to a modeling error of a few (<10) percent and significantly hampering the ability to differentiate closely related elemental compositions, whose theoretical isotope profiles may differ by far less than this modeling error. For example, quadrupole and magnetic sector MS are typically considered as having Gaussian-type peak shape functions when the actually measured peak shape function may be modified Gaussians with unsymmetric peak shapes. Similarly, in the case of FT-ICR MS or Orbitrap, the peak shape may nominally be Lorentzian but specifically dependent on the processing method used, involving various apodization functions and zero-filling, all part of a Fourier transform experiment, for which a good reference is available on its impact on quantitative peak height and area measurement [24].

6.2.3 With Profile Data with Peak Shape Calibration

The uniqueness of this method is that the best possible match can be achieved between experimental and theoretical spectra for unknown identification because both measured and calculated spectra are described by exactly the same mathematical functions generated by the peak shape calibration, or lineshape calibration. With the mass spectral peak shape calibration [20,25], the peak shape can be accurately known and mathematically defined. When the most likely elemental composition needs to be determined from numerous possible formula candidates, the same mass spectral peak shape function is superimposed onto a discrete isotope abundance distribution to form a theoretical profile mass spectrum for each possible candidate formula. Since no assumption is made about the peak shape function used for both the calibrated and theoretical mass spectrum, the theoretical mass spectrum should exactly match the calibrated experimental mass spectrum, except for a difference in scaling between the two due to ion abundance, assuming (1) no systematic or random measurement error and (2) that a given candidate formula is indeed the correct formula.

6.3 THE IMPORTANCE OF LINESHAPE CALIBRATION

It has been recognized that obtaining a well-defined mass spectral peak shape is key to achieving accurate mass measurements, peak integration, and isotope pattern matching for unknown identification by both unit mass resolution [26,27] and high-resolution [13,14,28] mass spectrometers. An experimental approach to achieving symmetric peak shape was reported by carefully tuning instrumentation using standards developed before accurate mass measurements [13,14], while various mathematical algorithms or computing methods were developed for effectively calculating isotope distribution [29–33], accurately performing peak centroiding [28,34], and investigating peak shape changes at different resolution settings [35]. In 2008 Kuehl [36] described (in sections of 6.3–6.4) a novel lineshape calibration approach that simultaneously corrects both mass shift and peak shape deviations. Mass spectrometry is unique among instrumental techniques in that the physical property being measured, the mass-to-charge ratio (*m/z*), is a discrete value that can be calculated very accurately on the basis of well-characterized physical constants (the sum of the exact mass of its constituent elements). Since the mass value is discrete, the measured lineshape and position error are due solely to the inherent characteristics of the instrument. The measured lineshape is simply a convolution of the theoretical spectrum (a discrete distribution) with the instrument response function. The instrument response function is the result of imperfections in the source and detector geometries and electronics, deviation from ideal electromagnetic fields, and a variety of other factors. Some examples of typical lineshape distortions are shown in Figure 6.3 for several different MS analyzers. Figure 6.4 shows an example of a theoretical mass spectrum of an ion at the position of the M + 2 isotope peak. If a perfect instrument of infinite resolving power and infinite signal-to-noise

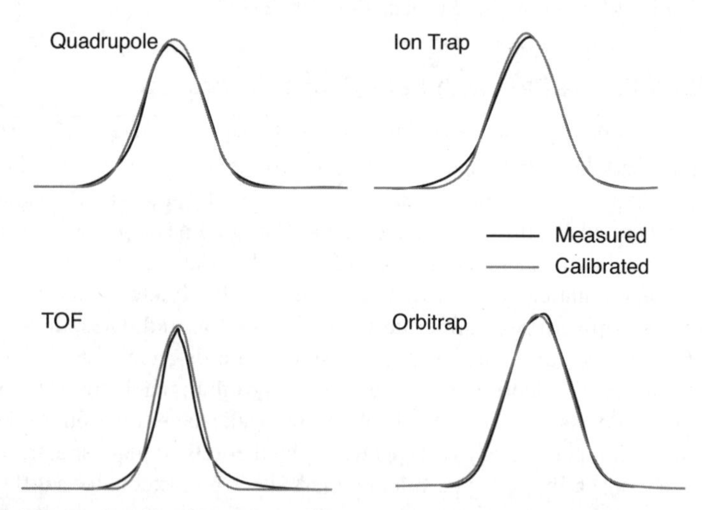

FIGURE 6.3 Typical lineshapes from different MS analyzers compared to mathematically defined symmetric functions. (reprinted from Ref. [36], with permission from International Scientific Communications, Inc.).

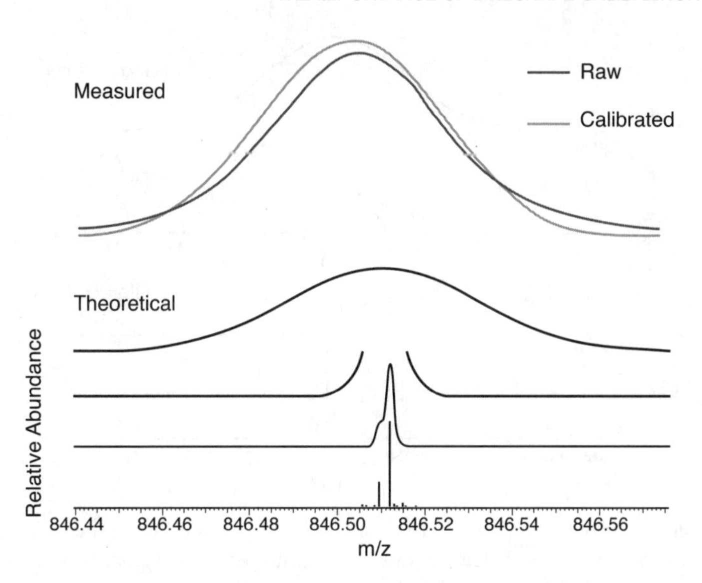

FIGURE 6.4 *Bottom tracings*: Theoretical spectrum for the M + 2 peak in the ion $C_{43}H_{74}NO_{15}$ at different resolutions. *Top tracings*: Raw and self-calibrated spectrum measured on a MicroTOF (Bruker Daltonics, Billerica, MA). While the unique isobar pattern for this ion is not apparent in the TOF data, the search results based on spectral accuracy using only the M + 2 peak reveal the correct formula for the compound. (Reprinted from Ref. [36], with permission from International Scientific Communications, Inc.).

ratio were used, the measured spectrum would match the theoretical spectrum exactly. This would allow unique identification of the ion, based on both its mass and isotope pattern. However, the measured spectrum is a convolution of the instrument lineshape with this spectrum. While it is not possible to create the perfect instrument, if one could correct the lineshape to a mathematically defined function, for example, a pure Gaussian form of known peak width, it would be possible to accurately match the measured spectrum to the theoretical spectrum by simply convolving it with the exact same mathematically defined function. Figure 6.5 illustrates the theoretical spectrum generated using a Gaussian function of different widths representative of the typical linewidths of commercially available systems. While the isobar pattern is not apparent to the naked eye in all except the highest-resolution measurements, accurate matching of the pattern should still be possible, with sufficient signal-to-noise ratios, even for the time-of-flight (TOF) spectrum shown in Figure 6.6.

6.3.1 Lineshape Calibration Using Standards

This calibration is similar to the conventional one in terms of the standards usage requirements. By correcting on both *m/z* values and peak shape, this calibration is most often applied to mass spectral data obtained from unit mass resolution instruments. High-resolution mass spectra usually are calibrated by lineshape self-calibration, discussed below for the purpose of formula identification, unless

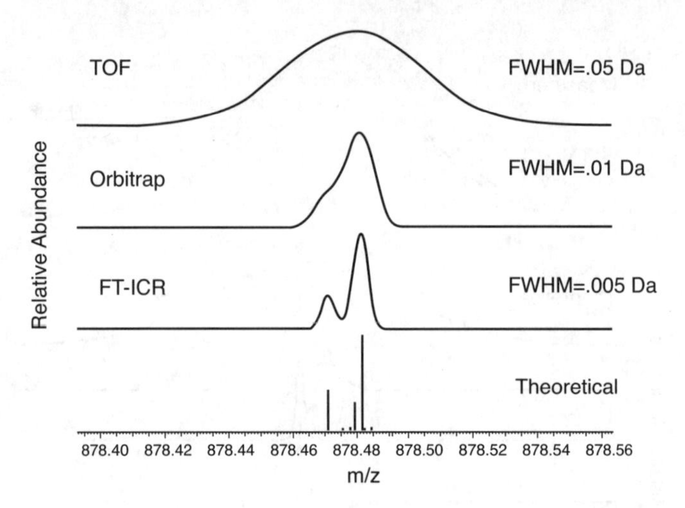

FIGURE 6.5 Theoretical mass spectrum of $C_{43}H_{74}NO_{15}S$ at different linewidths typical of high-resolution systems (reprinted from Ref. [36], with permission from International Scientific Communications, Inc.).

large mass errors from the spectra need to be calibrated. In unit mass resolution spectra, by definition, the monoisotope peak M and its M + 1 and M + 2 peaks overlap. To calibrate the spectra, the entire isotope pattern of the standard ions is used to generate the calibration function. This calibration function, also sometimes

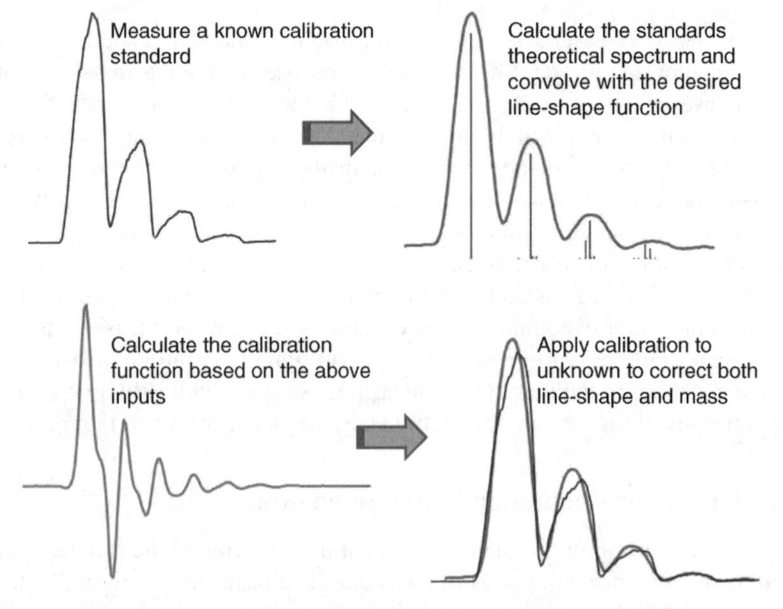

FIGURE 6.6 Flow diagram of calibration process that uses a known standard ion; note that both lineshape and mass axis are calibrated (reprinted from Ref. [36], with permission from International Scientific Communications, Inc.).

called *a lineshape filter*, is then applied to unknown spectra as illustrated in Figure 6.6. As a result, the calibrated spectra not only have symmetric peak shape, but more importantly, the peak shape is mathematically defined so that the overlapped monoisotope and its ^{13}C satellite peaks can be mathematically separated to drastically improve the mass accuracy of unit mass resolution mass spectra to ~5 mDa level [37–43]. Furthermore, the calibrated spectra and theoretically calculated spectra obey the same line-shape function allowing highly accurate isotope pattern matching for unknown identification.

6.3.2 Lineshape Self-Calibration

The basic requirement for this calibration is that the resolution of mass spectra be high enough for resolution of the monoisotope peak and its satellite ^{13}C peak. With self-calibration, the isotopically pure monoisotopic peak of any ion is used as the model lineshape for calibration, which is then applied to the entire ion isotope pattern; hence the term *self calibration* [44]. The lineshape self-calibration process is illustrated in Figure 6.7. The advantage of this approach is that the lineshape calibration is nearly perfect. First, the calibration peak is very close in mass to the ion of interest, because it is contained within the ion of interest! This minimizes any differences in peak shape as a function of mass. Also, since both the calibrant ions (M) and the sample ions (M+1, M+2, M+3) were measured almost simultaneously, any effects due to

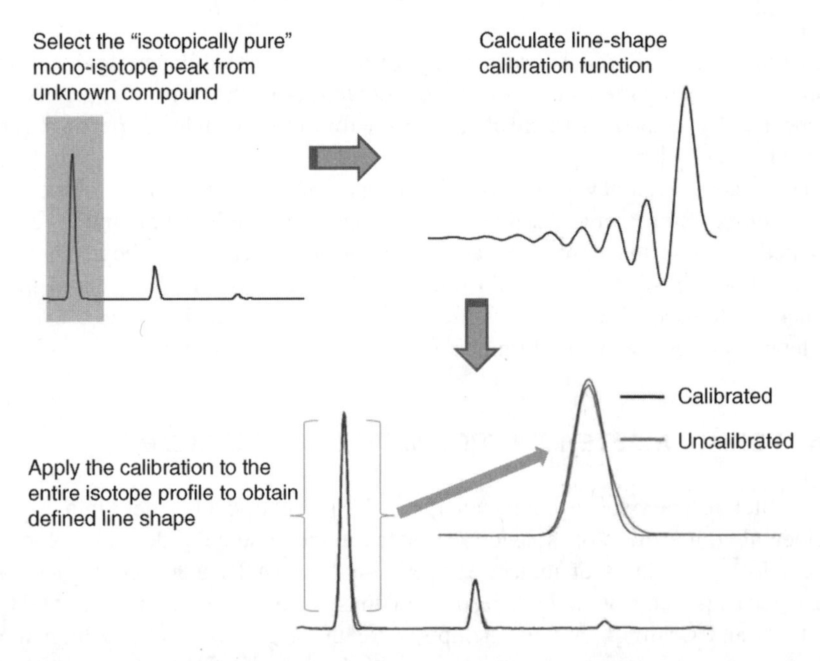

FIGURE 6.7 Flow diagram of the lineshape self-calibration process; note that only the lineshape is calibrated (reprinted from Ref. [36], with permission from International Scientific Communications, Inc.).

instrument drift as a function of time were minimized. The other advantage of this calibration is that the highly accurate isotope pattern matching with an additional metric, spectral accuracy, dramatically improved the ability to identify unknown formulas so that the requirement for high mass accuracy in formula determination can be relaxed. Since accurate mass calibration can be a tedious task on some instruments, this additional metric can make it easier to attain high mass accuracy in more time-consuming and complex calibrations.

6.4 SPECTRAL ACCURACY

Formula determination without isotope pattern matching usually uses mass accuracy to rank possible formulas. This ranking does not provide meaningful information for determining which formula is most likely the correct one, because within a given mass tolerance (e.g., 5 ppm), all calculated formulas have similar probabilities of being correct. On the other hand, ranking based on spectral accuracy reveals the intrinsic connection between measured spectra and calculated formulas. This spectral accuracy (SA) can be readily calculated as follows to describe the congruence between the calibrated and theoretical isotope profile data:

$$\mathrm{SA} = \left(1 - \frac{\|\mathbf{e}\|_2}{\|\mathbf{r}\|_2}\right) \times 100$$

Here **e** is the fitting residual or spectral error vector, **r** is the calibrated isotope profile vector, and $\|\cdot\|_2$ represents the 2-norm (or square root of the sums of squares of all elements) of a vector. As an equivalent measure for isotope matching, spectral error is defined as $\|\mathbf{e}\|_2/\|\mathbf{r}\|_2$.

This spectral accuracy metric is used to evaluate all possible formulas whose exact monoisotope masses come within a user-defined mass tolerance window of the reported accurate mass from the actual monoisotope peak and whose elemental compositions satisfy the user-defined chemical constraints. The formulas with the higher spectral accuracy or the lower spectral error are ranked higher and considered as the more likely candidate formulas for the unknown ion of interest.

6.5 FORMULA DETERMINATION WITH QUADRUPOLE MS

In qualitative mass spectrometry analysis, the ultimate goal is determination of elemental composition of molecular ions or fragments to help identify unknowns or elucidate structures of molecules. This goal is usually achieved by accurate mass measurements with high-resolution mass spectrometers such as Q-TOF, FT-ICR, and Orbitrap. A few attempts were made earlier to obtain high mass accuracy on unit mass resolution GC-MS [45] and LC-MS [27], requiring chemical derivatization for internal calibration and a much high sampling rate at 128 data points per atomic mass unit (amu), respectively. With the innovative mass spectrometry calibration technology discussed above, achieving high mass accuracy and DEC with a unit mass resolution quadrupole mass spectrometer has proved to be feasible [20,37–43].

6.5.1 Impurity Identification with LC-MS

In a recent report [46], the author described the application of this calibration technology to identify impurities of simvastatin with a single-quadrupole LC/MS system. Simvastatin is a hypolipidemic drug belonging to the class of pharmaceuticals called *statins*. It is used to control hypercholesterolemia (elevated cholesterol levels) and to prevent cardiovascular disease. The simvastain obtained from a commercial source and its impurities generated according to USP (US Pharmacopoeia) procedures were separated by ultra-high-pressure liquid chromatography (UPLC) (Waters, Milford, MA) and detected by a Waters Acquity SQD single-quadrupole mass spectrometer (Waters, Milford, MA). All mass spectral data were acquired in profile mode with a scan rate of 1000 amu/s and a mass range from 380 to 520. The mass spectral data files in MassLynx format were read directly by MassWorks software (Cerno Bioscience, Danbury, CT) for accurate mass calibration and formula identification.

Separated by UPLC in a 9-min chromatographic analysis, the parent drug simvastatin appears at a retention time of 4.4 min as the most abundant peak, while the minor peaks appearing up before or after simvastatin are the impurities related to simvastatin (Figure 6.8). As shown in Figure 6.9, simvastatin was ionized in three formations as the protonated $[M + H]^+$, ammonium adducts $[M + NH_4]^+$, and potassium adduct $[M + K]^+$ ions, observed at *m/z* 419, 436, and 457, respectively. Since the elemental composition for the three ions is known, these ions can be used as standards to perform an internal calibration for the best mass accuracy and spectral accuracy.

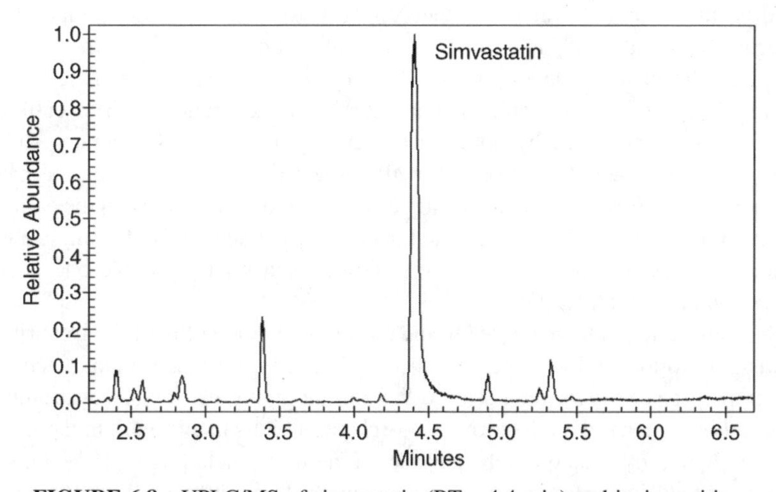

FIGURE 6.8 UPLC/MS of simvastatin (RT ~ 4.4 min) and its impurities.

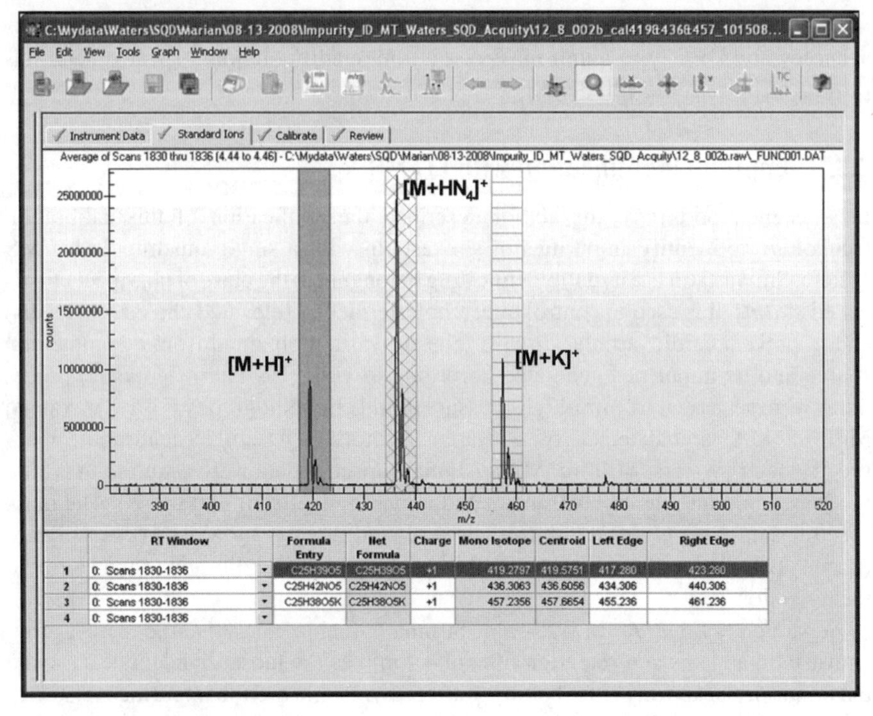

FIGURE 6.9 The calibration ions from parent drug simvastatin.

The MassWorks calibration requires the input of the molecular formula of the calibration standards instead of *m/z* values as in the classical mass-only calibration. For example, in Figure 6.9 the entire isotope profile of the calibration standard simvastatin is highlighted in gray and selected for mass and peak shape calibration, based on the exact *m/z* value and theoretical isotope distribution provided by the molecular formula of $C_{25}H_{39}O_5$. After the calibration, not only is the *m/z* value corrected; the peak shape (Figure 6.10, calibrated spectra shown in red) is also calibrated to a symmetric and mathematically defined function. The calibration performance is measured by both mass accuracy and spectral accuracy. Relative mass accuracy is less than 5 ppm for all three calibration standards. A spectral accuracy $= \geq 98\%$ is also achieved, suggesting a good match between the calibrated spectra and the theoretically calculated spectra of the standards. The high spectral accuracy attained from this calibration allows for a highly reliable and accurate isotope pattern search for DEC.

This calibration is then applied to the entire LC-MS data file of the impurities of simvastatin followed by averaging the calibrated spectra across of each chromatographic peak for formula search. Without applying particular constraints, the author has summarized all the formula search parameters in this work in Figure 6.11. The search employed a comprehensive set of elements including C, H, N, O, K, Cl, and S with their lower limits set to zero and upper limits set very high. Prior results

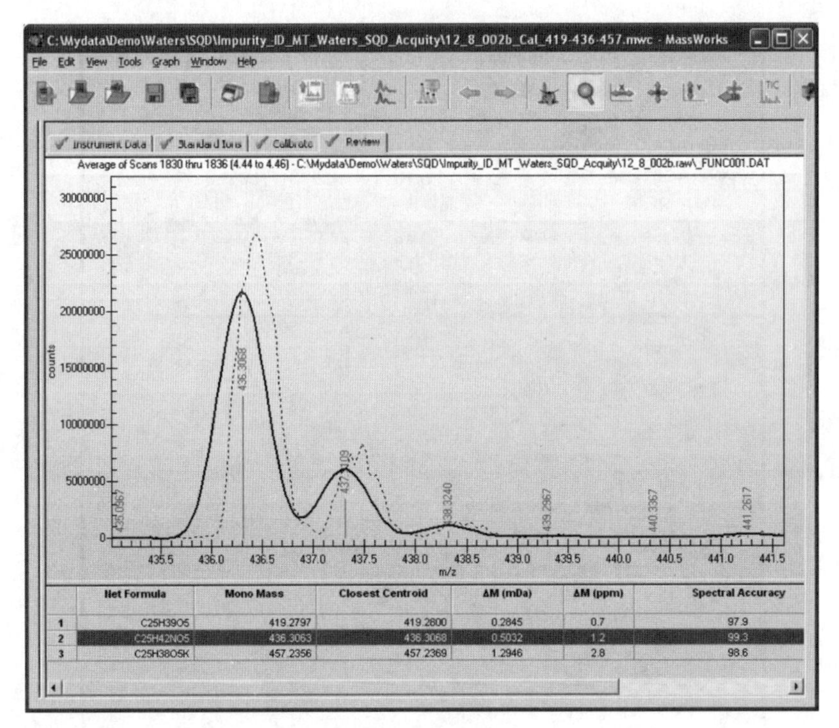

FIGURE 6.10 The calibration report summary and overlay of the uncalibrated raw data (black) and the accurate mass and peak shape calibrated data (red). Note the substantial improvement in signal-to-noise ratio as well.

showed the instrument's mass accuracy to be better than 10 mDa using the MassWorks calibration, and as such a value of 10 mDa for mass tolerance was used for the formula search. Since the mass spectra of most impurities are similar to the parent drug's spectra, having three different ion formations such as $[M + H]^+$, $[M + NH_4]^+$, and $[M + K]^+$, only the most abundant ions of either $[M + NH_4]^+$ or $[M + H]^+$ were selected for formula search.

As summarized in Table 6.1, the results show that most of impurities were identified as number 1 or 2 hit and have spectral accuracy better than 97.2% except for the ions at *m/z* 422, which eluted at a retention time of 4.0 min, due to poor signal-to-noise ratio (SNR). It is evident that using spectral accuracy can effectively distinguish correct molecule formula from among many possible candidates.

The results from this LC-MS analysis clearly demonstrate that formula identification by the combination of mass accuracy and spectral accuracy is much more effective than by mass accuracy alone. For example, simvastatin acid, one of the impurities of simvastatin, was identified as the top hit from 49 possible candidates with mass accuracy of 4 ppm and spectral accuracy of 99.4% [Table 6.2, upper (spectral accuracy) portion]. This high spectral accuracy of 99.4% allowed calibrated spectrum of simvastatin acid to be superimposed almost perfectly with the theoretically calculated spectrum, leading to confident identification of the compound by the

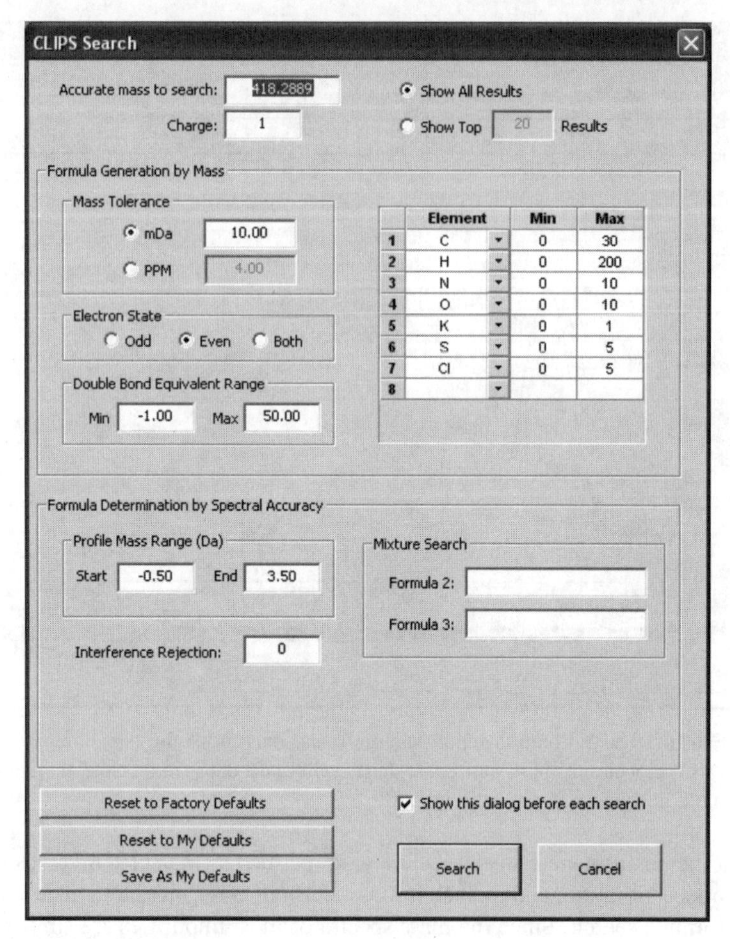

FIGURE 6.11 The formula search parameters used for all impurity searches.

TABLE 6.1 Summary for DEC for Simvastatin Impurities

Formula	Monoisotope	Mass Error (mDa)	Mass Error (ppm)	Spectral Accuracy	R_t	Rank	Total Formula Searched
$C_{25}H_{44}NO_5$	438.3219	4.4485	10.1491	98.4	5	1	34
$C_{25}H_{40}NO_4$	418.2957	5.7338	13.7077	99.1	5.3	1	28
$C_{25}H_{41}O_6$	437.2903	1.814	4.1484	99.4	3	1	49
$C_{24}H_{40}NO_5$	422.2906	8.3484	19.7697	94.5	4.1	4	34
$C_{24}H_{40}NO_5$	422.2906	7.1484	16.928	98.1	4	3	32
$C_{26}H_{44}NO_5$	450.3219	−0.3515	−0.7805	97.3	5.3	1	33
$C_{25}H_{40}NO_5$	434.2906	−1.3516	−3.1122	97.2	4.2	2	36

TABLE 6.2 Top 10 Formulas from Search for *m/z* 437

Row	Formula	Monoisotope	Mass Error (mDa)	Mass Error (ppm)	Spectral Accuracy	RMSE[a]	DBE[b]
			Ranked by Spectral Accuracy				
1	$C_{25}H_{41}O_6$	437.2903	1.814	4.1484	99.4	35,330	5.5
2	$C_{22}H_{33}N_{10}$	437.289	0.4661	1.0659	98.8	71,877	11.5
3	$C_{21}H_{37}N_6O_4$	437.2876	−0.8713	−1.9925	98.0	115,843	6.5
4	$C_{26}H_{37}N_4O_2$	437.2917	3.1515	7.2068	97.8	132,404	10.5
5	$C_{27}H_{37}N_2O_3$	437.2804	−8.0819	−18.4819	97.5	146,570	10.5
6	$C_{23}H_{41}N_4O_2S$	437.295	6.5223	14.9153	97.1	174,600	5.5
7	$C_{24}H_{41}N_2O_3S$	437.2838	−4.7111	−10.7735	96.9	183,226	5.5
8	$C_{20}H_{37}N_8OS$	437.2811	−7.3965	−16.9144	96.7	197,386	6.5
9	$C_{19}H_{37}N_{10}S$	437.2923	3.8369	8.7744	96.6	200,644	6.5
10	$C_{20}H_{41}N_2O_8$	437.2863	−2.2087	−5.0509	96.1	231,489	1.5
			Ranked by Mass Accuracy				
1	$C_{22}H_{33}N_{10}$	437.289	0.4661	1.0659	98.8	83,184	11.5
2	$C_{29}H_{41}OS$	437.2878	−0.6584	−1.5742	95.7	295,740	9.5
3	$C_{21}H_{37}N_6O_4$	437.2876	−0.8713	−1.9925	98.2	124,081	6.5
4	$C_{21}H_{42}N_4O_3K$	437.2894	0.8981	2.0539	94.0	415,173	2.5
5	$C_{20}H_{42}N_4O_4CI$	437.2895	0.9586	2.1921	73.0	1,868,435	1.5
6	$C_{21}H_{45}N_2O_3S_2$	437.2872	−1.3403	−3.065	93.0	437,894	0.5
7	$C_{16}H_{38}N_{10}O_2CI$	437.2868	−1.7268	−3.9488	72.8	1,881,594	2.5
8	$C_{17}H_{38}N_{10}OK$	437.2867	−1.7872	−4.087	93.1	476,649	3.5
9	$C_{25}H_{41}O_6$	437.2903	1.814	4.1484	99.5	37,173	5.5
10	$C_{20}H_{41}N_2O_8$	437.2863	−2.2087	−5.0509	96.4	252,309	1.5

[a]Root mean squire error.
[b]DBE.

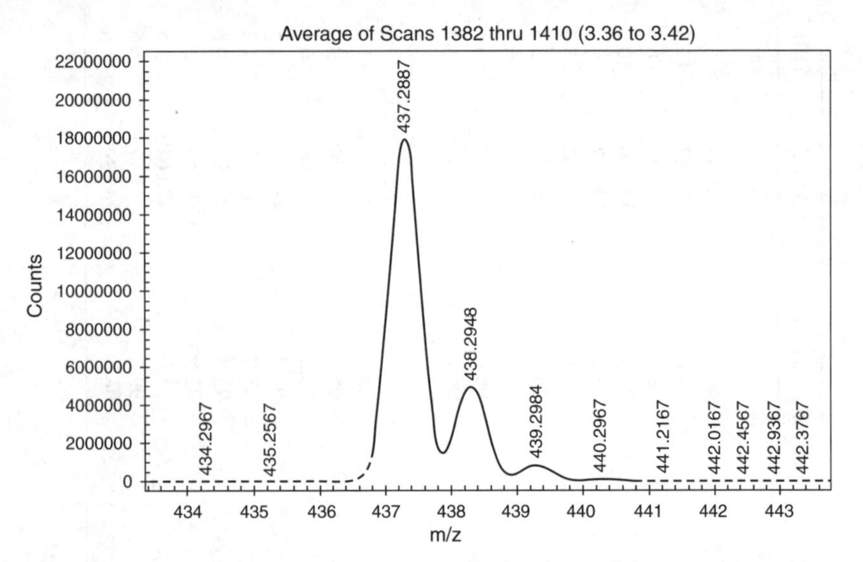

FIGURE 6.12 Overlay of calibrated spectra (red) and theoretically calculated spectra (green) of the impurity simvastatin acid.

great match (Figure 6.12). On the other hand, with the same search parameters except for mass tolerance selected within 5 ppm, simvastatin acid ranked only 9 on the basis of mass accuracy (Table 6.2, lower portion).

6.5.2 Impurity Identification with GC-MS

For accurate mass measurements and formula identification of small volatile organic compounds (VOCs) or semivolatile organic compounds (SVOC), high-resolution magnetic sector field [45] and GC-time-of-flight (GC-TOF) [47] instruments have been used, but these might not be the most user-friendly forms of MS, or be able to routinely deliver satisfactory performance, as the sector instrument often suffers from lack of sensitivity and GC-TOF requires the use of lock mass as internal calibration for high mass accuracy.

As an alternative approach enabled by the lineshape calibration, DEC with quadrupole-based GC-MS is becoming popular. Numerous groups [37,39,40,43,48] have performed formula identification with GC-MS systems, including the analysis of impurities observed in the pharmaceutical raw material of 4-fluorophenylethyl alcohol as reported by Mick and Gillespie [39]. Using an Agilent GC-MS system, they (Mick and Gillespie) acquired both the calibration standard of perfluoro-tri-*n*-butylamine (PFTBA) and the raw-material sample in profile mode (raw scan mode) with threshold set to zero and a sampling rate of $\sim$10 points per amu as opposed to the centriod mode (scan mode) traditionally used by most GC-MS investigators. The lineshape calibration in this system can be obtained by acquiring a separate data file of PFTBA for external calibration or acquiring PFTBA before or after GC separation in the same data file of the sample as pseudointernal calibration since the PFTBA does

TABLE 6.3 Overview of Impurities Observed in the Software-Processed Pharmaceutical Raw Material (4-Fluorophenylethyl Alcohol)

Impurities	Ranking of Correct Formula	Formula	Theoretical	Measure	Mass Error (mDa)	Mass Error (ppm)
Fluorobenzyl alcohol	2nd of 24	C_7H_7OF	126.0481	126.0428	−5.3	−42.0
Fluorobiphenyl	2nd of 45	$C_{12}H_9F$	172.0688	172.0774	8.6	49.8
Difluorobiphenyl	2nd of 57	$C_{12}H_8F_2$	190.0594	190.0624	3	15.7
Fragment of impurity *m/z* 184[a]	1st of 22	C_8H_7F	122.0532	122.056	2.8	23.1
Fragment of impurity *m/z* 184[a]	1st of 16	C_7H_6F	109.0454	109.0521	6.7	61.9
Fragment of impurity *m/z* 184[a]	1st of 16	C_8H_7	103.0548	103.0551	0.3	3.2
Impurity *m/z* 156[b]	1st of 38	$C_8H_9O_2F$	156.0587	156.0678	9.1	58.6
Impurity *m/z* 166[c]	1st of 40	$C_{10}H_{11}OF$	165.0794	166.0343	4.9	29.5
Fragment of impurity *m/z*166[c]	1st of 42	$c_{10}h_{10}of$	165.0716	165.0814	9.8	59.6

[a]Fragment ions were observed for this impurity rather than the molecular ion *m/z* 184. The compound was synthesized to confirm its structure and formula.
[b]Believed to be the correct formula based on known chemistry of the original material.
[c]Believed to be the correct formula based on known chemistry of the original material.

not actually appear in the spectra of the sample. In analysis of the impurities of 4-fluorophenylethyl alcohol, although an external mass spectral calibration was built by the fragment ions from PFTBA, a mass accuracy of >10 mDa was achieved for 4-fluorophenylethyl alcohol and related impurities as listed in Table 6.3. All correct formulas were determined and ranked according to spectral accuracy as either a 1 or 2 hit. This high-ranking performance is comparable to what can be obtained from a high-resolution GCT instrument.

6.5.3 Pros and Cons of Determination of Elemental Decomposition (DEC) with Quadrupole MS

The unit mass resolution quadrupole based instrument as one of the most mature mass spectrometers that has been traditionally used for the simple analysis of molecular weight (MW) confirmation at nominal mass accuracy. For DEC applications, they are still in an early developing stage. In addition to the most obvious advantages of being user-friendly and cost-effective, quadrupole-based MS systems do have technological advantages over high-resolution mass spectrometers for DEC, at least for some applications. First, they have better dynamic range compared with high-resolution instruments such as Q-TOF, Orbitrap, and FT-MS, and therefore can readily generate accurate isotope distribution that carries real fingerprint information of molecules to ensure successful matching for formula identification. On the other hand, distorted isotope patterns are commonly observed in TOF-type instruments due to signal

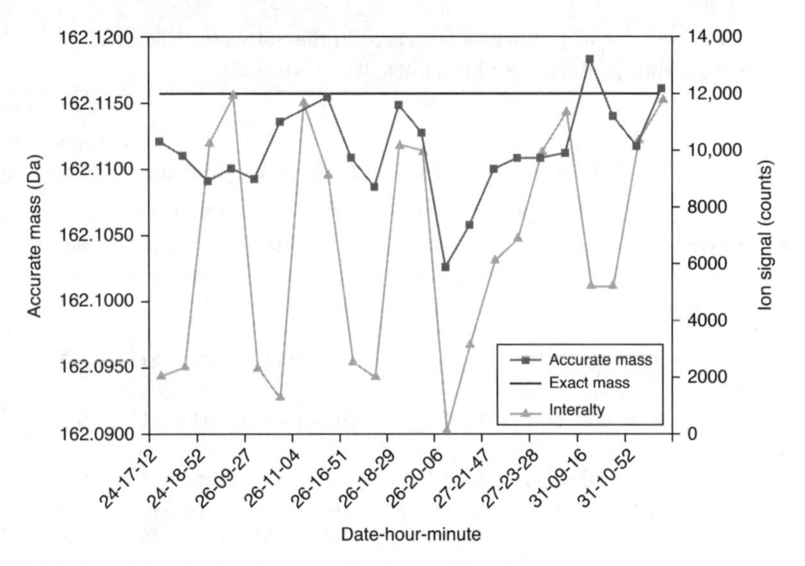

FIGURE 6.13 Reported accurate mass, monoisotopic peak ion count, and known exact mass for the ion at *m/z* 162.

saturation [23,49] and in Orbitrap and FT-ICR because of isotopic beat patterns [12,50,51]. Moreover, quadrupole MS systems usually have great instrument stability and can maintain a minimum mass shift for a long time. As shown in a week-long test, the same PFTBA calibration was applied to 22 GC-MS runs for the accurate mass measurements of nicotine ($C_{10}H_{14}N_2$) at *m/z* 162. For ions with intensities above 8000 counts, all reported accurate masses of nicotine (Figure 6.13) come within a few millidaltons of the exact mass. This result was confirmed with separated experiments that showed an overall mass shift ~7 mDa for all the calibration ions of between two PFTBA runs measured one week apart. This suggested the PFTBA calibration was indeed applicable to all 22 GC/MS runs acquired afterward for accurate mass measurements and DEC.

However, unit mass resolution MS systems lack the resolving power to separate ions of interest from complex background interference, in particular for biological samples from applications such as proteomics and in vivo drug metabolism. In these applications, all high-resolution instruments definitely hold indispensable advantages over unit mass resolution quadrupole systems for accurate mass measurements and DEC. In cases where the samples are very clean or have relatively low background interference, the quadrupole MS can be utilized for DEC as effectively as can the high-resolution instruments. For example, compound confirmation for high-throughput organic synthesis [52] by quadrupole MS has yielded results comparable to those obtained by Q-TOF. In the area of impurity or degradant identification, possible background interference can be effectively separated by UHPLC so that the molecular formulas of real impurities can be determined by the unit resolution quadrupole instrument [46].

6.6 FORMULA DETERMINATION WITH HIGH-RESOLUTION MS

Unique determination of formulas for unknowns based on mass accuracy alone is seldom feasible even at a mass accuracy of >1 ppm [1]. To obtain a conclusive unknown identification, it is necessary to utilize additional chemistry constraints [2] available to help avoid false positives. In the identification of drug-related impurities or degradants, for example, the elemental composition of the parent drug can provide useful information for limiting elements and their upper and lower bounds for a more restrictive search. As a general approach, isotope pattern recognition can effectively facilitate formula determination, as described earlier. Kuehl [53] has reported the use of DEC with data from high-resolution quadrupole, Q-TOF and Orbitrap using innovative lineshape self-calibration methodology. Of all the measurements made, spectral accuracy ranked the correct compound as the most (rank 1) correct match 14 out of 15 times. For example, ketoconazole at *m/z* 531 acquired from Orbitrap was one of the 14 best matches by spectral accuracy. Without the spectral accuracy, using the elements C, H, N, O, Cl, and S in an elemental composition search would require a mass accuracy of <200 ppb to uniquely identify the compound. Even at a mass accuracy of 1 ppm, over 40 formula candidates must be evaluated.

Another example clearly demonstrated that the usual requirement for mass accuracy of <5 ppm can be relaxed when spectral accuracy is used as an additional metric for DEC. Seven compounds, including acetaminophen, promethazine, buspirone, terfenadine, loperamide, Tyr–Tyr–Tyr, and reserpine, were measured using high-resolution quadrupole Quantum Ultra, and their formulas were determined using elements of C, H, N, O, Cl, and S and a mass tolerance of 30 ppm. Six of them scored a spectral accuracy of >99% and ranked as hit 1. Although the compound Tyr–Tyr–Tyr ranked as only hit 5, the selectivity provided by spectral accuracy was evident since a total of 694 candidate formulas were evaluated for Tyr–Tyr–Tyr. The correct formula determination with a high mass tolerance can be further illustrated with compound caffeine measured by microTOF with accuracy of ~20 ppm. When searched with 25 ppm mass tolerance, the formula of caffeine appeared as a top hit with spectral accuracy of 99.7%. As shown in Figure 6.14, the match between calibrated and calculated spectra of caffeine is nearly perfect. To compare the top hit with the second and third best matches, the major differences in isotope patterns were found in their M + 1 peaks. The formulas of both $C_7H_{15}O_6$ and $C_4H_7N_{10}$ contributed less to the M + 1 peak because they contained carbon atoms or fewer combinations of carbon and nitrogen atoms, respectively, than did the correct formula $C_8H_{11}N_4O_2$ (caffeine); therefore, their calculated abundances at M + 1 are lower than that measured. In this case, new measurements must be made for effective DEC without spectral accuracy.

As molecular weight increases, spectral accuracy may play an even more dominant role than mass accuracy in DEC. On the basis of the lineshape self-calibration, Erve and et al. authors [12] investigated formula identification with spectral accuracy for 11 compounds with a mass range of 639–1664 measured by Orbitrap. Their findings (summarized in the Table 6.4) indicated that spectral

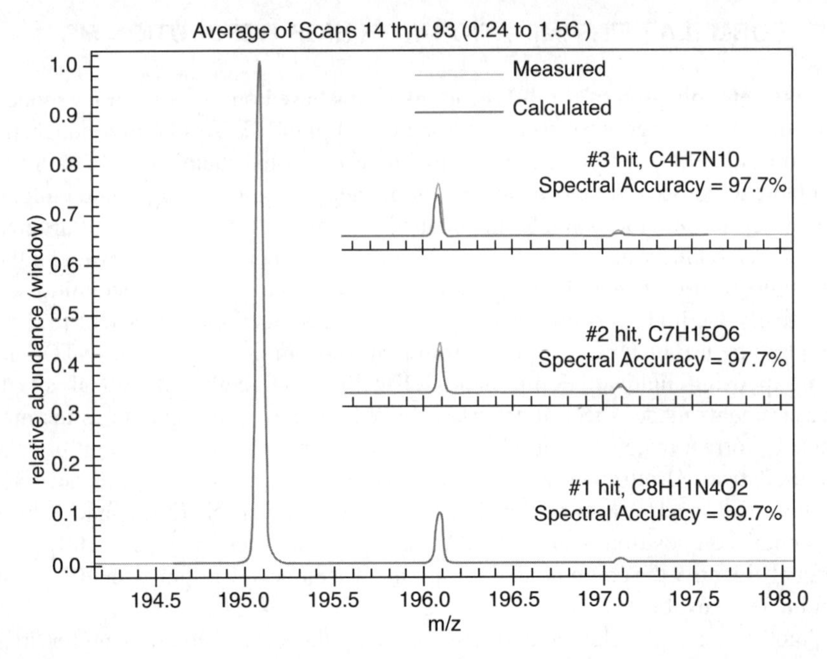

FIGURE 6.14 Overlay of calibrated spectra and theoretically calculated spectra for top three matches.

accuracy proved to be a powerful discriminator for formulas that fall within the 2 ppm mass tolerance range. For example, moxidectin at *m/z* 639 ranked second according to spectral accuracy by removing 94% false-positive formulas. For higher-molecular-weight compounds such as cyclosporine (1202 Da) or thiostrepton (1664 Da), spectral accuracy ranked the correct elemental composition within the top 0.5% of compounds, thereby eliminating >99% of false-positive candidates. Consistent with findings in work on impurity identification of simvastatin (Table 6.2), the results from the Orbitrap data confirmed that spectral accuracy not only helps eliminate incorrect formula candidates but also adds confidence to the formula candidates included for consideration. Thus, a higher spectral accuracy provides greater support to an elemental formula compared to the same elemental formula associated with a lower spectral accuracy. This reasoning, however, cannot be applied for mass accuracy, since it is often observed that a correct elemental formula can have a lower mass accuracy compared to incorrect elemental formulas with higher mass accuracy.

It is well known that mass accuracy is inversely proportional to mass resolution [54], which means that the higher the mass resolution, the higher the mass accuracy. However, this theory is not valid for spectral accuracy as found by Erve et al. [12]. They investigated the effect of mass resolution on spectral accuracy of various compounds measured at resolving powers of 7.5 K, 15 K, 30 K, 60 K, and 100 K (where K = thousand) on the Orbitrap. They observed that spectral accuracy

TABLE 6.4 Ranking of Elemental Formulas Based on Spectral Error or Mass Error[a]

Compound	Rank	Orbitrap Resolving Power (K)				
		7.5	15	30	60	100
Moxidectin	Spectral error	2	2	2	2	2
	Mass error	4 (42)	8 (34)	8 (33)	7 (31)	4 (32)
Erythromycin	Spectral error	1	1	1	1	1
	Mass error	23 (48)	16 (45)	16 (45)	16 (45)	16 (46)
Digoxin	Spectral error	1	3	3	3	4
	Mass error	10 (97)	16 (99)	22 (96)	37 (99)	22 (99)
Rifampicin	Spectral error	1[b]	3/2	3/1	3/3	1/1
	Mass error	60 (147)	13 (147)	46 (144)	34 (151)	24 (148)
Amphotericin B	Spectral error	3/2	1	1	1	2
	Mass error	30 (186)	45 (188)	40 (187)	26 (191)	28 (189)
Rapamycin	Spectral error	1	2	1	2	1
	Mass error	67 (281)	30 (275)	12 (278)	12 (278)	21 (278)
Gramicidin S	Spectral error	1	1	1	3	4
	Mass error	31 (570)	194 (577)	79 (584)	108 (577)	80 (584)
Gramicidin S[c]	Spectral error	1	1	1	3	3
	Mass error	183 (369)	76 (303)	97 (297)	97 (297)	54 (302)
Cyclosporin A	Spectral error	1	2	2	3	1
	Mass error	134 (1,089)	200 (1,089)	245 (1,089)	222 (1,099)	270 (1,090)

(*Continued*)

TABLE 6.4 *(Continued)*

		Orbitrap Resolving Power (K)				
Compound	Rank	7.5	15	30	60	100
Vancomycin	Spectral error	6	6	7	8	31
	Mass error	400 (1,523)	86 (1,515)	81 (1,529)	392 (1,529)	391 (1,529)
Vancomycin[c]	Spectral error	6	5	6	7	7
	Mass error	314 (769)	315 (769)	314 (769)	234 (774)	234 (774)
Thiostrepton	Spectral error	5	2	3	6	7
	Mass error	193 (1,908)	529 (1,919)	218 (1,908}	448 (1,912)	356 (1,912)
Thiostrepton[c]	Spectral error	1	2	4	4	3
	Mass error	309 (971)	309 (971)	71 (978)	135 (973)	135 (973)

[a]Based on number of formulas with mass errors in ppm less than or equal to the mass error of the correct formula. Numbers in parentheses are the total numbers of possible formulas consistent with search criteria.

[b]Second number shows rank when treating rifampicin as a mixture. See text for further explanation.

[c]Doubly charged ion.

Source: Reprinted from Ref. 12, with permission from Elsevier.

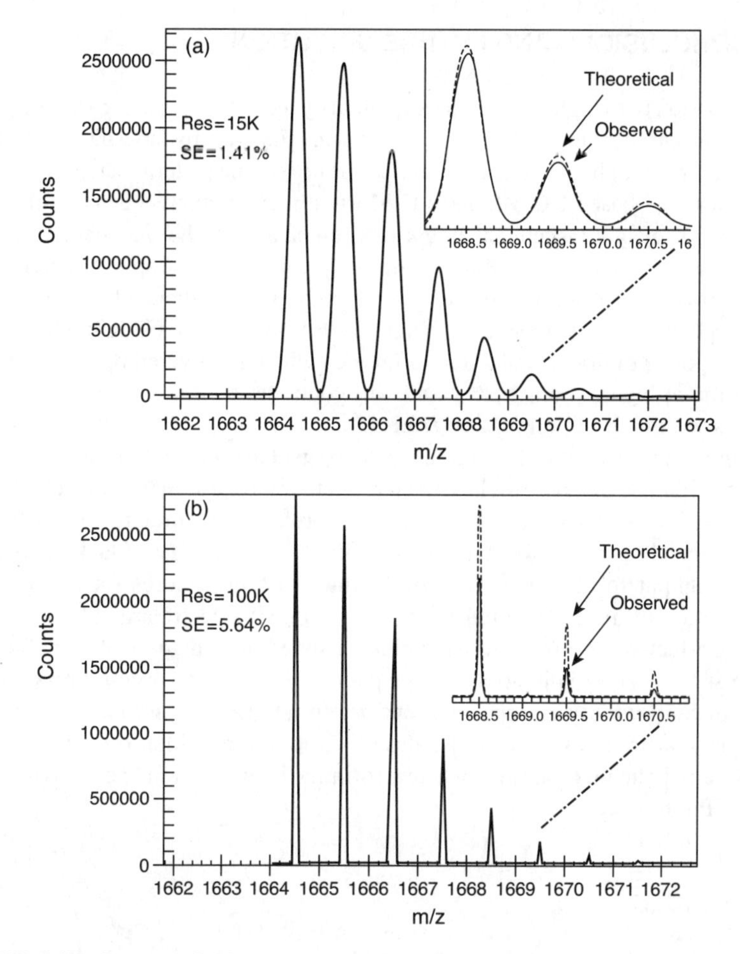

FIGURE 6.15 Spectra of thiostrepton at 7.5 K (a) and 100 K (b). The insets illustrate the intensity deficiency from the experimental data on ions below 1,000,000 counts, which is greater at 100 K than at 7.5 K. (reprinted from Ref. [12], with permission from Elsevier.)

tended to decrease with increasing resolving power. As an example, the spectra of thiostrepton acquired at resolving powers of 15 K and 100 K showed marked differences in spectral accuracy (Figure 6.15). High spectral accuracies of 98.59% (i.e., spectral errors of 1.41%) were observed for the spectra measured at a resolving power of 15 K, while at resolving power of 100,000, the spectral accuracy was found to be only 94.36%, with isotope peaks M + 4 and M + 5 having relative errors greater than 100%. This phenomenon can be explained by *isotopic beat patterns*, in which closely spaced yet unresolved frequencies cause constructive (or destructive) interference, especially for those compounds having rich isotope fine structure due to their high molecular weight and elemental composition [50,51].

6.7 CONCLUSIONS AND FUTURE DIRECTIONS

In combination with high mass accuracy, impurity identification using isotope pattern recognition has become the method of choice. Enabled by lineshape calibration technology with high mass accuracy and accurate isotope pattern matching capabilities, quadrupole-based GC-MS and LC-MS will play an unprecedented role on the identification of pharmaceutical impurities, especially in the manufacture process support area, where a large number of single-quadrupole mass spectrometers reside and unknown identification with short turnaround time is critical. On the other hand, highly sophisticated instruments such as Q-TOF, Orbitrap, and FT-ICR will certainly enjoy the power of drastic reduction of false candidates delivered by isotope pattern recognition at high mass accuracy and will continue to be used to tackle complex problems with high background interference.

During many decades of evolution, isotope pattern recognition technology has advanced significantly, but still has much room to improve to meet many challenging applications. In spite of providing great performance for impurity identification, current manual operation to analyze one unknown at a time has a lack of speed for high-throughput formula determination. Automatic procedures for simultaneous determination of multiple formulas in MS or MS/MS spectra are in high demand. Ensuring effective isotope pattern matching also requires instrument manufacturers to address the issues of distorted isotope patterns [12,23] and produce mass spectra data with high-fidelity isotope distribution. In addition to isotope pattern, more chemistry constraints such as those described in seven golden rules for heuristic filtering should be incorporated in a new formula determination approach to further improve the methodology.

REFERENCES

1. Kind, T.; Fiehn, O. (2006), Metabolomic database annotations via query of elemental compositions: mass accuracy is insufficient even at less than 1 ppm, *BMC Bioinform. 7*, 234–243.
2. Kind, T.; Fiehn, O. (2007), Seven golden rules for heuristic filtering of molecular formulas obtained by accurate mass spectrometry, *BMC Bioinform. 8*, 105–124.
3. Rock, S. M. (1951), Qualitative analysis from mass spectra, *Anal. Chem. 23*, 261–268.
4. Evans, J. E.; Jurinski, N. B. (1975), Program ELAL: An interactive minicomputer based elemental analysis of low and medium resolution mass spectra, *Anal. Chem. 47*, 961–963.
5. Tenhosaari, A. (1988), Computer-assisted composition analysis of unknown compounds by simultaneous analysis of the intensity ratios of isotope patterns of the molecular ion and daughter ions in low-resolution mass spectra, *Org. Mass Spectrom. 23*, 236–239.
6. Roussis S. G.; Proulx, R. (2003), Reduction of chemical formulas from the isotopic peak distributions of high-resolution mass spectra, *Anal. Chem. 75*, 1470–1482.
7. Darland, E.; McIntyre, D.; Weil, D.; Kuhlmann, F.; Li, X. (2008), *Superior Molecular Formula Generation from Accurate-Mass Data*, Agilent Application Notes, 5989-7409 EN.

8. Verena Tellstroem, V.; Ralf Dunsbach, R. (2008), *SmartFormula 3D—the New Dimension in Substance Identification—from Mass Spectrum to Chemical Formula*, Bruker Technical Note TN-26.
9. Wang, Y.; Gu, M. (2010), The Concept of Spectral Accuracy for MS, *Anal. Chem. 82*, 7055–7062.
10. Hobby, K.; Gallagher, R. T.; Caldwell, P.; Wilson, I. D. (2009), A new approach to aid the characterisation and identification of metabolites of a model drug, partial isotope enrichment combined with novel formula elucidation software, *Rapid Commun. Mass Spectrom. 23*, 219–227.
11. McLafferty, F. W.; Turecek, F. (1993), *Interpretation of Mass Spectra*, 4th ed. , University Science Books, Mill Valley, CA.
12. Erve, J. C. L.; Gu, M.; Wang, Y.; DeMaio, W.; Talaat, R. E. (2009), Spectral accuracy of molecular ions in an LTQ/Orbitrap mass spectrometer and implications for elemental composition determination, *J. Am. Soc. Mass Spectrom. 20*, 2058–2069.
13. Grange, A. H.; Donnelly, J. R.; Sovocool, G. W.; Brumley, W. C. (1996), Determination of elemental compositions from mass peak profiles of the molecular ion (M) and the M + 1 and M + 2 ions, *Anal. Chem. 68*, 553–560.
14. Grange, A. H.; Brumley, W. C. (1997), A mass peak profile generation model to facilitate determination of elemental compositions of ions based on exact masses and isotopic abundances, *J. Am. Soc. Mass Spectrom. 8*, 170–182.
15. Amirav, A.; Alon, T. (2008), Mass Spectrometric based method for sample identification, US Patent 7,345,275 B2.
16. Alon, T.; Amirav, A. (2006), Isotope abundance analysis methods and software for improved sample identification with supersonic gas chromatography/mass spectrometry, *Rapid Commun. Mass Spectrom. 20*, 2579–2588.
17. Zweigenbaum, J. A.; Thurman, E. M.; Ferrer, I. (2007), Agilent Technologies, US Patent Application.
18. Mather, J.; Goshawk, J.; Rao, R.; O'Malley, R.; McMillan, D.; Oldham, N. (2006), *Use of a Novel Exact Mass Isotopic Matching Algorithm and LDI for Rapid Identification of Dyes and Pigments*, Waters Application Notes, Lit. Code No. 720001754EN.
19. Hobby, K. (2005), *A Novel of Isotope Prediction Applied to Elemental Composition Analysis*, Waters Application Notes, Lit. Code No. 720001345EN.
20. Wang, Y.; Gu, M. (2007), PCT International Patent Application (Cerno Bioscience LLC, USA), US Patent 6,983,213.
21. Fernandez-de-Cossio, J.; Gonzalez, L. J.; Satomi, Y.; Betancourt, L.; Ramos, Y.; Huerta, V.; Besada, V.; Padron, G.; Minamino, N.; Takao, T. (2004), Automated interpretation of mass spectra of complex mixtures by matching of isotope peak distributions, *Rapid Commun. Mass Spectrom. 18*, 2465–2472.
22. Koester, C. (1999), (Bruker Daltonik G.m.b.H., Germany), Br. Patent Application.
23. Bristow T. C. J.; Harrison, M.; Cavoit, F. (2008), Performance optimisation of a new-generation orthogonal-acceleration quadrupole-time-of-flight mass spectrometer, *Rapid Commun. Mass Spectrom. 22*, 1213–1222.
24. Goodner, K. L.; Milgram, K. E.; Williams, K. R.; Watson, C. H.; Eyler, J. R. (1998), Quantitation of ion abundances in Fourier transform ion cyclotron resonance mass spectrometry, *J. Am. Soc. Mass Spectrom. 9*, 1204–1212.

25. Gu, M.; Wang, Y.; Zhao, X. G.; Gu, Z. M. (2006), Accurate mass filtering of ion chromatograms for metabolite identification using a unit mass resolution liquid chromatography/mass spectrometry system, *Rapid Commun. Mass Spectrom. 20*, 764–770.
26. Feng, F.; Konishi, Y. (1992), Analysis of antibodies and other large glycoproteins in the mass range of 150000–20000 Da by electrospray ionization mass spectrometry, *Anal. Chem. 64*, 2090–2095.
27. Thomas, S.; Claudia, H.; Thorsten, R. A.; Martin, J. (2001), Exact mass measurements online with high-performance liquid chromatography on a quadrupole mass spectrometer, *Anal. Chem. 73*, 589–595.
28. Savitski, M. M.; Ivonin, I. A.; Nielsen, M. L.; Zubarev, R. A. (2004), Shifted-basis technique improves accuracy of peak position determination in fourier transform mass spectrometry, *J. Am. Soc. Mass Spectrom. 15*, 457–461.
29. Yergey, J. A. (1983), A general approach to calculating isotopic distributions for mass spectrometry, *Int. J. Mass Spectram. Ion Phys. 52*, 337–349.
30. Yergey, J. A.; Heller, D.; Hansen, G.; Cotter, R. J.; Fenselau, C. (1983), Isotopic distributions in mass spectra of large molecules, *Anal. Chem. 55*, 353–356.
31. Yergey, J. A.; Cotter, R. J.; Heller, D.; Fenselau, C. (1984), Resolution requirements for middle molecule mass spectrometry, *Anal. Chem. 56*, 2262–2263.
32. Rockwood, A. L.; Van Ordent, S. L.; Smith, R. D. (1995), Rapid calculation of isotope distributions, *Anal. Chem. 67*, 2699–2704.
33. Rockwood, A. L.; Van Ordent, S. L.; Smith, R. D. (1996), Ultrahigh resolution isotope distribution calculations, *Rapid Commun. Mass Spectrom. 10*, 54–59.
34. Roussis, S. G. (1999), Exhaustive determination of hydrocarbon compound type distributions by high resolution mass spectrometry, *Rapid Commun. Mass Spectrom. 13*, 1031–1051.
35. Werlen, R. C. (1994), Effect of resolution on the shape of mass spectra of proteins: Some theoretical considerations, *Rapid Commun. Mass Spectrom. 8*, 976–980.
36. Kuehl, D. (2008), The importance of line-shape calibration in mass spectrometry, *Am. Lab. Online* (Jan. 2008).
37. Wang, H.; Press, H. (2006), Accurate mass measurement on real chromatographic time scale with a single quadrupole mass spectrometer, *Chromatograph. (Japan) 27*, 135.
38. Wang, J.; Gu, M.; Wang, Y. (2006), Accurate MS/MS measurements for metabolite identification on unit mass resolution mass spectrometers, *Proc. 54th ASMS Conf. Mass Spectrometry and Allied Topics*, Seattle, May 28–June 1, 2006.
39. Mick, J.; Gillespie, T. (2007), Accurate mass measurement using single quadrupole GC/MS for structure elucidation of unknowns, *Proc. 55th ASMS Conf. Mass Spectrometry and Allied Topics*, Indianapolis, June 3–7, 2007.
40. Dancle, M. C.; Gu, M.; Powell, D. H. (2007), Improving mass accuracy on a unit resolution quadrupole mass spectrometer, *Proc. 55th ASMS Conf. Mass Spectrometry and Allied Topics*, Indianapolis, June 3–7, 2007.
41. Zhang, M.; Gu, M.; Kagan, N.; Ratnayake, A. (2008), Metabolite identification using a unit mass resolution liquid chromatography/mass spectrometry with accurate formula identification and mass defect filtering, *Proc. 56th ASMS Conf. Mass Spectrometry and Allied Topics*, Denver, June 1–5, 2008.
42. Garrett, T. J.; Dawson, W. W.; Gu, M.; Powell, D. H.; Richard, A.; Yost, R. A. (2008), Identifying lipids and other small molecules from imaging mass spectrometry experiments

using tandem mass spectrometry and exact mass, *Proc. 56th ASMS Conf. Mass Spectrometry and Allied Topics*, Denver, June 1–5, 2008.

43. Chen, J.; Sparkman, O. D.; Gu, M. (2009), Confident unknown identification of SVOC compounds by combining NIST library search with elemental composition determination, *Proc. 57th ASMS Conf. Mass Spectrometry and Allied Topics*, Philadelphia, May 31–June 4, 2009.
44. Kuehl, D.; Wang, Y. D. (2007), Self-calibration of mass spectral line-shapes for improving the formula identification of unknown compounds, *Spectrosc. Suppl. Mass Spectrom.* (Nov. 1, 2007).
45. Fiehn, O.; Kopka, J.; Trethewey, R. N.; Willmitzer, L. (2000), Identification of uncommon plant metabolites based on calculation of elemental compositions using gas chromatography and quadrupole mass spectrometry, *Anal. Chem. 72*, 3573–3580.
46. Gu, M. (2008), *Identification of Pharmaceutical Impurities by UPLC and a Fast Scanning Quadrupole MS*, Cerno Bioscience, Application Note 105.
47. Hancock, H. (2006), *Application of Elevated Resolution of GC-TOF-MS for the Multi-residue Analysis of Pesticides in Food*, Waters Application Note 720001607EN.
48. Sparkman, D.; Jones, P. R.; Curtis, M. (2009), Anatomy of an ion's fragmentation after electron ionization, *Spectroscoy* (Part I, special issue, May 1).
49. Chernushevich, I. V.; Loboda, A. V.; Thomson, B. A. (2001), An introduction to quadrupole–time-of-flight mass spectrometry, *J. Mass Spectrom. 36*, 849–865.
50. Hofstadler, S. A.; Bruce, J. E.; Rockwood, A. L.; Anderson, G. A.; Winger, B. E.; Smith, R. D. (1994), Isotopic beat patterns in Fourier transform ion cyclotron resonance mass spectrometry: Implications for high resolution mass measurements of large biopolymers, *Int. J. Mass Spectrom. Ion Process. 132*, 109–127.
51. Easterling, M. L.; Amster, I. J.; van Rooij, G. J.; Heeren, R. M. A. (1999), Isotope beating effects in the analysis of polymer distributions by Fourier transform mass spectrometry, *J. Am. Soc. Mass Spectrom. 10*, 1074–1082.
52. Capka, V.; Gu, M. (2009), Feasibility and reliability of low and high-resolution MS approaches for accurate mass and molecular formula determination in drug discovery, *Proc. 57th ASMS Conf. Mass Spectrometry and Allied Topics*, Philadelphia, May 31–June 4, 2009.
53. Wang, Y.; Gu, M.; Kuehl, D. (2007), Beyond mass accuracy: The neglected role of spectral accuracy in mass spectrometry, *Proc. 55th ASMS Conf. Mass Spectrometry and Allied Topics*, Indianapolis, June 3–7, 2007.
54. Blom, K. R. (2001), Estimating the precision of exact mass measurements on an orthogonal time-of-flight mass spectrometer, *Anal. Chem. 73*, 715–719.

PART II

APPLICATION

CHAPTER 7

Practical Application of Very High-Pressure Liquid Chromatography Across the Pharmaceutical Development–Manufacturing Continuum

BRENT KLEINTOP and QINGGANG WANG

Bristol-Myers Squibb, Analytical Research and Development, New Brunswick, NJ 08903

7.1 INTRODUCTION

In the pharmaceutical industry, scientists are increasingly challenged to shorten development timelines to rapidly address unmet medical need. Accordingly, development scientists are challenged to investigate new instrumentation and techniques that could provide for increased productivity. However, this challenge is often accompanied by limited potential to employ customized instrumentation, due to the need to maintain strict GMP/GLP compliance and the need to transfer methodology to worldwide manufacturing sites and contract vendors.

High-performance liquid chromatography (HPLC) is probably the most widely used analytical technique in pharmaceutical industry for characterizing new pharmaceutical entities. A common trend to increase the productivity for HPLC analyses has been the development of analytical columns packed with smaller-diameter particles. More recently, the availability of columns employing <2-µm particles has shown promise to provide faster analyses without compromising chromatographic efficiency. Columns packed with <2-µm particles provide greater efficiency over a wider range of linear velocities than do columns using 3- and 5-µm particles. This allows the use of shorter columns and higher flow rates to provide shorter analysis

Characterization of Impurities and Degradants Using Mass Spectrometry, First Edition.
Edited by Birendra N. Pramanik, Mike S. Lee, and Guodong Chen.

times, while still maintaining the same efficiency as methods developed on conventional analytical columns [1–7].

However, the increased backpressure derived from using such columns may preclude the use of conventional HPLC systems that typically have backpressure limitations of around 5000 (psi). Fortunately most chromatographic instrument vendors now offer commercially available equipment that can accommodate backpressures ranging from 9000 to 17,000 psi, as well as redesigned injectors, detector flow cells with significantly decreased system volume, in order to handle smaller-i.d. columns, compared with traditional 4.6-mm i.d. columns. The commercialization of this LC equipment has also been accompanied by many column vendors providing a much wider range of column chemistries and dimensions becoming available with <2-μm particles. These two factors have contributed to a much more widespread use of this technique for many applications within the pharmaceutical industry. With a wide variety of equipment and columns available, a variety of terminology exists that is used by vendors and in literature. In this chapter, we will use the generic term *very high-pressure liquid chromatography* (VHPLC) to reference the equipment and applications that utilize <2-μm particle columns.

Within pharmaceutical development, the use of VHPLC has shown particular promise for applications such as impurity profiling and in-process analysis. In-process applications are often time-sensitive in that large-scale syntheses are often dependent on rapid turnaround to decide whether reactions are complete and can proceed to the next synthetic step. Impurity profiling applications often require HPLC runtimes greater than 30 min to afford resolution of all the components of interest. For several of these applications, we have been able to rapidly develop VHPLC profiling methods that provide 3–9 times shorter analysis time with no significant compromise in chromatographic efficiency. For some more challenging separations, HPLC profiling methods may require shallow gradients and therefore long runtimes to separate critical components. In these cases, scientists may prefer to maximize the chromatographic efficiency provided by VHPLC to ensure that adequate separation can be routinely achieved. Although these applications may not produce significant time-saving benefits, the increased efficiency of VHPLC may provide development scientists more rugged methods and greater confidence to successfully transfer methods to worldwide quality control (QC) and contract labs.

Development scientists must also consider that implementation of newer techniques such as VHPLC across the pharmaceutical development–manufacturing continuum faces regulatory, operational, and technical challenges that may limit or preclude widespread utilization. As processes are transferred from development labs to worldwide manufacturing facilities and contract vendors, development scientists must also ensure that methods and equipment are rugged and that local QC labs are equipped and trained to run these methods. This approach may invoke large-scale implementation costs in terms of capital equipment budgets, time, and personnel training. Scientists may face questions from collaborators, quality assurance, and regulatory agencies regarding the equivalence of results when compared to more widely utilized conventional HPLC methods.

Another good example can be the resistance to updating "legacy" HPLC methods implemented within manufacturing. Legacy HPLC methods are methods that were developed in the early–mid-1990s to support products that have been marketed for many years. Many of these methods may have analysis times greater than 60 min largely because column technology was not nearly as advanced at the time when methods were developed. Through the years, many companies have resisted developing new methods that would provide only small increases in analysis time and/or efficiency. The main barrier to this has been the significant costs incurred to update worldwide regulatory filings with the new methodology. This cost can be significant when considering that some products may be marketed in more than 50 countries worldwide, requiring a significant investment in filing costs, and also in time spent amending the existing regulatory filings.

In this chapter we will highlight some practical applications and strategies that can be considered when employing VHPLC within the pharmaceutical development–manufacturing continuum.

7.2 THEORY AND BENEFITS OF VHPLC

The relationship between column efficiency and particle size can be derived from the van Deemter equation:

$$H = A + \frac{B}{u} + Cu \tag{7.1}$$

Here H represents the plate height or column efficiency; u is the linear velocity of the mobile phase; and A, B, and C are the coefficients for eddy diffusion, longitudinal diffusion, and mass transfer, respectively. The maximum efficiency (H_{min}) is achieved at optimum linear velocity (u_{opt}), and can be expressed as follows:

$$u_{opt} = \sqrt{\frac{B}{C}} \propto \frac{1}{d_p} \tag{7.2}$$

$$H_{min} = A + 2\sqrt{BC} \propto d_p \tag{7.3}$$

Since A is directly proportional to particle size (d_p), C is proportional to d_p^2; therefore, H_{min} is directly proportional to d_p, and u_{opt} is inversely proportional to d_p.

On the other hand, the relationship between separation time and particle size can be expressed as follows:

$$t_R = \frac{L}{u}(1+k) = \frac{(1+k)Nh}{D_m \nu} d_p^2 \tag{7.4}$$

Here t_R and k are the retention time and retention factor of the analyte, respectively; L is column length; h and ν are the reduced plate height and linear velocity, respectively; and D_m is the diffusion coefficient of the analyte in the mobile phase. Therefore, for separations having the same efficiency (N), separation time is proportional to the square of the particle size.

From a practical standpoint, the use of smaller particles can provide two important advantages that analytical scientist can exploit: higher efficiency or higher speed. Decreasing the particle size by a factor of 2 (e.g., from 3.5 to 1.7 μm) may provide twice the column efficiency if the column length is kept the same. Alternatively; the separation time may be reduced by 4 times by reducing the column length and doubling the linearity velocity of the mobile phase to provide the same efficiency.

In our estimation, most scientists who currently utilize VHPLC leverage the capability to provide decreased runtimes to increase throughput. This is typically accomplished by employing shorter columns packed with smaller particles and higher mobile-phase flow rates. This provides shorter runtimes while maintaining roughly equivalent chromatographic efficiencies provided by existing HPLC methods using 3 and 5 μm particle sizes. An example of this is provided in Figure 7.1, which illustrates a decrease in chromatographic runtime from 60 to 8 min employing a shorter column at a higher flow rate. Examination of the impurity profile in Figure 7.1 also indicates that this was achieved without sacrificing chromatographic efficiency to resolve the impurity peaks in the sample. In our practical experience, a decrease in runtimes of approximately 3–9 times can be achieved while maintaining the necessary efficiency to separate all impurity peaks of interest.

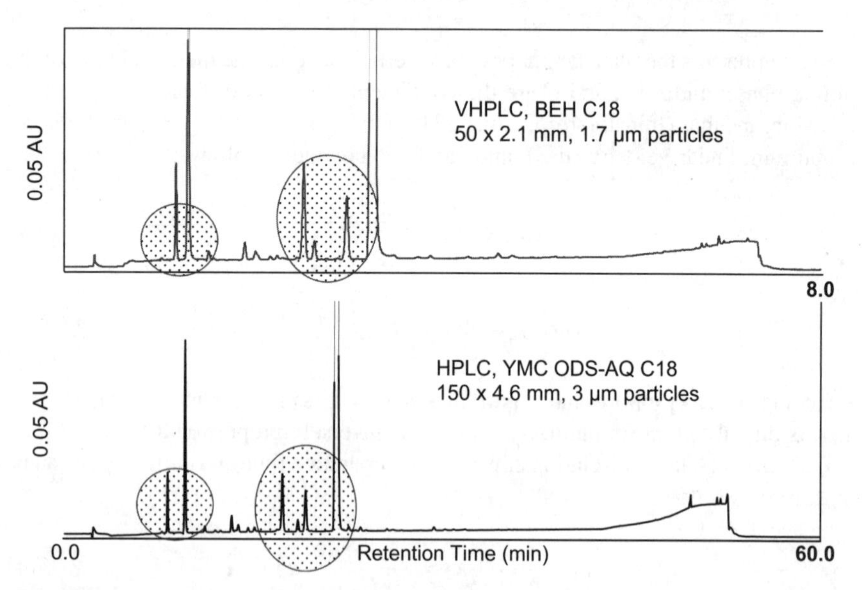

FIGURE 7.1 Chromatograms of an impurity mixture analyzed using HPLC and VHPLC method conditions. The figure illustrates that the chromatographic runtime was decreased from 60 min to 8 min while maintaining the chromatographic efficiency needed to resolve the impurity peaks in the sample.

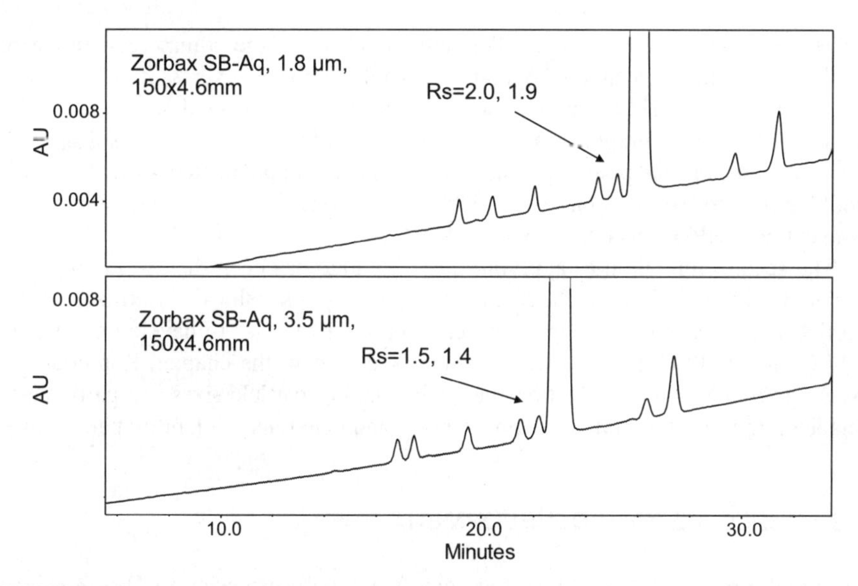

FIGURE 7.2 Chromatograms of an impurity mixture comparing results obtained using columns packed with 1.8- and 3.5-μm particles, otherwise similar conditions. In this application, the 1.8-μm-particle column provided increased chromatographic efficiency. Although no decrease in runtime was achieved, the increased resolution may provide for increased method robustness and confidence so that the method could be successfully transferred to worldwide QC labs.

During HPLC method development, analytical scientists frequently encounter impurities that are difficult to chromatographically resolve from other components in a sample. In these situations, scientists can also exploit the increased chromatographic efficiency provided by smaller particles to improve the resolution between a critical pair of impurities without necessarily achieving a decrease in runtime. An example of this application is provided in Figure 7.2. The two chromatograms in the figure compare the resolution achieved on a critical pair of impurities using 15-cm columns packed with either 3.5- or 1.8-μm particles of the same column chemistry. The increased efficiency provided by the smaller particles resulted in improved resolution between the critical components in the sample. Although no significant decrease in runtime was achieved, the increased resolution may provide for increased method robustness and confidence that the method could be successfully implemented and transferred to other labs around the world.

Traditionally there have been two main practical considerations that have limited widespread implementation of this technique. The obvious drawback has been the backpressure limitations of HPLC equipment (typically ~5000 psi), which have limited the column lengths and flow rates that could be employed when using <2-μm columns. In the late 1990s, however, several academic labs developed and applied HPLC systems capable of achieving higher pressures [8–12]. With this capability they were able to demonstrate the high efficiency and speed benefits that could be routinely achieved using <2-μm particles. This ultimately led to the commercialization of VHPLC equipment capable of operating reliably at backpressures of ≤15,000 psi. The market for VHPLC instrumentation has currently evolved such that major vendors of chromatographic instrumentation now offer VHPLC systems. Improvements in detectors and pumping systems are still being implemented. Currently, a lab could expect to pay a 15–30% premium for a VHPLC system when compared to conventional HPLC systems.

The second limitation to applying this technique has been the availability of a variety of different column chemistries and column dimensions that utilize <2-μm particles. Concerns with using different column chemistries to develop equivalent HPLC and VHPLC methods will be addressed later in the chapter. Reproducibly packing columns on a production scale with smaller particle sizes is a particularly challenging task from both a packing process and materials availability perspective.

7.3 VHPLC METHOD DEVELOPMENT

Very high-pressure liquid chromatography is a natural extension of HPLC, which possesses the same chromatographic theory and principles. As a result, scientists can not only develop VHPLC methods "from scratch" for new projects and needs but also adapt existing HPLC methods to provide more rapid or rugged VHPLC versions. In the following sections, we will provide some examples and practical considerations for both scenarios.

7.3.1 Adapting Existing HPLC Methods to VHPLC

There are many literature references sources that describe the transfer of conventional HPLC methods to VHPLC equivalent methods. Briefly, the most straightforward approach to developing VHPLC conditions for existing HPLC methods is to maintain most of the existing HPLC conditions (e.g., mobile-phase composition, column chemistry, column temperature, gradient steps) and simply geometrically scale the gradient step times and injection volumes to account for the different column volumes of the HPLC and VHPLC columns. Most vendors provide simple software programs to facilitate this conversion. Although a scaling approach can provide a good starting point to developing VHPLC separations, several other factors should be considered prior to finalizing the method. In the next few paragraphs we will outline some considerations for method parameters that should be considered.

1. *Sample Preparation* Sample preparation parameters include sample concentration, diluent, and extraction procedures. Since these parameters will likely have minimal effects on the VHPLC separation, keeping them the same will simplify the method validation. However, additional consideration should be given to makeup of the sample diluent if methods are being scaled down to narrowbore VHPLC columns (e.g., <3 mm i.d.), which are currently more readily available.

Effective sample loading onto the column should be verified if high levels of organic are used in the sample diluent.

2. *Mobile Phase* Mobile-phase composition is the major factor that will affect selectivity (a), as well as retention factor (k). Therefore it should be kept the same to reserve the same separation profile.

3. *Column Chemistry* Column chemistry is another major parameter that affects selectivity (a); therefore, it should also be kept the same. Since the silica particle properties also play an important role in column selectivity, a practical consideration for column selection is to use the same brand of column (e.g., from the same manufacturer), which differs only in particle size and column dimensions to minimize selectivity differences. The chromatograms provided in Figure 7.1 illustrate the variations in selectivity that can result from using different column chemistries. Close examination of the three impurity peaks immediately preceding the major component reveal a somewhat different profile between the two columns. Although all the impurities are still separated in this example, such selectivity change could result in peak coelution on method conversion. In Figure 7.3, the same sample was analyzed using the same column chemistry under both HPLC and VHPLC conditions. This resulted in virtually identical elution profiles being obtained for both method conditions. When such VHPLC version columns are not available, the overall selectivity should be as similar as possible between two columns. Currently, software programs are commercially available that can be used to assess column equivalence.

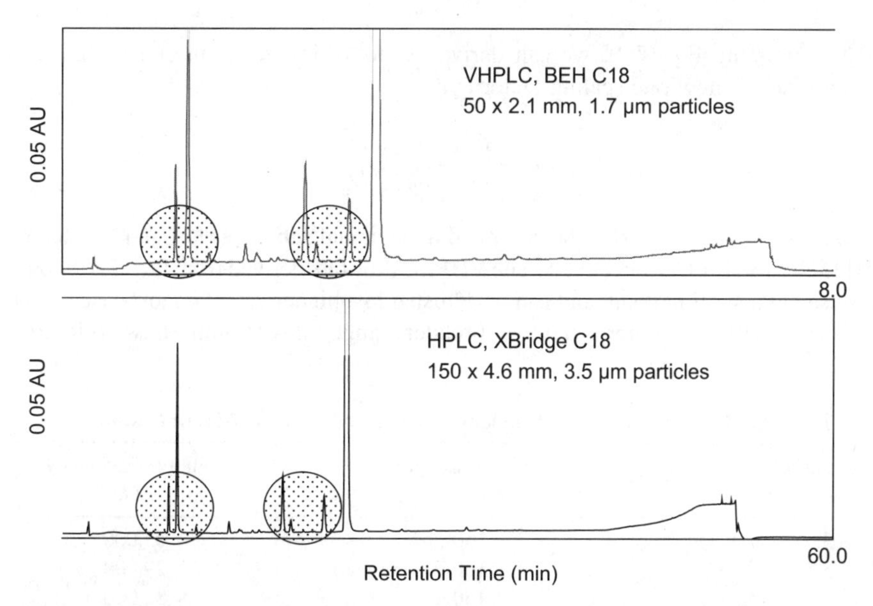

FIGURE 7.3 Chromatograms of an impurity mixture analyzed using HPLC and VHPLC columns with similar column chemistries. Note that the elution profiles are more similar to the chromatograms shown in Figure 7.1.

4. *Column Dimensions* The relationship between column length (L), particle size (d_p), and efficiency (N) can be roughly expressed as

$$N \propto \frac{L}{d_p} \tag{7.5}$$

Therefore, to ensure the same column efficiency, the ratio of column length to particle size needs to be kept the same. In Table 7.1, we compare the efficiency values typically achieved using different column dimensions. From this table, a 10-cm-long VHPLC column packed with 1.7-μm particles will provide ~30% more efficiency than will a 15-cm-long HPLC column packed with 3.5-μm particles. From a practical standpoint, the uniformity of either the particle size or the packing bed of a <2-μm particles is not as good as that of columns with 3.5-μm particles [4]. As a result, the real column efficiency for <2-μm columns is a little less than the theoretically predicted values. Therefore, a 10-cm-long VHPLC column packed with <2-μm particles will be expected to provide roughly the same efficiency as that of a 15-cm-long HPLC column packed with 3.5-μm particles. However, from a practical standpoint, if the resolution between critical impurities is sufficient, a 5-cm-long VHPLC column may be used to further increase the speed.

5. *Flow Rate* As shown in Eq. (7.2), the mobile-phase linearity velocity should be inversely proportional to particle size; therefore

$$\frac{u_v}{u_h} = \frac{F_v/D_v^2}{F_h/D_h^2} = \frac{d_{p,h}}{d_{p,v}} \tag{7.6}$$

By rearranging Eq. (7.7), we can derive the following equation and utilize it to determine the flow rate scaling factor (f_r):

$$f_r = \frac{F_v}{F_h} = \frac{d_{p,h}}{d_{p,v}} \cdot \frac{D_v^2}{D_h^2} \tag{7.7}$$

Here F is the flow rate, D is the column diameter, and subscripts "h" and "v" denote HPLC and VHPLC, respectively. The VHPLC flow rate derived from Eq. (7.8) should be used as a starting point, and can be adjusted to a higher value in most cases since <2-μm VHPLC columns have a broader range of optimum flow velocities

TABLE 7.1 Comparison of Efficiencies Expected Using Different Columns

Particle Size d_p (μm)	Column Length L (mm)	Relative Efficiency L/d_p
3.5	150	43,000
3.5	100	29,000
1.7	150	88,000
1.7	100	59,000
1.7	50	29,000

according to the van Deemter equation. The broader optimum flow velocities also afford reduction of the flow rate if system backpressure limitations are exceeded, without a significant decrease in chromatographic efficiency.

6. *Gradient Profile* To convert the gradient profile between HPLC and VHPLC conditions, we need to understand the gradient steepness (b) [13]:

$$b = \frac{V_m \, \Delta\phi S}{t_G F} \tag{7.8}$$

Here V_m is the column dead volume, $\Delta\phi$ is the change in B solvent change during the gradient step, S is the slope of the relationship between $\ln(k)$ and ϕ, which is independent of particle size, and t_G is the time for the gradient step. To keep the same gradient steepness between VHPLC and HPLC, we need to keep the same gradient step ($\Delta\phi$) while adjusting gradient time to ensure $V_m/t_G F$ constant between VHPLC and HPLC. From the discussion above, we can derive the following equation to determine the gradient scaling factor (f_g) which can then be applied to the time for each gradient step:

$$f_g = \frac{t_{G,v}}{t_{G,h}} = \frac{L_v}{L_h} \cdot \frac{D_v^2}{D_h^2} \cdot \frac{F_h}{F_v} \tag{7.9}$$

7. *Injection Volume* Although injection volume has minimal effect on separation, it also needs to be scaled according to column volume to avoid overloading or detector saturation. The following equation can be used to determine the injection volume scaling factor (f_i):

$$f_i = \frac{V_{inj,v}}{V_{inj,h}} = \frac{L_v}{L_h} \cdot \frac{D_v^2}{D_h^2} \tag{7.10}$$

8. *Column Temperature* Since column temperature also affects selectivity and retention, it should be kept the same between VHPLC and HPLC. However, for columns packed with $<$ 2-μm particles, both longitudinal and radial temperature gradients have been found to exist within the column as a result of the frictional heating between the mobile phase and small particles. While the radial temperature gradient is rarely a concern for columns sitting in a forced-air oven, a significant temperature increase can occur along the longitudinal direction (up to 20°C higher in the column outlet than in the inlet) [14,15]. This longitudinal temperature gradient can cause a decrease in retention, and may also change selectivity if it is sensitive to temperature. To accommodate this effect, the column temperature in VHPLC may need to be decreased to provide equivalent efficiency. This effect is illustrated in the chromatograms in Figure 7.4. At a 0.5 mL/min flow rate, the impurity in front of the main peak was well separated using a column temperature of 55°C. However, increasing the flow rate to 1.0 mL/min caused the impurity to coelute with the main peak. However, adequate separation was achieved by reducing the column temperature to ~40°C.

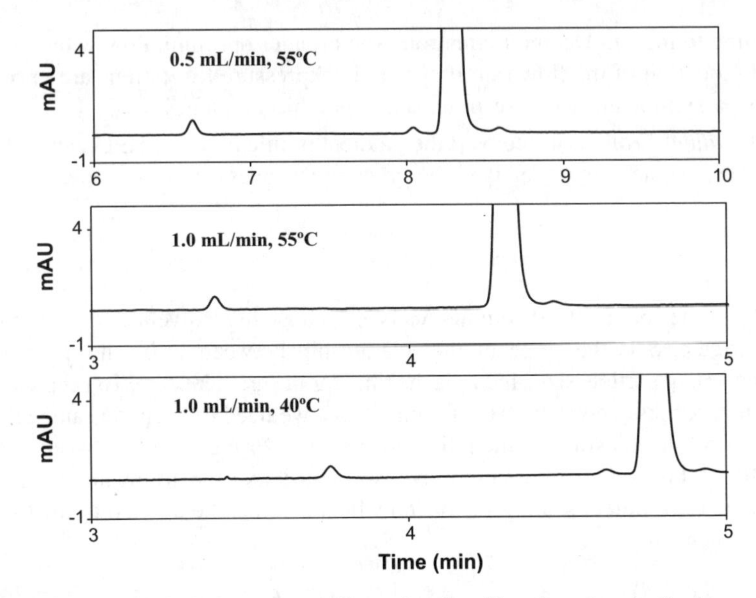

FIGURE 7.4 VHPLC chromatograms illustrating the effect that a longitudinal temperature gradient may have on retention. At a 0.5 mL/min flow rate, the impurity in front of the main peak was well separated using a column temperature of 55°C. However, increasing the flow rate to 1.0 mL/min caused the impurity to coelute with the main peak. Adequate separation was achieved by reducing the column temperature to ~40°C.

9. *Detection* UV detection, including variable UV-visible and photodiode array (PDA), is the most common detection mode used in most HPLC applications. As detection has no effect on separation, detection parameters, such as detection wavelength, should be kept the same. However, some other detector settings, such as sampling rates and filter time constant, may need to be adjusted to accommodate the narrower peak widths encountered when using VHPLC. Typically, chromatography software needs at least 10–15 points across the peak to ensure a reproducible integration. Since VHPLC peak widths are typically only a few seconds, a sampling rate of at least 5 points per second is generally required. The same consideration should also be used when applying other detection techniques, such as mass spectrometry.

7.3.2 Developing New VHPLC Methods

As more VHPLC equipment is routinely use in pharmaceutical labs, analytical scientists are becoming much more confident with the ruggedness of these relatively new systems. Accordingly, we are seeing a rapid increase in the use of VHPLC as methods are initially developed as well. The ability to initially develop faster separations provides some distinct advantages during method development. The obvious benefit is that a greater number of individual experiments (e.g., chromatographic runs) can be performed during a workday, which may lead to a more effective set of unattended, overnight investigations being utilized. While method development can now be performed in a few days rather than a couple of weeks, we would recommend tempering any management expectations that effective VHPLC method development can always be completed in a fraction of the time usually needed for developing HPLC methods. In practice, effective method development is somewhat limited by the appropriateness of samples that are provided to the method development scientist. These samples should contain a suitable level and number of potential impurities to effectively guide the method development. As new projects enter pharmaceutical development from discovery or in-licensed activities, there is usually very limited information on the related impurities of the active pharmaceutical ingredient.

The common LC parameters that can affect selectivity, discussed in Section 7.3.1, include column chemistry, organic solvent, mobile-phase pH and modifiers, temperature and solvent strength for isocratic separations and gradient profile for gradient separations. We have found it useful to incorporate a screening strategy to be able to screen some of these important parameters to provide an important and rapid first step in method development.

In practice our initial screening steps to investigate the effect of mobile-phase pH, organic composition, and column chemistry. Mobile-phase pH screening is performed over a range from pH 2 to 9. In choosing a pH for method develop we consider the separation profile and the sample solution stability at different pH values. We will then perform screening of different columns and organic solvents to determine appropriate starting conditions for method development. More recently, more < 2-μm column chemistries have become available, including a wide variety of columns with polar-embedded groups and hydrophilic interaction (HILIC) columns. These columns may provide different selectivity compared with regular C18, C8, and phenyl columns. After this step, one or two combinations may be picked for next-step screening. We then optimize other parameters such as column temperature and gradient profile screening, often using method development software packages, such as DryLab.

Figure 7.5 illustrates a schematic diagram of a home-built automatic VHPLC screening system. this system is capable of screening up to six different pH or mobile-phase modifiers, four columns, and two solvents. Following the screening strategy described in last paragraph, a complete set of screening can be finished in roughly one day.

An additional consideration for VHPLC method development is the variation between commercially instruments. In general, the configuration of HPLC instruments has become largely standardized over the years. However, with the rapid improvements we are seeing, VHPLC instruments from different vendors have somewhat different configurations, such as pressure rating, system delay volume, and injector design. Scientists need to consider such differences during method development stage to avoid future method transfer issues. For example, some VHPLC instruments are available only as binary pumps; therefore, a tertiary gradient should be avoided. Pressure ratings with current VHPLC instruments from different vendors

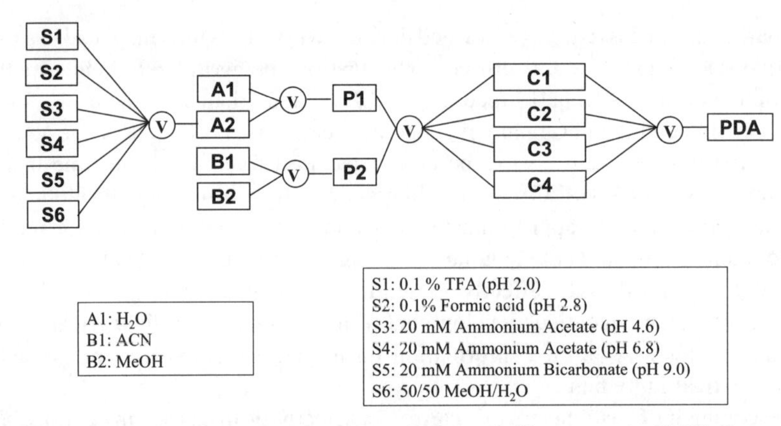

FIGURE 7.5 A schematic diagram of an in-house-built automatic VHPLC screening system. This system is capable of screening up to six different pH or mobile phase modifiers (designated S1–S6), four columns (designated C1–C4), and two solvents (designated B1–B2).

range from 9000 to 17,000 psi, and currently, columns have different pressure ratings as well. Systems from different vendors also have different delay volumes, which can provide subtle differences in elution profiles.

7.4 OTHER PRACTICAL CONSIDERATIONS

Today's pharmaceutical development organizations often leverage domestic and offshore contract labs to perform analytical characterization, utilize chemical vendors to synthesize intermediate compounds and manufacture drug products, and often reassign existing projects within the organization to more effectively utilize available resources. As a result, scientists often need to ensure that methods can be successfully implemented in different labs throughout the development lifecycle. Until such time when VHPLC enjoys widespread instrument availability, pharmaceutical development scientists will need to consider mechanisms to provide comparable methods that can yield equivalent analytical results when only conventional HPLC systems are available.

To address this issue, it may be practical to provide equivalent VHPLC and HPLC methods [16,17] and allow either method to be used for analytical testing. The VHPLC version can be transferred to the labs where such instruments are available; otherwise, the HPLC version can be transferred. To develop equivalent VHPLC/HPLC methods, the same principles discussed in Section 7.3.2 can be followed.

An example of results from method equivalence experiments using this approach are shown in Figure 7.6, and some batch analysis results are shown in Table 7.2. From these results, it is evident that relative retention times (RRT) are very similar between HPLC and VHPLC results, as well as the quantitative impurity results. This suggests

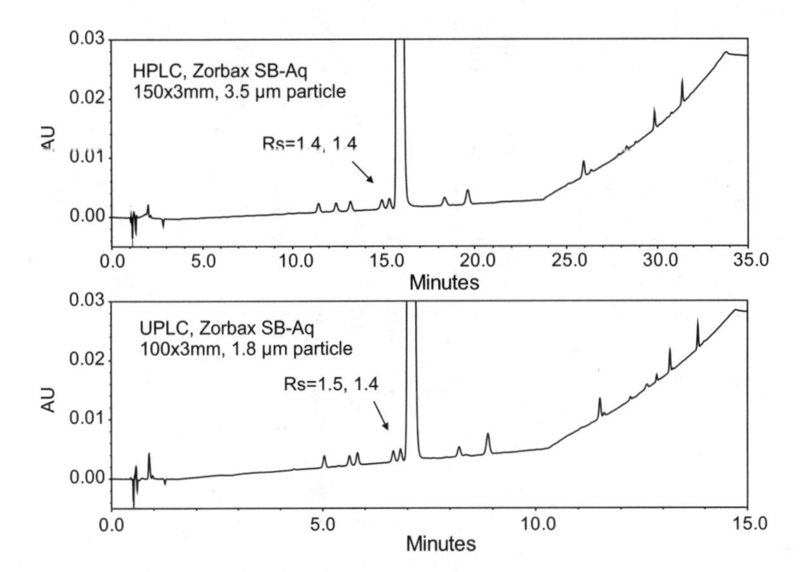

FIGURE 7.6 Chromatograms illustrating an example of results from method equivalence experiments between VHPLC and HPLC methods. From these chromatograms, it is evident that relative retention times (RRTs) are very similar between HPLC and VHPLC results, as well as the quantitative impurity results. This suggests that the two methods are indeed equivalent.

that the two methods are indeed equivalent. Validation requirements are briefly discussed in the next section.

7.5 VHPLC METHOD VALIDATION

Validation of methods used in pharmaceutical development is required for all GMP/GLP testing by different health authority guidelines. Depending on the stage of development, the validation can take a few days or a few weeks, and involves a significant amount of work. Given this, the advantage of using faster VHPLC separations are obvious. For methods where only VHPLC conditions exist, validations should be performed according to standard ICH guidelines.

However, versions of both VHPLC and HPLC methods may have been developed to allow worldwide QC labs to utilize methods that are compatible with whatever version of LC equipment is available in those labs. In these situations, performing all validation tests on both methods is not scientifically warranted. If one of the method versions is fully validated, then sound scientific judgment should be utilized to determine which validation tests should be performed to validate method equivalence.

TABLE 7.2 Batch Analysis Results from Equivalent HPLC and VHPLC Methods

	Impurity												
	1	2	3	4	5	6	API	7	8	9	10	11	Total (Number of) Impurities
HPLC RRT	0.13	0.73	0.78	0.84	0.94	0.97	1.00	1.16	1.24	1.62	1.86	1.96	
VHPLC RRT	0.13	0.72	0.80	0.83	0.95	0.97	1.00	1.17	1.26	1.64	1.89	1.98	
HPLC AP Batch 1	0.05	0.04	0.05	0.07	0.18	0.09	99.03	0.14	0.26	<0.03	0.09	<0.03	0.97(9)
VHPLC AP Batch 2	0.05	0.04	0.06	0.07	0.19	0.09	99.03	0.12	0.25	<0.03	0.10	<0.03	0.97(9)
HPLC AP Batch 2	0.04	<0.03	0.05	0.06	0.19	0.15	99.15	0.10	0.14	<0.03	0.09	<0.03	0.85(9)
VHPLC AP Batch 2	0.04	0.03	0.06	0.06	0.19	0.16	99.11	0.09	0.13	<0.03	0.09	<0.03	0.89(9)

AP = Area Percent

7.6 SUMMARY

In this chapter, we briefly discussed some practical applications and strategies that can be considered when employing VHPLC within the pharmaceutical development–manufacturing (PDM) continuum. VHPLC has shown promise to provide faster analyses without significantly compromising chromatographic efficiency. However, implementation of VHPLC across the PDM continuum faces operational and technical challenges which should be considered.

REFERENCES

1. Wu, N.; Clausen, A. M. (2007), *J. Sep. Sci. 30*, 1167.
2. Jerkovich, A. D.; LoBrutto, R.; Vivilecchia, R. V. (2005), *LCGC North Am.* 15.
3. Dong, M. W. (2007), *LCGC North Am.* 89.
4. De Villiers, A.; Lestremau, F.; Szucs, R.; Gelebart, S.; David, F.; Sandra, P. (2006), *J. Chromatogr. A 1127*, 60.
5. Jerkovich, A. D.; Mellors, J. S.; Jorgenson, J. W. (2003), *LCGC North Am. 21*, 600.
6. Wren, S. A. C. (2005), *J. Pharm. Biomed. Anal. 38*, 337.
7. Wren, S. A. C.; Tchelitcheff, P. (2006), *J. Chromatogr. A 1119*, 140.
8. MacNair, J. E.; Lewis, K. C.; Jorgenson, J. W. (1997), *Anal. Chem. 69*, 983.
9. MacNair, J. E.; Patel, K. D.; Jorgenson, J. W. (1999), *Anal. Chem. 71*, 700.
10. Lippert, J. A.; Xin, B.; Wu, N.; Lee, M. L. (1999), *J. Microcolumn Sep. 11*, 631.
11. Wu, N.; Lippert, J. A.; Lee, M. L. (2001), *J. Chromatogr. A 911*, 1.
12. Tolley, L.; Jorgenson, J. W.; Moseley, M. A. (2001), *Anal. Chem. 73*, 2985.
13. Synder, R. L.; Dolan, W. J. (2007), *High-Performance Gradient Elution*, Wiley, Hoboken, NJ, p. 17.
14. de Villiers, A.; Lauer, H.; Szucs, R.; Goodall, S.; Sandra, P. (2006), *J. Chromatogr. A 1113*, 84.
15. Gritti, F.; Guiochon, G. (2008), *Anal. Chem. 80*, 5009.
16. Dai, J. (2009), *Am. Pharm. Rev. 12*, 12.
17. Neue, D. U.; McCabe, D.; Ramesh, V.; Pappa, H.; DeMuth, J. (2009), *Pharmacop. Forum 35*, 1622.

CHAPTER 8

Impurity Identification for Drug Substances

DAVID W. BERBERICH, TAO JIANG, JOSEPH McCLURG, FRANK MOSER, and R. RANDY WILHELM

Covidien, Ltd., Pharmaceutical R&D, St. Louis, MO 63147

8.1 INTRODUCTION

Ensuring the safety and efficacy of active pharmaceutical ingredients (API's) by controlling the levels of both related and unrelated substances is required by agencies that regulate pharmaceutical products. In addition, quality manufacturers of such products are committed to controlling impurities, including degradants and extractable container components, to ensure that their products are of high quality and safety for their customers. Related substances generally originate from the process as side products, residual solvents, or reactants, or from degradation of the drug substance [1]. Unrelated substances would usually have packaging components as their sources [1–5].

The identification of these substances is essential to develop the means to sufficiently reduce, eliminate, or remove them from the API. Such identification is also required when these substances exceed thresholds, such as those defined by the International Conference on Harmonization (ICH). The ICH definition of the thresholds is based on the daily dose of the API: 0.05% for doses $>$ 2 g/day and 0.10% for doses less than or equal to that level, although lower thresholds may be applicable if the substance is highly toxic [6]. When the thresholds are exceeded, identification of the substance in the API is required.

The focus of this chapter is on the identification of such impurities in API's, primarily by mass spectrometry. The great advances in the sensitivity, speed, ability to interface to separation systems, and other technical aspects of mass spectrometers that has occurred since the 1970s have made MS one of the primary technologies

Characterization of Impurities and Degradants Using Mass Spectrometry, First Edition.
Edited by Birendra N. Pramanik, Mike S. Lee, and Guodong Chen.

employed in these identifications. However, it should be noted that many other techniques, such as spectroscopic (NMR, IR, UV–Visible, etc), chromatographic (HPLC, UHPLC, GC, etc), and wet-chemical methods (titrations, derivatizations, colorimetric tests, etc.) are also often employed to complete these identifications in an effective and thorough fashion. Three case studies are presented that illustrate the identification process and some of the tools employed in our laboratories.

8.2 CASE STUDIES

8.2.1 Identification of Impurities in Each Synthetic Step of Drug Substance during Process Development

Various impurities were identified during the process development of oxymorphone. The analytical tools that are routinely used to identify the impurities are LC-MS and LC/UV.

Oxymorphone is a powerful semisynthetic opioid analgesic with approximately 6–8 times the potency of morphine. Oxymorphone is usually manufactured by *O*-demethylation of oxycodone using a variety of *O*-demethylation reagents such as BBr_3 and HBr (route 1 in Figure 8.1). The yield for this type of *O*-demethylation ranges from 30% to 80%. The synthesis starts with an expensive material, oxycodone. A much more economic approach for manufacturing oxymorphone is oxidation of oripavine, followed by hydrogenation reduction (route 2 in Figure 8.1). This method is similar to the method for the synthesis of oxycodone from thebaine [7].

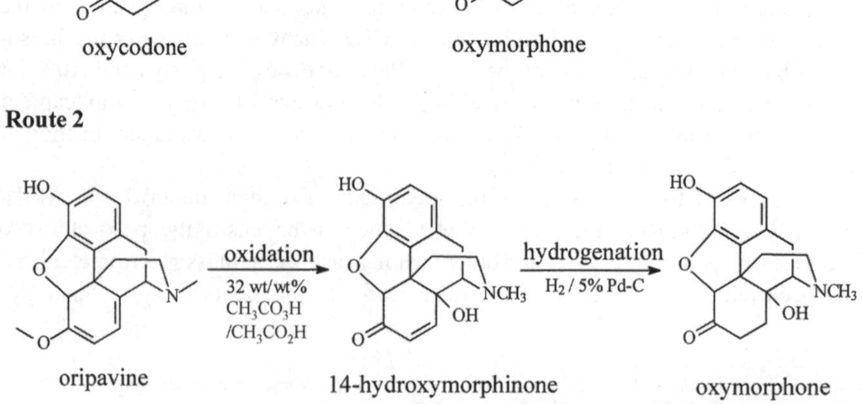

FIGURE 8.1 Synthetic routes for oxymorphone.

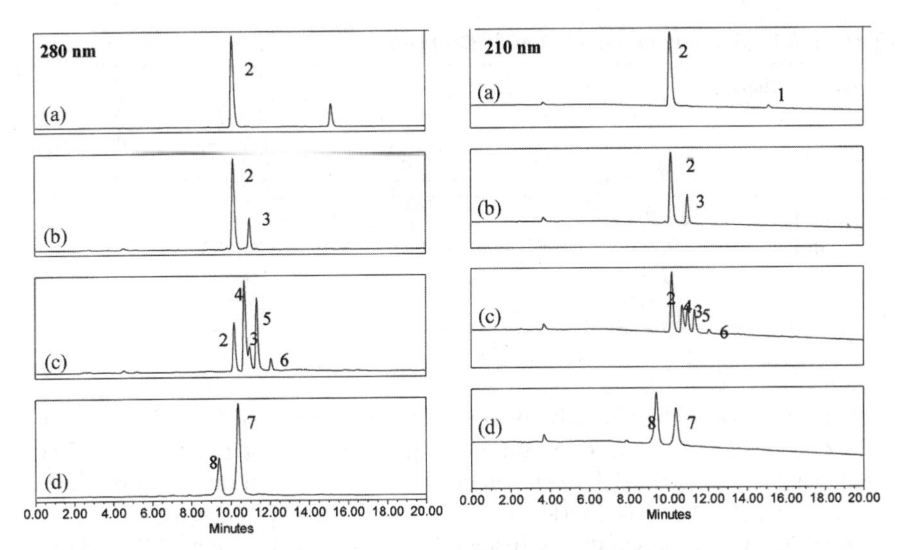

FIGURE 8.2 Representative chromatograms of the samples taken from the route 2 reactions: (a) oripavine oxidation for 1 h; (b) after completing oxidation reaction; (c) after charging catalyst and before starting hydrogenation; (d) after completing hydrogenation reaction. Compounds: 1—starting material (SM), oripavine; 2—desired product of oxidation, 14-hydroxymorphinone (14-OHM); 3–6—byproducts of oxidation; 7—byproduct of hydrogenation; 8—desired product of hydrogenation and final product, oxymorphone.

However, the oxidation of oripavine proved somewhat problematic using percarboxylic acid as an oxidant. The highest reported yield was 62% for the desired product, 14-hydroxymorphinone, and only minor yields of 14-hydroxymorphinone could be isolated [8,9]. The oxidation of oripavine is much more sensitive to the reaction conditions than thebaine [8,9]. The major impurities of the reaction were not separated and identified; and, the root cause for the formation of the impurities in the final product remained unknown. Therefore, the route 2 method for manufacturing oxymorphone had not been industrialized until 2008 [10].

An HPLC method was developed to monitor the reactions of route 2 in Figure 8.1 at the early stage of oxymorphone process development. Figure 8.2 shows the typical chromatograms for the samples taken during the process at wavelengths 280 and 210 nm. Several peaks were observed: starting material (peak 1), desired products (peaks 2 and 8), and five other byproduct impurities (peaks 3–7). impurity 3 has similar UV responses at both wavelengths, while the UV absorbance of impurities 4, 5, 6 and 7 at 280 nm is about twice the values at 210 nm, which indicates that compounds 4–7 have extensive conjugation. The major impurity (peak 7) in the final product (peak 8) was about 40%, which was very difficult to remove by crystallization. Historically, purifying the final product has been a problem for the route 2 chemistry. The stability of a sample lot is shown in Table 8.1.

To identify the unknown impurities, a series of LC-MS experiments were performed. Figure 8.3 contains the mass spectra of five unknown impurities (3–7)

TABLE 8.1 Stability of Sample Lot 0682B05889

Timepoint under Accelereated Stability Conditions	Step 5, RT 4.6 min, Area 226 nm	Step 6, RT 7.4 min, Area 226 nm	MW 705, RT 7.2 min, Area 226 nm	MW 1048, RT 6.8 min, Area 565 nm
1 month H2	26,394 (140 ppm)	12,924 (90 ppm)	0	0
1 month H4	19,269	11,190	21,023	905
2 months H4	8,533	9,770	21,678	3,855
6 months H4	6,882	7,101	16,827	19,502

in samples (b)–(d). The respective protonated molecular ions ($[M + H]^+$) are at *m/z* 597 for compound 4, *m/z* 316 for compound 3, *m/z* 613 for compound 5, *m/z* 629 for compound 6, and *m/z* 601 for compound 7. Each can be confirmed by their adduct ions, including $[M + NH_4]^+$ (M + 18), $[M + Na]^+$, and $[2M + H]^+$.

Impurity 3 was formed during oxidation. The molecular weight (MW = 315) is 16 u higher than the desired oxidation product 14-OHM (compound 2, MW = 299). Thus, impurity 3 can result from the addition of an oxygen atom to compound 2. When excess peracetic acid exists in the reaction solution, 14-OHM can be further oxidized to 14-OHM-*N*-oxide (14-OHM-NO, MW = 315, Figure 8.4). The LC-MS^4 experiments indicate that standard 14-OHM-NO undergoes the same fragmentation and has the identical chromatographic retention time as does compound 3, which confirms that compound 3 is 14-OHM-*N*-oxide. During hydrogenation, 14-OHM-NO can be reduced to oxymorphone without generating any new impurities.

The relatively high odd-number molecular weights at ~597–629 for impurities 4–7 suggest that these four compounds are the dimers of 14-OHM related compounds. The strong UV absorbance at high wavelength (Figure 8.2) implies that two aromatic rings in dimers are conjugated. As shown in the chromatogram of sample (b) in Figure 8.2, there are two major compounds, 14-OHM and 14-OHM-NO, in the reaction solution after the oxidation reaction completes. Therefore, the possible dimers are 2, 2′-bis-14-OHM (A, MW = 596), 2-14-OHM-2′-14-OHM-NO dimer (B, MW = 612), and 2, 2′-bis-14-OHM-NO (C, MW = 628), as proposed in Figure 8.4. Their molecular weights match exactly with compounds 4, 5 and 6, respectively. Therefore, compounds 4, 5, 6 were proposed to correspond to A, B and C. The fragmentation data and chromatographic retention behavior also corroborate this conclusion. Furthermore, after hydrogenation, dimers 4, 5, and 6 were reduced to the same compound, 2,2′-bis-oxymorphone (D in Figure 8.4) with MW = 600, which is peak 7 in chromatogram of sample (d) in Figure 8.2. The highest- level impurity in oxymorphone resulted from the dimerization that occurred before hydrogenation.

To understand the mechanism of dimerization, various different reaction conditions in route 2 were investigated. Results suggest that dimerization was triggered by Pd/C catalyst with the existence of peracetic acid. For instance, in reaction (b) of Figure 8.5, in which the molar ratio of peracetic acid to oripavine was 3:2, about 45% area of dimers was formed after charging catalyst in oxidation reaction

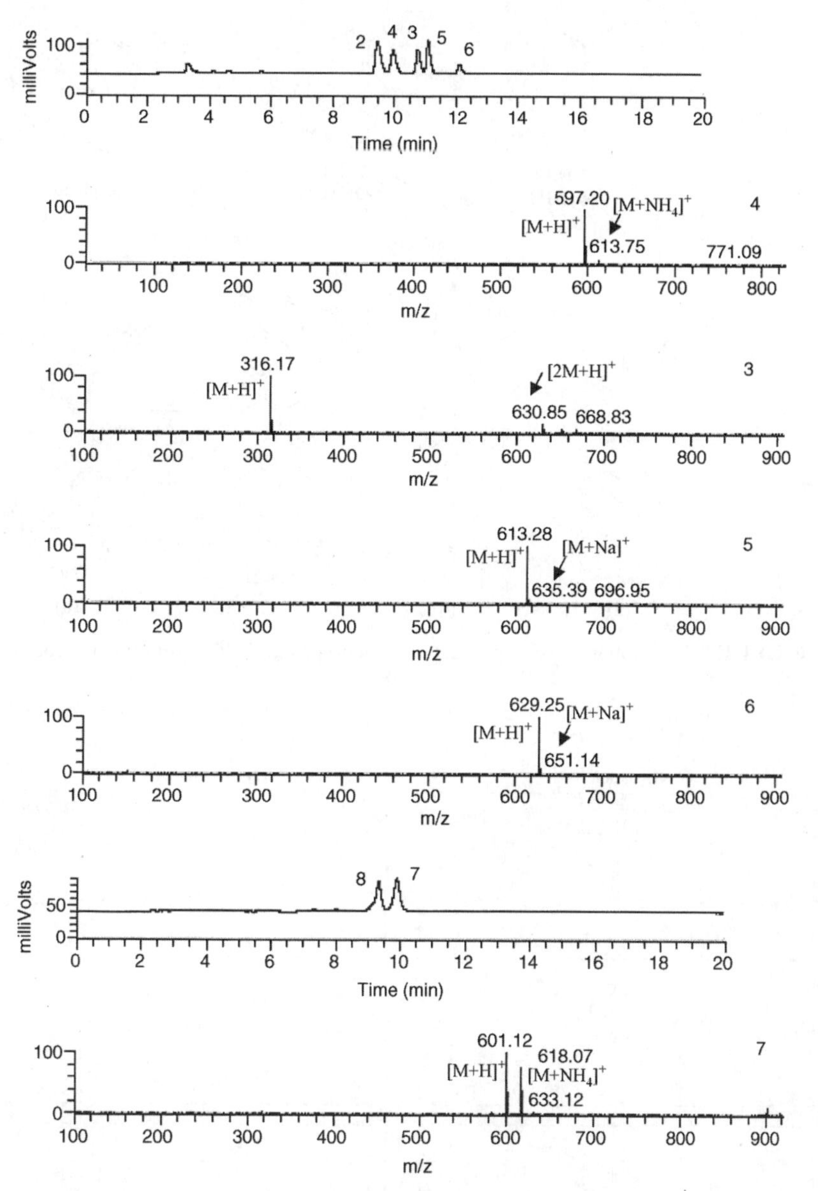

FIGURE 8.3 ESI mass spectra of byproducts 3–7 in samples (c) and (d).

solution. However, in reaction (a) in Figure 8.5, where there was no excess peracetic acid (molar ratio of peracetic acid to oripavine charged is less than 1), and reaction (c), where excess peracetic acid (~20%) was consumed by adding a reducing reagent, such as ascorbic acid before charging catalyst, the dimers were not formed. To ensure the completeness of oxidation, a slight excess of peracetic acid was used for the oripavine oxidation. The excess peracetic acid was then reduced by ascorbic

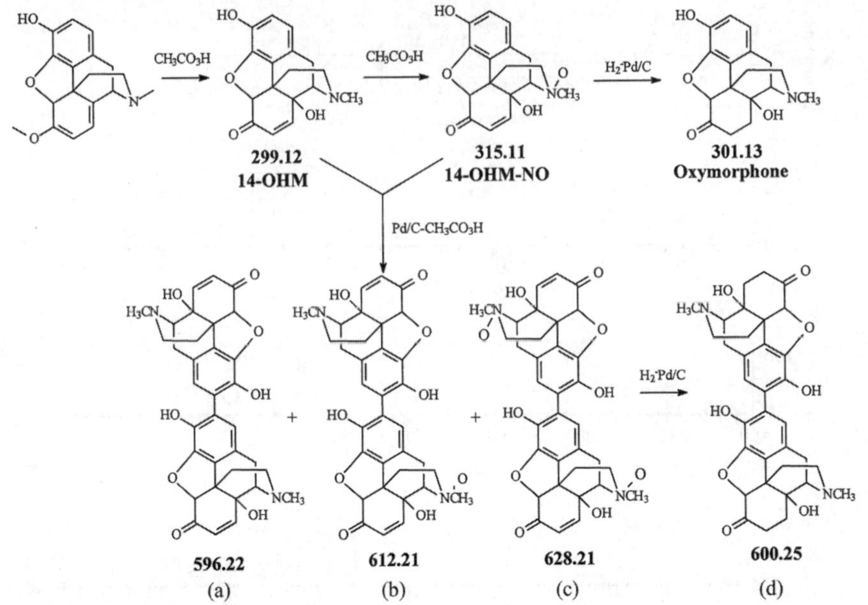

CH_3CO_3H → 299.12 14-OHM → CH_3CO_3H → 315.11 14-OHM-NO → H_2-Pd/C → 301.13 Oxymorphone

Pd/C-CH_3CO_3H

596.22 (a) + 612.21 (b) + 628.21 (c) → H_2-Pd/C → 600.25 (d)

FIGURE 8.4 Formation of byproducts in reaction route 2 of oxymorphone process.

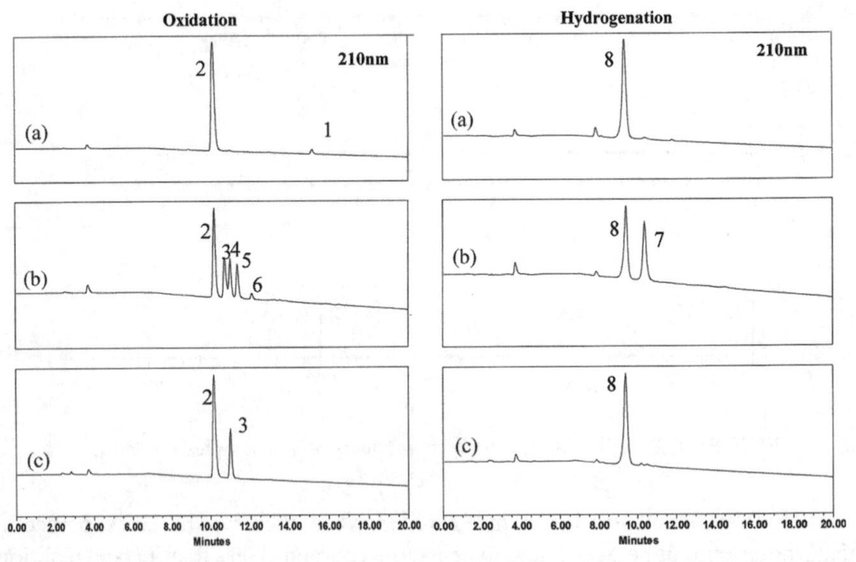

FIGURE 8.5 Representative chromatograms of the reactions with different reaction conditions: reaction (a) molar ratio of peracetic acid to oripavine is less than 1; reaction (b) molar ratio of peracetic acid to oripavine is 3 : 2; reaction (c) molar ration of peracetic acid to oripavine is 3 : 2, ascorbic acid was added before charging catalyst.

acid prior to the addition of Pd/C, and the formation of dimers could be eliminated. This process has been successfully scaled up and transferred to production to replace the costly *O*-demethylation process. The yield of new process is about 85–95%.

This case study demonstrated that by identifying the structures of byproducts, the origin of the impurities was pinpointed, which made elimination/reduction of the impurity achievable. Understanding the mechanisms by which impurities are formed expands the knowledge of the process chemistry and enables control of the impurities.

8.2.2 Impurity ID by LC/MS during Exploratory Chemistry: Evaluation of New Raw Materials

Gadoversetamide, a polyaminocarboxylate gadolinium complex, is an effective agent for creating contrast in magnetic resonance imaging (MRI). This agent is useful for imaging abnormalities in brain, spinal, and hepatic tissues by creating contrast through a decrease in T1 relaxation times.

A laboratory batch of gadoversetamide, made for research purposes using new raw materials, was determined to contain a new impurity when examined by HPLC. A typical approach to the identification of such an impurity would be to first examine the material by LC-MS to determine whether the impurity could be identified from those data alone. The next step, which is often required to complete or verify the identification, would involve NMR analysis of the impurity. The identification in this study was particularly challenging: (1) the HPLC method in which the impurity was observed utilized a nonvolatile buffer in the mobile phase and was, therefore, not MS compatible; (2) gadolinium complexes often provide little to no diagnostic structural information by MS^n analysis since the complex features a tightly bound core; and (3) finally, since the impurity contained gadolinium, it was not suitable for analysis by NMR without significant modification.

The mass spectral analyses in this study were performed in positive ESI mode on linear ion trap (LIT) and time-of-flight (TOF) mass spectrometers. The HPLC analyses and isolation were performed on an analytical-scale HPLC system. Solid-phase enrichment was performed on C18 cartridges, while NMR analysis was performed on a 500 MHz NMR equipped with a cryoprobe.

An MS-compatible HPLC method was developed to acquire molecular weight information for the impurity (see Figure 8.6). The molecular weight of the unknown impurity was determined to be 660 u (based on the most abundant isotope peak), and the isotopic pattern indicated the presence of one gadolinium atom per molecule.

MS^n experiments were performed to obtain structural information for the impurity. The gadolinium complex form generated few significant structural details. A rapid, small-scale analytical isolation yielded a few micrograms of the unknown that were employed in LC-MS experiments with low-pH mobile phases. This technique was employed to dechelate the isolated unknown complex online, due to instability of the complex at pH of approximately ≤ 2. The experiments allowed the acquisition of MS^n data on the impurity in its free ligand form [*m/z* 506 observed,

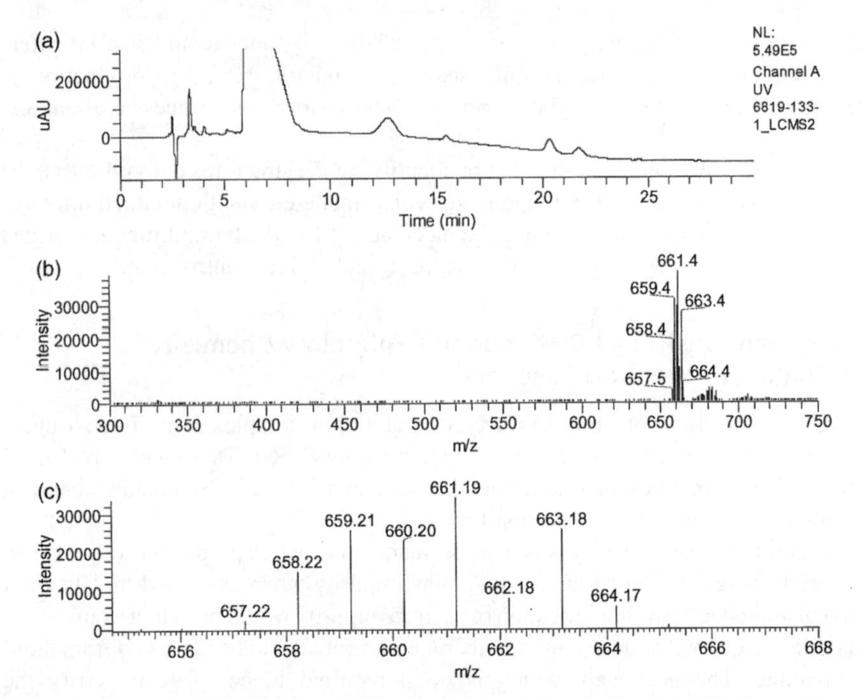

FIGURE 8.6 LC-MS with compatible ammonium acetate buffer: (a) UV chromatogram showing impurity; (b) full-scan MS; (c) ultrazoom scan—Gd isotope pattern.

corresponds to loss of gadolinium (157.9) and addition of 3 protons] (see Figures 8.7 and 8.8). Significant structural detail was obtained. These data enabled elucidation of more than half of the molecule; however, more information was necessary to complete the identification. Figure 8.9a shows the proposed portion of the structure

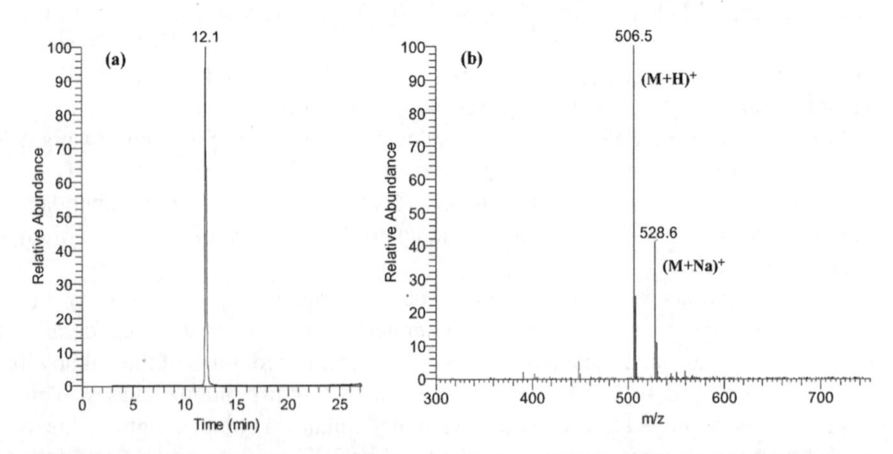

FIGURE 8.7 LC-MS with online dechelation by low pH: (a) extracted ion chromatogram (EIC) of the dechelated form; (b) mass spectrum of the dechelated form.

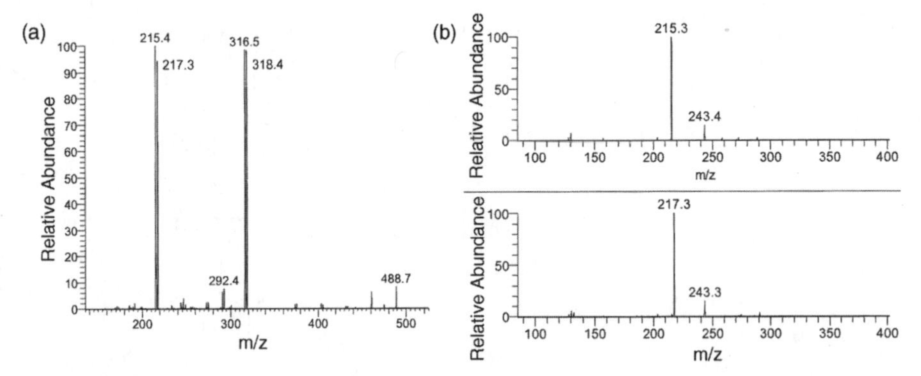

FIGURE 8.8 LC-MS^n with online dechelation by low pH: (a) MS^2 of dechelated form; (b) MS^3 of *m/z* 316 and 318 product ions.

(dechelated form); the R group has a mass of ~ 187 (see Figure 8.9b gadoversetamide, for comparison).

Accurate mass data was also acquired on both the complex form of the unknown and the dechelated form (from low-pH LC-MS; see Table 8.2 for both). These results led to formulas for both the gadolinium complex and dechelated forms, which, when combined with the MS^n data, defined ~85% of the structure. In Figure 8.9c, R_1 has a best fit of $C_4H_{10}N$ from accurate mass and MS^n data. The accurate mass data showed measured values of *m/z* 506.2820 and 661.1824, respectively, for the dechelated and gadolinium complex forms of the impurity. On the basis of the MS^n fragmentation and the isotope pattern, certain constraints could be applied to the calculations of formulas from these data. For instance, it was clear from the isotope pattern that one and only one gadolinium atom had to be present in the complex. Additionally, from the fragmentation of the dechelated form, the numbers of nitrogens and oxygens had to be a minimum of 3 and 6, respectively (Figure 8.9a). Therefore, the following constraints and ranges of atoms within formulas were employed: C from 0 to 30, H from 0 to 60, N from 3 to 10, O from 6 to 15, and for the complex only, gadolinium from 1 to 1. The maximum allowed error was $\leq$ 5 ppm, as TOF instruments consistently provide data within this range. On the basis of the formulas generated (Table 8.2) and knowledge of the chemistry, $C_{21}H_{40}N_5O_9$ and $C_{21}H_{37}N_5O_9Gd$ were selected as best fits for the protonated molecular ions of the free ligand and gadolinium complex forms, respectively.

While these experiments provided valuable information that defined much of the molecule, a fully elucidated structure had not been derived yet. NMR data would be essential to complete the identification. However, the presence of even trace levels of gadolinium would prevent the acquisition of NMR data. Therefore, a procedure was developed to quantitatively dechelate the complex.

The impurity was enriched by a factor of approximately 33, and the gadoversetamide reduced significantly, with a simple solid-phase extraction (Figure 8.10b). The impurity was rapidly isolated (on a mg scale) from this highly enriched material by analytical HPLC (Figure 8.10c). The isolated impurity was then prepared for NMR by quantitatively removing the gadolinium by taking advantage of the

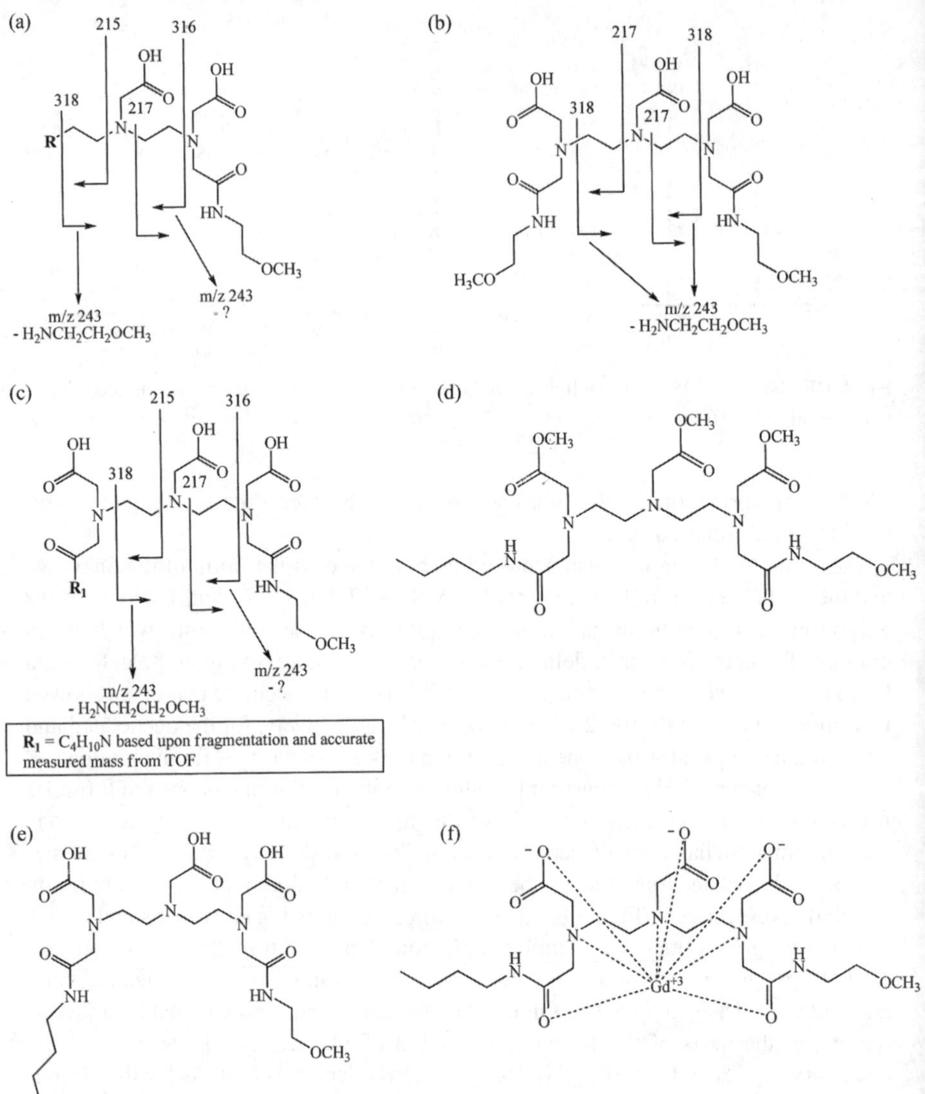

FIGURE 8.9 Structures of various forms of the impurity: (a) unknown (dechelated form) based on MS^n; (b) gadoversetamide (dechelated form) with known MS^n fragments; (c) unknown (dechelated form) based on MS^n and measured accurate mass; (d) impurity—dechelated, methylated form (chemical formula, $C_{24}H_{45}N_5O_9$, exact mass 547.3217); (e) impurity—dechelated form (chemical formula $C_{21}H_{39}N_5O_9$, exact mass 505.2748); (f) impurity structure (chemical formula $C_{21}H_{36}GdN_5O_9$, exact mass 660.1754).

TABLE 8.2 Best Formulas for Unknowns Based on TOF Data

Theoretical MW (u)	Δ (ppm)	Formula
Dechelated Form		
506.2821	−0.1	$C_{21}H_{40}N_5O_9$
506.2807	2.5	$C_{19}H_{38}N_8O_8$
Gadolinium Complex Form		
661.1827	−0.4	$C_{21}H_{37}N_5O_9Gd$
661.1813	1.6	$C_{19}H_{35}N_8O_8Gd$
661.1800	3.6	$C_{18}H_{39}N_4O_{12}Gd$
661.1854	−4.5	$C_{24}H_{35}N_6O_6Gd$

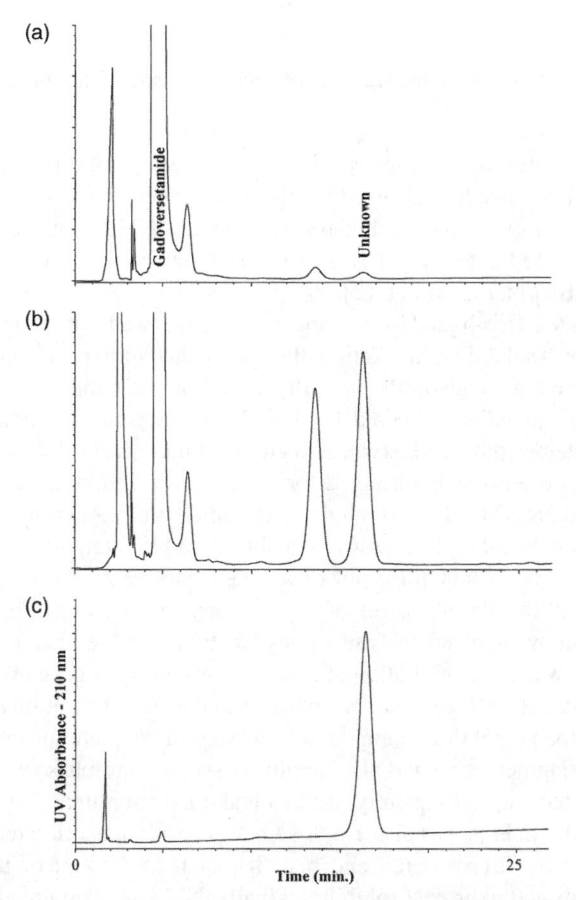

FIGURE 8.10 LC-UV of the gadoversetamide and isolates: (a) gadoversetamide as received; (b) after SPE enrichment; (c) after LC isolation of SPE-enriched products.

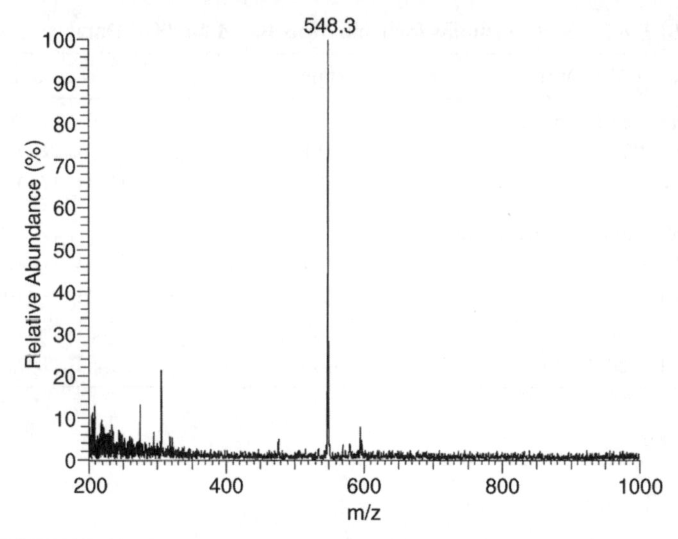

FIGURE 8.11 Mass spectrum of the dechelated isolate after methylation.

instability of gadolinium complexes at low pH and the insolubility of certain salts of gadolinium. This procedure allowed for the acquisition of the pure ligand form of the impurity at a quantity suitable for 1D and 2D NMR analyses. Finally, the isolated, dechelated free ligand form of the impurity was then methylated by reaction with BF_3/methanol [11,12].

The methylated, free-ligand form of the impurity was well suited for NMR analysis and had improved solubility. In addition, this methylated form provided a count of the number of carboxylic acids in the impurity, based on the number of methyl groups (units of $+14$) added as observed by MS. These experiments indicated that the compound contained three carboxylic acid groups (Figure 8.11), due to the addition of 42 to the *m/z* 506 previously observed for the underivatized, dechelated form.

Both 1D and 2D NMR data revealed that the unidentified R_1 group in the structure (Figure 8.9c) was *N*-(*n*-butyl). This was valuable since MS fragmentation and accurate mass data alone could not fully pin down the structure of this $C_4H_{10}N$ portion. The structures of the various forms of the unknown are shown in Figure 8.9d–f.

This case study involved the use of several tools for the characterization of a complex unknown. The exploitation of chemical instability (e.g., acid) and the use of the accurate mass data to generate probable formulae added additional information. Utilization of the parent drug compound to serve as a template for interpretation of mass spectral fragmentation was also helpful [13–17]. Solid-phase extraction (SPE) enrichment, a technique frequently used to make isolation unnecessary or at least more rapid and efficient, was also employed. Methylation, another tool used in this study, provided additional detail and benefits, such as a count of the number of carboxylic acids and improved solubility. Finally, NMR was employed to define the last small portion of the molecule that could not be elucidated by mass spectral analysis.

8.2.3 Impurity Identification during Accelerated Stability Studies

The development process to produce alfentanil hydrochloride (Figure 8.12) drug substance (MW 453) required accelerated stability testing. At a 2-month timepoint under the harshest conditions (H4 = 40°C/75% relative humidity) the drug substance had a pinkish/purple hue. In order to investigate the source of the coloration, the sample was analyzed via HPLC/PDA utilizing the previously developed stability indicating HPLC method. This coloration had not been previously observed during development of the stability indicating methodology.

The HPLC/PDA analysis was performed on the discolored drug substance with a reversed-phase gradient method that featured aqueous trifluoroacetic acid (0.1% v/v)/acetonitrile on a 150 × 4.6 mm C18 (5-μm) column. The data were collected at 190–700 nm. The results are shown in Figure 8.13.

The UV–visible spectrum of reference standard alfentanil hydrochloride is shown in Figure 8.14. It was obtained under the same chromatographic conditions as those utilized to analyze the discolored drug substance.

Comparison of the spectrum index plot in Figure 8.13 with the spectrum in Figure 8.14 illustrates the huge bathochromic shift associated with the formation of color bodies in aromatic amine and phenolic drug substances. In many cases, this shift occurs through the oxidative formation of "quinone"-type species conjugated to aromatic rings. Exactly how the oxidative formation of the quinone would take place in an arylamide-type molecule such as alfentanil hydrochloride was unclear.

Comparison of the data also demonstrates that the three main species responsible for the color in the drug substance had very little UV absorbance (0.01% by area). Previous experience in our laboratory with "color bodies" in drug substances suggested that coloration can be caused by substances present at the ppm level due to excellent

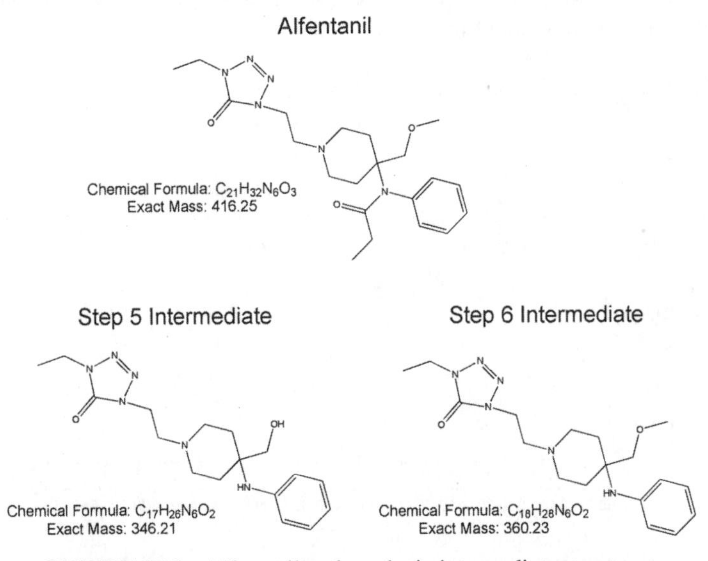

FIGURE 8.12 Alfentanil and synthetic intermediate structures.

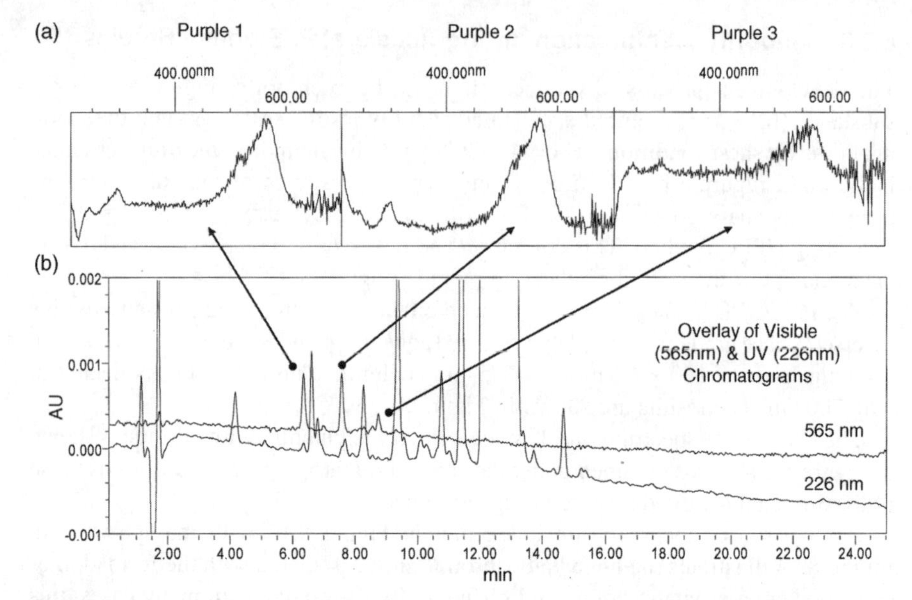

FIGURE 8.13 (a) Spectrum index plot chromatogram showing UV/visible spectra of the colored species; (b) overlaid chromatograms at UV (226 nm) and visible (565 nm) wavelengths.

sensitivity of the human eye at visible wavelengths. This observation is consistent with the expected increase in molar extinction coefficient that arises with the extended conjugation necessary to provide absorbance in the visible wavelength range.

The next step involved the use of the existing chromatographic conditions with an LC-MS system. An ion trap mass spectrometer with an ESI interface operated in the positive-ion mode was used to obtain the molecular weight and as much structural information as possible about the colored species.

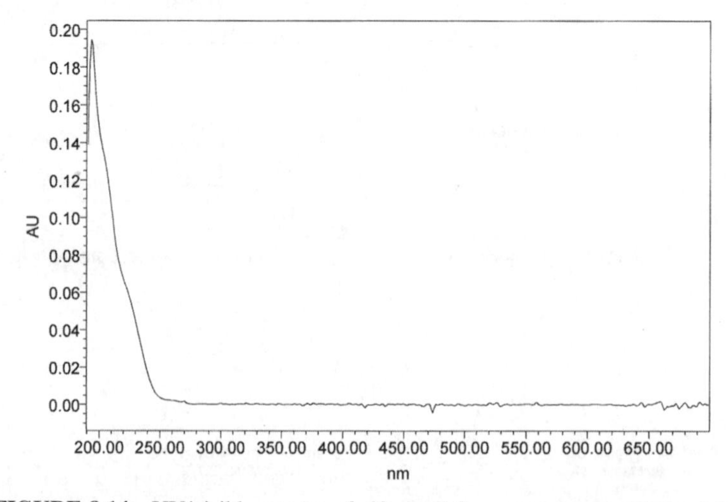

FIGURE 8.14 UV/visible spectra of alfentanil for comparison to Figure 8.13.

Under normal LC-MS operating conditions, the sensitivity for the colored species was not adequate to obtain structural information about the molecules. In fact, even under overload chromatographic conditions (1 mg alfentanil hydrochloride injected), which began to compromise the chromatography, only the molecular weights of the species could be obtained.

A rather unique approach was applied in order to increase the concentration of the colored species introduced to the mass spectrometer. The separation was scaled up to a 250×10-mm-i.d. HPLC column (C18, 5 μm) and a loop injector capable of 1 mL injections was installed. This setup allowed for injections of $\leq$150 mg of alfentanil HCl without sacrificing chromatographic efficiency when run at a modest flow rate (5.0 mL/min) with appropriate gradient modification. The chromatographic peaks that corresponded to the colored species in question were 30 times greater in concentration than previously obtained. The post-UV/visible detector eluant was minimally split (4 : 1) prior to introduction to the mass spectrometer ESI interface. The chromatograms are shown in Figure 8.15 with the mass ranges of interest extracted.

The chromatograms shown in Figure 8.15 contain the actual *m/z* values of the colored species, which are 2.5–2.6 times the exact mass of the alfentanil HCl base (416 u). The increased concentration afforded by the semipreparative chromatography allowed the collection of MS^6 fragmentation spectra. An example of this experiment is shown in Figure 8.16. A surprising result was the discovery of uncolored "dimeric"-type species in the 700–800 u range. These species had *m/z* values of approximately 1.7 times the exact mass of the alfentanil HCl base. This discovery, along with the MS^5 fragmentation data acquired for the "dimers," allowed the tentative identification of both the "dimeric" and "trimeric" species leading to the color formation (Figure 8.17).

A conjugated quinone-type structural moiety appeared to be responsible for the visible coloration of the drug substance. With these tentative structures the color issue associated with the drug substance was proposed to be due to the oxidative coupling of

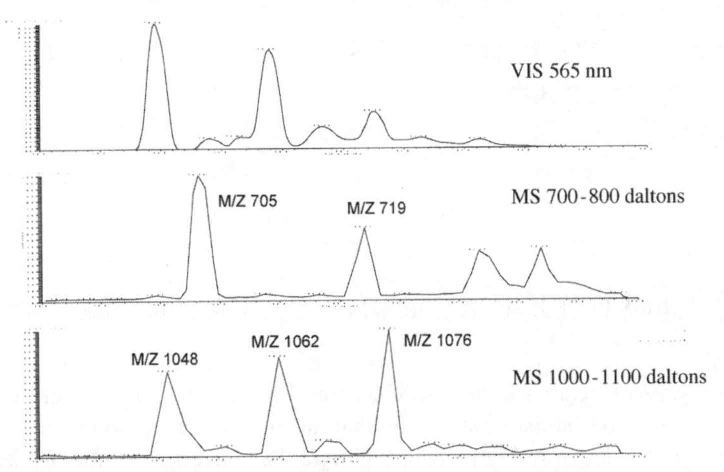

FIGURE 8.15 Representative semipreparative LC/MS chromatogram showing visible wavelength and two extracted mass ranges of interest.

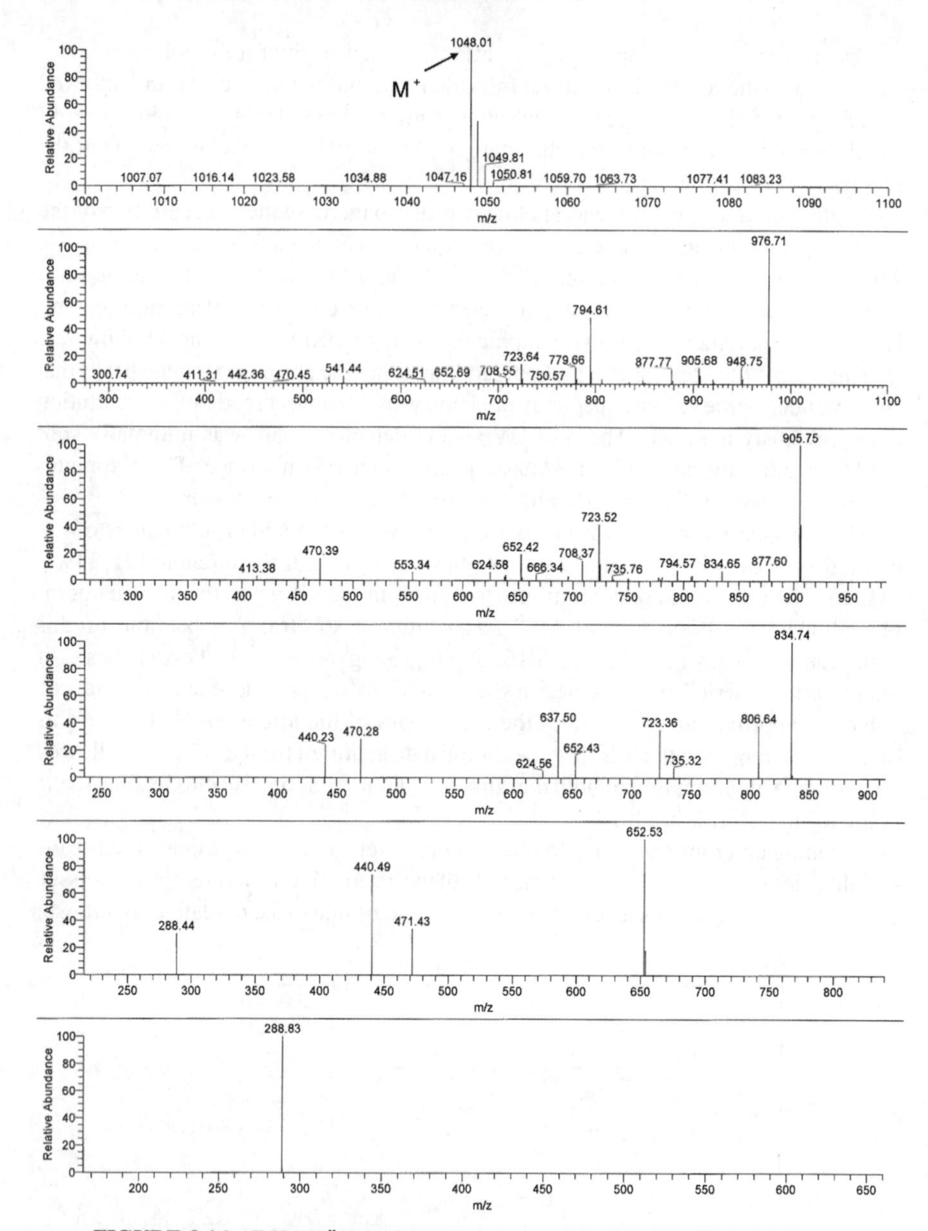

FIGURE 8.16 ESI MS^n data-dependent spectra of m/z 1048 colored species.

trace amounts of synthesis intermediates (see Figure 8.12) present on the crystal surfaces via a mechanism similar to that in aniline dye production [18]. The chromatograms shown in Figure 8.18 highlight the conversion of the two synthesis intermediates into the "dimers" – labeled Unknown 1 and 2, and the three main colored compounds – labeled Purple 1-3.

MWt's 704 & 718

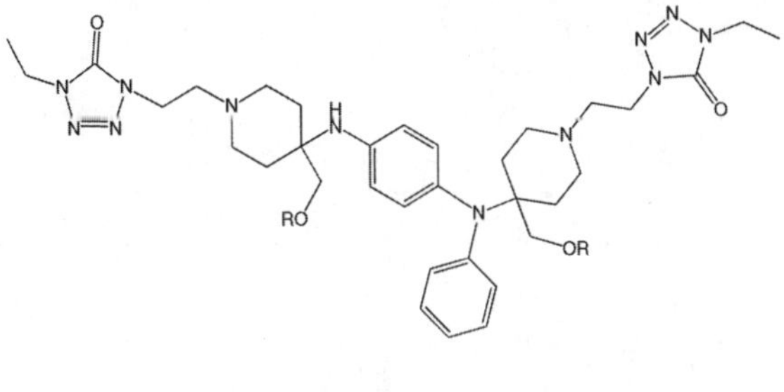

R=H,CH_3

MWt's 1048, 1062 & 1076

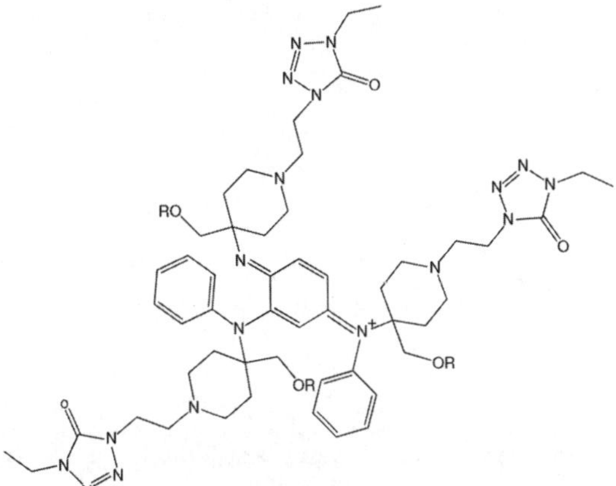

R=H,CH_3

FIGURE 8.17 Proposed structures of the tentatively identified species.

The data contained in Table 8.3 and Figure 8.19 illustrate the initial low levels of process intermediates present. These data also illustrate the subsequent coupling over time to generate one of the uncolored "dimers" and further oxidative coupling to generate one of the "trimeric" colored species. The same trends were obserbed for the other "dimeric" and "trimeric" species in this lot of alfentanil HCl. The same overall stability profile was also observed for two other lots of alfentanil HCl.

Finally, further experiments were performed to determine the composition of the alfentanil HCl at or near the surface of the product particles where the oxidation was occurring. These experiments involved the preparation of a slurry of alfentanil HCl in a solvent mixture at a calculated concentration such that only 2% (w/w) of the material would be soluble. The slurry was filtered after 10 min. The solids and filtrate were analyzed by HPLC and compared to initial values. The dissolved material, presumed to be primarily from the surface of the particles, was enriched in steps V and VI intermediates by 5 and 8 times, respectively.

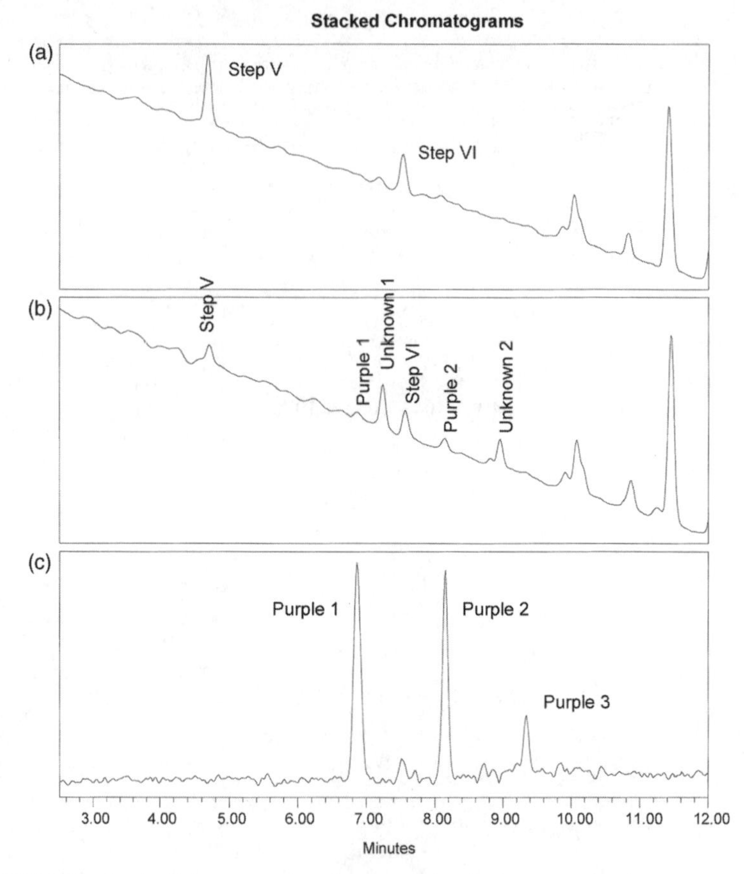

FIGURE 8.18 Stacked chromatograms demonstrating the drop in steps 5 and 6 intermediates over time in the UV (226 nm) with corresponding increase in the "dimeric" and "trimeric" species in the UV (226 nm) and visible (565 nm) ranges, respectively: (a) sample 0682B05889, 1 month H2, PDAD 226 nm; (b) same sample, 6 months H4, PDAD 226 nm; (c) same sample, 6 months H4, PDAD 565 nm.

TABLE 8.3 Data Showing Reactions of Intermediates 5 and 6 over Time to Form One of the Dimers and Subsequently a Trimer (Stability Sample Lot 0682B05889)

Timepoint under Accelereated Stability Conditions	Step 5, RT 4.6 min, Area 226 nm	Step 6, RT 7.4 min, Area 226 nm	MW 705, RT 7.2 min, Area 226 nm	MW 1048, RT 6.8 min, Area 565 nm
1 month H2	26,394 (140 ppm)	12,924 (90 ppm)	0	0
1 month H4	19,269	11,190	21,023	905
2 months H4	8,533	9,770	21,678	3,855
6 months H4	6,882	7,101	16,827	19,502

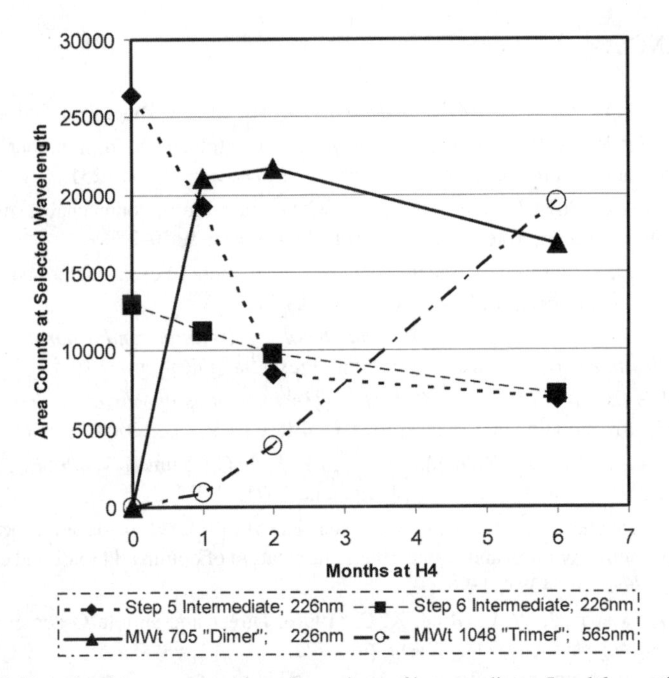

FIGURE 8.19 Graphic representation of reactions of intermediates 5 and 6 over time to form one of the dimers and subsequently a trimer.

The results of this experiment provided compelling information that the process intermediates were the source of this accelerated stability color problem. Process changes were made to reduce both synthesis intermediates to undetectable levels in the drug substance, and no further stability failures due to color have been noted.

8.3 CONCLUSIONS

Pharmaceutical research and development will continue to rely heavily on analytical methods that are capable of delivering accurate results in an efficient manner. Especially in cases where products are ingested by humans, there is no room for error in the identification of impurities and related substances, because the risks to life and health are too great. Other tangible costs drive the need for proper identification as well, including manufacture time; rework costs; internal and external investigations; and government fines, levies, or bans.

In each of the aforementioned case studies, mass spectrometry played a critical role in the identification of impurities in drug substances. The results were used to enhance the research, development, process manufacturing, and formulation of potentially important active pharmaceutical ingredients. Since the 1980s, mass spectrometry has evolved into one of the primary techniques used in the pharmaceutical industry. As it continues to develop, it will undoubtedly remain a vital tool in laboratories for many years to come.

REFERENCES

1. Smith, R. J.; Webb, M. L. (2007), *Analysis of Drug Impurities*, Blackwell, Oxford, pp. 2–3.
2. Paskiet, D. M. (1997), Strategy for determining extractables from rubber packaging materials in drug products, *PDA J. Pharm. Sci. Technol. 51*, 248–251.
3. Markovic, I. (2006), Challenges associated with extractable and/or leachable substances in therapeutic biologic protein products, *Am. Pharm. Rev. 9*, 20–27.
4. Kauffman, J. S. (2006), Identification and risk-assessment of extractables and leachables, *Pharm. Technol.* (Suppl.), S14, S16–S18, S20–S22.
5. Ahuja, S.; Alsante, K. M. (2003), *Handbook of Isolation and Characterization of Impurities in Pharmaceuticals*, Academic Press, San Diego, p. 249.
6. USDHHS (2008), *Guidance for Industry, Q3A Impurities in New Drug Substances,* US Dept. Health and Human Services, FDA-CDER, CBER, June 2008, revision 2, p. 11.
7. Francis, C.A.; Lin, Z.; Kaldahl, C.A.; Antczak, K.G.; Kumar, V. (2006) *Process for Manufacturing Opioid Analgesics*, US Patent 7,071,336B2.
8. Coop, A.; Janetka, J. W.; Lewis, J. W. L.; Rice, K. C. (1998), Selectride as a general reagent for the O-demethylation and N-decarbomethoxylation of opium alkaloids and derivatives, *J. Org. Chem. 63*, 4392–4396.
9. Coop, A.; Lewis, J. W. L.; Rice, K. C. (1996), Direct and simple O-demethylation of thebaine to oripavine, *J. Org. Chem. 61*, 6774.
10. Wang, P.; Jiang, T.; Cantrell, G.; Berberich, D. (2008), *Improved Preparation of Oxymorphone from Oripavine*, Int. Patent WO 2008118654.
11. Bryant, W.; Mitchell, J.; Smith, D. M. (1940), Analytical procedures employing Karl Fischer reagent. III. The determination of organic acids, *J. Am. Chem. Soc. 60*, 4–6.
12. Johler, J.; Niebruegge, L.; Miller, D.; Wilhelm, R. R. (2005), Use of methylation with electrospray ionization mass spectrometry to assist in the identification of polyaminocarboxylate impurities, *Proc. 53rd ASMS* (Am. Soc. Mass Spectrometry) *Conf.*
13. Kerns, E. H.; Rourick, R. A.; Volk, K. J.; Lee, M. S. (1997), Buspirone metabolite structure profile using a standard liquid chromatographic-mass spectrometric profile, *J. Chromatogr. B 698*, 133–145.
14. Perchalski, R. J.; Wilder, B. J.; Yost, R. A. (1982), Structure eludication of drug metabolites by triple quadrupole mass spectrometry, *Anal. Chem. 54*, 1466–1471.
15. Lee, M. S.; Yost, R. A.; Perchalski, R. J. (1986), Tandem mass spectrometry for the identification of drug metabolites, *Annu. Rep. Med. Chem. 21*, 313–321.
16. Lee, M. S.; Yost, R. A. (1988), Rapid identification of drug metabolites with tandem mass spectrometry, *Biomed. and Environ. Mass Spectrom. 15*, 193–204.
17. Fernadez-Metzler, C. L.; Subrumanian, R.; King, R. C. (2005), Application of technological advances in biomtransformation studies, in Lee, M. S., ed., *Integrated Strategies for Drug Discovery Using Mass Spectrometry*, Wiley, Hoboken, NJ, pp. 261–287.
18. James, T. H. (1977), *The Theory of the Photographic Process*, Eastman Kodak Co., New York, pp. 339–353.

CHAPTER 9

Impurity Identification in Process Chemistry by Mass Spectrometry

DAVID Q. LIU, MINGJIANG SUN, and LIANMING WU
GlaxoSmithKline, Analytical Sciences, 709 Swedeland Road, King of Prussia, PA 19406

9.1 INTRODUCTION

According to the ICH Q3A (R2), impurities in drug substances can be classified into three categories: organic, inorganic, and residual solvent impurities [1]. Organic impurities can arise during manufacturing process (i.e., synthesis, purification, and storage), which can be the starting materials, intermediates, reagents/ligands/catalysts, reaction by-products, degradation products, or other substances. The *process impurities* discussed in this chapter refer to those organic impurities formed during the chemical synthesis as reaction byproducts. These unwanted reaction byproducts usually arise from uncontrolled reactions among the starting materials, intermediates, and reagents, as well as low-level impurities from the chemicals used in the synthesis.

The presence of process impurities in a drug substance usually provides no benefit to patients. Therefore, designing robust manufacturing processes that produce highly pure drug substances is a key objective in process chemistry. In order to achieve impurity controls, the structures and origin of impurities must be determined. Therefore, rapid impurity identification in process chemistry becomes a crucial task in drug development [2]. Sophisticated recrystallization or washing procedures are usually adopted to remove unwanted impurities. Nonetheless, because of the structural similarities between the impurities (viz., drug-related substances) and the active pharmaceutical ingredient, effective purifications are not always readily achievable. Therefore, proactive modification of the chemical process including changing the reagents or reaction conditions would have to be implemented to provide ultimate impurity control whenever feasible [3]. Knowing the mechanism of formation of

Characterization of Impurities and Degradants Using Mass Spectrometry, First Edition.
Edited by Birendra N. Pramanik, Mike S. Lee, and Guodong Chen.

impurities is crucial to improving the synthetic processes with the aim of minimizing or preventing formation.

The coupling between atmospheric-pressure ionization (API) mass spectrometry, including electrospray ionization (ESI) and atmospheric-pressure chemical ionization (APCI) with liquid chromatography (LC), has facilitated the impurity identification enormously [4]. This is due primarily to the superior speed, sensitivity, and selectivity of such "hyphenated" mass spectrometric techniques. Most of the reaction mixtures can be analyzed directly, and structural information of process impurities can be obtained very quickly. This chapter discusses several case studies of identification of unknown impurities encountered in process chemistry with respect to employing LC-MS and GC-MS methodologies. On identification of the impurity, its mechanism of formation can then be elucidated, which subsequently enables chemists to design a control strategy to reduce or sometimes completely eliminate the impurity in the drug substance.

Typical mass spectrometry support in process chemistry is provided in two tiers: *open access* (or "walkup") and *expert support*, respectively. Open-access LC-MS and GC-MS are routine tools maintained by mass spectrometry experts but operated by chemists themselves for monitoring impurities in reaction mixtures. A generic "fast LC" method employing a short C18 column eluted by a gradient of water–acetonitrile fortified with 0.05% TFA as the mobile phases is typically satisfactory for separations (see Section 9.2 for details). Single-quadrupole mass spectrometry such as an Agilent mass selective detector (MSD) is used as the detector. This setup satisfies majority of impurity identification needs of process chemists. Similarly, for open-access GC-MS, chemical ionization single-quadrupole MSD offers a convenient means for examining impurities of nonpolar small-molecule starting materials. In many circumstances, however, impurity peaks are not readily identified, due either to poor ionization of the unanticipated structure (the chemical nature of the impurity) or the extremely low levels. These samples will have to be analyzed by expert mass spectrometrists who can design targeted experiments to investigate the impurity structures. The purpose of this chapter is to provide selected examples on this subject with an attempt to discuss the usefulness of hyphenated mass spectrometric approaches for structural identification in reaction mixtures. Following the positive identification of impurity structures, proper impurity control strategies can then be implemented.

9.2 EXPERIMENTATION

9.2.1 Liquid Chromatography Conditions

For cases 1, 2, and 3, chromatographic separations were achieved by using a generic "fast LC" method where a Phenomenex (Torrance, CA, USA) Luna C18 (2) column (50×2 mm, 3 μm) was used. The mobile phases were delivered by an Agilent binary HP1100 system (Palo Alto, CA, USA) with a linear ramp from 0% B to 100% B in 8 min. Mobile phases A and B were HPLC-grade water with 0.05% TFA and acetonitrile with 0.05% TFA, respectively. For cases 5 and 6, project-specific columns and chromatographic conditions were used for separations.

9.2.2 LC-MS Systems

For cases 1 and 2, mass spectra were acquired on an LTQ ion trap mass spectrometer (Thermo Scientific, San Jose, CA, USA) coupled with a Thermo Scientific Ion Max electrospray ionization source in the positive-ion mode. The source temperature was set at 300°C, and the spray voltage was 4 kV. One-fifth of the 1-mL/min LC effluent was flowed into the mass spectrometer and nebulized using nitrogen as the sheath gas at a flow of 30 au (arbitrary units). The auxiliary gas and sweeping gas flows were both set to 5 au. Full-scan spectra were collected from *m/z* 100 to 1200. Collision-induced dissociation (CID) experiments were performed using helium as the collision gas with typical relative collision energy of 25–30%. All CID spectra were collected using the data-dependent MS^n function. Xcalibur 1.4 was used for instrument control and data reduction. For cases 3 and 5, mass spectra were collected on an Agilent LC-MSD ion trap mass spectrometer model G2445D (Agilent, Palo Alto, CA, USA) operated in the ESI positive-ion mode. The source voltage was set to 4.5 kV with a temperature of 350°C. An LC flow of 1 mL/min was flowed into the ion source directly. The nebulizer gas pressure was 60 psi while the dry-gas flow was set to 12 L/min. LC-MSD trap software 5.3 was used for instrument control and data analysis.

9.2.3 GC-MS System

For case 4, the analysis was carried out using an Agilent GC-MS system (Palo Alto, CA, USA) consisting of a 6890A GC and a 5973 inert mass detector. An HP-5MS (30 m × 250 μm × 0.25 μm) GC column was used. The oven temperature gradient was started at 50°C for 2 min and then ramped to 250°C at 25°C/min and held for 2 min at 250°C. A 4-mm-i.d. liner containing glass wool was used. Helium was used as the carrier gas with a constant flow rate of 1.2 mL/min. The injector temperature was kept at 200°C in split mode (5 : 1). The mass detector was operated in chemical ionization mode where methane was used as the reagent gas. The source temperature and quad temperature were set to 230°C and 150°C, respectively. The MSD transfer line temperature was set at 250°C.

9.2.4 Accurate Mass

For case 6, LC-MS experiments were performed on a Q-TOF Premier quadrupole orthogonal acceleration time-of-flight mass spectrometer (Waters, Manchester, UK) coupled to an Agilent 1100 HPLC system controlled by MassLynx 4.1 software. The electrospray ionization source was operated in the positive-ion mode with a spray voltage of 3.5 kV. The source and desolvation gas temperatures were set to 120°C and 300°C, respectively. The desolvation gas flow rate was 600 L/h, and the sample cone voltage was set to 30 V. Argon was used as the collision gas at 0.45 mL/min with a collision energy of 10 eV for MS and 30 eV for MS/MS experiments. A solution of leucine–enkephalin at *m*/*z* 556.2771 was used as the lockmass, which was introduced via a Lockspray™ device.

9.2.5 Online H/D Exchange LC-MS

Deuterium oxide was employed as the LC-MS mobile phase directly, which can be referred to as *on-column H/D exchange* [5]. On-column H/D exchange is advantageous since it generally produces complete exchanged mass spectra. A trace amount of TFA serves as a good choice among acidic modifiers in terms of maintaining the chromatographic integrity and ionization efficiency in the positive-ion mode if a pure D_2O/ACN system is inadequate. One severe drawback of adding TFA, however, is its limitation for use in the negative-ion mode. In such cases, acetic or formic acids are preferable choices. The organic mobile phase is ACN, used either directly or with a low level of the same modifier present in the aqueous mobile phase. Acetonitrile is an ideal choice as the organic phase since it contains no exchangeable hydrogens that can compromise the completeness of H/D exchange otherwise.

9.3 APPLICATIONS

9.3.1 Identification of Reaction Byproducts by Data-Dependent LC-MS^n (Case 1)

One of the main purposes of identifying process impurities is to understand their mechanism of formation and design solutions for their reduction or elimination. An investigational compound (**A**) [6] was prepared by coupling **B** (an amine) with **C** (a thionyl chloride) as shown in Scheme 9.1. The HCl salt of **B** was originally used; however, it led to a low yield and the resulting reaction mixture was difficult for workup. In order to enhance the reactivity of **B** in attempting to improve the yield, the free base was tested. As expected, the reaction yield improved greatly. However, the new process generated two new impurities, **Imp 1** and **Imp 2**, as shown in Figure 9.1, and they could not be removed readily by solvent washing procedures. In order to reduce or eliminate the two impurities, it was imperative to understand how the

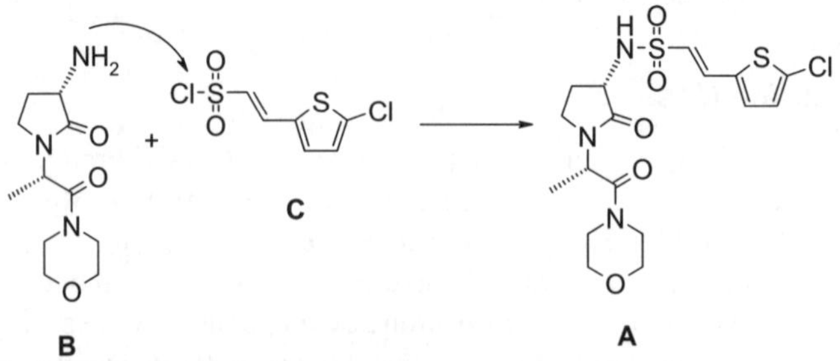

SCHEME 9.1 The reaction scheme for the synhesis of compound **A**.

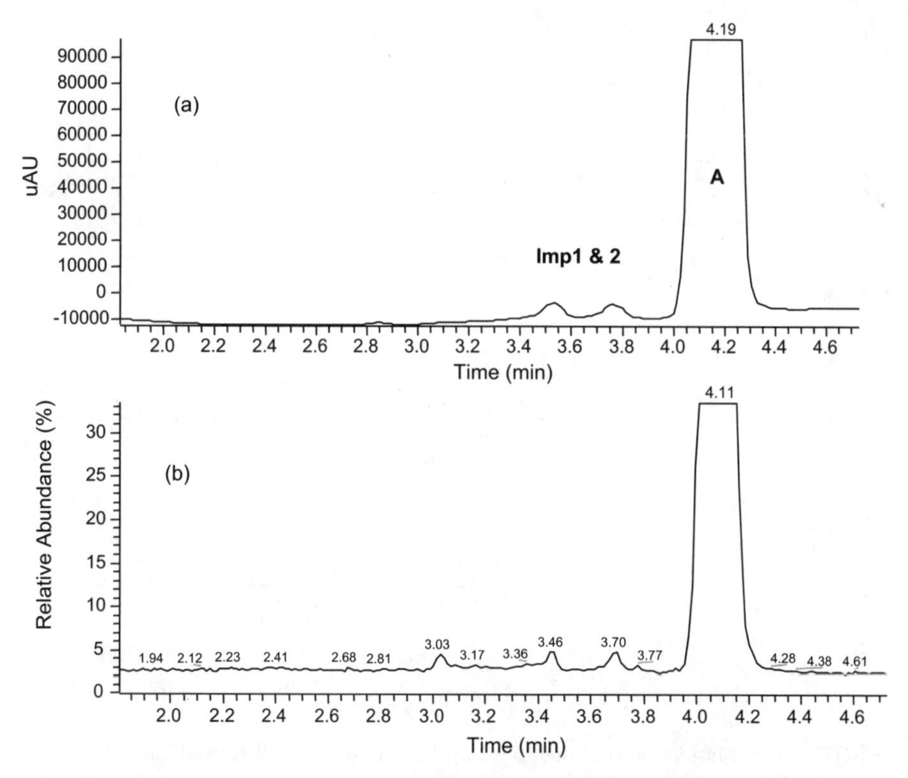

FIGURE 9.1 LC-MS analysis of a synthetic reaction mixture of compound **A** showing the presence of **Imp 1** and **Imp 2**: (a) UV trace; (b) MS total-ion chromatogram (TIC).

impurities were formed. Thus, LC-MS analysis was performed on an LTQ mass spectrometer operated in the data-dependent analysis mode. As a result, full-scan MS, MS^2, and MS^3 (or MS^n if necessary) spectra of the main peak and the two impurities can be collected simultaneously during a single injection [7]. In other words, the selected ion (most intense in this case) in the full-scan MS spectrum (Figure 9.2a) is automatically isolated for MS^2 analysis by obtaining the spectrum as shown in Figure 9.2b; likewise, the selected ion generated by MS^2 is subject to further fragmentation and generates a MS^3 spectrum (Figure 9.2c). Generation of further product ion spectra from the fragment ion produced by the previous stage leads to MS^n data. This continuous multistage MS^n analysis provides a convenient way for structural elucidation of unknown process impurities in reaction mixtures.

The full-scan spectra of the two impurities are identical, containing ions at *m/z* 689 $[M + H]^+$ and *m/z* 711 $[M + Na]^+$ (Figure 9.2a). The MS^2 and MS^3 of both impurities (Figure 9.2b,c) were also identical suggesting that they are structural isomers. Fragmentation of *m/z* 689 gave rise to a number of fragment ions, including *m/z* 545, 448, 361, 333, and 242 in the MS^2 spectrum (Figure 9.2b). The assignments of these ions are shown in Scheme 9.2. Coincidentally, the two ions at *m/z* 448 and *m/z* 242 appeared to be complementary fragments having the same mass as those of

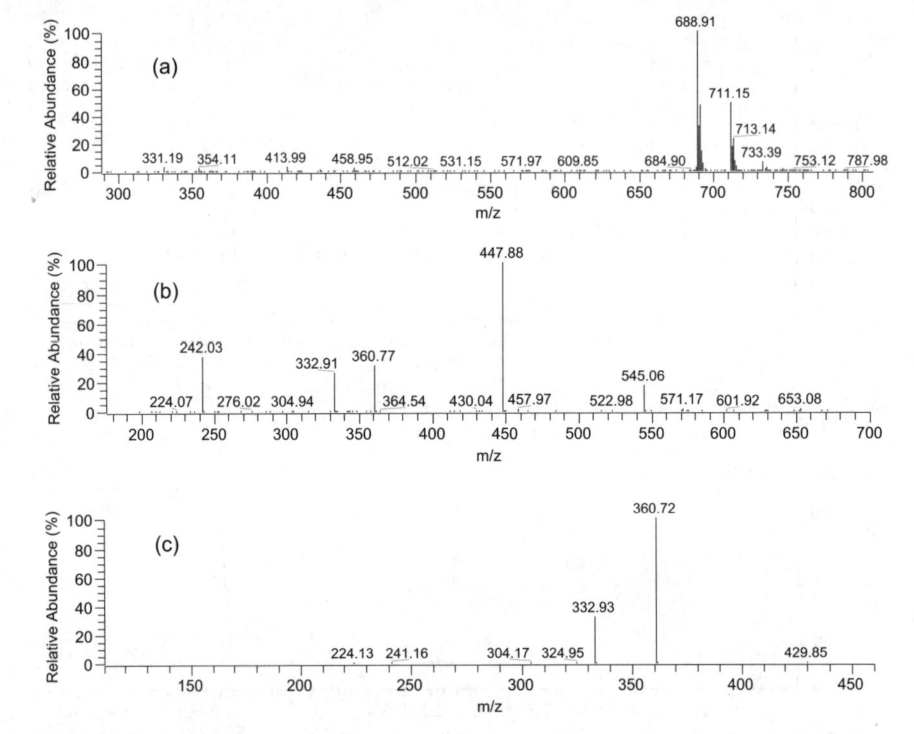

FIGURE 9.2 Mass spectra of impurities **Imp 1** and **Imp 2**: (a) full-scan MS; (b) MS^2 of *m/z* 689; (c) MS^3 of *m/z* 689 → 448.

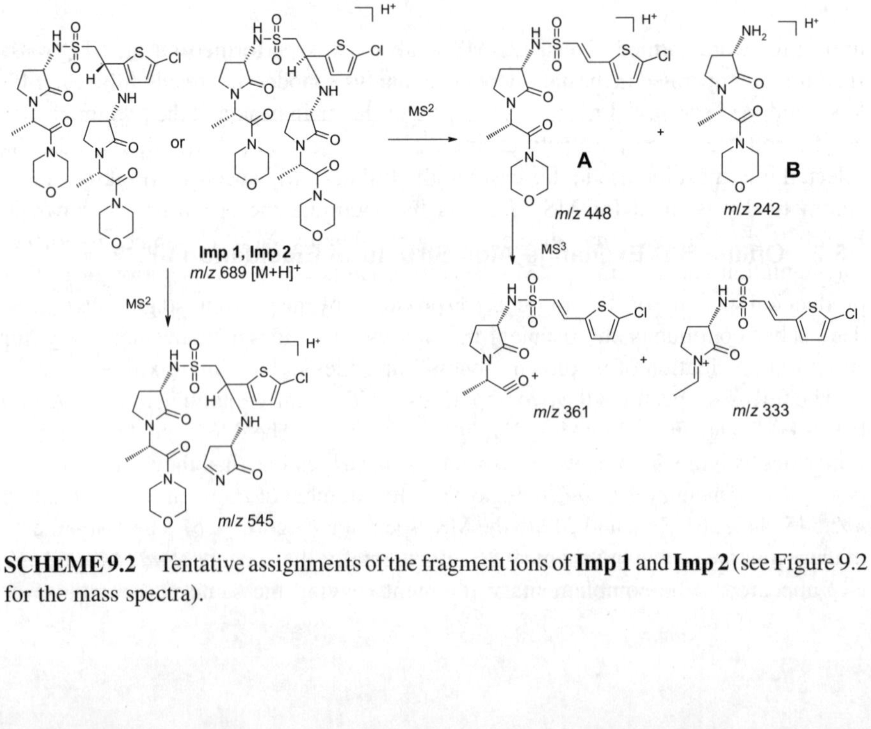

or

Imp 1, Imp 2
m/z 689 $[M+H]^+$

MS^2

A
m/z 448

B
m/z 242

MS^3

m/z 361

m/z 333

MS^2

m/z 545

SCHEME 9.2 Tentative assignments of the fragment ions of **Imp 1** and **Imp 2** (see Figure 9.2 for the mass spectra).

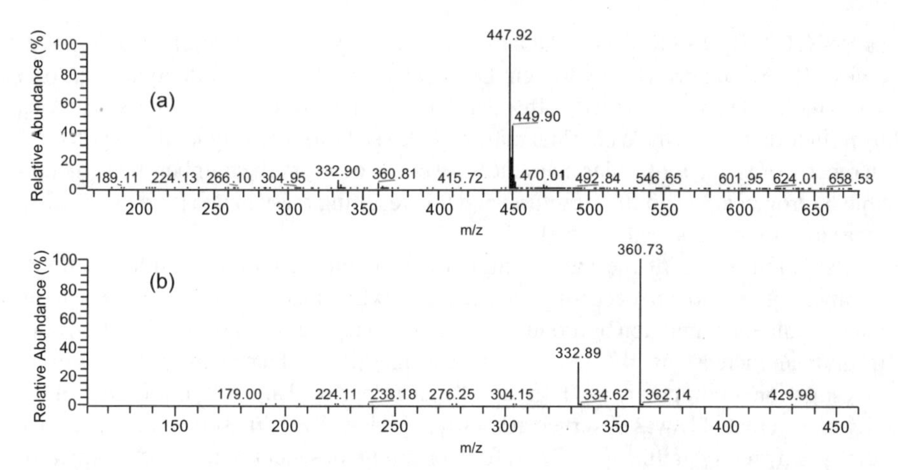

FIGURE 9.3 Mass spectra of the main peak, compound **A**: (a) full-scan MS; (b) MS^2 of *m/z* 448.

A and **B**, respectively. The full-scan MS and MS^2 spectra of the main peak **A** are shown in Figure 9.3. The fragmentation pattern of *m/z* 448 in the MS^2 spectrum of **Imp1** (Figure 9.2c) is identical to that of **A** (Figure 9.3b). This strongly suggests that a second molecule of **B** was added to the target reaction product **A** as a side reaction. Therefore, the Michael addition of a second molecule of **B** to the olefin double bond of **A** was proposed as the mechanism for the two impurities where formation of a pair of diastereoisomers was expected. This was confirmed by LC-NMR data (not discussed here). On identification of the structures and the mechanism of formation of the impurities, a strategy was designed and implemented to reduce the generation of the impurities. The strategy involves addition of a small quantity of pyridine HCl into the solution of **B** to attenuate the nucleophilicity of the primary amine before encountering compound **C**. Indeed, this procedure helped minimize the formation of **Imp 1** and **Imp 2**. This case study exemplifies that data-dependent LC-MS^n is a quick means for identification of process impurities in reaction mixtures.

9.3.2 Online H/D Exchange Aids Structural Elucidation of Process Impurities (Case 2)

During identification of process impurities, isomeric or isobaric structures can often be proposed. Distinguishing between the isomeric or isobaric structures solely on the basis of fragmentation data cannot be easily achieved. Hydrogen/deuterium (H/D) exchange is a useful strategy for distinguishing impurity structures and for structural confirmation. Conducting H/D exchange analysis online using LC-MS made it possible to aid impurity structural elucidation with no need to isolate the impurity, thus greatly facilitating impurity identification [5,8,9]. When organic molecules are exposed to D_2O, heteroatom-bonded labile hydrogens (H) will be replaced by deuterium (D) in the presence of a high concentration of D. This can be described as R–XH + D_2O → R–XD + DOH (where R = partial structure of organic molecules; X = N, O, or S; H = hydrogen; D = deuterium). When *n* labile hydrogens of a given molecule are exchanged with *n* deuterium atoms, the molecular mass increases by *n* since deuterium has a nominal mass of 2 Da (Dalton), 1 Da higher than hydrogen. In other words, the molecular mass increase corresponds to the number of exchangeable hydrogen atoms in the structure, and this resulting mass increase can be readily measured by mass spectrometry.

Oxidation is one of the most commonly encountered undesired side reactions. Methylation is another occurring side reaction when methyl donors are present as either a solvent or reaction byproduct. Oxidation to form a ketone or methylation both result in an increase of 14 Da in molecular weight (MW). During the synthesis of an investigational compound **D** (Figure 9.4), an impurity (**Imp 3**) with an *m/z* of 14 higher than that of **D** was observed at a level as high as 2%. **Imp 3** could not be readily removed by re-crystallizations. Therefore, unambiguous identification of its structure

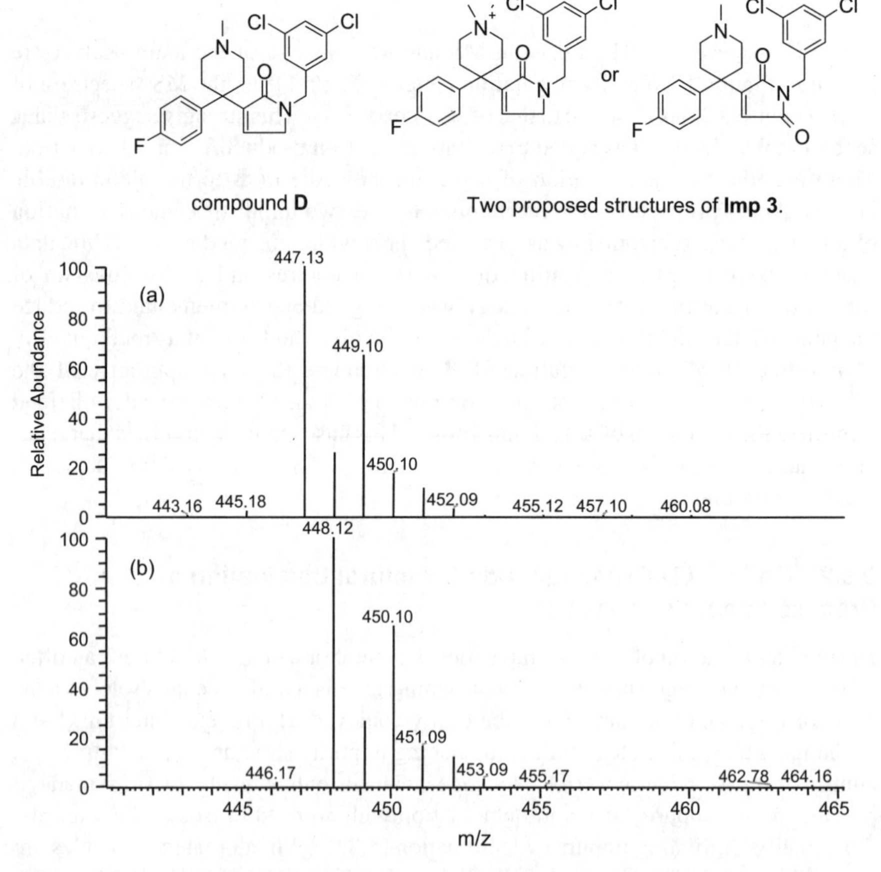

FIGURE 9.4 Full-scan MS spectra of compound **D** in (a) H_2O and (b) D_2O. The structures of compound **D** and **Imp 3** are shown on the top.

by LC-MS was necessary. The desired product (**D**) gave a protonated molecule $[M + H]^+$ at *m/z* 433 (MW 432, spectrum not shown), while the impurity gave an *m/z* of 447 (MW 446) as shown in Figure 9.4a. On the basis of the molecular weight information (plus 14 Da) and the fact that methylation was performed in the previous reaction step, one of the plausible proposals for the impurity structure is a methylated structure (see Figure 9.4). However, modification of the reaction condition to minimize the methyl donor could not reduce or eliminate this impurity. Therefore, on-column H/D exchange LC-MS was performed for structural elucidation of this impurity. When analyzed in D_2O mobile phase, the deuterated molecule $[M_D + D]^+$ of the impurity was detected at *m/z* 448 (Figure 9.4b, bottom spectrum), which was 1 Da higher than that obtained in regular H_2O (Figure 9.4a, top spectrum). This indicates that no exchangeable hydrogen atoms were present in the neutral molecule (i.e., the single deuterium belongs to the charge). If the quaternary methyl structure was correct, the molecular ion would remain unchanged since it is already positively charged. Therefore, an oxidation structure was proposed for **Imp 3**. The fragmentation data shown in Figure 9.5 confirm this proposal. Tentative assignments of the fragment ions of **Imp 3** are described in Scheme 9.3. Unambiguous identification of this impurity was critical for establishing that oxygen was the root cause of the formation of this impurity. On the basis of this finding, exclusion of oxygen from the reaction vessel was implemented as a measure to successfully control the impurity level in the final product. Unambiguous identification of the impurity structure is critical to controlling process impurities, while H/D exchange is a useful approach for distinguishing isomeric or isobaric structures.

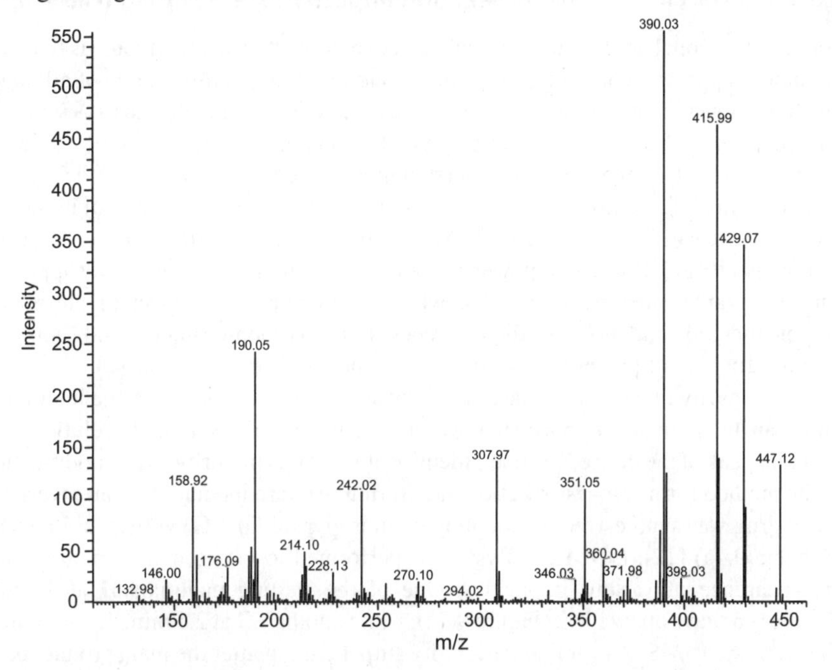

FIGURE 9.5 MS/MS spectrum of the protonated molecule of **Imp 3** at *m/z* 447.

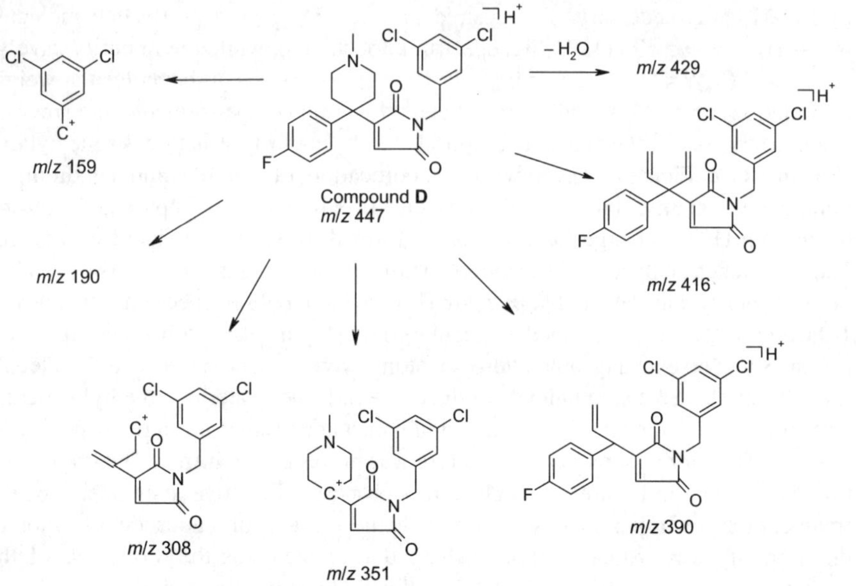

SCHEME 9.3 Tentative assignments of the fragment ions of **Imp 3** (see Figure 9.5 for the mass spectrum).

9.3.3 LC-MS for Chemical Reaction Impurity Fate Mapping (Case 3)

For investigational drugs in the late stage of development where final route is fixed for regulatory approval, comprehensive understanding of the origin and fate of all major reaction impurities at each stage is crucial. This activity is often referred to as *impurity fate mapping*. Comprehensive impurity fate mapping of all stages of chemical reactions is part of the process understanding that ensures successful technology transfer to manufacturing facilities in terms of impurity control. Because of its superb specificity, sensitivity, and speed, LC-MS has become an important tool for impurity mappings [10,11]. It is also a powerful tool for examining peak purity in support of chromatographic method specificity, which is an important aspect of chromatographic method validation. A valid in-process impurity monitoring method should be able to separate all potential impurities from the main product peak.

The impurity fate mapping case study presented here demonstrates the identification of an unanticipated impurity that was inevitably missed because of coelution with the main peak of the desired product. Identification of this impurity led to modification of the method for in-process reaction monitoring. An intermediate **F** of an investigational drug was synthesized by coupling dichloropyrimidine (**G**) with an aniline (**H**) (Scheme 9.4a) [12]. LC-MS analysis was performed to monitor the formation and disappearance of various impurities. One of the starting materials **H** at 1.7 min (Figure 9.6) gave an *m/z* 162 (Figure 9.7a), while another, **G** at 2.5 min, did not ionize by electrospray MS. A prominent impurity **Imp 4** eluting after the main product peak gave a protonated molecule at *m/z* 274 (Figure 9.7b), which was the same as that of the

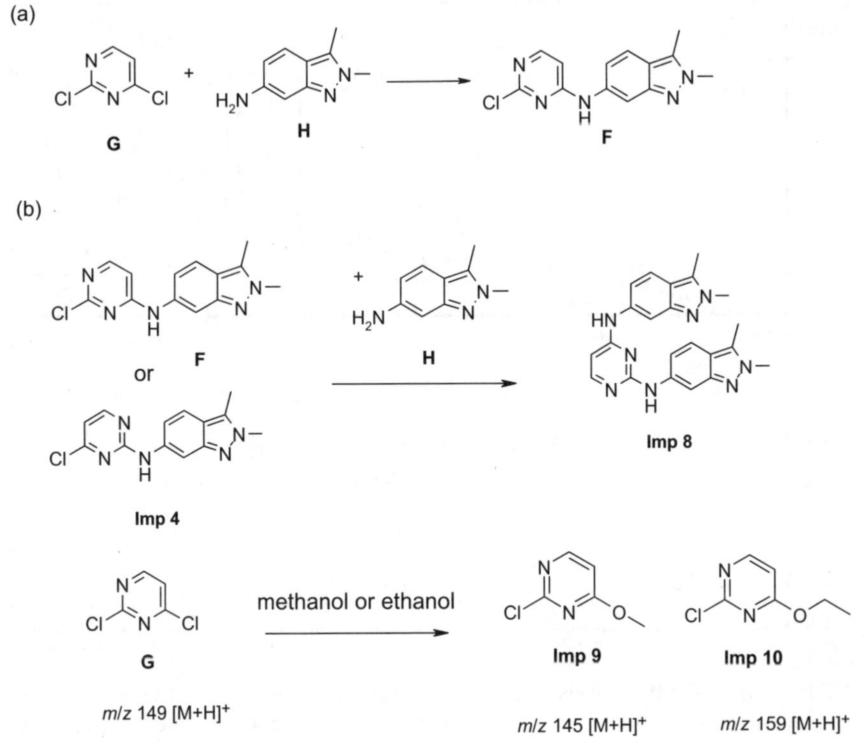

(a)

G + H → F

(b)

F or Imp 4 + H → Imp 8

G → (methanol or ethanol) → Imp 9, Imp 10

m/z 149 [M+H]+   *m/z* 145 [M+H]+   *m/z* 159 [M+H]+

SCHEME 9.4 (a) The reaction scheme for the synthesis of compound **F** and (b) the impurity structures of **Imp 4, Imp 8–10,** and their formation pathways (**Imp5–7** structures are not shown).

desired reaction product **F** (Figure 9.7c). It was readily identified as the isomeric byproduct that was presumably formed by coupling the aniline with the other chlorine-substituted carbon. Several additional impurities, **Imp 5–7**, were identified in a similar manner. Impurities **Imp 5, Imp 6,** and **Imp 7** gave protonated molecules at *m/z* 335 (Figure 9.7d), *m/z* 511(Figure 9.7e), and *m/z* 636 (Figure 9.7f), respectively. The possibility of them being carried into the next stage of reaction or converted into new impurities was examined. This led to an impurity fate map that cannot be disclosed because of proprietary information.

Furthermore, the impurity fate mapping of this stage of chemistry led to the identification of a coeluting impurity **Imp 8** under the main peak **F** when a generic "fast LC" method was used (see Section 9.2). **Imp 8** gave an *m/z* of 399 $[M + H]^+$ under the main peak (Figure 9.7c). On the basis of the chemistry and the MS^2 and MS^3 data shown in Figure 9.8, its structure was proposed as shown in Scheme 9.4b. Mass spectrometric fragmentations of **Imp 8** were tentatively assigned as shown in Scheme 9.5. The formation of this impurity was proposed where a second molecule of **H** appears to react with **Imp 4** or product **F** itself giving rise to **Imp 8**

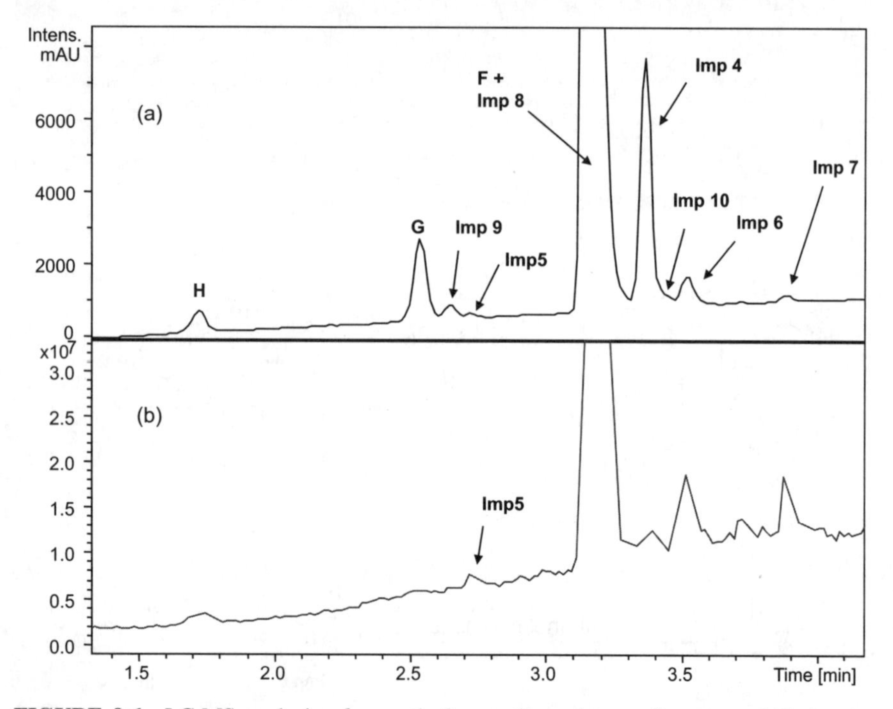

FIGURE 9.6 LC-MS analysis of a synthetic reaction mixture of compound **F** showing the detection of the starting materials **G** and **H** as well as **Imp 4–10**: (a) UV trace; (b) MS TIC.

(Scheme 9.4b). Should one rely solely on the conventional HPLC-UV methodology in absence of LC-MS for process impurity monitoring, **Imp8** would have not been discovered readily. Actually this impurity was present in the isolated product, which was unnoticed prior to the use of LC-MS for the impurity fate mapping. As a consequence, an improved in-process monitoring method was developed (not discussed here) that was able to resolve **Imp 8** from **F** effectively. As demonstrated here, LC-MS has become an essential tool for monitoring *peak purity* during LC method development and validation [13] to ensure that the impurity method is able to adequately separate or accurately determine the process impurities as the method is intended for, namely, method specificity.

One of the starting materials, dichloropyrimidine (**G**), and **Imp 9** and **Imp 10** are not amenable to ESI MS analysis; thus GC-MS was employed for their identification or confirmation (see case study 4 below).

9.3.4 GC-MS for Impurity Profiling of Small-Molecule Starting Materials (Case 4)

In process chemistry, with regard to identification of small-molecule starting materials and their impurities, GC-MS is the tool of choice. During the same stage synthesis

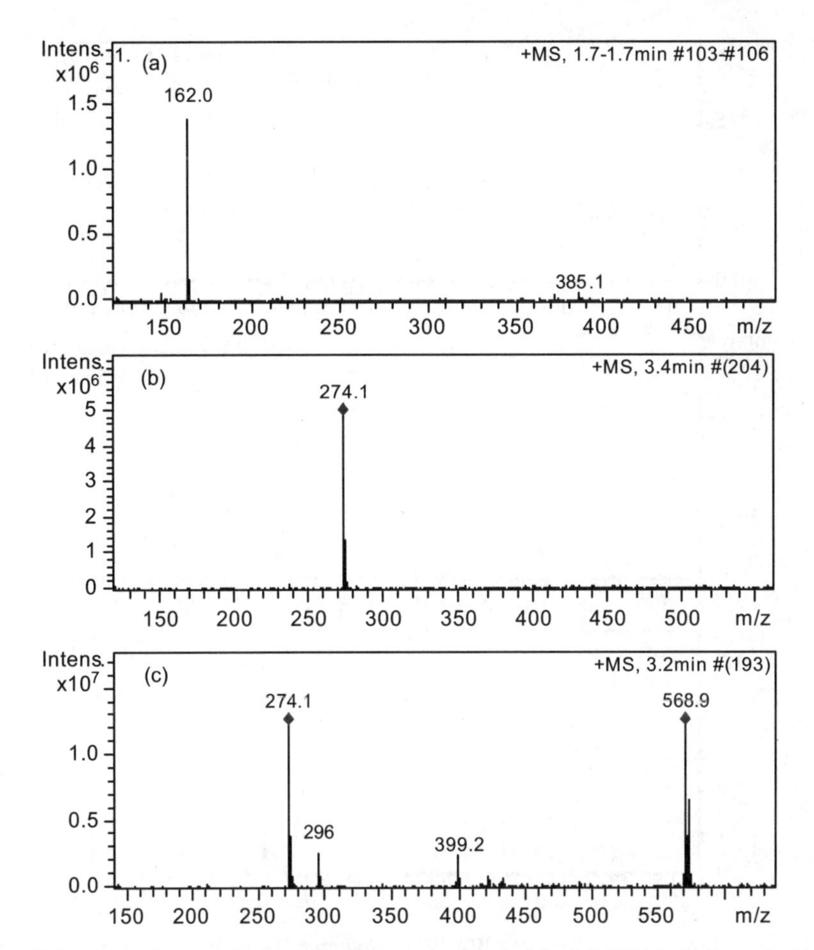

FIGURE 9.7 Full-scan MS spectra of the peaks detected by LC-MS analysis as shown in Figure 9.6: (a) compound **H**; (b) **Imp 4**; (c) the main peak, **F** plus coeluting **Imp 8**; (d) **Imp 5**; (e) **Imp 6**; (f) **Imp 7**. Compound **G**, **Imp 9**, and **Imp 10** did not ionize well under the ESI MS conditions.

of **F** as described above (see case study 3), one of the starting materials, dichloropyrimidine (**G**), when encountering alcoholic solvents such as ethanol and methanol, was anticipated to generate methoxychloropyrimidine (**Imp 9**) and ethoxy-chloropyrimidine (**Imp 10**) impurities, respectively (Scheme 9.4b). By the generic fast LC ESI-LC-MS method, peaks **G**, **Imp 9,** and **Imp 10** in the UV trace (Figure 9.6a) did not give apparent peaks in the total-ion chromatogram (TIC); thus the molecular ions cannot be determined readily (Figure 9.6b). Chemical ionization GC-MS, on the other hand, afforded intense signals for these impurities (Figure 9.9). As shown in Figure 9.10, peak **G** and the two impurities, **Imp 9** and **Imp 10**, gave intense molecular ions at *m/z* 149 (with an isotope pattern of two chlorines), *m/z* 145 (with an isotope pattern of one chlorine), and *m/z* 159 (with an

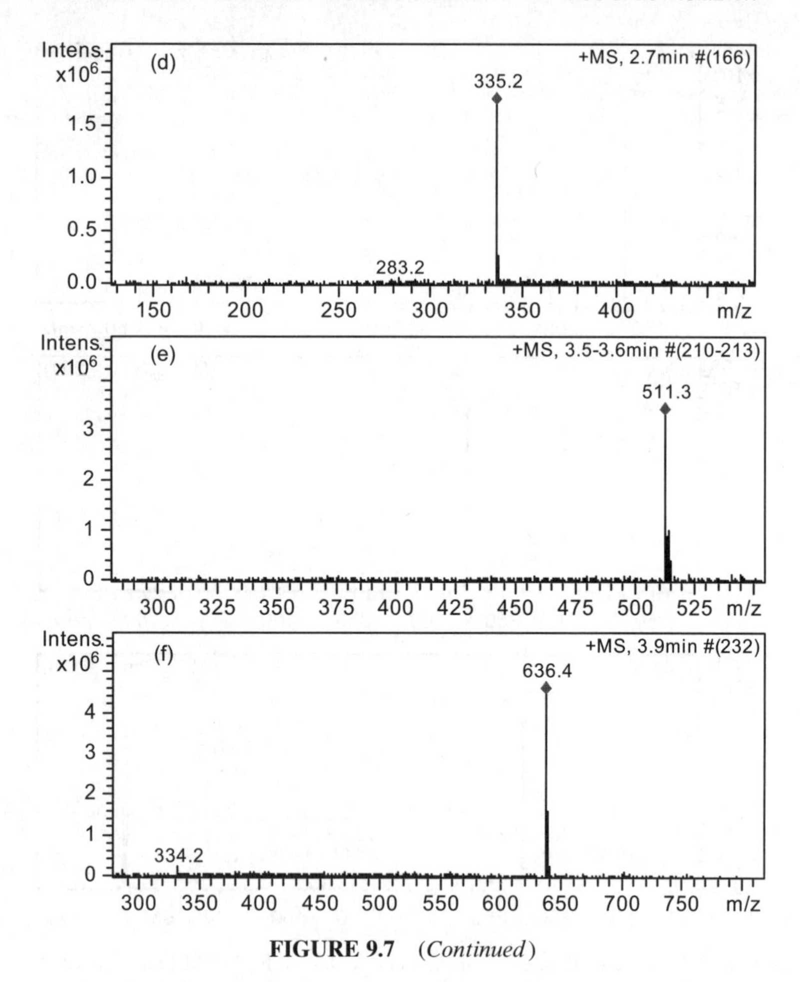

FIGURE 9.7 (*Continued*)

isotope pattern of one chlorine), respectively. Once the structures of **Imp 9** and **Imp 10** were established, corresponding ions at *m/z* 145 and *m/z* 159 at trace levels were extractable from the ESI LC-MS data, allowing correlation of the GC-MS peaks with the LC-UV peaks. In addition, authentic materials were prepared chemically later on for use as retention time markers as a means of confirmation of the LC-UV peaks.

In process chemistry, organic solvents are often used as reaction media or for equipment cleanup and thus may end up as impurities in drug substances. In addition, drug substances can sometimes become contaminated in process equipments (e.g., lubricants for mechanical parts) or packaging materials. These small molecules are often nonpolar in nature and may not always be amenable to atmospheric-pressure ionization LC-MS methods. As such, GC-MS has become the method of choice for identification of volatile small-molecule impurities or contaminants [3,4].

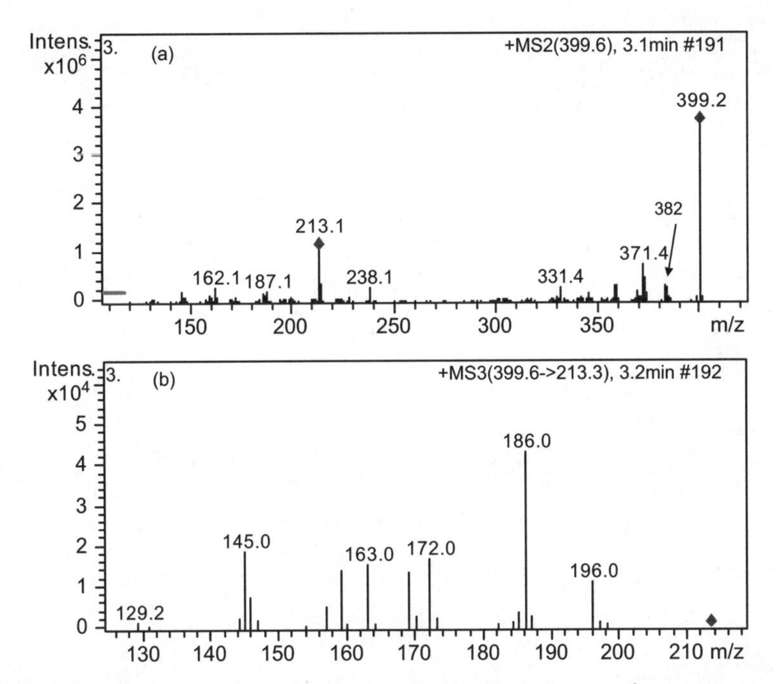

FIGURE 9.8 Mass spectra of the protonated molecule of **Imp 8**: (a) MS^2 of *m/z* 399; (b) MS^3 of *m/z* 299 → 213.

9.3.5 Identification of a Process Impurity that Impacts Downstream Formulation (Case 5)

This case study discusses the identification of a process impurity that impacts downstream formulation (especially injectable formulations). During an intravenous formulation of compound **I** (viz., Ispinesib, Toronto Research Chemicals Inc., North York, Ontario, Canada), fine floating particulates were observed in the vials during product inspection. This led to a quality investigation of the root cause responsible for the product failure. The floating particulates were collected by filtration and subjected to LC-MS identification. A prominent late-eluting impurity (**Imp 11)** was detected at 8.7 min by reversed-phase HPLC indicating its high hydrophobicity (Figure 9.11). This impurity gave an *m/z* of 460, which was 57 Da less than that of compound **I** itself, suggesting that was the desaminopropyl impurity (see Scheme 9.6 for the structures of **I** and **Imp 11**). This hypothesis was supported by the MS/MS data (Figure 9.12). The fragmentation data of compound **I** and **Imp11** were tentatively assigned as shown in Scheme 9.6a,b, respectively.

This impurity was present in the particulates (collected by filtration) at a level as high as 3% in the formulation solution (Figure 9.11a), while it was only 0.07% in the original batch of drug substance (Figure 9.11b). LC-MS analysis showed that all batches of drug substances contain this impurity. However, the highest level detected was 0.07%. Typically an impurity below 0.1% does not constitute a great concern for

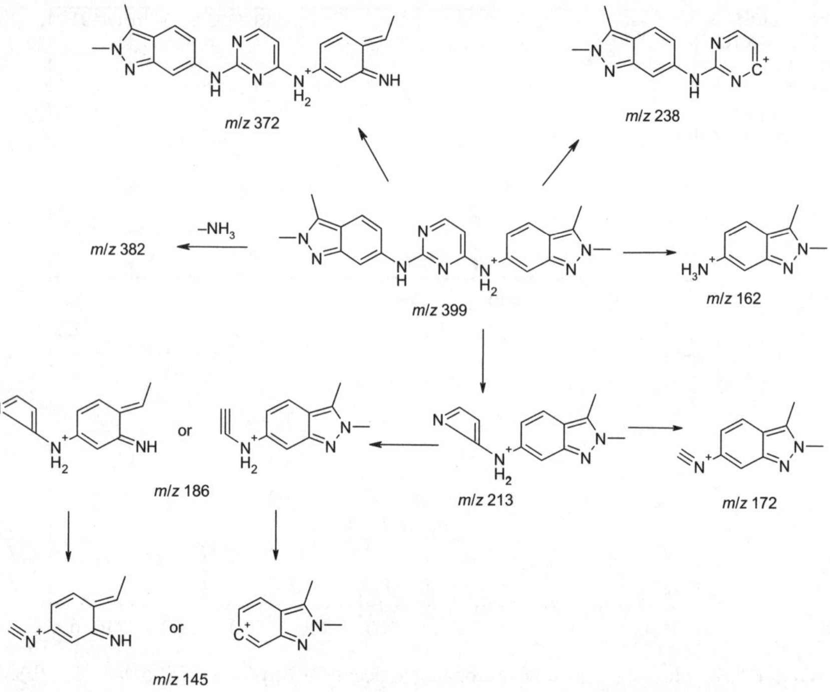

SCHEME 9.5 Tentative assignments of the fragment ions of **Imp 8** (see Figure 9.8 for the mass spectra).

drug development, and its identification is not required when the dose is below one gram. In this case, however, it was believed that the impurity was partially solubilized during compounding of the bulk solution and passed through the filter at a very low level (only ~0.03%). On storage at a low temperature, it precipitated out as a result of the poor water solubility, presumably due to the absence of the primary amine functional moiety in the API molecule that is ionizable in the formulation buffer. Although present at trace levels, water-insoluble impurities can constitute a critical quality attribute of drug substances intended for injectable formulations. It's imperative to consider this possibility during synthetic route development. Since a trace level of this impurity was unavoidable for the current process, an extra filtration procedure had to be implemented during formulation manufacturing to remove this impurity completely. Alternatively, in order to ensure the successful formulation of the injectable solution of compound **I** that is free from forming particulates of **Imp 11**, a new synthetic route might have to be developed. LC-MS plays a critical role in monitoring and identification of impurities at this trace level.

It is worth noting that both compound **I** and the impurity (**Imp11**) afforded water-loss fragment ions at *m/z* 499, and 442, respectively, from their respective protonated molecules $[M + H]^+$ on collision-induced dissociation. However, they may arise from two different mechanisms. Elimination of water at the quinazolinone is unlikely

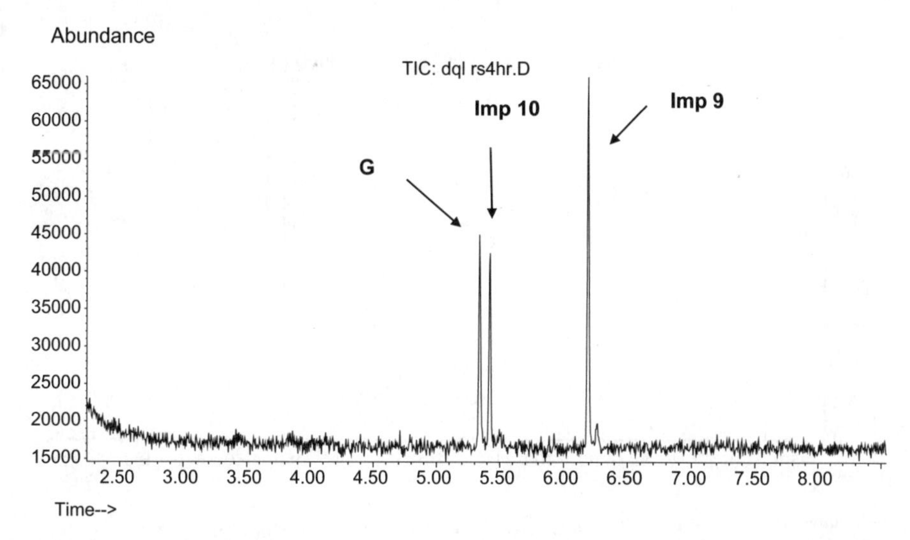

FIGURE 9.9 GC-MS chromatogram of a synthetic reaction mixture of compound **F** showing the detection of the starting material **G** and two impurities, **Imp 9** and **Imp 10**.

due to the lack of neighboring proton. Therefore, dehydration must have occurred at the benzamide carbonyls. Two distinct dehydration mechanisms (1 and 2), could be proposed for compound **I** as illustrated in Scheme 9.7a. Cyclization between the primary amine of the sidechain and the benzamide carbonyl would give a stable six-membered ring imine structure at *m/z* 499 (1). Alternatively, tautomerization involving a double-bond shift (becoming an enamine) followed by cyclization between the quinazolinone nitrogen and the benzamide carbonyl would give rise to a conjugated five-membered ring imidazole cation (2). The former mechanism resembles the water loss from cysteine-containing peptides proposed by Reid et al. [14], where it involves an intramolecular nucleophilic sidechain thiol attack on the carbonyl of the amide bond. As for **Imp 11**, on the other hand, the retro-Ritter process (3) shown in Scheme 9.7b could be one of the possible water-loss processes [15] that forms a nitrilium product ion at *m/z* 442. Alternatively, it could follow the "tautomerization/cyclization forming the imidazole" process (4). The nitrilium product ion may have higher relative energy compared to the toluene imidazole structure, as demonstrated by Reid et al. [14] in the case of *N*-formylcysteine. In considering the product ion stability, the imidazole pathway (4) is more likely the favored process. Further evaluation by ab initio calculations would be helpful in confirming the true underlying mechanisms.

9.3.6 Differential Fragmentation between Sodiated and Protonated Molecules as a Means of Structural Elucidation (Case 6)

Protonated molecules $[M + H]^+$ are the most commonly observed ions in ESI mass spectrometry. Sodiated molecules $[M + Na]^+$, however, are also common for some

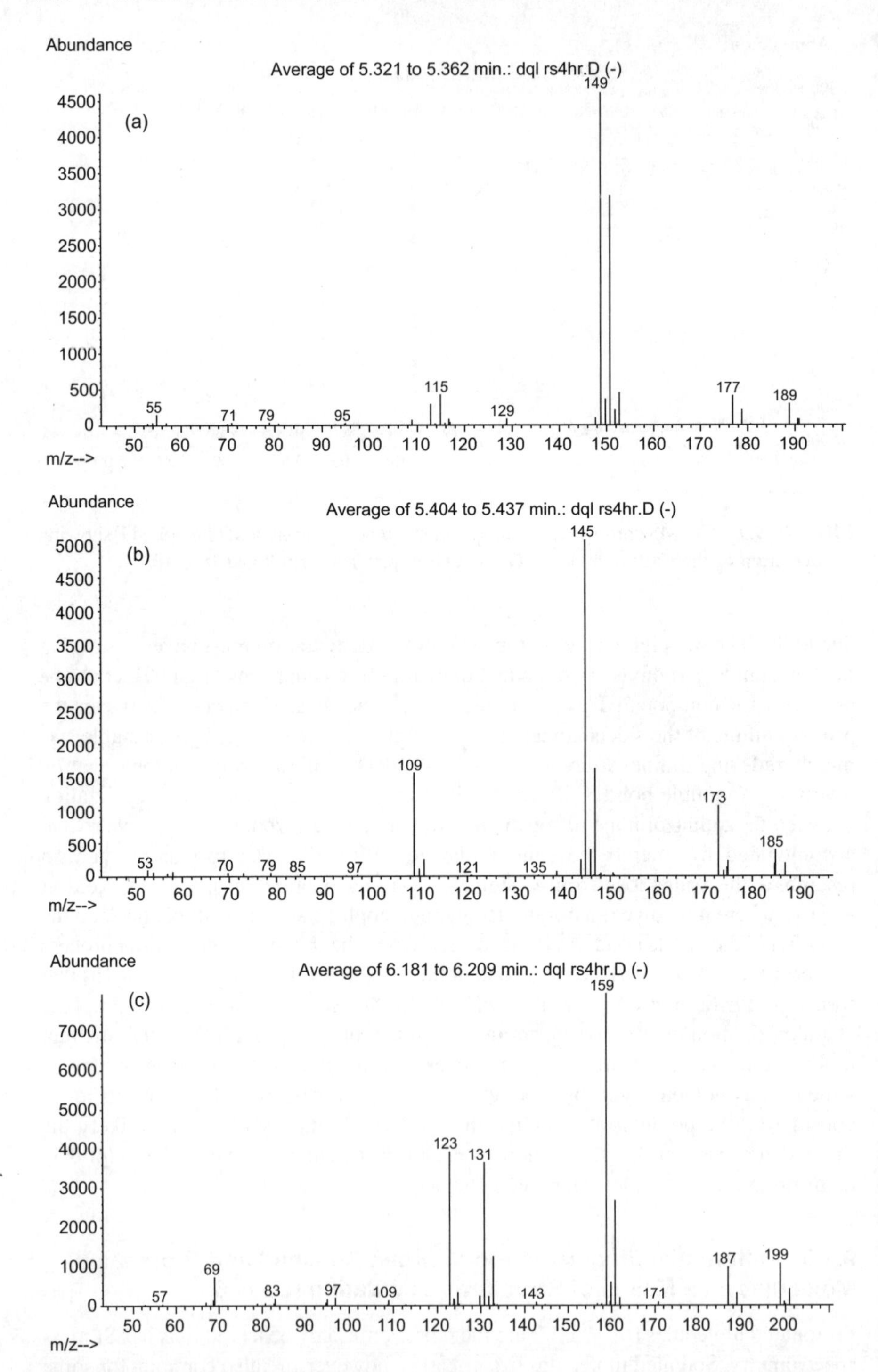

FIGURE 9.10 Chemical ionization (CI) mass spectra of the peaks **G** (a), **Imp 9** (b), and **Imp 10** (c) detected by the GC-MS analysis illustrated in Figure 9.9.

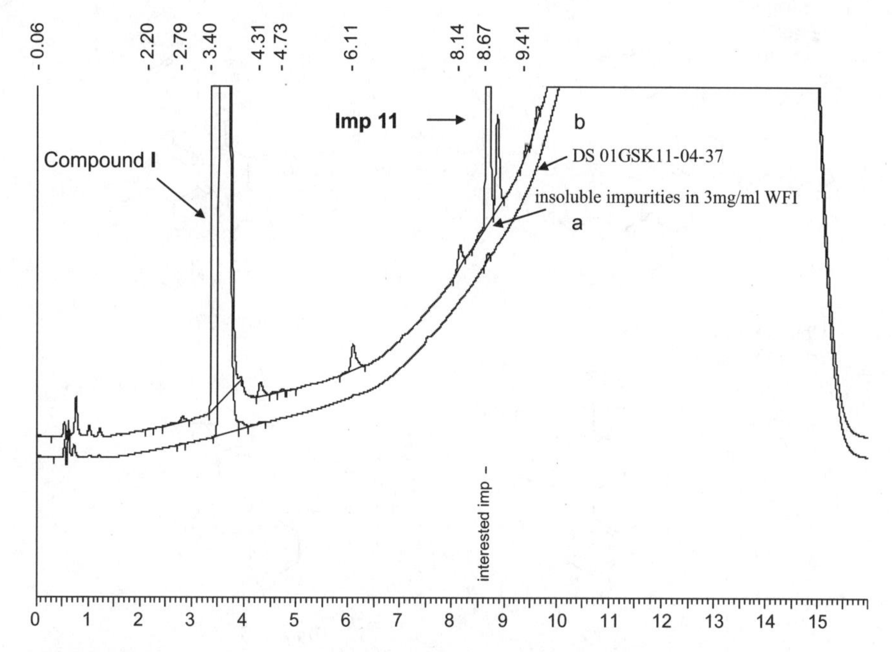

FIGURE 9.11 LC-UV chromatograms of (a) the insoluble particles collected by filtration during formulation of compound **I**, which contains enriched **Imp 11**, and (b) the original drug substance of compound **I** containing a very low level of **Imp 11**.

molecules such as the pleuromutilin class of antibiotics. Compound **J** (Scheme 9.8), a synthetic intermediate of an investigational pleuromutilin antibiotic [16], tends to give intense sodiated adduct $[M + Na]^+$. The fragmentation pathways of the sodiated molecules can be very different from those of the protonated species as a result of different bonding affinity, that is, attaching H^+ or Na^+ charge at different sites within a molecule. Thus, examining fragmentation of the sodiated molecules can provide complementary information for structural elucidation of unknown process impurities. We demonstrate here the use of such a strategy for the structural elucidation of **Imp 12** during the synthesis of compound **J** (Scheme 9.8). The protonated molecules of **Imp 12** showed that it had a mass of 14 Da higher than **J**. Accurate mass measurement using a Q-TOF suggested that this mass difference was attributed to a methylene group (CH_2). Furthermore, MS/MS of protonated molecules of compound **J** and **Imp 12** indicated that the mutilin core was intact as evidenced by the common ions at *m/z* 317 and *m/z* 285, respectively (Scheme 9.8a). The MS/MS spectra of the protonated molecules, however, did not afford any fragment ions in relating to the sidechain where the structure modification occurred. Since the sidechain contains multiple oxygen and nitrogen atoms, it provides a good chelating moiety for sodium ion. Indeed, compound **J** and **Imp 12** afforded intense signals corresponding to the sodiated molecule at *m/z* 641 and *m/z* 655, respectively. Performing MS/MS fragmentation

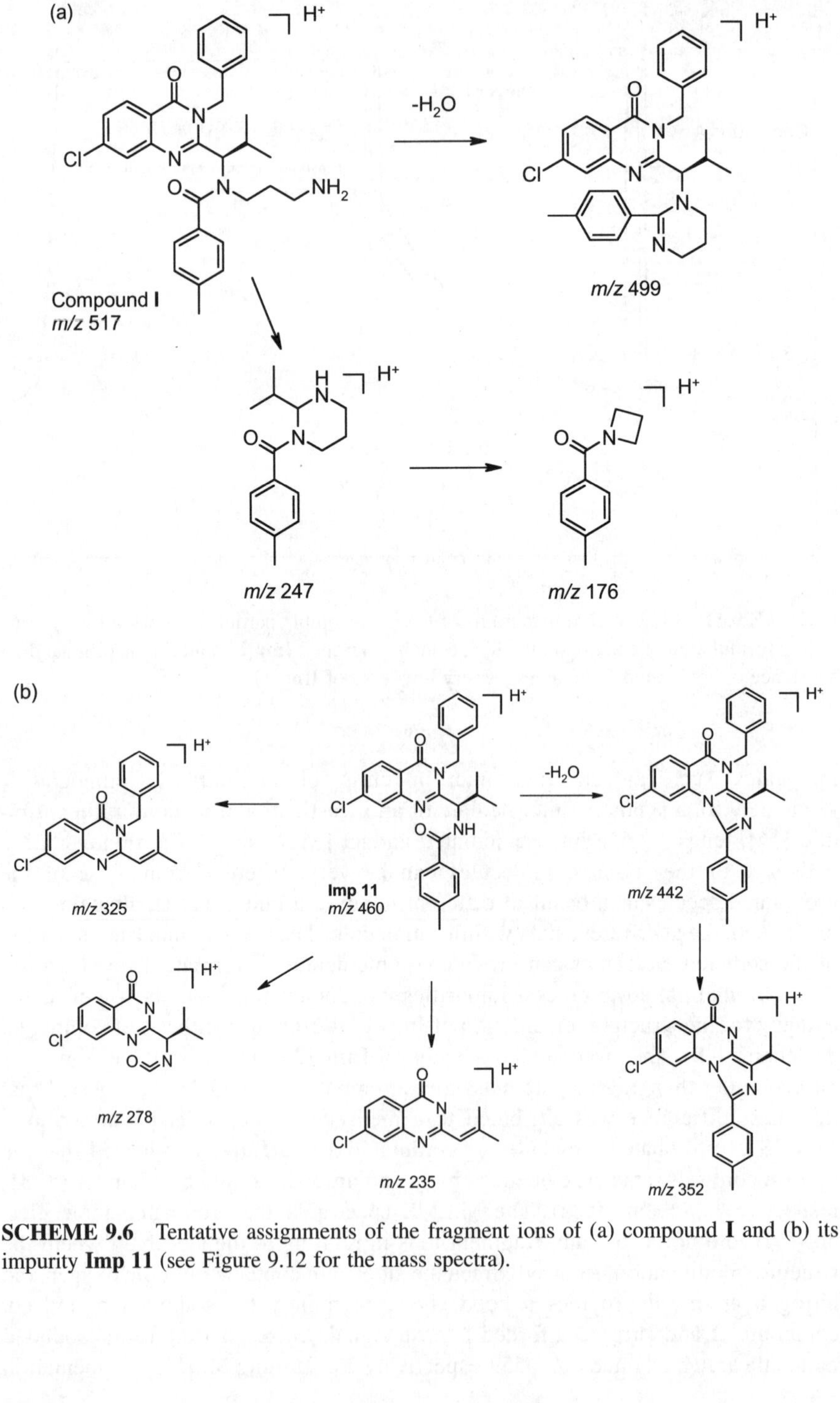

(a)

Compound **I**
m/z 517

$-H_2O$

m/z 499

m/z 247

m/z 176

(b)

m/z 325

Imp 11
m/z 460

$-H_2O$

m/z 442

m/z 278

m/z 235

m/z 352

SCHEME 9.6 Tentative assignments of the fragment ions of (a) compound **I** and (b) its impurity **Imp 11** (see Figure 9.12 for the mass spectra).

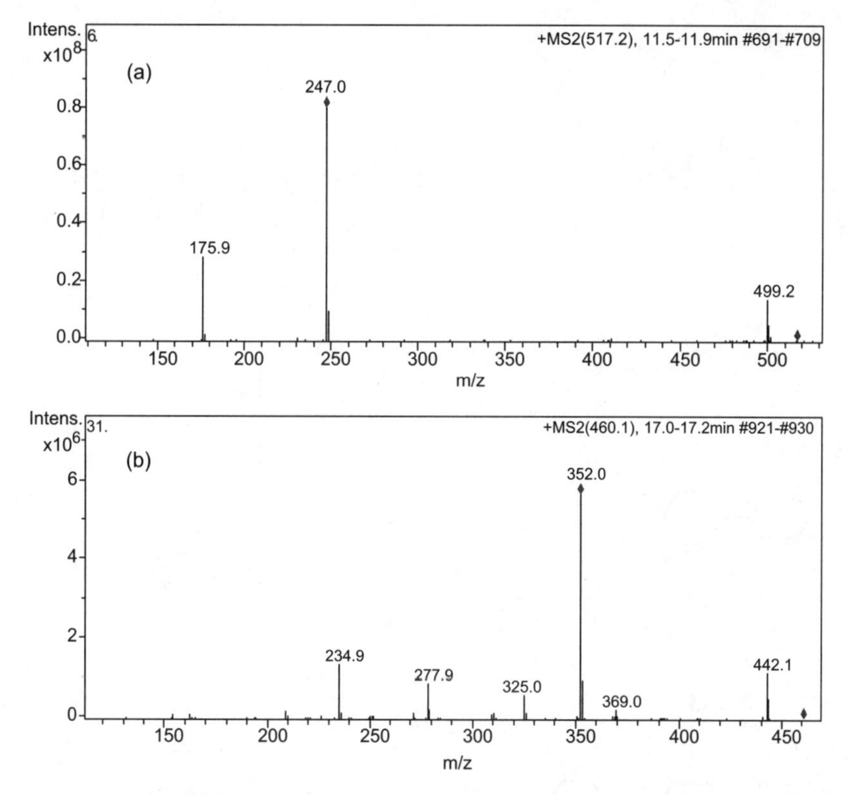

FIGURE 9.12 MS/MS spectra of the protonated molecules of: (a) compound **I** at *m/z* 517, and (b) **Imp 11** at *m/z* 460.

on the sodiated molecule of **Imp 12** did shed light on its structure with regard to the site of the methylene moiety.

Accurate mass measurements were used to make unambiguous structural assignments of the product ions. By comparing the three product ions of **Imp 12** at *m/z* 339.1524, *m/z* 295.1633, and *m/z* 234.1470 (Figure 9.13b) with those of compound **J** at *m/z* 325.1371, *m/z* 281.1484, and *m/z* 220.1317 (Figure 9.13a), we can easily deduce that **Imp 12** contained an additional methylene group. There were two product ions at *m/z* 225.0845 and *m/z* 164.0690 that are the same for **Imp 12** and **J**, respectively, within 2.2 ppm, indicating that this portion of the structure was unchanged. The structures of the major fragment ions are assigned in Scheme 9.8b. Combining all the evidence, it allowed unambiguous determination of the structure of **Imp 12**, which contains a *tert*-pentoxycarbonyl instead of a *tert*-butyloxycarbonyl (*t*-BOC). This impurity was presumably introduced via the use of sodium *tert*-pentoxide as the base in the process. Indeed, reducing the

(a)

(1)

Compound **I**, *m*/*z* 517

H_2O

m/*z* 499

(2)

(R = $CH_2CH_2CH_2NH_2$)

Compound **I**, *m*/*z* 517

H_2O

m/*z* 499

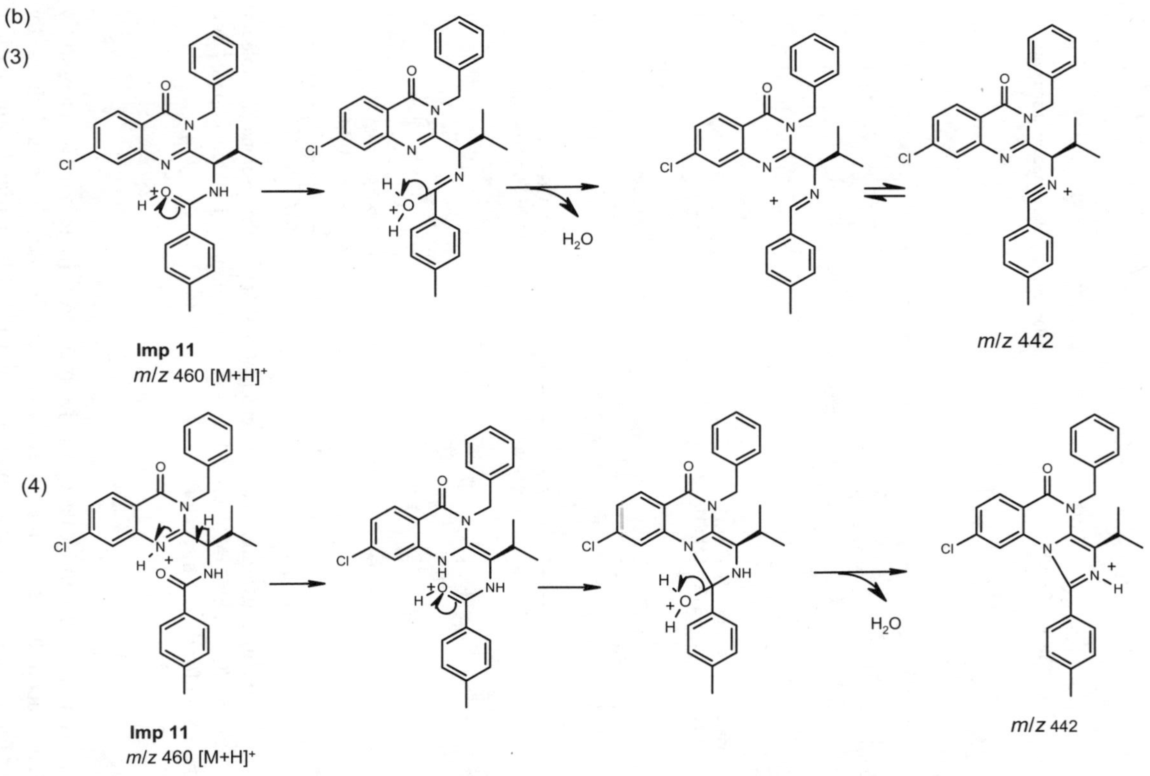

(b)

(3)

Imp 11
m/z 460 $[M+H]^+$

H_2O

m/z 442

(4)

Imp 11
m/z 460 $[M+H]^+$

H_2O

m/z 442

SCHEME 9.7 Proposed dehydration mechanisms for (a) compound **I** at m/z 517 and (b) **Imp 11** at m/z 460 in gas phase in the ESI MS.

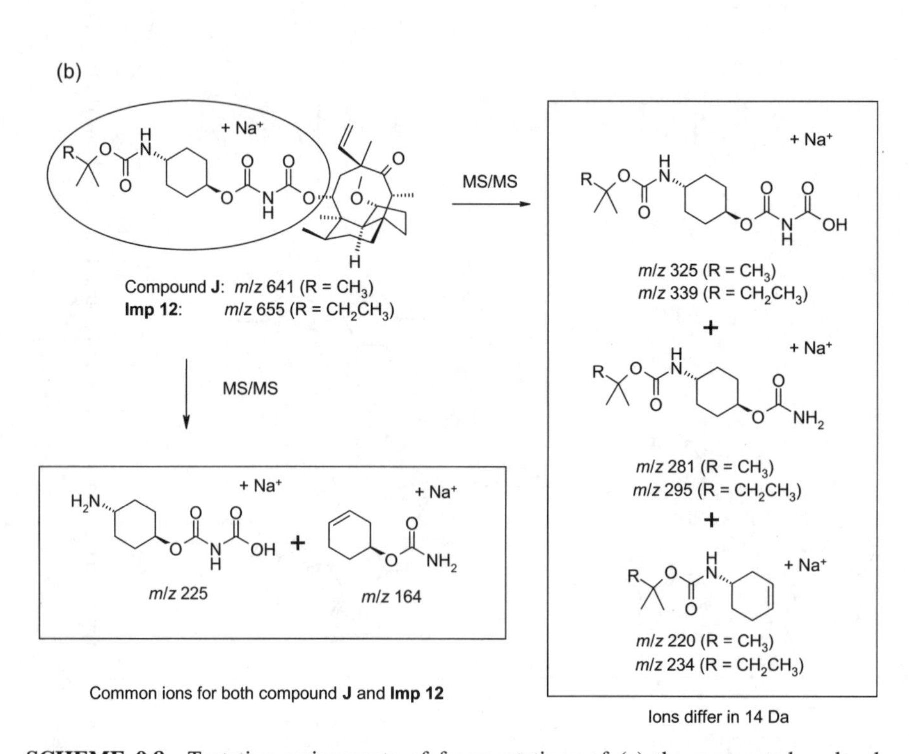

SCHEME 9.8 Tentative assignments of fragmentations of (a) the protonated molecule $[M + H]^+$ and (b) the sodiated molecule $[M + Na]^+$ of compound **J** and its impurity **Imp 12** (see Figure 9.13 for the mass spectra).

amounts of sodium *tert*-pentoxide was able to control the impurity successfully. In summary, accurate mass measurements were helpful in establishing the molecular formulas of unknown impurities, and the differential fragmentations of protonated and sodiated molecules are complementary for structural elucidations. Different ion adducting sites lead to different fragment pathways, which afford fine structural details in the relevant part of the unknown impurity under investigation.

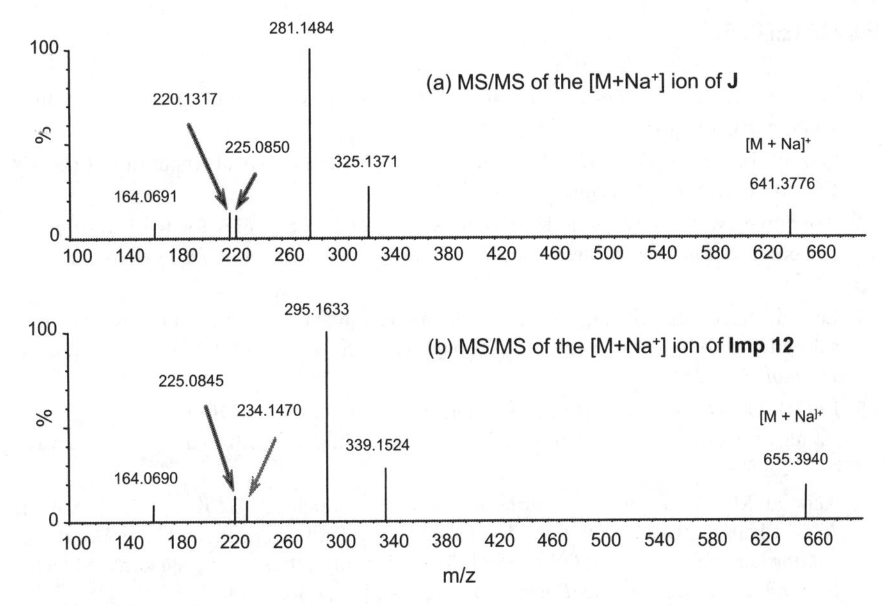

FIGURE 9.13 MS/MS spectra of the sodiated molecules $[M + Na]^+$ of compound **J** at *m/z* 641 and its impurity **Imp 12** at *m/z* 655.

9.4 CONCLUDING REMARKS

Identification of chemical impurities at low levels in the presence of a high concentration of main product is a crucial part of the process chemistry in drug development. Establishing the origin of the impurity enables the design of proper control strategy for controlling process impurities. Six case studies of identification of various process impurities discussed in this chapter demonstrate the selected application of several tandem mass spectrometric techniques. This is by no means a comprehensive review of the full utility of these hyphenated mass spectrometric techniques; instead, it represents a few selected case studies only. Great strides have been made in the development of online LC-MS and GC-MS techniques capable of providing insight into impurity structures. Despite the utility and convenience of these hyphenated techniques, they do not eliminate the need for impurity isolation sometimes for structural identification or confirmation. When definitive structures are desired, preparative isolation followed by NMR analysis is a powerful approach for impurity identification [17]. Alternatively, confirmation of impurity structures can be achieved by chemical synthesis of reference standards.

ACKNOWLEDGMENTS

The authors are grateful to Drs. Rennan Pan, Ravinder Sudini, John Guo, and Mohanmed Mokhallalati of GlaxoSmithKline for their contributions.

REFERENCES

1. USDHHS (2008), *Guidance for Industry, Q3A Impurities in New Drug Substances*, June 2008, ICHQ3A, revision 2 (R2).
2. Qiu, F.; Norwood, D. L. (2007), Identification of pharmaceutical impurities, *J. Liquid Chromatogr. Related Technol. 30*, 877–935.
3. Argentine, M. D.; Owens, P. K.; Olsen, B. A. (2007), Strategies for the investigation and control of process-related impurities in drug substances, *Adv. Drug Deliv. Rev. 59*, 12–28.
4. Lee, H.; Shen, S.; Grinberg, N. (2008), Identification and control of impurities for drug substance development using LC/MS and GC/MS, *J. Liquid Chromatogr. Related Technol. 31*, 2235–2252.
5. Liu, D. Q.; Wu, L.; Sun, M.; MacGregor, P. A. (2007), On-line H/D exchange LC-MS strategy for structural elucidation of pharmaceutical impurities, *J. Pharm. Biomed. Anal. 44*, 320–329.
6. Abboud, M. A.; Needle, S. J.; Burns-Kurtis, C. L.; Valocik, R. E.; Koster, P. F.; Amour, A. J.; Chan, C.; Brown, D.; Chaudry, L.; Zhou, P.; Patikis, A.; Patel, C.; et al. (2008), Antithrombotic potential of GW813893: A novel, orally active, active-site directed factor Xa inhibitor, *J. Cardiovasc. Pharmacol. 52*, 66–71.
7. Liu, D. Q.; Hop, C. E. C. A. (2005), Strategies for characterization of drug metabolites using liquid chromatography-tandem mass spectrometry in conjunction with chemical derivatization and on-line H/D exchange approaches, *J. Pharm. Biomed. Anal. 37*, 1–18.
8. Olsen, M. A.; Cummings, P. G.; Kennedy-Gabb, S.; Wagner, B. M.; Niool, G. R.; Munson, B. (2000), The use of deuterium oxide as a mobile phase for structural elucidation by HPLC/UV/ESI/MS, *Anal. Chem. 72*, 5070–5078.
9. Novak, T. J.; Helmy, R.; Santos, I. (2005), Liquid chromatography-mass spectrometry using the hydrogen/deuterium exchange reaction as a tool for impurity identification in pharmaceutical process development, *J. Chromatogr. B, 825*, 161–168.
10. Chen, Y.; Brill, G. M.; Benz, N. J.; Leanna, M. R. R.; Dhaon, M. K.; Rasmussen, M.; Zhou, C. C.; Bruzek, J. A.; Bellettini, J. R. (2007), Normal phase and reverse phase HPLC-UV-MS analysis of process impurities for rapamycin analog ABT-578: Application to active pharmaceutical ingredient process development, *J. Chromatogr. B (Anal. Technol. Biomed. Life Sci.) 858*, 106–117.
11. Dorottya, B.; Sandor, G. (2008), Recent advances in the impurity profiling of drugs, *Curr. Pharm. Anal. 4*, 215–230.
12. Sorbera, L. A.; Bolós, J.; Serradell, N. (2006), Pazopanib hydrochloride: Oncolytic angiogenesis inhibitor VEGFR-2 tyrosine kinase inhibitor, *Drugs Future 31*, 585–589.
13. Zhou, L. Z. (2005), Applications of LC/MS in pharmaceutical analysis, in Ahuja, S.; Dong, M. eds., *Handbook of Pharmaceutical Analysis by HPLC*, Elsevier, San Diego, pp. 499–568.
14. Reid, G. E.; Simpson, R. J.; O'Hair, R. A. J. A. (1998), Mass spectrometric and ab initio study of the pathways for dehydration of simple glycine and cysteine-containing peptide $[M + H]^+$ ions, *J. Am. Soc. Mass Spectrom. 9*, 945–956.

15. Ballard, K. D.; Gaskell, S. J. (1993), Dehydration of peptide $[M + H]^+$ ions in the gas phase, *J. Am. Soc. Mass Spectrom. 4*, 477–481.
16. Clawson, J. S.; Vogt, F. G.; Brum, J.; Sisko, J.; Patience, D. B.; Dai, W.; Sharpe, S.; Jones, A. D.; Pham, T. N.; Johnson, M. N.; Copley, R. C. P. (2008), Formation and characterization of crystals containing a pleuromutilin derivative, succinic acid and water, *Crystal Growth Design 8*, 4120–4131.
17. Huang, Y., Ye, Q., Guo, Z., Palaniswamy, V. A., Grosso, J. A. (2008), Identification of critical process impurities and their impact on process research and development, *Org. Process Res. Devel. 12*, 632–636.

CHAPTER 10

Structure Elucidation of Pharmaceutical Impurities and Degradants in Drug Formulation Development

CHANGKANG PAN, FRANCES LIU, and MICHAEL MOTTO

Pharmaceutical and Analytical Development, Novartis Pharmaceuticals Corporation, One Health Plaza, East Hanover, NJ 07936

10.1 IMPORTANCE OF DRUG DEGRADATION STUDIES IN DRUG DEVELOPMENT

A *pharmaceutical impurity* in a drug product, as defined by the International Conference on Harmonization Guidelines (ICH A3B), is any component of the drug product that is not the chemical identity, defined as the drug substance or an excipient in the drug product [1–5]. The safety of a drug product, or a dosage form, depends not only on the toxicological properties of the drug substance but also on the properties of those pharmaceutical impurities, including drug degradation products. As a result, the structures of drug degradation products over the identification threshold must be identified, and their levels must be carefully monitored and controlled during drug formulation development [2]. These are also the key information for regulatory assessment in submission [6].

Understanding drug degradation in the formulated product is critical in pharmaceutical development, as drug stability and degradation products could have significant impacts on formulation development, analytical method development, package development, storage conditions and shelf-life determination, and safety and toxicology concerns. During pharmaceutical drug development, identification of drug degradation products and assessment of their toxicity are required by regulatory authorities [2,5,6,7]. The identity of pharmaceutical impurities, including drug

Characterization of Impurities and Degradants Using Mass Spectrometry, First Edition.
Edited by Birendra N. Pramanik, Mike S. Lee, and Guodong Chen.

degradation products in the formulated dosage form, are an integral part of the information provided to regulatory authorities in registration application dossiers [ICH Q1A(R2)].

During preformulation development to select appropriate excipients for a particular drug substance, information on chemical stability of a drug substance is always critical. For example, if a drug substance is found to be highly hydroscopic and quickly forms hydrolysis products once exposed to the air, then special consideration in formulation development might be given to protect this dosage form from moisture. Similarly, this information is essential for packaging development so as to ensure full protection from moisture, thereby ensuring long-term storage stability of dosage forms. Hydrolysis of active pharmaceutical ingredients and excipients in a dosage form could create a stability issue for drug products. Hydrolysis mechanisms, analytical methods, and recommendations for formulation stabilization have been discussed in a review paper [9]. Similarly, if a compound is susceptible to degradation at elevated temperature, care must be taken in the selection and control of the temperature used during the formulation process. For example, a drug breakdown compound was detected using LC-MS for the drug substance under heat and pressure in the formulation process. The formulation conditions were then adjusted to reduce the heating temperature and heating time, by which the thermal breakdown compound can be minimized. Oxidation is another degradation mechanism for drug products. A guide for stabilization of pharmaceuticals to oxidative degradation was discussed in a review paper, where liquid and solid dosage forms were discussed with options for formulation changes, additives, and the use of antioxidants [8]. In another study, formation of hydroperoxides from PEG3350 in the coating caused electrophilic oxidation of drug to the sulfoxide degradation. An antioxidant BHT was then used in asymmetric membrane tablet coatings to stabilize the osmotic tablet core to the acid-catalyzed peroxide oxidation of a thioether drug [10]. Understanding the chemical properties of drug substances and excipients is very important for a formulator to select the appropriate formulation and to optimize manufacturing process, or to design alternative dosage forms for better drug stability.

Similarly, knowledge of drug chemical stability and degradation pathways also helps analytical chemists develop a good stability-indicating HPLC analytical method. Stress studies are normally conducted to study the chemical stability of the drug products (either tablets or capsules) under heat (thermal), acid and base, oxidation, and exposure to light [11–18]. Drug degradation products that form under these conditions can be identified using a variety of analytical technologies. LC-MS has been found to be a very powerful technique in the identification process. The stressed samples are often used for analytical method development so as to achieve good separation of these "potential" degradation products. Even though the degradation profiles observed in the stress studies may not be the same as those ultimately found in drug product stability samples, the use of a LC-MS-compatible analytical method helps monitor all new degradation products during various stages of drug development [19–29].

Additionally, the information obtained from drug degradation studies plays a crucial role in evaluating the drug safety profile that is associated with these degradation compounds. After the identity and the quantity of a degradant are positively confirmed, the safety profile of the degradation compound can be fully evaluated. The specification for this degradant can then be set if necessary. Furthermore, if drug degradation pathways are positively identified and fully understood, appropriate formulations or techniques can be applied to prevent or limit its formation in the final drug product. Synthetic chemists can even develop a "new" compound with a similar structure but with an improved intrinsic chemical stability.

10.2 DRUG DEGRADATION STUDIES IN FORMULATION DEVELOPMENT

Drug degradation in a formulated dosage form is often a very complicated and may be an unpredictable process. Depending on the intrinsic nature of the drug substance, degradation could be induced by either a single chemical reaction or a combination of multiple chemical reactions. The reaction(s) may result from a variety of factors, such as product composition (formulation) of a dosage form, the process by which the dosage form is prepared (formulation process), and the storage conditions (temperature, humidity, and light). Identification of drug substance degradation pathways helps better understand the drug product, and assists in determining how to control the stability of the drug product. Furthermore, it is important to understand how the physical form of a drug substance affects the nature of the degradation process. For a given drug substance, there may be different chemical activities if its physical property changes from a crystalline form to an amorphous form during storage conditions. In general, drug degradation in formulation development has some unique characteristics as discussed below.

10.2.1 Drug Substance–Excipient Interaction

For a formulated drug, reactions between drug substance and excipients can be common mechanisms in drug degradation. A variety of excipients with different pharmaceutical properties are used in a formulation process to combine with a drug substance, from which either tablets or capsules can be formed. Each of these excipients plays a specific role in the formulation, such as compressibility, binding, dissolution, and bio-availability. If these excipients are not optimally selected, they may react with a drug substance to form drug degradation products. The drug–excipient interaction in formulated dosage forms has been reported in a number of papers [30–48]. For example, lactose has been widely used as a filler or diluent in tablets and capsules. Since lactose is a reducing sugar, it can react with the primary amine of a drug substance to form degradation products through a well-known Maillard reaction mechanism [19,32]. The Maillard reaction can cause browning effects, especially with amorphous material. Therefore, if the drug substance contains primary or secondary amine(s) and lactose is used in the formulation, the Maillard reaction may be one of the most important chemical mechanisms in drug degradation.

In-depth knowledge of the chemical reaction mechanism is very helpful for a formulator to develop a good formulated dosage form.

10.2.2 Small Unknown Peaks (~0.1%) (Low-Dose Drugs <1 mg per Dose)

In current pharmaceutical industries, the drug load may be very low, even less than 1 mg per tablet or capsule. This creates additional challenges for the structure elucidation of unknowns in a dosage form. Compared with the unknown peak studied in chemical development/process development, where only drug substance is involved, identification of unknowns in formulated drug products becomes very challenging as the absolute content of an unknown compound may be quite low (μg or ng drug loading on a LC column). In some cases, the quality of mass spectra for trace-level impurities can be very poor. Identification of these low-level impurities requires special precautions since even the assignment of a molecular ion may not be evident in the analysis.

10.2.3 "Busy" LC Chromatogram with Multiple Peaks (Combination Drug Products)

When compared to the HPLC chromatograms for drug substance and related materials in chemical/process development, the HPLC chromatograms for the formulated dosage forms are usually quite complicated. In addition to observing the drug substance and its related compounds, one may also observe some of the excipients in the chromatograms, which may complicate the chromatogram. Furthermore, some dosage forms (combination drug products) contain multiple active pharmaceutical ingredients (APIs). As a result, the number of the peaks observed in the chromatograms is increased for these combination drug products. During stability studies of these combination drug products, the degradation product must be correctly assigned to one particular drug substance since the content (%) of the degradant relative to the drug substance is very critical for meeting the specification of a drug product. For a combination drug product containing multiple APIs with a great difference in percent content, the proper identification of a drug degradation product may determine either "pass" or "failure" of a dosage form under stability study.

10.2.4 Modification of Non-MS-Compatible LC Methods

Whenever at all possible, it is recommended that the analytical chromatographic method be compatible with LC-MS so that any extra peak appearing in the chromatogram can be readily studied using this technique. However, some analytical LC methods have to use non-LC-MS-compatible additives in the mobile phases, such as ion-pairing reagents or phosphate buffers, in order to accomplish the separation needed for a robust method. If an unknown peak found in a non-LC-MS-compatible condition need to be identified, a new LC-MS method must be "modified" or developed. Identification of the unknown peaks will be based on this "modified"

LC-MS method. It is critical to demonstrate that the peak identified in a "modified" method is the peak of interest found in the original non-LC-MS-compatible method. There are a number of ways to show the peak equivalence between two different analytical methods, which will be discussed in detail in Section 10.4.

10.2.5 Uncontrollable Multiple Chemical Reactions in Stability Samples

Drug degradation may occur when a dosage form is stored under the temperature and humidity conditions typical for stability programs. The stability samples, stored under either regular or accelerated stability conditions, indicate the physiochemical stability of the dosage form. Drug degradation or excipient degradation occurs if an extra peak (unknown) is found in the chromatogram, or if the peak area (or intensity) is increased over time. The unknown peak should be identified if it appears above the ICH identification thresholds, such as 1.0% for the drug product with maximum daily dosage (MDD) of <1 mg, 0.5% for the drug product with MDD 1–10 mg, 0.2% for the degradation product with MDD 10 mg -2 g, and 0.1% for the drug product with MDD >2 g [2]. Since the stability samples are stored under specific temperature and relative humidity, a drug substance may undergo multiple chemical reactions caused by heat, moisture, light, and oxidation. In addition, the physical forms (crystalline or amorphous form) of drug substances and excipients may vary under stability conditions, which could lead to different chemical activities of drug substances and excipients. Because of the lack of mobility of drug substance and excipients, the solid-state microenvironment of the drug product is highly localized. In addition, drug substance and excipients are such that it is difficult to ensure homogeneous distribution (heterogeneity) in a dosage form. Within this localized microenvironment (e.g., local pH and local concentration), the chemical reactions observed in the solid state might be more complex than those observed in homogeneous liquid solutions.

10.2.6 Separation Interference and Contamination Induced by Excipients

Even though most of the excipients in a dosage form do not have a UV chromophore and thus have little influence on the peak separation in the chromatograms detected by the UV detector, some excipients do exhibit absorbance (peaks) in the chromatogram if the detection wavelength is low enough. The presence of excipients in a dosage form could cause more difficulties for a MS detector than for a UV detector. The mass spectrometer can detect many excipients, which show little or no signals in the UV trace. In one of our studies, a small LC peak appeared to be fully separated (chromatographically clean) in the chromatogram detected by UV detector. However, the peak severely overlapped with other peaks in the total-ion chromatogram (TIC) detected by MS. An excipient with typical polymeric ions (mass difference = 44 Da) was detected at the same retention time of the peak of interest in the MS trace. It was very difficult to obtain a "clean" mass spectrum of the unknown compound in the presence of this excipient interference. Furthermore, injection of a drug product solution generally introduces much more excipients onto the LC-MS systems than does injection of a drug substance. This could subsequently cause severe instrument contamination at the capillary tip or on the sampling cone.

10.2.7 Peak Isolation and NMR Confirmation for Late-Phase Projects

Even though mass spectrometric study can provide structure-rich information on drug degradation products, the final structure of a degradation product may not be fully confirmed by mass spectrometric data alone. In this case, preparative LC peak isolation and enrichment are needed to collect a sufficient quantity of material for NMR confirmation [49–55]. This may also be achieved through LC-NMR analysis to confirm the identity of the degradation products for toxicity assessment [56,57]. After structure confirmation, authentic compounds can be synthesized and used for the quantitation of degradation products. Structure confirmation is especially important for late-phase projects as degradation products must be not only identified but also quantified in a drug dosage form, such as tablet or capsule formulation.

10.3 COMPLEXITY OF IMPURITY IDENTIFICATION IN DRUG DEVELOPMENT

As mentioned above, drug degradation in a dosage form could be influenced by multiple factors, such as formulation composition (e.g., excipients, pH), formulation process (temperature and pressure, wet or dry granulation, etc.), storage conditions (temperature, humidity, light), and package selection (containers). Drug degradation in a dosage form is very complex and often difficult to predict [11,19,36,37]. Since pharmaceutical impurities, including drug degradation products, could come from a variety of sources and channels, identification of these unknown impurities in the formulation development becomes quite challenging.

10.3.1 Drug Substance (DS) Degradation

Common drug degradation reactions include oxidation, hydrolysis, dehydration, photolysis, and dimerization. The potential degradation products and their degradation pathways can be found in a variety of stressing studies of a drug substance prior to or during formulation development. Typical pharmaceutical stress testing includes thermolytic, hydrolytic, oxidative, and photolytic degradation studies. The procedures of these tests have been described in detail in a number of review articles [11,12,16,35–37]. Some "new" degradation products, however, can still be found in stability samples, as stress-testing studies may not produce all potential degradation products because of the complexity of a drug substance and its environment. Care must be taken when identifying degradation mechanisms, as, in addition to the drug substance undergoing chemical degradation itself, secondary reactions may

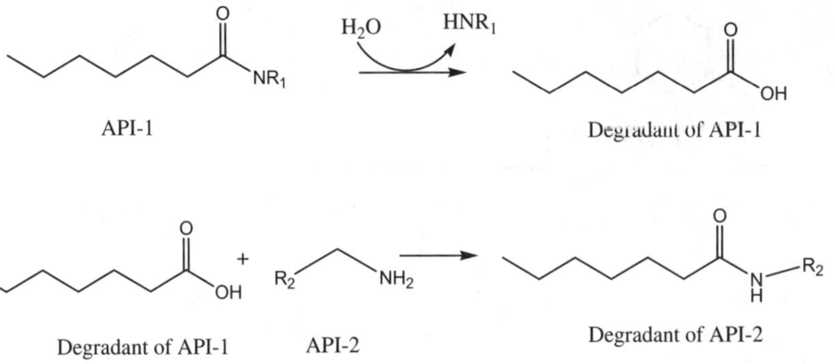

API-1

H_2O HNR$_1$

Degradant of API-1

Degradant of API-1 API-2 Degradant of API-2

FIGURE 10.1 A chemical reaction between two active pharmaceutical ingredients (APIs) within a combination drug product. A degradation product of API-1 reacts with API-2 to form a new degradation product of API-2.

occur between the drug substance and its degradation product to form additional degradation products.

The situation can be even more complex in a combination drug product containing more than one API in a dosage form. For example, one API may react with a degradation product formed from another API to form a "new" degradation product. In one of our previous studies, API-1 underwent hydrolysis to form a carboxylic acid degradant, which could then react with an amine group in API-2 to form an amide degradant of API-2. The reactions resulting from the degradation of two APIs are illustrated in Figure 10.1.

Since byproducts may have common substructures compared to their parent drug substance, drug byproducts can undergo similar degradation pathways as to the drug substance. Under certain circumstances, a drug substance could react with its byproduct to form a new degradation product, which is shown as "Deg.4" in Figure 10.2.

When conducting drug degradation studies, it is very important to have a synthetic scheme of the drug substance and to understand the chemical reaction on each of the synthesis steps.

10.3.2 DS–Excipient Interaction

A formulated pharmaceutical drug usually contains one drug substance and multiple excipients. Most of the excipients are chemically inert and thus cannot interact with a drug substance. In some cases, however, a drug molecule may interact with specific excipients to form drug degradation products. Different types of drug–excipient interactions in pharmaceutical drug development have been reported in the literature [30,35,36,38,41,43,45,46]. These interactions and reaction products can be characterized using LC-MS, GC-MS, or other analytical techniques. For example, cutina and magnesium stearate are lubricants frequently used in tablet or capsule formulation processing. They could be readily converted into carboxylic acid through

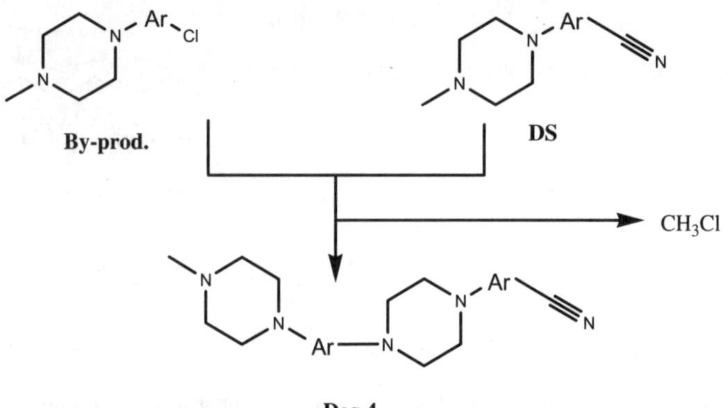

By-prod.

DS

CH_3Cl

Deg.4

FIGURE 10.2 A chemical reaction between a drug substance and its byproduct generated in synthesis.

hydrolysis, which may have chemical reactions with a variety of nucleophilic functional groups ($-NH_2$ or $-OH$) of drug substances. In a formulation development study, several unknown compounds were found in the stability samples for one drug product. LC-MS studies indicated that these unknowns were structurally related to the drug substance. The dosage form contained cutina as a lubricant, which is a natural product with a mixture of middle to long chain fatty acid glycerol esters (monoacylglycerols, diacylglycerols, triacylglycerols, and free fatty acids). Cutina was suspected to be the reaction precursor as it can be readily hydrolyzed to form fatty acids, which could react with amine and hydroxyl groups in the API to form new degradation products. The testing procedure of cutina material specification was to titrate free fatty acids with NaOH. As the composition of cutina varies considerably with different continents, areas, and soil conditions, there has been little information on cutina composition in the literature. Therefore, a GC-MS method was developed to determine the chemical composition of the cutina batch that was used in the formulation, from which the proposed structures of these degradation products can be justified and even the authentic compounds can be synthesized for structure confirmation.

Since direct GC-MS analysis of the resulting carboxylic acids in cutina did not work due to their high boiling points (>300°C), cutina was derivatized to convert all carboxylic acids into methyl esters prior to GC-MS analysis. The resulting ester compounds can be readily detected using GC-MS. The mass spectra of these esters matched very well (match factor >90%) with the reference mass spectra in the library database. The results are shown in Table 10.1.

Three strong peaks observed in the GC chromatogram were related to palmitic acid, stearic acid, and eiconsanic acid. The reactions between these carboxylic acids and the drug substance resulted in unknown peaks in the LC chromatogram of the stability samples. On the basis of GC/MS and LC/MS results, the relevant carboxylic acid reference materials were purchased and the targeted compounds were synthesized in an analytical lab. The molecular structures of these degradation compounds were quickly confirmed.

TABLE 10.1 Chemical Composition of Cutina and Magnesium Stearate

Components Detected by GC-MS	Cutina (Lot RGA353)	Mg Stearate (Lot 1740418)	Chemical Formula
Methylpalmitate	Strong	Strong	$C_{17}H_{34}O_2$
Methylstearate	Strong	Strong	$C_{19}H_{38}O_2$
Methyleicosanoate	Strong	Medium	$C_{21}H_{42}O_2$
Methyltridecanoate	Weak	—	$C_{15}H_{30}O_2$
Methylpentadecanoate	Weak	Weak	$C_{16}H_{32}O_2$
Methylheptadecanoate	Medium	Weak	$C_{18}H_{36}O_2$
Methylnonadecanoate	Medium	—	$C_{20}H_{40}O_2$
Methyldodecanoate	—	Weak	$C_{13}H_{26}O_2$
Methyltatradecanoate	—	Weak	$C_{15}H_{30}O_2$
Methyldocosanoate	—	Weak	$C_{23}H_{46}O_2$

10.3.3 DS–Residual Solvent Interaction

Typically organic solvents are used during the synthesis of drug substances. These residual solvents are the "fingerprint" of a particular drug substance in a dosage form. The interaction between a drug substance and its residual solvents may be observed especially in long-term stability samples where the reaction has progressed far enough over time to form a degradation product above the identification threshold. For example, ethyl acetate was an organic solvent used for the crystallization of one drug substance. It was present at low levels in the drug substance, and its reaction with the drug substance was not detectable at an early stage (initial time in stability program). For a tablet batch stored at 40°C/75% relative humidity (RH) for one month, an extra peak was found in the chromatogram compared with the same tablets at the initial time point. LC-MS studies revealed the molecular weight of this compound to be 28 Da higher than that of the drug substance. The first neutral loss detected in the MS/MS study was 46 Da instead of the 18 Da that was typically seen for the drug substance. The compound was identified as ethyl ester degradant, resulting from the reaction between one residual solvent (ethyl acetate) and the drug substance containing a carboxylic acid.

10.3.4 DS–Solvent Impurity Interaction

Even though analytical solvents used for HPLC analysis are relatively pure, normally HPLC-grade or American Chemical Society (ACS)-grade, they still contain some impurities at low levels [58]. These impurities may undergo chemical reactions with a drug substance to form degradation products. It is important to determine whether the observed "unknown peak" is introduced during sample preparation or is generated in a stability sample itself. Interaction between a drug substance and solvent impurities can be readily observed if a drug molecule is highly reactive.

In one of the drug development studies, the formation of a trace unknown peak was found to be related to the quality of the methanol being used in the sample preparation step. The unknown appeared when certain lots or brands of methanol were used but not were absent with the use of other lots/brands of methanol. Therefore, the extra peak was related to the impurity in the methanol solvent. LC-MS studies indicated that the molecular weight of this unknown was 12 Da higher than that of the drug substance. MS/MS results further indicated that the unknown compound had the same core substructure as did the drug substance. More importantly, it had a characteristic neutral loss of 29 Da, which could be related to the loss of methylene imine (CH_2=NH) from this amine-containing drug substance. Therefore, the degradation compound was identified as the reaction product between formaldehyde and the drug substance containing a primary amine group. In order to confirm the proposed structure, a laboratory synthesis was conducted by mixing formaldehyde and the drug substance at room temperature. The reaction occurred immediately with heat release. The sample solution was then injected into a LC column. The retention time and mass spectrum of this synthetic compound were consistent with the unknown peak of interest. Further studies showed that one particular batch of methanol contained multiple impurities, not only formaldehyde but also acetaldehyde and propionaldehyde. Three extra peaks were detected when this batch of methanol was used in sample preparation. These degradation compounds were related to the reactions between the reactive drug substance and three aldehyde impurities present in the methanol, as illustrated in Figure 10.3. These three degradation products were induced by solvent impurities during sample preparation and were not present in the drug product.

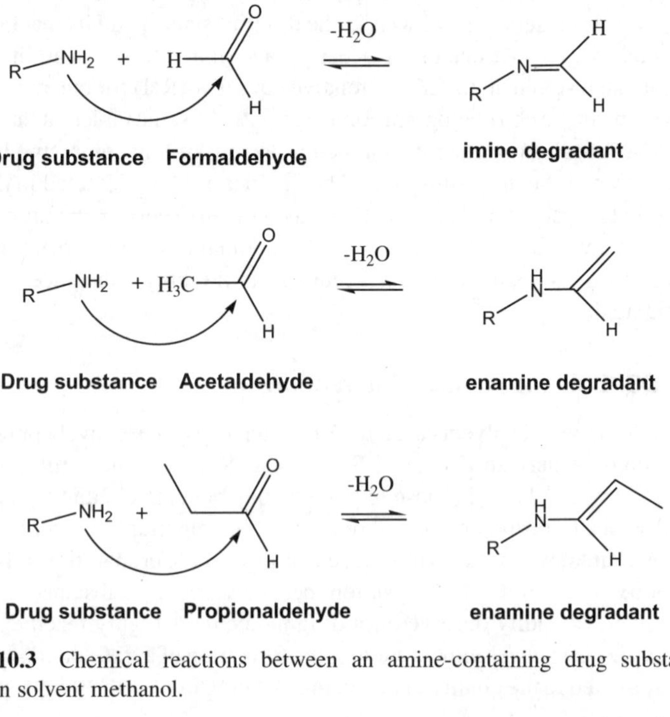

Drug substance Formaldehyde imine degradant

Drug substance Acetaldehyde enamine degradant

Drug substance Propionaldehyde enamine degradant

FIGURE 10.3 Chemical reactions between an amine-containing drug substance and aldehydes in solvent methanol.

10.3.5 Metal Ion–Catalyzed Reaction

Metal ions can play an important role as catalysts in organic and inorganic chemical reactions [59–61]. The presence of these trace metal ions may lead to various drug degradation pathways. One example is the metformin degradation catalyzed by a metal ion (M^{n+}) to form a drug degradation product as shown in Figure 10.4. This particular metformin degradation product induced by a metal catalyst has not been reported in the literature. The reactions catalyzed by metal ion are quite complex. The charge status of the metal ion and the types of metal are critical factors in the reactions.

An example was also observed during another drug product development, where the rate of dimerization of a drug substance increased by a factor of 4 when stored in amber vials when compared with the same sample solution stored in clear vials. Using inductively coupled plasma mass spectrometry (ICP-MS) as the analysis technique, the extraction of amber vials was found to contain a heavy-metal content higher than that found in the extraction of clear vials. The differences in metal content between amber vials and clear vials suggest that the metal ions lead to a different degree of dimerization for the same sample solution, even though the study cannot exclusively conclude the "true catalyst" for the dimerization.

10.3.6 DS–Excipient Impurity Interaction

Although the physical properties of many commonly used excipients in pharmaceutical dosage forms have been well characterized, the chemical impurity profiles of excipients have not been fully investigated. The impurities in excipients may have significant impacts on the chemical stability of drug product, especially for those reactive drug substances [62–72]. Although most of the excipients used in the formulation development are inert to chemical reactions, impurities in some excipients may induce chemical reactions, which then create drug efficacy and drug safety concerns [63,65,68,71]. For example, povidone and crospovidone are widely used in formulation development as tablet binder and tablet disintegrant, respectively. These excipients are not chemically active but contain certain levels (<0.2 % by PhEur

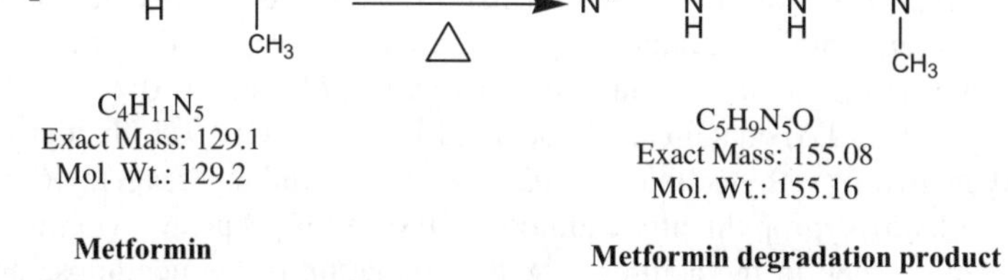

M^{n+}, Δ

$C_4H_{11}N_5$
Exact Mass: 129.1
Mol. Wt.: 129.2

Metformin

$C_5H_9N_5O$
Exact Mass: 155.08
Mol. Wt.: 155.16

Metformin degradation product

FIGURE 10.4 Metformin degradation induced by metal ion under heat, a metal ion–catalyzed chemical reaction for a drug substance.

and USPXXII) of aldehyde and formaldehyde [63], which are chemically active and may react with various functional groups of a drug substance. For example, it was observed in several cases during our formulation development that a drug substance containing amine functional groups reacted with aldehydes to form new degradation products.

Peroxides are present in a number of excipients in the form of either hydroperoxide (ROOH) or hydrogen peroxide (HOOH) [67–71]. These peroxide species are chemically active and can readily induce the oxidation of drug substances. A review paper has been published on the evaluation of hydroperoxide in common pharmaceutical excipients [65], where four excipients, povidone (PVP), polysorbate 80 (PS80), polyethylene glycol 400 (PEG400), and hydroxypropyl cellulose (HPC), were found to contain substantial contents of hydroperoxide with significant lot-to-lot and manufacturer-to-manufacturer variation. Peroxide generates free radicals that can react with a drug substance through a radical chain process (autoxidation), leading to severe drug degradation over time, especially for a low-dose drug molecule with a high ratio of excipient to drug substance. Peroxides can be found either as impurities in the excipient manufacturing process or as oxidation products from polymeric excipients [71]. Peroxides are present as initiators in polymerization processes for polymeric ethers, such as poly(ethylene glycol) [69], polysorbates, poly(oxyethylene alkyl ether)s, polyoxyethylene stearates, and other ethylene oxide–based materials [30,63,65]. Peroxides are also present as process impurities in polyvinylpyrrolidone (PVP)-based excipients, including povidone and crospovidone [63,65]. Peroxide can form by autoxidation of PVP, and the content may be increased under shear conditions in granulation and tableting processes [71].

In formulation development, we have observed in a number of studies where the oxidation of drug substances was induced by the presence of peroxide in the excipients. Drug molecules with different functional groups can readily react with peroxide to form various degradation products. During the development of a peptide formulation, for example, the presence of hydroperoxide in an excipient (crospovidone) had a significant impact on peptide stability. Oxidation of methionine in the peptide resulted in the formation of a sulfoxide on several methionine residues in the peptide. As the peroxide content varied from batch to batch for crospovidone, the degree of methionine oxidation could not be fully controlled. In order to develop a stable formulation for this peptide, a preventive action has been taken in the formulation development to minimize the methionine oxidation.

Impurities or excipient degradation products can also be found in many other pharmaceutical excipients. Detailed information can be found in a review article [71]. Cornstarch often contains hexamethylenetetramine as a preservative, which could be hydrolyzed to form ammonia and formaldehyde [72]; many dye and flavoring reagents (e.g., vanillin) contain aldehyde impurities [63]. Glyoxal (dialdehyde) may be found in hydroxyethylcellulose (HEC) as a crosslinking reagent [62] and an impurity in hydroxypropylmethylcellulose (HPMC) [30]. Furfural (furan-2-carbaldehyde) can be seen in the acid-catalyzed degradation of hemicellulose and other sugar-based excipients such as lactose, maltose, and glucose [72]. Common excipients and their potential impurities are summarized in Table 10.2.

TABLE 10.2 Potential Impurities and Degradation Products in Pharmaceutical Excipients Commonly Used in Formulation Development

Excipient	Potential Impurity/Degradation Product
Povidone (PVP)	Peroxide
Crospovidone (crosPVP)	Peroxide
Lactose	Aldehydes, furfural
Poly(ethylene glycol) (PEG)	Peroxide, aldehydes
Poly(oxyethylene alkyl ether)s	Peroxides
Poly(oxyethylene stearate)s	Peroxides
Microcrystalline cellulose (Avicel)	Hemicellulose, furfural
Cornstarch	Hexamethylenetetramine, formaldehyde, ammonia
Stearate lubricants	Carboxylic acid
Hydroxyethylcellulose (HEC)	Glyoxal (dialdehyde)
Hydroxypropyl cellulose (HPC)	Hydroperoxide
Poloxamer	Hydroperoxide
Polysorbate 80 (PS80)	Hydroperoxide, aldehydes
Hydroxypropylmethylcellulose (HPMC)	Glyoxal (dialdehyde)
Glucose	Furfural

10.3.7 DS–Salt Interaction

In order to maintain or enhance the physical stability and chemical stability of a drug substance, a stable salt form is typically preferred. In these cases the "free acid or free base" of the drug substance is combined with a particular pharmaceutically suitable base or acid to form a salt. Although the salt form is generally stable, chemical reactions may occur under specific conditions, due to the chemical nature of the drug substance and its counterions. In our previous study, a formulated dosage form contained amlodipine and another API. It was found that under a specific set of stability conditions, amlodipine reacted with maleic acid, the counterion of the other API. From a chemical perspective, two reactions may occur between a primary amine in amlodipine and maleic acid; one undergoes nucleophilic attack at the carbonyl carbon with 1,2 addition, and the other attacks at the β carbon with 1,4 addition or Michael addition. The reactions are illustrated in Figure 10.5.

Similar reactions were also observed during storage between an amine-containing drug substance and its counterion of fumaric acid, *trans*-butenedioic acid. Michael addition seems to be one of the major reaction mechanisms in the drug–salt interaction where the molecular weight of the degradation product is the sum of the drug substance and the counterion.

10.3.8 DS–Preservative Interaction

Propylparaben, ethylparaben, and methylparaben are widely used as antimicrobial preservatives in pharmaceutical formulations. Paraben preservatives contain an ester

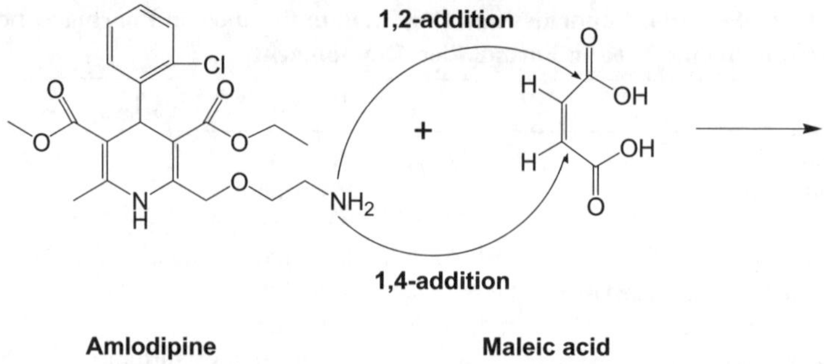

FIGURE 10.5 Amlodipine reacts with maleic acid through two reaction mechanisms: 1,2 addition and 1,4 addition (Michael addition).

functional group, which has the potential to react with carboxylic acid, hydroxyl group, or amine group from the drug substance. These reactions may not be detectable in the fresh formulated products but could be detected under stability conditions where temperature and humidity speed up the chemical reaction.

10.3.9 Preservative–Excipient Interaction

Paraben preservatives are ester compounds, which could react under certain conditions with excipients containing hydroxide groups to form extra "unknown" peaks in LC chromatograms. In one study, two unknown peaks were found in the stability samples. These two unknowns had the same UV spectra, which were different from those of the drug substance but similar to that of methylparaben. For both unknowns, a major ion of *m/z* 485 was detected in the positive-ion mode and a major ion of *m/z* 461 was detected in the negative-ion mode (see Figure 10.6). Therefore, the molecular weights of these two unknown compounds are 462 Da; the ion at *m/z* 485 was determined as a sodium adduct ion. The MS/MS data indicated that the $[M + Na]^+$ ion could lose 162 and 180 Da (Figure 10.6). Sucrose was used as an excipient in the formulation. It is a disaccharide of glucose and fructose, both of which have the same molecular weights of 180 Da. Therefore, these two unknowns were proposed as the reaction products of methylparaben and sucrose. To confirm this hypothesis, systematic stress studies were conducted. All stressed materials were heated at 80°C and then analyzed by LC-MS. The results are summarized in Table 10.3.

The results indicated that these two unknowns were the reaction products of methylparaben and sucrose. One of the reactions is illustrated in Figure 10.7. However, the MS data could not pinpoint the exact reaction site.

10.3.10 Excipient Degradation

Certain excipients may also undergo chemical degradation during storage or under stress conditions. Their degradation products may be detected in the chromatograms

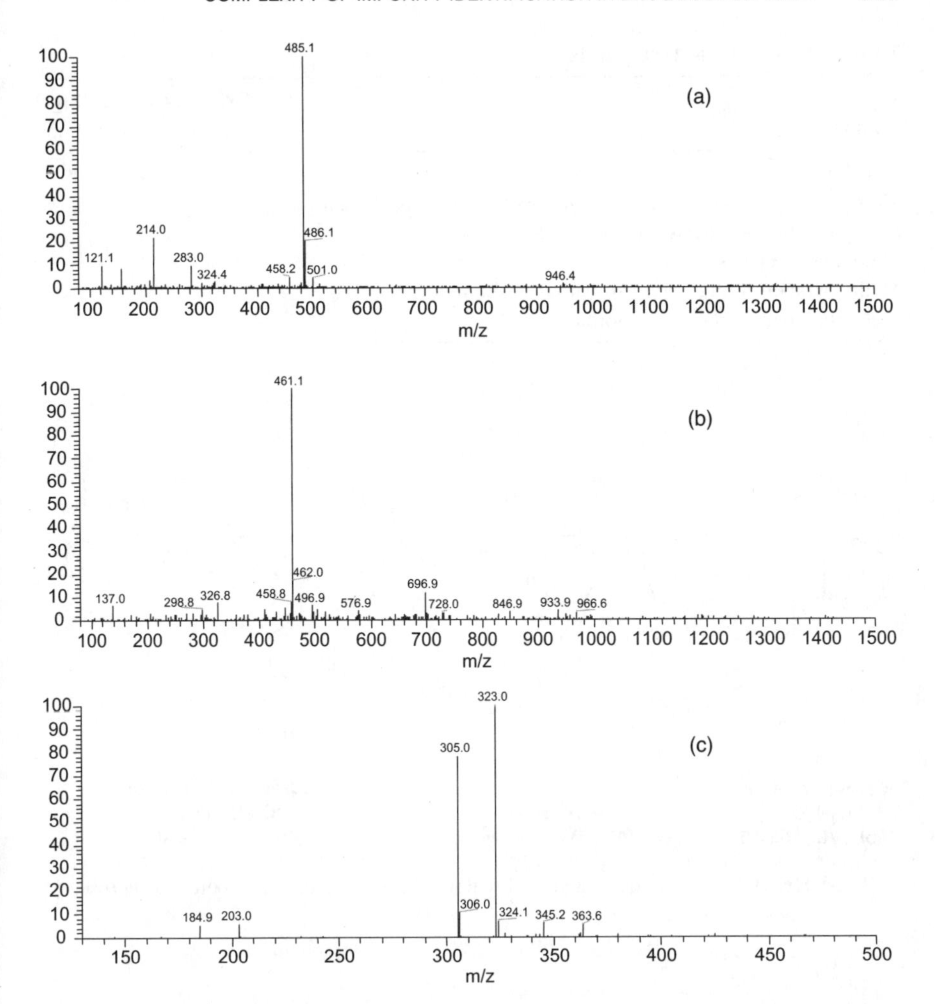

FIGURE 10.6 Positive-ion mass spectrum of the unknown peak (a); negative ion mass spectrum of the unknown peak (b); positive-ion MS/MS spectrum of *m/z* 485.1 (c).

by HPLC-UV or HPLC-MS. One example was the degradation of triethylcitrate (TEC) detected at a low wavelength in stability studies of tablets. The mass spectral pattern indicated the unknown peak was not structurally related to the drug substance. Instead, it was related to triethylcitrate, an excipient used in the dosage form. The TEC excipient was degraded to dimethylethylcitrate as described below and illustrated in Figure 10.8.

10.3.11 Leachables and Extractables

Leachable or extractable chemicals usually originate from packaging materials, such as glass, rubber stoppers, plastic tubing, and liquid solution bottles. These compounds

TABLE 10.3 Stress Test Results

Components	Existence of the Target Unknown Compounds
Placebo	+
Methylparaben + potassium sorbate + sucrose + sodium citrate	+
Methylparaben + potassium sorbate + sucrose	+
Methylparaben + sucrose	+
Methylparaben + potassium sorbate + sodium citrate	−
Methylparaben + potassium sorbate	−

Methyl Paraben
$C_8H_8O_3$
Mol. Wt.: 152.15

Sucrose
$C_{12}H_{22}O_{11}$
Mol. Wt.: 342.30

Degradation product
$C_{19}H_{26}O_{13}$
Mol. Wt.: 462.40

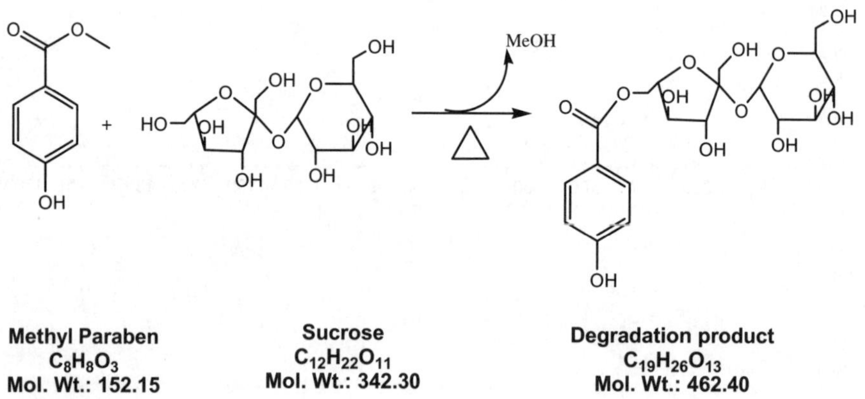

FIGURE 10.7 A proposed chemical reaction between methylparaben and sucrose.

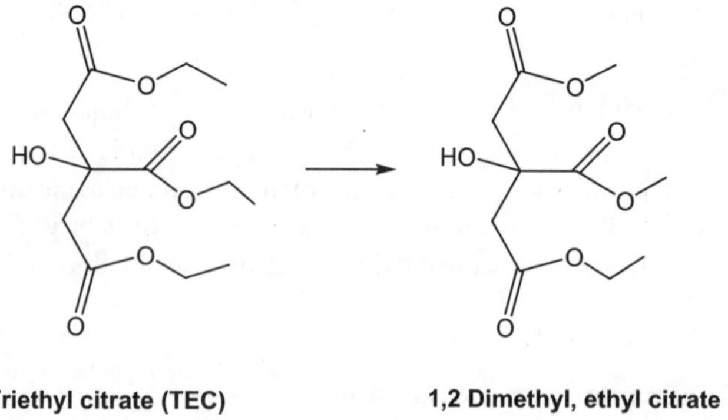

Triethyl citrate (TEC)

1,2 Dimethyl, ethyl citrate

FIGURE 10.8 Excipient degradation: tri(ethyl citrate) (TEC) to form 1,2-di(methylethyl citrate)

can be detected as "unexpected unknown" peaks in the chromatograms during HPLC-UV analysis of drug product samples [73–77]. Metal oxides are the major leachable components from glass, while manganese and iron can leach from amber glass vials. Different types of phthalate compounds can be extracted into sample solutions from rubber stoppers or plastic tubing materials. In the development of a liquid dosage form (injectable solution), an unexpected unknown peak was found in the sample solution. A MS/MS study indicated that the compound has a major fragment of *m/z* 149. The peak was therefore proposed as a phthalate compound since all phthalate compounds have a characteristic fragmentation ion of *m/z* 149. The identity of this unknown was then confirmed by running the reference phthalate. Further investigation indicated that this compound was an extractable originating from the use of a "new" type of carrier tubing. Different types of carrier tubing may leach different phthalate compounds, which elute at different retention times in the LC chromatogram. Leachable originating from the packaging materials may also penetrate through a semipermeable low-density polyethylene (LDPE) bottle and enter into the sample solution during storage, which will be discussed in the Section 10.6.2.

Unknown peaks may also originate from the contamination of external sources, container, and closure [76]. For example, one unknown peak was found to be above the identification threshold in the stability samples of a dosage form. LC-MS/MS studies indicated that the unknown peak was not structurally related to the drug substance. After a series of systematic experiments, the appearance of this peak was found to depend on how analysts shook the sample vials during sample preparation. For the same sample solution, the intensity of the unknown peak strengthened if the vial was shaken up and down. The LC vial septa were then extracted overnight using the same sample solvent. A LC-MS study indicated that the septa extract exhibited a peak that had the same retention time and mass spectrum as did the unknown. Therefore the unknown was confirmed as the extractable component from the vial septum. The extractable on the vial septum surface could enter into the sample solution if the solution had a good contact with the septum in the shaking step. Both Agilent and SUN SRI septa were found to contain the same extractable component, while the Agilent septum had a higher content than did that of the SUN SRI septum.

10.4 STRATEGY FOR STRUCTURE ELUCIDATION OF UNKNOWNS

10.4.1 Non-MS-Compatible Method versus MS-Compatible Method

Although an increasing number of HPLC methods are using LC-MS-compatible mobile phases, some analytical methods, due to the limitation of drug substances or excipients, are still employing non-MS-compatible mobile phases, such as ion pair reagents, surfactants, or nonvolatile additives. Development of a new LC-MS method is then necessary if the original LC method is not MS-compatible. In many cases, volatile additives, such as formic acid, acetic acid, or trifluoroacetic acid (TFA), can be directly used to replace a nonvolatile phosphoric acid. Ammonium counterion is preferred for use to adjust the buffer pH for any of these three acids. Trifluoroacetic acid has been reported in the literature as capable of reducing the sensitivity due to the ion-pairing effect. As the only volatile acid that can provide very low pH ranges (pH~2) [71], TFA is still widely used for the modification of a non-MS-compatible LC method. Triethylamine (TEA) has often been used as a mobile-phase additive in HPLC methods to improve the peak shape in the LC chromatogram. As TEA could quench the positive ionization process in an API source and therefore reduce the sensitivity, it should be avoided in a modified LC method for LC-MS with a positive ionization mode. After modification, the resulting LC chromatographic pattern in a new method should be similar to that obtained from a non-MS-compatible mobile phase. The use of UV spectral comparison can also help determine the peak of interest even though UV spectral comparison has its own limitation.

For a mobile phase containing an ion-pairing reagent, a completely new LC method must be developed for LC-MS analysis. The mobile phases used will be completely different, and the gradient may not necessarily be the same. In this case, it is impossible to compare the relative retention time (RRT) of the peak between two completely different LC methods. Therefore, it is very challenging to determine the peak of interest in the new chromatogram and to correlate the LC-MS results with the unknown found in the original non-MS-compatible method. In one study, an unknown peak was found in the stability sample for a combination drug containing multiple APIs. Identification of this unknown became very important since its content relative to one of the drug substances could have a significant impact on the status (pass or fail) of this stability batch. Since the original ion-pairing LC method was not MS-compatible, a new LC-MS method with volatile mobile phases was developed for the identification purpose. Because of the incompatibility of these two methods, the retention times of the unknown peak in these two methods were completely different. The traditional RRT comparison cannot be applied in this study. In order to determine the unknown peak in the LC-MS method, a control sample (a sample without this unknown) was tested using the LC-MS-compatible method together with the requested samples containing this unknown. One extra peak in the requested sample was found in the chromatogram compared to the control sample. This "suspected" peak was identified using LC-MS, and its structure was tentatively proposed. The compound may be due to the chemical reaction between one of the drug substances and one of the excipients in the formulation. On the basis of the chemical nature of this compound, synthesis was conducted in an analytical laboratory. The drug substance and the excipient were mixed together under several stressing conditions, such as heating time, heat temperature, solution pH, and various ratios of the drug substance versus the excipient. The ideal sample should have the major targeted compound but minimal other reaction products. Among these multiple testing samples, the best set of conditions was determined. The synthetic sample was then tested using LC-MS. The identity of this compound was confirmed by mass spectral analysis and retention time match. Then the same synthetic sample solution was tested using the original non-MS-compatible method. As the synthetic compound had the same retention time and UV spectrum as did the unknown in the stability sample, the identity of the unknown peak in the stability sample was fully confirmed. A general illustration of this identification processes is shown in Figure 10.9.

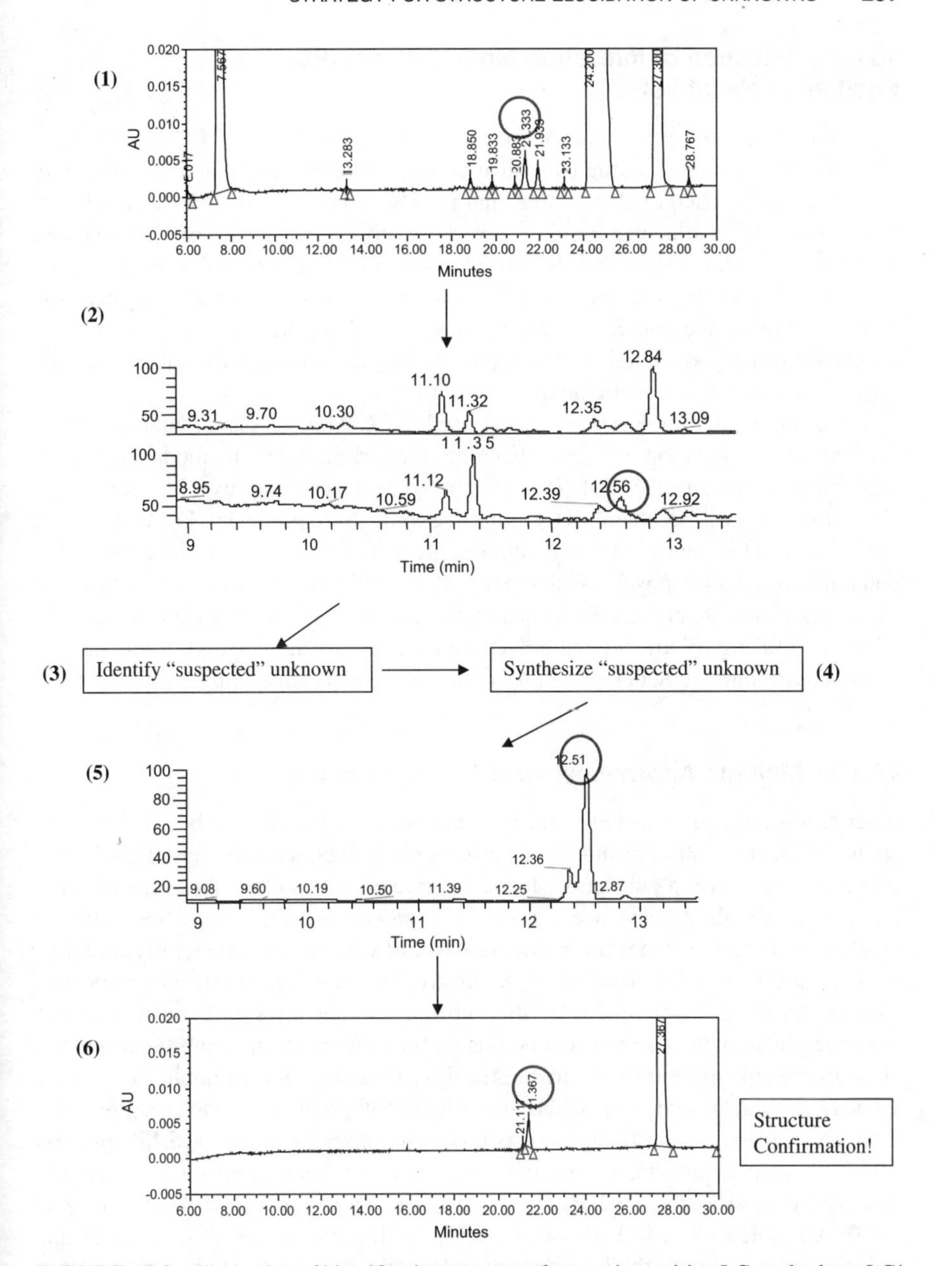

FIGURE 10.9 Illustration of identification processes from an ion-pairing LC method to a LC/MS-compatible method for structure elucidation of an unknown: (1) found an unknown at 21.3 min in a stability sample detected with an ion-pairing LC method; (2) ran control sample (top) and requested sample (bottom) and determined the "suspected" unknown at 12.6 min; (3) identified the structure of the "suspected" unknown by LC-MS/MS; (4) synthesized the "suspected" unknown; (5) analyzed the synthetic sample by LC-MS method to confirm the molecular structure; (6) tested the synthetic sample with the original LC method. The unknown peak identity was confirmed by matching the retention time and UV spectrum of the synthetic compound.

10.4.2 Selection of Ionization Mode (ESI or APCI, Positive or Negative)

Selection of an appropriate ionization mode is very important for LC-MS analysis. The chemical nature of a drug substance is such that either positive and/or negative ionization can be used to obtain molecular ion $[M + H]^+$ or $[M - H]^-$ information. If a "suspected" molecular ion does not show any adduct ions, both positive and negative ionization modes can be used to determine the molecular weight of an unknown. The mass difference of 2 Da between the positive-ion and negative-ion modes confirms the molecular weight assignment of the unknown compound. Ionization in a negative mode can be suppressed significantly if the employed mobile phase contains trifluoroacetic acid.

Postcolumn addition has been widely used in LC-MS analysis of unknowns as it can improve ionization without changing the original LC methods and chromatographic patterns [78–80]. Postcolumn addition can improve MS sensitivity since it can adjust the eluant pH to improve ionization efficiency. For example, if the column pH is neutral or basic, introducing a strong acidic solution after an LC column can enhance positive-ion signals. If the column pH is acidic, addition of a basic solution after an LC column can improve sensitivity for negative ionization. In addition, infusing volatile organic solvent, such as isopropanol, to highly aqueous LC eluant could improve nebulization, desolvation and ion evaporation efficiency [81].

10.4.3 Multiple Approaches for Structure Elucidation

After determination of a molecular ion, tandem LC-MS/MS will be conducted to study the fragmentation of this ion to determine its chemical structure. Usually the fragments of the drug substance should be studied first so that the fragmentation pathway of the drug substance can be fully understood. Then the same collision conditions are applied to the unknown to determine whether it is structurally related to the drug substance. In general, the core substructure of a drug substance can be also found in the degradation product of that substance. Sometimes the core substructure of the degradation product may not be exactly the same as the drug substance, but the changes are still structurally related to the drug substance. For example, addition of 16 or 14 Da may be due to the addition of a hydroxyl group or ketone group onto the core substructure, respectively. The neutral loss observed in the MS/MS process provides valuable information on the structure of the molecule. Some commonly observed neutral losses are NH_3 (−17), H_2O (−18), CO and C_2H_4 (−28), HCHO (−30), CH_3OH (−32), CO_2 (−44), C_2H_5OH (−46), and so on. Proper use of this information could help the investigator understand the structural change in a drug degradation product.

In addition to the tandem MS^n studies, a high-resolution mass spectrometer can be used to determine the elemental composition (chemical formula) of an unknown. On the basis of the combination of MS/MS analysis and accurate MS measurement, together with the background information on the unknown formation, a tentative structure could be proposed.

In some cases where both LC-MS/MS and high-resolution MS cannot provide a definite answer, hydrogen/deuterium exchange LC-MS experiments may provide additional structure information. Pharmaceutically active ingredients normally contain multiple exchangeable hydrogen atoms. Determination of the number of exchangeable hydrogen atoms helps define the chemical structure of a drug degradation product; the process is usually termed *hydrogen/deuterium (H/D) exchange LC-MS*. The exchangeable hydrogen atoms, also called *labile hydrogen*, are the hydrogen atoms associated with heteroatoms such as oxygen (O), nitrogen (N), and sulfur (S) in the forms of amine (–NH, or $-NH_2$), hydroxyl (–OH), thiol (–SH), and carboxylic acid (–COOH). When organic molecules are exposed to deuterium oxide (D_2O), heteroatom-bonded labile hydrogen (H) could be replaced by deuterium (D). This subsequently increases the molecular weight of the compound due to the replacement of $[M_H + H]^+$ with $[M_D + D]^+$, from which the change in particular functional groups can be determined. The H/D exchange LC-MS method has been used for structure elucidation of unknowns. It generates valuable information allowing the complete identification of unknown compounds in several areas, such as metformin oxidation endproducts [82], process impurities [83], and proteins/peptides [84–86]. In some cases where neither LC/MS^n nor accurate mass measurement can provide a definite structure, the H/D exchange experiment may generate a clear answer to the question. A more recent review paper described the application of H/D exchangeable LC/MS in some case studies, including differentiation between ketone and methylation (MW + 14 Da); differentiation between *N*-oxide formation and *C*-hydroxylation (MW + 16 Da); differentiation between desfluoro and dehydration (MW – 18 Da); and differentiation between alcoholysis and Michael addition (MW + 60 Da) [87].

One example is shown on Figure 10.10, where two proposed dimer-type structures had matching ion fragmentation pathways and exhibited little difference in accurate mass analysis because of its high molecular weight. However, the numbers of active protons in these two structures were different. In a H/D experiment, the molecular ion of this unknown was found to increase from *m/z* 872 $[M + H]^+$ to *m/z* 878 $[M + D]^+$, indicative of the presence of five exchangeable hydrogen atoms. The results were consistent with the proposed structure 1, on the top. The structure was further confirmed by LC-NMR results.

10.4.4 Structure Confirmation

For late-phase projects, the identity of major or minor drug degradation products must be positively confirmed. If mass spectral data alone are not sufficient for structure confirmation, NMR analysis becomes necessary. One approach is to isolate sufficient material through preparative LC separation, enrich the isolation, and then submit it for NMR analysis. The other approach is to directly use LC-NMR with a modified LC method. LC-NMR analysis provides direct structure confirmation within a short period of time. If feasible, the use of LC-NMR could avoid time-consuming and labor-intensive isolation, and thus facilitate and accelerate the identification. In-house synthesis is another quick way of confirming the identity of an unknown. This could

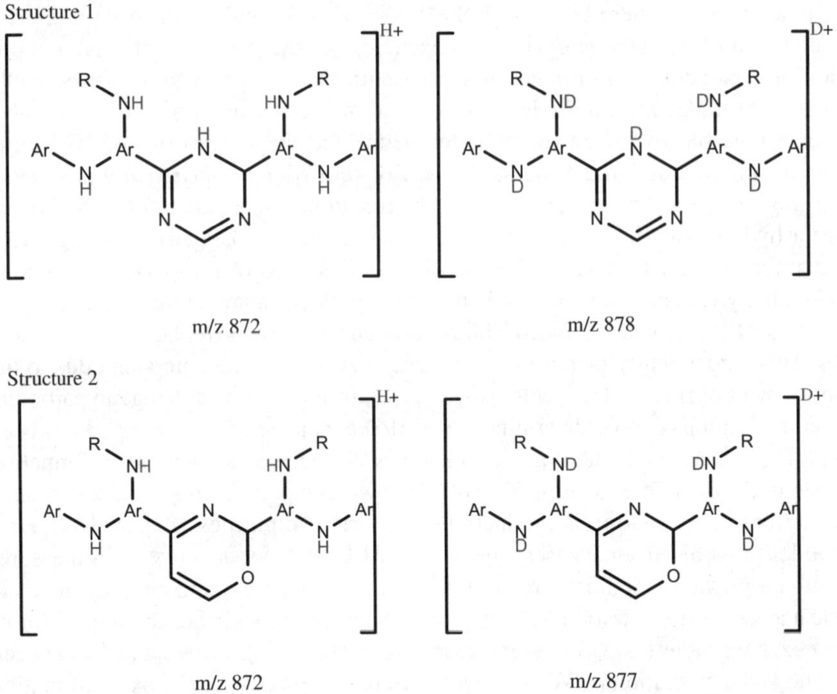

Structure 1

m/z 872 m/z 878

Structure 2

m/z 872 m/z 877

FIGURE 10.10 Two proposed molecular structures detected by hydrogen/deuterium exchange LC-MS experiment. Structure 1 indicates five exchangeable protons; structure 2 shows four exchangeable protons.

be carried out especially for oxidation and hydrolysis products as these reactions may possibly be conducted in an analytical laboratory. Figure 10.11 illustrates the general strategy for identification of unknowns in drug development.

10.5 HYPHENATED ANALYTICAL TECHNIQUES USED IN DRUG DEVELOPMENT

"Hyphenated" analytical techniques, in which a chromatographic separation is coupled online with one or more information-rich detectors, such as liquid chromatography/mass spectrometry (LC-MS), gas chromatography–mass spectrometry (GC-MS), and liquid chromatography–nuclear magnetic resonance (LC-NMR), have quickly become powerful tools for the identification or confirmation of low- or trace-level impurities [88–99]. These techniques, having complementary selectivity, are often required to completely define an unknown molecular structure. The LC-MS technique has been widely used in the pharmaceutical industries because of its high sensitivity, selectivity, dynamic range, and ruggedness as described in several research papers [71,100–105]. The technique has excellent sensitivity for the

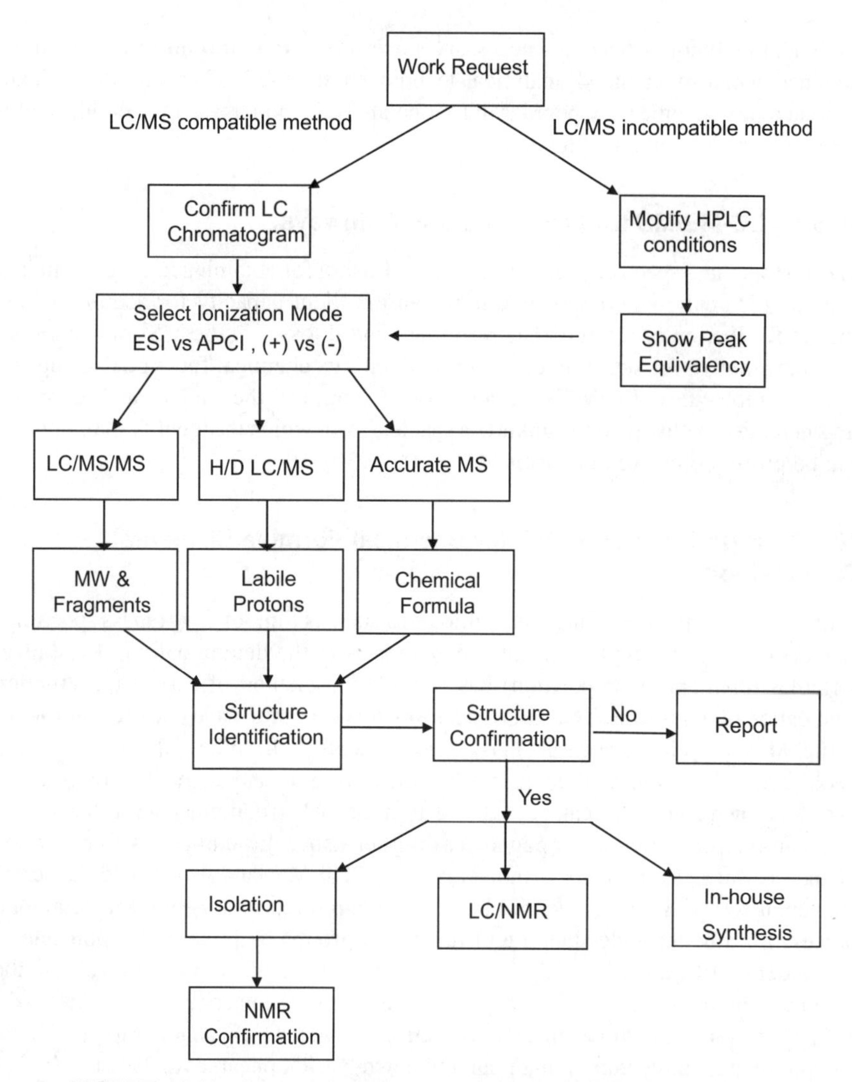

FIGURE 10.11 Strategy for unknown identification in drug development.

detection of trace-level impurities and degradation products observed in pharmaceutical drug development and manufacturing processes. In certain complex situations, however, LC-MS alone cannot come up with the final structure. The combination of MS and NMR techniques are therefore necessary for structure confirmation. With the development of new technology, LC-NMR has become a useful technique for structure elucidation in the pharmaceutical industry [93,94,96,97,106,107].

Liquid chromatography–mass spectrometry is the major tool in structure elucidation of drug degradation products or impurities. Although LC-MS provides high sensitivity and selectivity on the detection of unknowns at a fast pace, a detailed structure sometimes cannot be derived from LC-MS data alone. The use of other analytical techniques becomes necessary since more structural information can be obtained when hyphenated analytical techniques are used. This multidisciplinary approach has become very popular in routine analysis to increase the confidence for the structure assignments [88–105].

10.5.1 LC-MS/MS for Fragmentation Pathways

The ion trap mass spectrometer is a very powerful tool for obtaining ion fragmentation patterns for unknown compounds. It has an excellent capacity for acquiring MS^n spectra for the ion of interest. The resulting neutral losses in the MS^n processes are very informative for structure elucidation of various chemical/functional groups on targeted molecules. LC/MS^n studies greatly facilitate the understanding of ion fragmentation pathway for an unknown species, from which the identity of unknowns can be proposed or even confirmed [101,103,105–110].

10.5.2 High-Resolution MS for Chemical Formula/Elemental Composition

With the increasing use of high-resolution MS such as time-of-flight mass spectrometry (TOF-MS), the application of accurate mass to the determination of chemical formulas (elemental composition) has become the method of choice for structure elucidation of unknowns. TOF mass spectrometers generally have mass resolutions of >10,000, which enable accurate mass measurements with a minimal mass deviation. With accurate mass measurement, the chemical formula and elemental composition of an unknown compound can be clearly determined when combined with all essential information from a parent compound or a drug substance. In addition to determination of the chemical formula for a molecular ion, TOF-MS can also provide accurate elemental composition of a neutral loss molecule produced in the ion fragmentation process for a better understanding of functional groups in the targeted molecule.

Tandem MS and TOF-MS are mutually complementary. In many cases, the combination of both techniques could increase identification confidence [98,99,102–105]. One example is shown in a mass spectrum acquired from an ion trap MS, where the protonated molecular ion of an unknown peak appeared to be at *m/z* 214 (Figure 10.12). However, the *m/z* 214 ion was also present as a strong background signal observed in LC-MS analysis. Since the ion trap MS had only a unit mass resolution in normal operations, it cannot differentiate the mass difference between analyte ion and background ion.

In this case, a high-resolution TOF-MS was used to obtain accurate mass data. The $[M + H]^+$ ion of the unknown was determined as *m/z* 214.2544, while the background ion was at *m/z* 214.0899. With a mass resolution of 11,000, the background ion at 214.0899 was fully resolved from the $[M + H]^+$ ion (*m/z* 214.2544) of the unknown even when their mass difference was 0.16 Da. Furthermore, the proposed molecular ion (*m/z* 214) was confirmed by the presence of *m/z* 427, an adduct ion of $[2M + H]^+$. Figure 10.13 shows the accurate mass spectra of the background and the unknown.

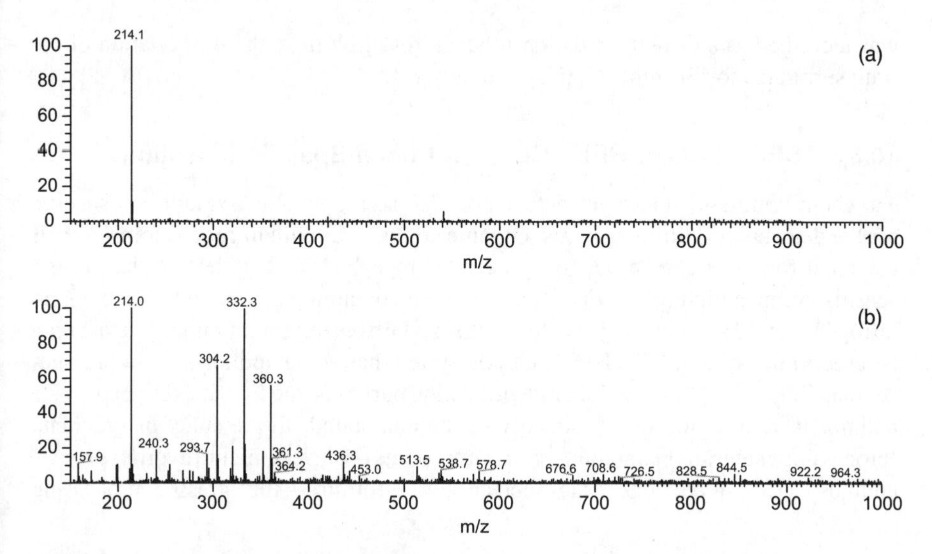

FIGURE 10.12 Mass spectra of an unknown (a) and background ion on baseline (b), obtained from an ion trap MS with a unit mass resolution.

The measured accurate mass of 214.2544 Da for this unknown corresponded exclusively (only one formula matched this ion within a 5 mDa tolerance range) to a chemical formula of $C_{14}H_{32}N$ ($[M + H]^+$ ion) with a mass deviation of 0.9 mDa. On the basis of the accurate mass results and the drug substance structure, the unknown

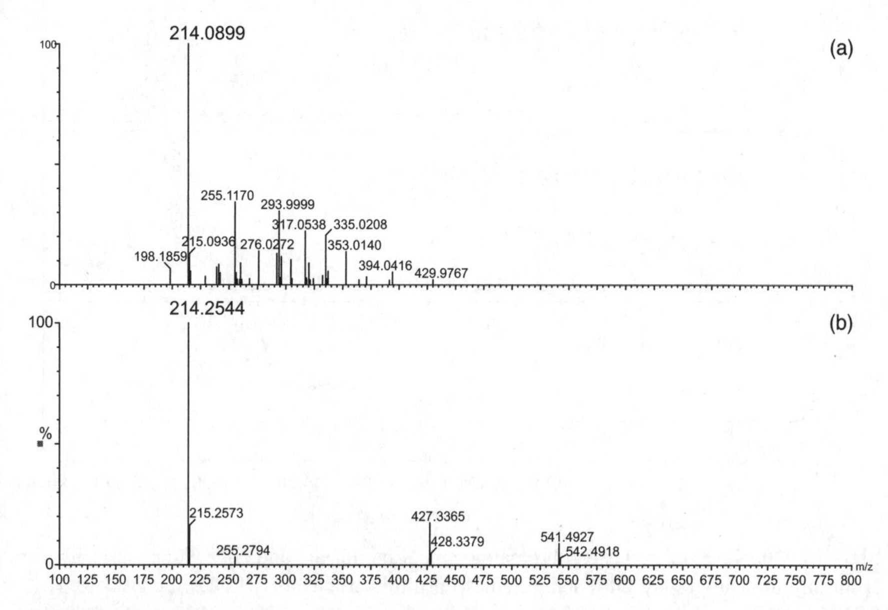

FIGURE 10.13 Mass spectra of background ion on baseline (a) and an unknown (b), obtained from a TOF MS with a mass resolution of 11,000.

was identified as a drug degradation product, resulting from the dealkylation of the drug substance to form this $C_{14}H_{31}N$ moiety.

10.5.3 SEC/CLND or HPLC/CLND: Nitrogen-Specific Detection

The chemiluminescent nitrogen detector (CLND) is highly selective and very specific for the detection of nitrogen atoms. Organic compounds with nitrogen atoms in their chemical formulas can be positively detected by a CLND. The detector has a high sensitivity and a high selectivity for all nitrogen-containing compounds, although no nitrogen-related solvents (e.g., acetonitrile) or additives (e.g., ammonium acetate) can be used in the system. The HPLC-CLND system has some applications in pharmaceutical industries [111–114]. For identification purposes, the LC-CLND can provide additional information on the identity of the compound. In the study of granulate (processing materials) containing two components—one excipient [hydroxypropylcellulose (HPC)] and one drug substance (metformin)—the stressed processing

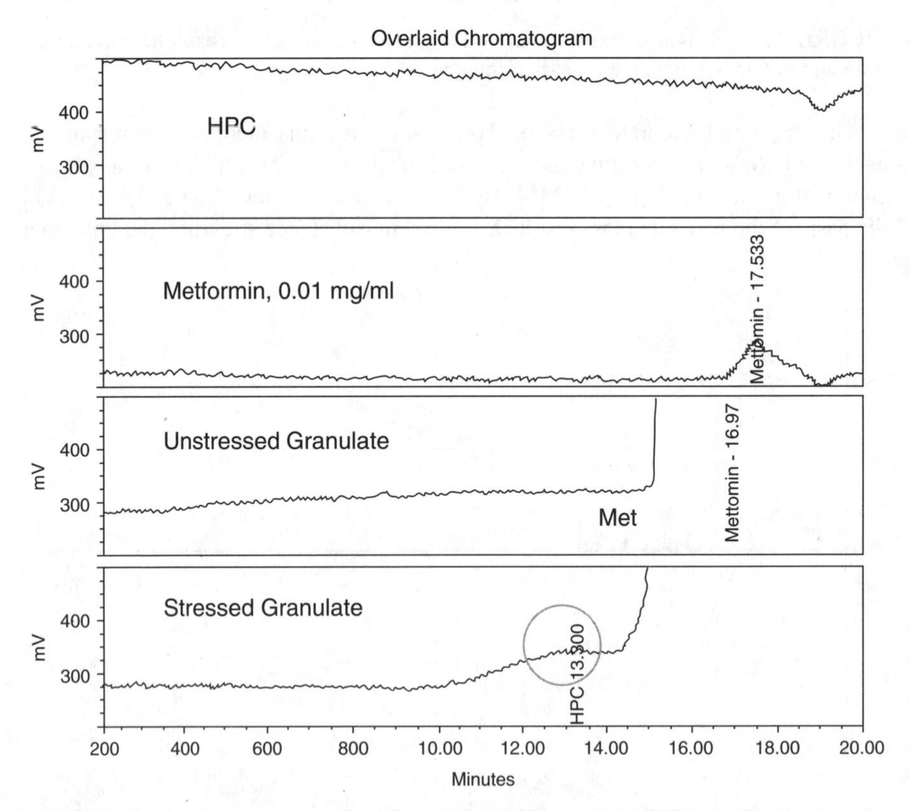

FIGURE 10.14 Size exclusion chromatograms of metformin and HPC excipient using size exclusion chromatography with chemiluminescent nitrogen detector (CLND), model 8060. TSK-GEL GMPWXL column, 7.8 × 300 mm, mobile phase, 0.1% formic acid in water; flow rate, 0.7 mL/min; column temperature, ambient; detector oxidative furnace temperature, 1050°C.

material was found to have an extra peak in the chromatogram using size exclusion chromatography interfaced with a nitrogen detector (SEC-CLND) (Figure 10.14). HPC contains no nitrogen atoms and should have no signal in CLND. Metformin contains several nitrogen atoms and has a peak at ~17.5 min in the chromatogram. In the unstressed granulate, only one peak of metformin was observed in the chromatogram. Since HPC cannot directly react with metformin, the HPC must first degrade, most likely by oxidation to form unsaturated moieties (aldehydes or ketone). This degraded HPC can then react with metformin to form a degradation product that contains nitrogen atom(s). The resulting nitrogen-containing compound can then be detected by CLND. The results from SEC-CLND analysis reveal the difference between the stressed and unstressed granulates in the formulation process. This particular information will not be readily obtained using HPLC-MS or SEC-MS.

10.5.4 GC-MS with EI-CI Combination

When some degradation products or impurities are not ionizable in atmospheric-pressure ionization (API) sources, either ESI or APCI, the use of alternative ionization sources becomes necessary. LC fraction can be collected and injected into a GC-MS. A combination of chemical ionization (CI) and electron impact (EI) in GC-MS provides detailed information on the molecular ion and fragmentation pattern under the "hard" ionization process. For a complete unknown (impurity or leachable), API-LC/MS may have molecular weight information but insufficient information for structure elucidation since the unknown is not related to the drug substance. In this case, GC-MS combined with EI and CI may provide additional information. Since the molecular weight of an unknown has been determined by API-LC/MS, this compound can be tracked by CI with the targeted $[M + H]^+$ ion, from which the peak of interest can be found in the gas chromatogram. Then GC/EI-MS experiments can be conducted to obtain an EI mass spectrum of the peak of interest. The unknown can be identified through a NIST (National Institute of Standards and Technology) library search if the compound is present in the database. The value of this approach is described in detail in Section 10.6.2.

10.5.5 Headspace GC-MS: Volatile Compounds

Headspace GC-MS is a powerful tool for studying residual solvents in tablets or capsules [115–120]. Residual solvents are the solvents that are used for the synthesis of a drug substance, and thus are considered as a "fingerprint" for a particular drug product containing this drug substance. Since residual solvents may play an important role in drug degradation, detection of residual solvents in tablets or capsules becomes necessary if this information is not available by LC-MS.

Furthermore, any volatile compounds present in solid materials can be readily detected with headspace GC-MS. The presence of small and volatile molecules (e.g., formaldehyde in povidone) in excipients can be studied with this approach [120]. Figure 10.15 shows the headspace gas chromatograms of multiple batches of povidone excipient manufactured by different vendors. As shown in the chromatograms, various batches of povidone have different volatile impurity profiles. The presence of these impurities in povidone may cause drug substance degradation in the formulation process and stability conditions. Therefore, the results from headspace GC-MS can be used to justify the source of the reactant involved in the drug degradation.

10.5.6 NMR and LC-NMR

For late-phase projects in formulation development, it is always required that the identity of degradation products be confirmed and the authentic compounds be

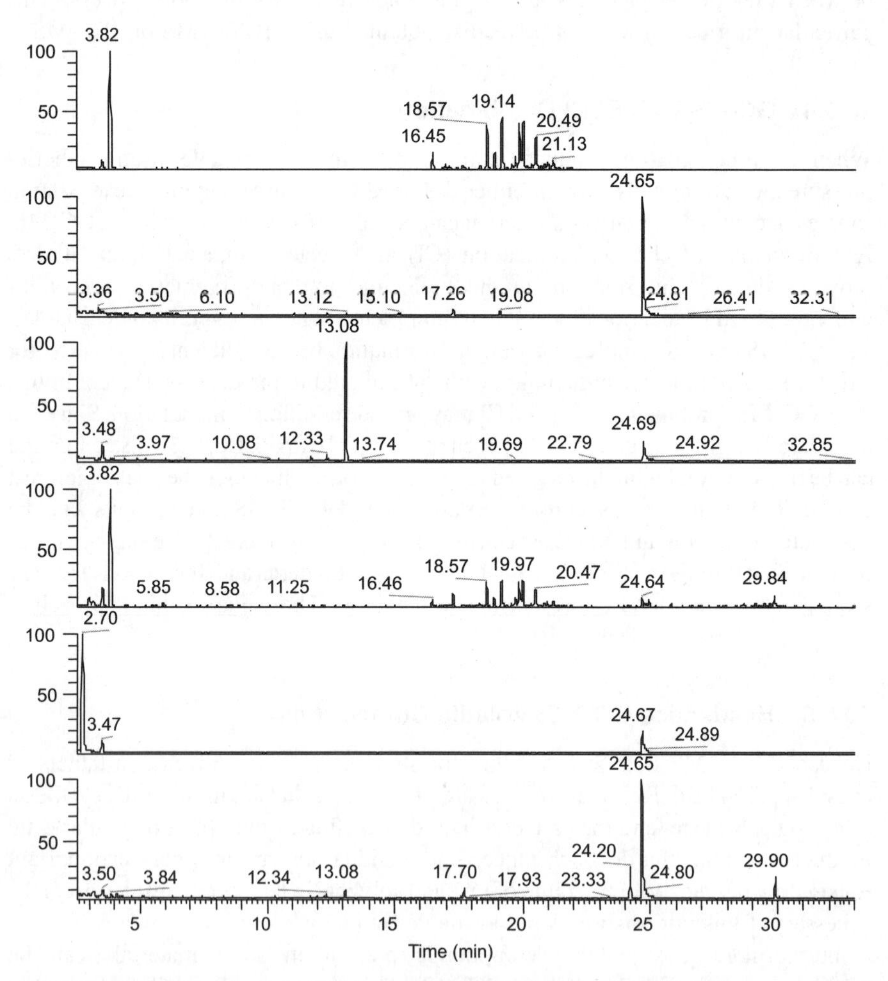

FIGURE 10.15 Headspace GC-MS analysis of volatile impurities in different batches of excipient (povidone) made by different vendors. Headspace, ThermoElectron TriPlus Autosampler Model TP0321131; GC, ThermoElectron Trace GC2000; MS, Finnigan DSQ MS; column, Phenomenex ZB-624, 30 m × 0.25 mm × 1.4 μm.

available through synthesis. Therefore, the structures proposed through LC/MS often need confirmation using isolation/NMR or LC-NMR [121–126]. While LC-PDAD and LC-MS instruments are well-developed laboratory tools now widely available in the pharmaceutical industry, LC-NMR has only relatively recently overcome the sensitivity limitations historically impeding its widespread application in the pharmaceutical industry [93,94,96,97,106].

10.5.7 TD-GC/MS: Chemical Reactions Attributing to Weight Loss in TGA

Thermal desorption gas chromatograph mass spectrometry (TD-GC/MS) has been used for polymer sciences and pharmaceutical industries [127–131]. The technique could be used to study weight loss in thermogravimetric analysis (TGA). TGA measures the weight loss as a function of temperature and is a useful analytical tool in pharmaceutical development as it characterizes the thermal behavior of drug substances and drug products. When multiple species are released during a single weight-loss event in a TGA, only the total weight loss is observed from the thermogram. It is therefore interesting to know how the weight loss is distributed among the various components and to understand the chemistry behind the weight loss. In addition, it is important to know in formulation development whether a drug substance has bound or unbound solvent molecules and at what temperature solvent molecules are released. TGA can provide partial answers but cannot indicate positive identity. In TD-GC/MS, the technique of thermal desorption utilizes the same temperature heating rate as the TGA to thermally desorb volatiles from solid sample matrices. Volatiles were cryotrapped at −60°C. After thermal desorption is complete, the trapped volatiles are separated by a GC capillary column and identified by MS.

In one example, TD-GC/MS experiments were performed to understand chemical reactions attributed to the weight loss in the thermal decomposition of two dicarboxylic acid salts of a drug substance. These two salts exhibited different thermal stabilities in TGA. The thermally induced chemical reactions obtained from these two salts included dehydration and decarboxylation. Thermal degradation compounds were identified, and reaction pathways for decomposition were proposed. The stability of the salts is dependent on the identity of the dicarboxylic acids from which they were generated. The information obtained from TD-GC/MS experiments can facilitate better understanding of the weight-loss process in thermogravimetric analysis [131].

10.6 CASE STUDIES

10.6.1 LC-MS, GC-MS, and LC-NMR Studies of a Drug Degradation Product

TCH346 is a propargylamine compound with a pK_a at 6.3. TCH346 drug substance is chemically stable; however, a number of degradation products were observed in the forced degradation at high pH of a TCH346 tablet formulation. One of the unknown degradation products observed during the storage conditions (25°/60%RH and 40°/75% RH) of the TCH346 drug product grew over time. Identification of this unknown is particularly challenging because of its poor ionization property and difficult separation from other basic drug impurities in HPLC. Stress studies indicate that this degradation product can be produced in the basic condition under xenon light, indicating that it may be a photocatalytic oxidation product of the TCH346 drug substance. Several approaches were used for structural elucidation of this degradant.

10.6.1.1 LC-MS Analysis

Figure 10.16 shows the chromatograms of a TCH346 tablet stability sample obtained from four different HPLC methods. Under the acidic mobile phase of 0.1% TFA, the peak of interest (labeled as "unknown") was eluted at 11 min (Figure 10.16a). No protonated molecular ion was observed for this unknown peak using either ESI (+) mode or APCI (+) mode. A negative ESI mode was then investigated under neutral and basic mobile phase conditions (Figure 10.16c,d). No ionization was detected for this peak, either. The results imply that this compound has no ionizable functional groups as it could not be ionized in either a positive or a negative ionization mode. The retention time of this unknown peak remains almost constant under all four different mobile-phase conditions from pH 2.5 to 8.7, indicating that its charge status is not changed during this acid–base transition. In contrast, the retention time of the TCH346 drug substance shows a strong dependence on the mobile-phase pH. The drug substance is a basic compound with a pK_a of 6.3. When the mobile phase was changed from acidic pH to basic pH, the drug substance was changed from a protonated form to a neutral state, and therefore retained longer in the column.

10.6.1.2 GC-MS Analysis

Since the unknown could not be ionized by API techniques (ESI and APCI), the isolate was analyzed by chemical ionization, a much less selective ionization mode. Because the TCH346 drug substance is semivolatile and can be detected by GC-MS, the unknown may be amenable to this technique as it is a degradation compound of the drug substance. Figure 10.17 displays the gas chromatogram of the isolated material. The major peak elutes at 16.9 min and is presumed to be the unknown compound. A minor peak at 18.1 min is due to the TCH346 drug substance resulting from the peak tailing in the chromatogram during the isolation process. The methane CI mass spectrum illustrated in Figure 10.17 suggests that the molecular weight of the unknown compound is 222 Da, which is confirmed by the presence of $[M + H]^+$ ion at *m/z* 223 and an adduct ion $[M + C_2H_5]^+$ at *m/z* 251. The CI-MS/MS mass spectrum of the protonated pseudomolecular ion is also exhibited in Figure 10.17.

10.6.1.3 LC-NMR Analysis

The isolate was dissolved in acetonitrile and injected onto a C18 LC column for purification and to focus the material into the LC-NMR flowprobe. The concentrated

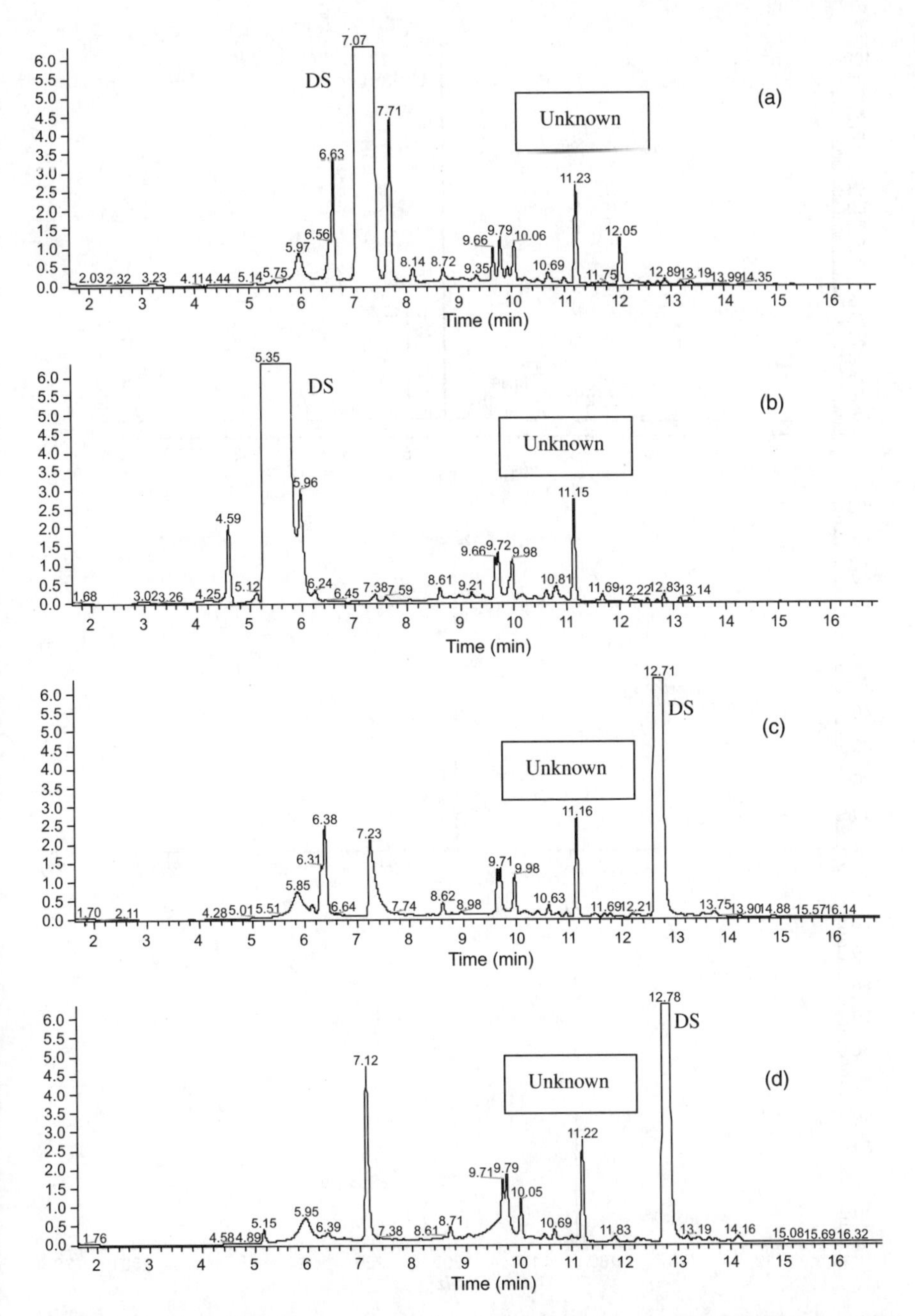

FIGURE 10.16 HPLC chromatograms of the sample solution of a TCH346 0.25-mg tablet stored at 40°C/75% RH for 6 months. Mobile phase A: (a) 0.1% trifluoroacetic acid (TFA); (b) 0.1% formic acid (HCOOH); (c) 20 mM ammonium acetate (CH_3COONH_4), pH 5.8; (d) 20 mM ammonium carbonate $(NH_4)_2CO_3$, pH 8.7. Mobile phase B: acetonitrile in each case. The same gradient condition was used for all four cases. (Reprinted from Ref. 94, with permission of Elsevier Science, Inc; copyright 2005 Elsevier B.V.)

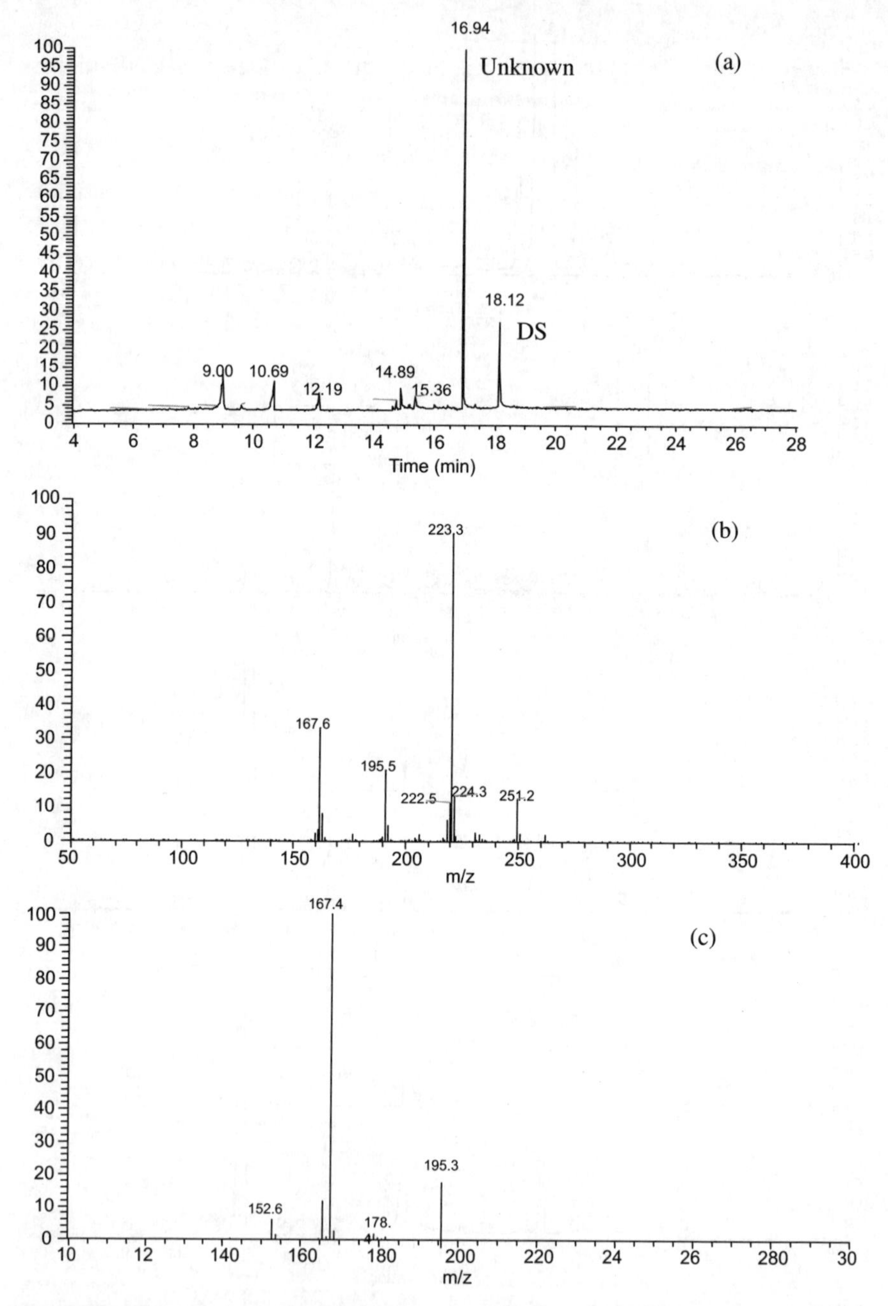

FIGURE 10.17 Chemical impact (CI) total-ion chromatogram (TIC) of isolated material (a); mass spectrum of the unknown (b) and CI MS/MS spectrum of *m/z* 223, the protonated molecular ion of the unknown (c). TRACE GC2000/PolarisQ ion trap mass spectrometer equipped with both El and CI sources. Capillary column: J&W DB-5MS, 30 m × 0.25 mm i.d. × 0.25 μm film thickness. Methane as a CI reagent gas at a flow rate of 1.5 mL/min. (Reprinted from Ref. 94, with permission of Elsevier Science, Inc; copyright 2005 Elsevier B.V.)

analyte in the flowprobe was of sufficient purity and quantity to provide both ^{1}H and ^{13}C data, as well as 2D homonuclear and heteronuclear polarization transfer spectra. Figure 10.18 shows the proton ^{1}H spectrum obtained. There is a single resonance for the aldehyde proton at 9.4 ppm, and nine aromatic protons. The integration for the aromatics is slightly high because of the presence of the residual chloroform signal under the analyte signal. The ^{13}C spectrum was also shown in Figure 10.18. 2D multinuclear experiments were used to make tentative assignments. The six upfield protons (three sets of two) correspond to the six-membered ring protons

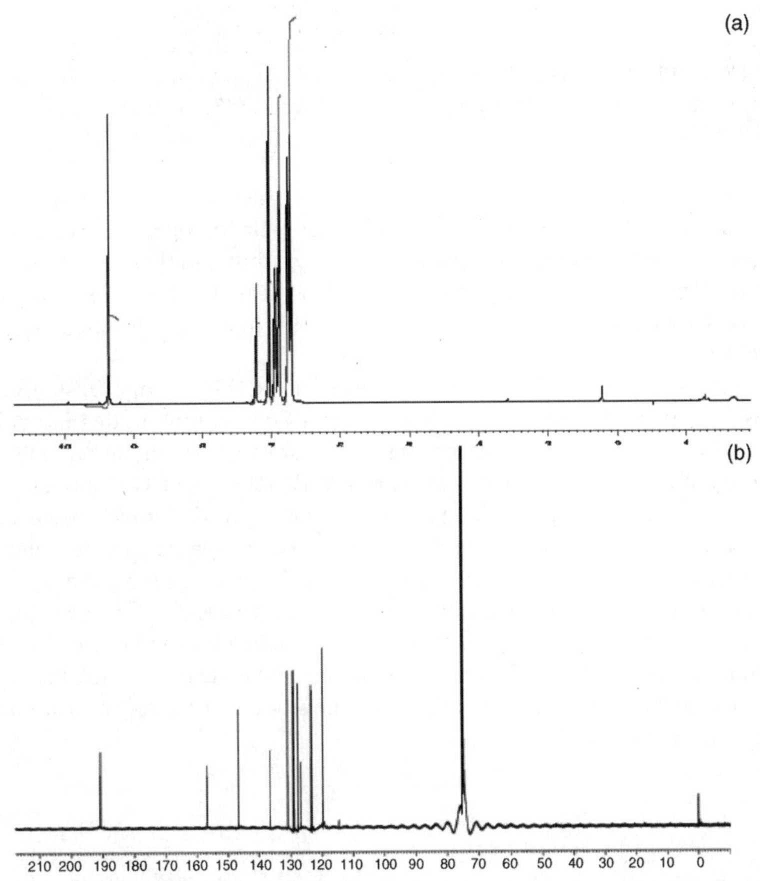

FIGURE 10.18 The ^{1}H NMR spectrum of the isolate purified by a second, automated LC/SPE step (a); ^{13}C NMR spectrum of the isolate purified by a second, automated LC/SPE step (b). NMR spectra were recorded in at 300 K using a Bruker Avance 600 MHz (125 MHz for ^{13}C) spectrometer equipped with a Bruker CryoPlatform and a 3 mm cryofit dual inverse probe fitted with a 60 flow cell (30 μL active volume). The temperature of the probe and amplifier was 15 K. Chemical shifts were referenced to the respective $CDCl_3$ resonances. (Reprinted from Ref. 94, with permission of Elsevier Science, Inc; copyright 2005 Elsevier B.V.)

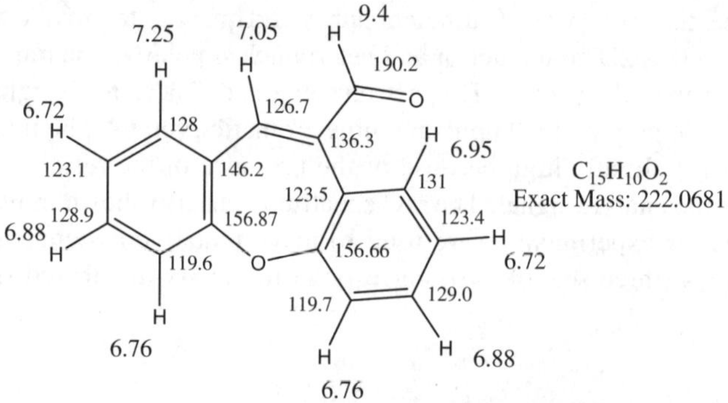

$C_{15}H_{10}O_2$
Exact Mass: 222.0681

FIGURE 10.19 Confirmed chemical structure of the unknown based on LC/NMR results (reprinted from Ref. 94, with permission of Elsevier Science, Inc; copyright 2005 Elsevier B.V.)

closest to the ether. The resonances within each pair (on opposite rings) are not resolved, so assignments of proton or carbon shifts to particular rings are tentative. The three downfield protons correspond to the three protons closest to the aldehyde. The spectra are consistent with aldehyde structure shown in Figure 10.19.

The compound is a degradation product of TCH346 drug substance. The proposed structure is consistent with the previous observation in that it manifests no response in ESI and APCI in either positive or negative ionization mode. The formation of aldehyde conjugates the carbonyl double bond to the aromatic ring, resulting in a *redshift* in its UV spectrum compared to the drug substance. The aldehyde structure is also in agreement with the gas-phase ion fragmentation obtained by MS/MS, as shown in Figure 10.20. The degradation pathway for the formation of this compound was proposed in Figure 10.21. A reference compound was synthesized, and the authentic compound matched the "unknown" in the LC retention time and UV and MS spectra. With this authentic compound, the relative response factor was obtained and the peak can now be quantitatively measured at the wavelength of 281 nm.

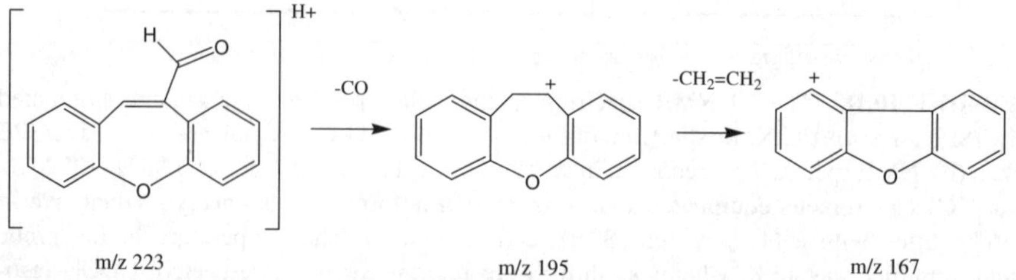

m/z 223 —(-CO)→ m/z 195 —(-CH_2=CH_2)→ m/z 167

FIGURE 10.20 Proposed ion fragmentation pathways for this unknown (reprinted from Ref. 94, with permission of Elsevier Science, Inc; copyright 2005 Elsevier B.V.)

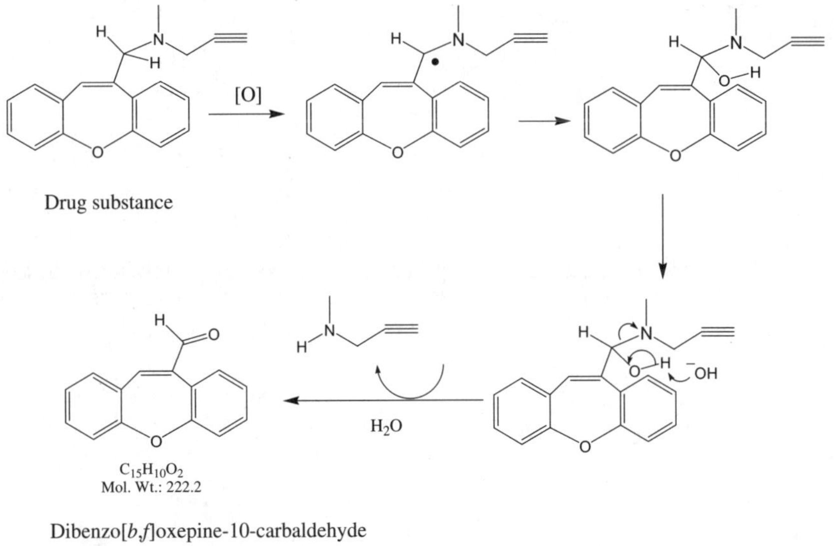

Drug substance

$C_{15}H_{10}O_2$
Mol. Wt.: 222.2

Dibenzo[*b*,*f*]oxepine-10-carbaldehyde

FIGURE 10.21 Proposed degradation pathway for the formation of this unknown (reprinted from Ref. 94, with permission of Elsevier Science, Inc; copyright 2005 Elsevier B.V.)

10.6.2 Strategy for Identification of Leachables in Packaged Liquid Formulation

This investigation involved the identification of an unknown leachable found in a registration stability sample (40°C/20% RH for 6 months) of an ophthalmic solution stored in a semipermeable low-density polyethylene (LDPE) bottle. Initial LC-MS/MS analysis of the unknown indicated that the unknown was not structurally related to the active ingredient. The unknown had a molecular weight of 196 Da and fragmentation ions at *m/z* 179, 135, 109, and 89, as shown in Figure 10.22. Hence, the unknown was identified as a leachable from plastics or labels on the semipermeable bottles. An investigation was initiated to determine both the source and the identity of the unknown impurity, followed by an assessment of the maximum daily exposure.

Controlled extraction studies were conducted to determine the source of this impurity using reversed-phase HPLC-MS/MS. All packaging components, including bottle, plug, cap, label, ink, and varnish were extracted and then tested by LC-MS. Results indicated that varnish extract contained the same peak as the unknown in terms of retention time, MS/MS spectrum, and UV spectrum. As this is a complete unknown, LC-MS alone cannot reveal its full structure. In order to obtain further structural information, GC-MS with CI was performed on the varnish extract. As shown in Figure 10.23, the CI mass spectrum of the peak at 15.6 min showed ions similar to those in the ESI MS/MS spectrum. Clearly, both spectra exhibit $[M+H]^+$ ion as *m/z* 197 and some common fragmentation ions, such as *m/z* 179, 135, 109, and 89. This indicated that the peak that eluted at 15.6 min in the gas chromatogram corresponded to the unknown peak observed at ~9.2 min in the liquid chromatogram. Even though CI in the gas phase and ESI in the liquid phase are two different ionization processes, they

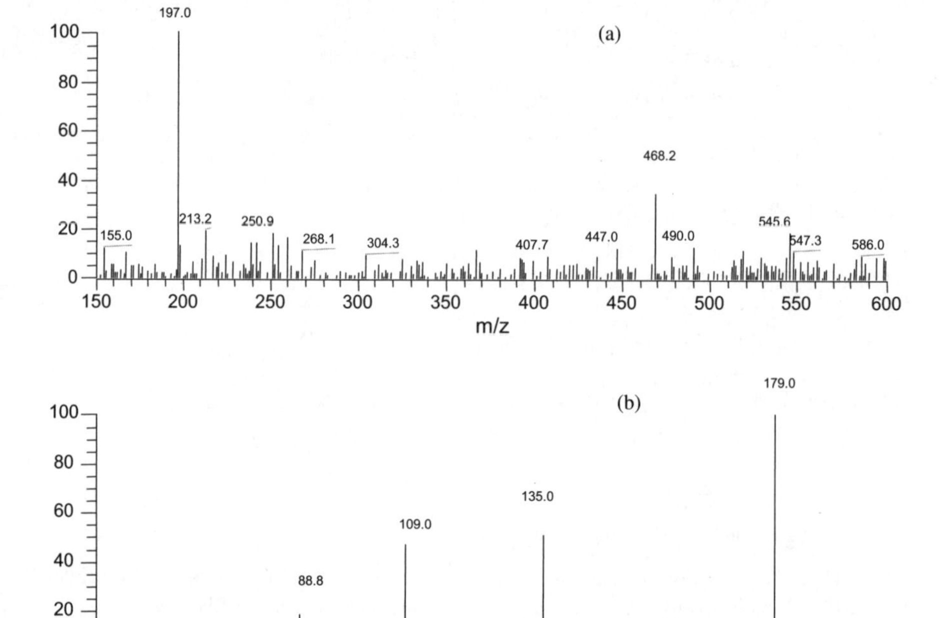

FIGURE 10.22 Mass spectrum of the unknown peak at 9.2 min found in a registration stability sample (40°C/20% RHumidity for 6 months) of an ophthalmic solution (a); MS/MS spectrum of *m/z* 197, the protonated ion of the unknown (b) (reprinted from Ref. 73, with permission of Elsevier Science, Inc; copyright 2007 Elsevier B.V.)

both generate the same protonated molecular ions $[M+H]^+$, which could be used to establish a peak correlation between gas chromatograms and liquid chromatograms.

On the basis of CI-MS results, EI-MS experiments were performed to generate the EI mass fragmentation of the peak at 15.6 min, from which the compound could be identified by the NIST library search. Figure 10.23 shows the EI mass spectral data, where the peak eluting at 15. 6 min has an odd-electron ion at *m/z* 196 (a radical cation $M^{+\cdot}$ ion). This observation further confirms that the compound eluting at 15.6 min in the gas chromatogram has a molecular weight of 196 Da. The EI and CI mass spectral data are fully complementary to each other in the molecular weight determination. From the EI-MS library search, the unknown in the stability samples was identified as a monomethyl derivative of mephenesin with a good matching factor. The structure is consistent with the fragmentation pattern observed in Figure 10.22. The compound has a neutral loss of water (197 → 179), indicating that one hydroxyl group is present in the structure. The substructures of this molecule match well with the fragmentation pathway.

In summary, this approach in conjunction with systematic data analyses provided the following conclusions: (1) the unknown was most likely a leachable, as it was not

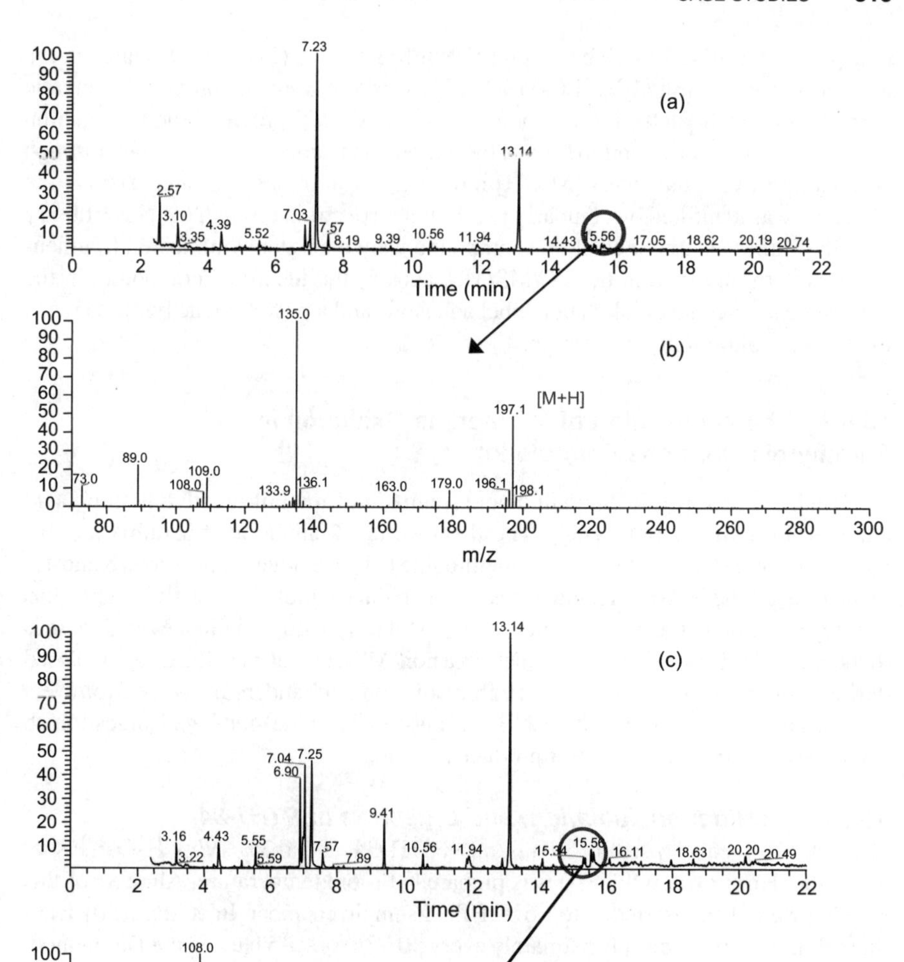

FIGURE 10.23 Total-ion chromatogram of varnish extraction detected by GC-MS with CI (a); CI mass spectrum of the peak at 15.6 min (b); total-ion chromatogram of varnish extraction detected by GC-MS with EI (c); EI mass spectrum of the peak at 15.6 min (d) (reprinted from Ref. 73, with permission of Elsevier Science, Inc; copyright 2007 Elsevier B.V.)

related to the drug substance based on LC/MS/MS results; (2) the exact source of the unknown was identified by LC-MS analysis as the varnish through systematic extraction of each packaging component; (3) GC/CI-MS provided the information needed to identify the retention time of the unknown in the GC chromatogram through correlation of the unknown's [M + H] ion observed in the LC-MS analysis; (4) the unknown was identified by comparing its EI mass spectrum through the NIST library search; (5) the substructures of this compound were consistent with its ion fragmentation pattern observed in the LC/MS/MS; and (6) the identified compound in the varnish penetrated label ink, label, label adhesive, and a polyethylene bottle to reach the sample solution during storage.

10.6.3 Characterization of Methionine Oxidation in Parathyroid Hormone Formulation

PTH1-34 is a therapeutic 34–amino acid fragment of the full-length 84 amino acid parathyroid hormone (PTH1-84). The therapeutic 34–amino acid parathyroid hormone fragment (PTH1-34) contains methionine (Met) residues at positions 8 and 18. Oxidation of these MET residues results in reduced biological activity and thus efficacy of the potential drug product [132,133]. The practical challenge in this case is to identify PTH1-34 oxidation products in a non MS-compatible HPLC method used during drug product development. High-resolution and tandem mass spectrometers were used in conjunction with CNBr (cyanogen bromide)-mediated digestion to accurately identify the oxidation products.

10.6.3.1 Oxidation, Isolation, and Digestion of PTH1-34

A solution of equal parts of 10 mg/mL PTH1-34 and 0.3% (v/v) H_2O_2 (Fisher Scientific, Fair Lawn, NJ, USA) was prepared at room temperature. Aliquots of this solution were then inserted in to a 5°C HPLC sample chamber. Injections (50 μL) of this solution were made approximately every 30 min for <2.5 h using the DS method. All observed peaks were manually collected from replicate injections and labeled as fractions 1–4 in order of increasing retention time (Figure 10.24) (fraction 1—7.5 min; fraction 2—9.5 min; fraction 3—10.9 min; fraction 4—13.0 min). CNBr (Acros Organics, Morris Plains, NJ) (20 μL, 100 mM) was added to 500 μL of each fraction and allowed to react at room temperature, protected from light, for ~40 h. The digested fractions were then re-injected onto the MS-compatible LC method.

PTH1-34 is susceptible to oxidation at residues Met8 and Met18, resulting in the following three variants: two monooxygenated peptides with a methionine sulfoxide at either Met8 or Met18 and one dioxygenated peptide with a methionine sulfoxide at both residues. The biological activity of these variants decreases as follows: oxidation at Met18 > oxidation at Met8 > oxidation at both residues [133]. It is therefore important to identify and control these variants during manufacture and stability due to their impact on the efficacy of the drug product. Notably, differentiation of the two mono-oxygenated variants requires site-specific digestion with subsequent LC-MS analysis. CNBr selectively hydrolyzes peptide bonds at the *C* terminus of Met [134]. The sulfoxide form of methionine is not cleaved by CNBr because of the reduced

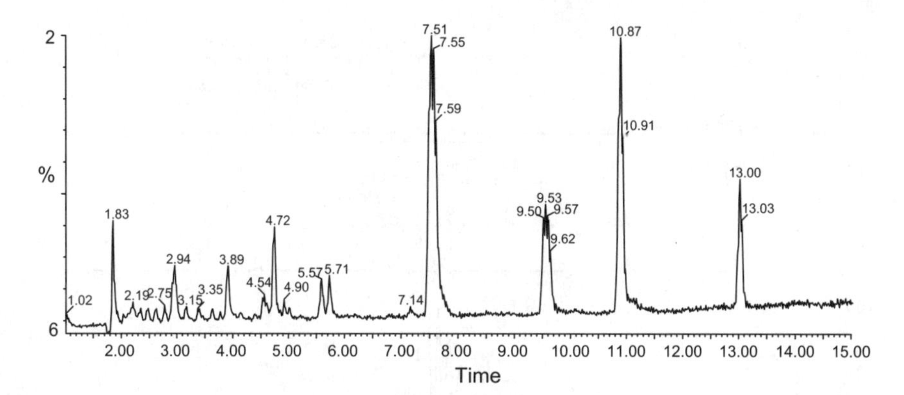

FIGURE 10.24 Chromatogram of PTH1-34 peptide (RT 13.0 min) and its three major oxidation products at RT = 7.5, 9.5, and 10.9 min.

nucleophilicity of the sulfur. This specificity greatly reduces the number of digestion fragments. The expected mass of the modified *N*-terminal fragment is 31 Da less than that of the native sequence fragment owing to the losses of S atom and $-CH_3$ and a gain of O atom ($-32-15+16=-31$) during the conversion of the Met to a homoserine lactone.

10.6.3.2 Mass Assignment of PTH1-34 Oxidized Variants

In order to ensure the mass accuracy for each fragment, a Lockspray of leucine enkephalin (0.75 ng/μL, 10 μl/min) in a ACN/H_2O (1 : 1) matrix was used to facilitate the mass measurement during the analysis. The use of a high-resolution TOF mass spectrometer enabled the identification of many multiply charged species from which molecular weights were positively calculated. Since the TOF mass spectrometer has a high resolution (11,500), the multiple charges on each ion can be readily detected, from which peptide molecular weights can be calculated. For example, the peak at 13.0 min (PTH1-34 peptide) has two major ions: one at *m/z* 1030.28 Da with a split of 0.25 Da, indicative of four charges; and the other, at *m/z* 1373.40 Da with a split of 0.33 Da, indicative of three charges. The mass of this peptide is thus calculated as 4117.17 Da, which is highly consistent with the theoretical molecular weight of 4117.80 Da for the PTH1-34 peptide. This high-mass-accuracy measurement was applied to the determination of oxidized and digested peptide fragments as well.

Figure 10.25 shows the mass spectra of four peaks obtained from TOF-MS. The peak at 7.5 min contains an ion of 1038.28 Da with a split of 0.25 Da and an ion of 1384.06 with a split of 0.33 Da, indicative of four and three charges on these ions, respectively. The mass of this peptide is 4149.1492 Da, which is 32 Da higher than that of PTH1-34 peptide. This could be due to the addition of two oxygen atoms. Two peaks at 9.5 and 10.9 min have almost identical mass spectra. The mass split of 0.25 Da on the ion *m/z* 1034.28 indicates that both peptides have the same molecular weight of 4133.16 Da, which is 16 Da higher than that of the PTH1-34 peptide. These two peptides could result from the oxidation of either Met8 or Met18 residue.

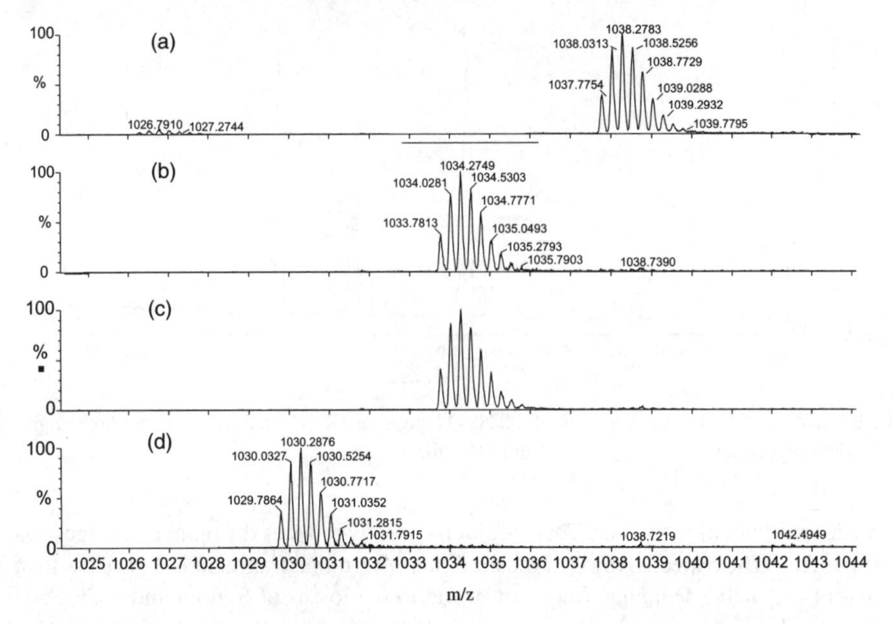

FIGURE 10.25 High-resolution mass spectra of dioxygenated PTH (RT 7.5 min) (a); monooxygenated PTH (RT 9.5 min) (b); monooxygenated PTH (RT 10.9 min) (c); PTH1-34 peptide (RT 13.0 min) (d).

However the mass spectral data without CNBr digestion cannot positively define which Met in PTH1-34 has been oxidized.

10.6.3.3 Mass Assignment of CNBr Digested Peptide Fragments

10.6.3.3.1 PTH Met8(O)Met18(O)

The chromatogram of fraction 1 after digestion with CNBr contains only one major peak at 7.5 min, very similar to that of the undigested fraction (Figure 10.26a). This fraction contained an ion of 1038.28 Da with a split of 0.25 Da, indicating that its molecular weight was 4149 Da, which is 32 Da higher than that of PTH1-34 peptide. The absence of other peptide fragments after CNBr digestion along with the mass assignment is consistent with the formation of the doubly oxidized variant, PTH Met8(O)Met18(O), as illustrated below:

SVSEIQLM(O)HNLGKHLNSM(O)ERVEWLRKKLQDVHNF
1 2 3 4 5 6 7 8 9 10 11 12 13 14 15 16 17 18 19 20 21 22 23 24 25 26 27 28 29 30 31 32 33 34

The chromatographic results also rule out the possibility of a methionine sulfone formation. If this were the case, the characteristic fragments resulting from CNBr digestion at the unoxidized methionine residue would have been observed. Therefore, the peak at 7.5 min is due to the formation of a sulfoxide on both methionine residues of the PTH1-34 peptide.

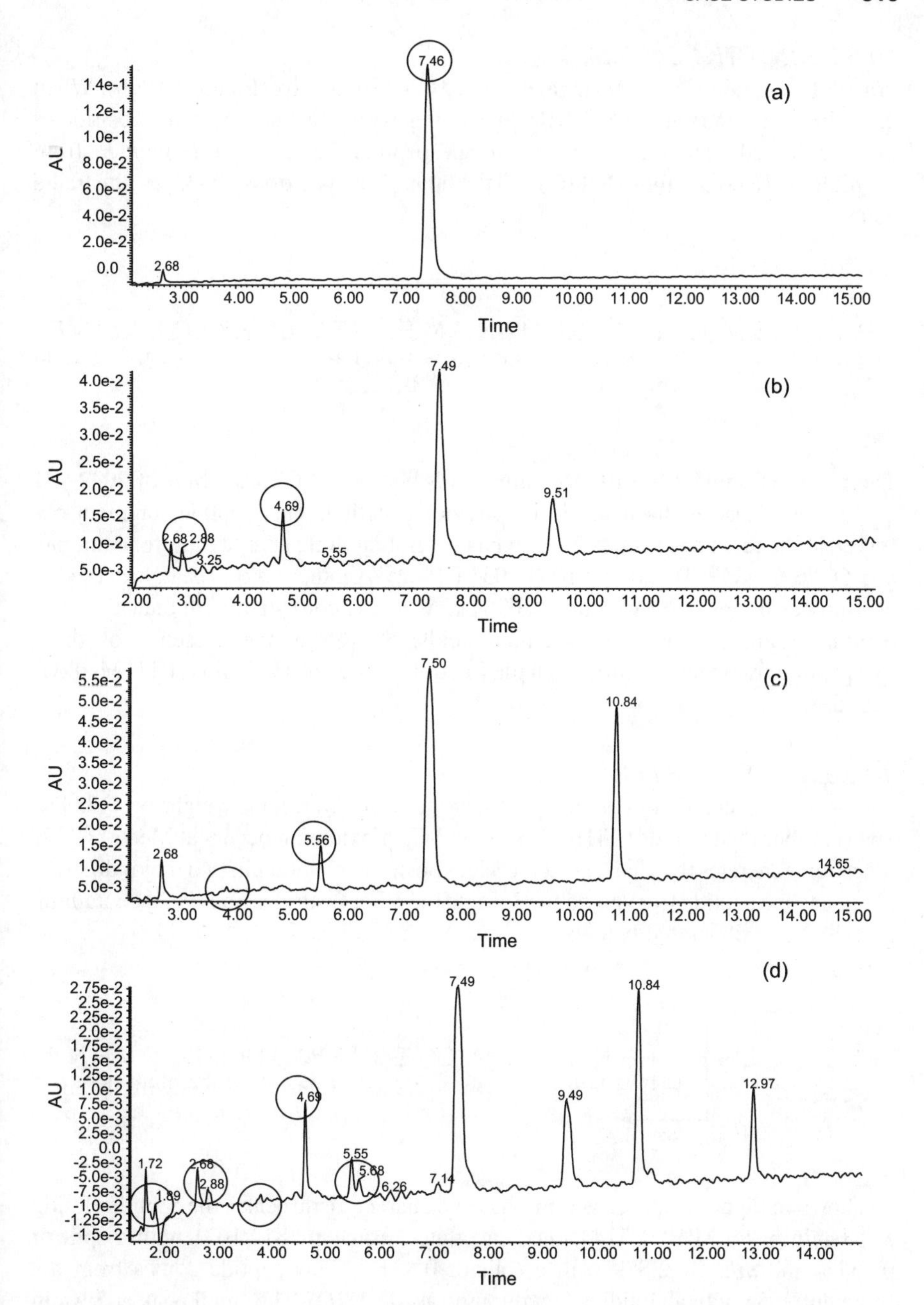

FIGURE 10.26 Chromatograms of isolated fractions 1 (a), 2 (b), 3 (c), and 4 (d) after CNBr digestion.

10.6.3.3.2 PTH1-34/Met8(O)

Fraction 2 contained several fragments after CNBr digestion (Figure 10.26b). When the oxidation occurs at Met8, CNBr digestion generates a cleavage at the position of the unoxidized Met18 to form two major peptide fragments: one peptide from positions 1–18 containing Met8(O) and the other, from positions 19–34, as illustrated below:

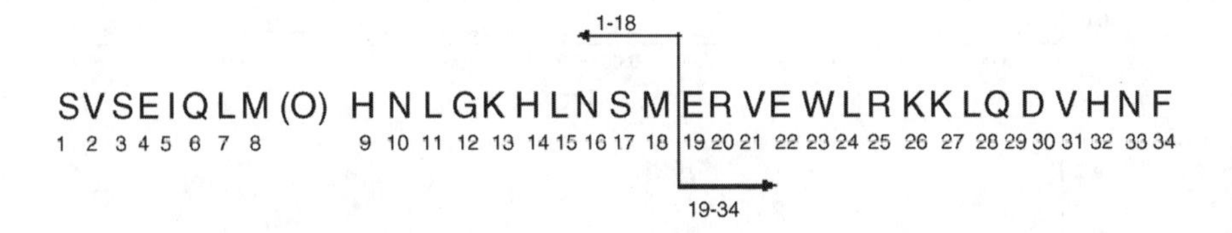

The peptide fragment from 1–18 amino acids with Met8(O) has a MW of 2022−31 + 16 = 2007 Da, as discussed in the previous section. This peptide fragment observed at the retention time of 2.9 min has several multiple charge ions, for example, *m/z* 669.6721 (Z = 3) and *m/z* 1004.0031 (Z = 2). Another unique peak observed at 4.7 min has a molecular weight of 2096 Da, which is consistent with the calculated peptide fragment from 19–34 amino acids. Therefore, the presence of these two peptide fragments confirms that the fraction collected at 9.5 min is PTH Met8(O) peptide.

10.6.3.3.3 PTH1-34/Met18(O)

Fraction 3, collected at 10.9 min, also has the same molecular weight of 4133 Da, 16 Da higher than that of PTH1-34 peptide. When oxidation occurs at Met18, CNBr digestion generates the cleavage at the Met8 position, resulting in two major digested peptides: one peptide from positions 1–8 and the other from positions 9–34 containing Met18(O), as illustrated below:

As shown in Figure 10.26c, fraction 3 has two characteristic peaks: the peptide eluting at 5.6 min has a MW of 3244.7 Da, calculated from *m/z* 812.1671 with a split of 0.25 Da and *m/z* 1082.889 with a split of 0.33 Da. This peptide comes from the fragment 9–34 with an oxidized methionine at Met18(O). The small peak at 3.9 min has an ion of *m/z* 858.5 Da with no split, which is consistent with the calculated mass of peptide fragment 1–8. As a result, the presence of these two anticipated peptide fragments and lack of Met8(O) confirm that the fraction collected at 10.9 min is PTH Met18(O) peptide.

10.6.3.3.4 Unoxidized PTH1-34 Peptide

The unoxidized PTH1-34 peptide could have several peptide fragments resulting from CNBr digestion, as illustrated in Figure 10.26d:

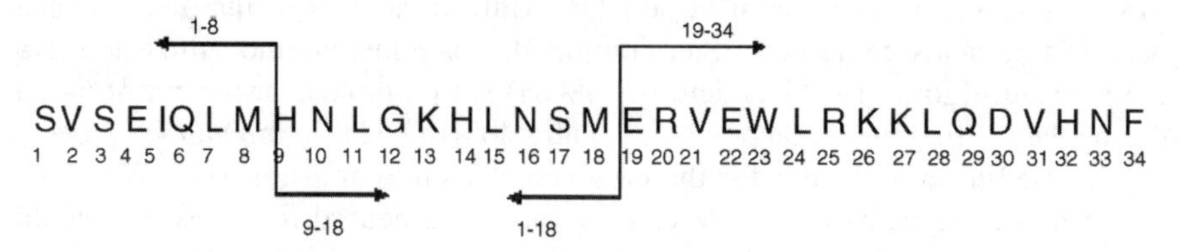

The peptide fragment from 1–8 amino acids with a MW of 858 Da was found at 3.9 min. The peak at 4.7 min is due to the peptide fragment from 19–34. The peak at 5.7 min has *m/z* 664.3372 with $Z=3$ and *m/z* 995.0149 with $Z=2$. This peptide has a molecular weight of 1990.0299 Da, resulting from the peptide fragment from 1–18. The peak at 1.9 min has *m/z* 551.7884 with $Z=2$ and *m/z* 1102.5903 with $Z=1$. The molecular weight of this peptide is 1102.0800 Da, resulting from the PTH fragment 9–18. Additional PTH1-34 oxidation products were observed in these experiments that were generated as by-products of the CNBr-mediated digestion. Therefore, the two peaks at 2.9 and 5.6 min are due to the oxidized peptide fragments 1–18 [containing Met8(O)] and 9–34 [containing Met18(O)], respectively. Table 10.4 summarizes the PTH1-34 peptide fragments after CNBr digestion observed in this LC-MS study.

TABLE 10.4 Summary of PTH-134 CNBr Digested Peptide Fragments Detected by LC-MS

Retention Time (min)	Measured Ion (m/z)	Charge	Calculated Mass (MW)	Peptide Fragment Assignment
1.9	551.7844	2	1101.5688	—
	1102.5903	1	1102.5903	9–18
2.9	669.6721	3	2006.0163	—
	1004.0031	2	2006.0113	1–18(M8(O))
3.9	858.4572	1	857.4572	1–8
4.7	700.0454	3	2097.1362	—
	1049.0581	2	2096.6262	19–34
5.6	812.1671	4	3244.6684	—
	1082.8892	3	3245.1680	9–34(M18(O))
5.7	664.3372	3	1990.0116	—
	995.0149	2	1990.0299	1–18
7.5	1038.2802	4	4149.1208	—
	1384.0592	3	4149.1492	PTH M8(O)M18(O)
9.5	1034.2823	4	4133.1292	—
	1378.7303	3	4133.1600	PTH M8(O)
10.9	1034.2805	4	4133.1220	—
	1378.7291	3	4133.1547	PTH M18(O)
13.0	1030.2876	4	4117.1504	—
	1373.3971	3	4117.1709	PTH1-34

10.6.3.4 LC-MS/MS Studies of Ion Fragments from Oxidized Peptides

In addition, a tandem mass spectrometer (ion trap MS) was used to study the LC/MS^n fragmentation patterns of these PTH-oxidized variants so that the assigned peptide fragments could be further confirmed. Since methionine sulfoxide has a unique neutral loss of sulfinic acid (CH_3SOH) with a molecular weight of 64 Da during the ion fragmentation process, a neutral loss of 64 Da in the MS/MS process can be used as an indicator for the presence of methionine sulfoxide. When the detected peptide carries multiple charges on it, the neutral loss observed could change proportionally through an empirical relationship of $64/z$ (where z denotes charges on the parent ion).

Figure 10.27 shows the MS/MS spectra of the multiply charged peptides that are related to methionine oxidation. The ion of m/z 1003.9 is related to fragment 1–18 [Met8(O)]. It has two charges (Z = 2), and its fragmentation generates a neutral

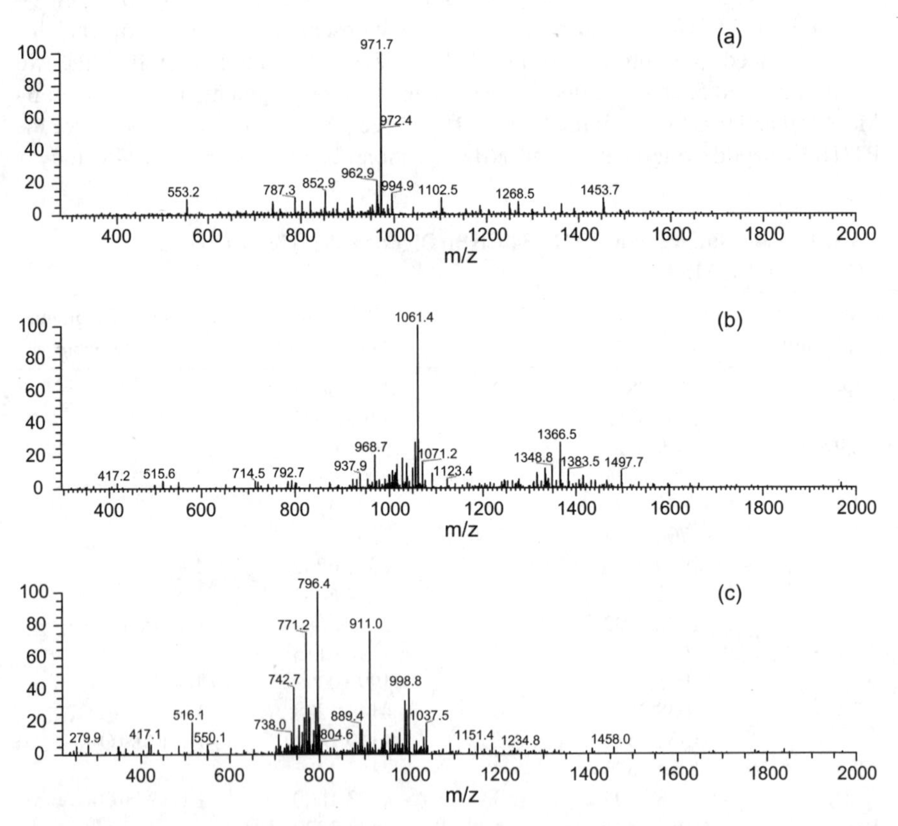

FIGURE 10.27 MS/MS spectra of multiply charged peptide species that are related to methionine oxidation: (a) MS/MS spectrum of m/z 1004 (+2) that is related to 1–18 [M8(O)]; (b) MS/MS spectrum of m/z 1082 (+3) that is related to 9–34 [M18(O)]; (c) MS/MS spectrum of m/z 812 (+4) that is related to 9–34 [M18(O)].

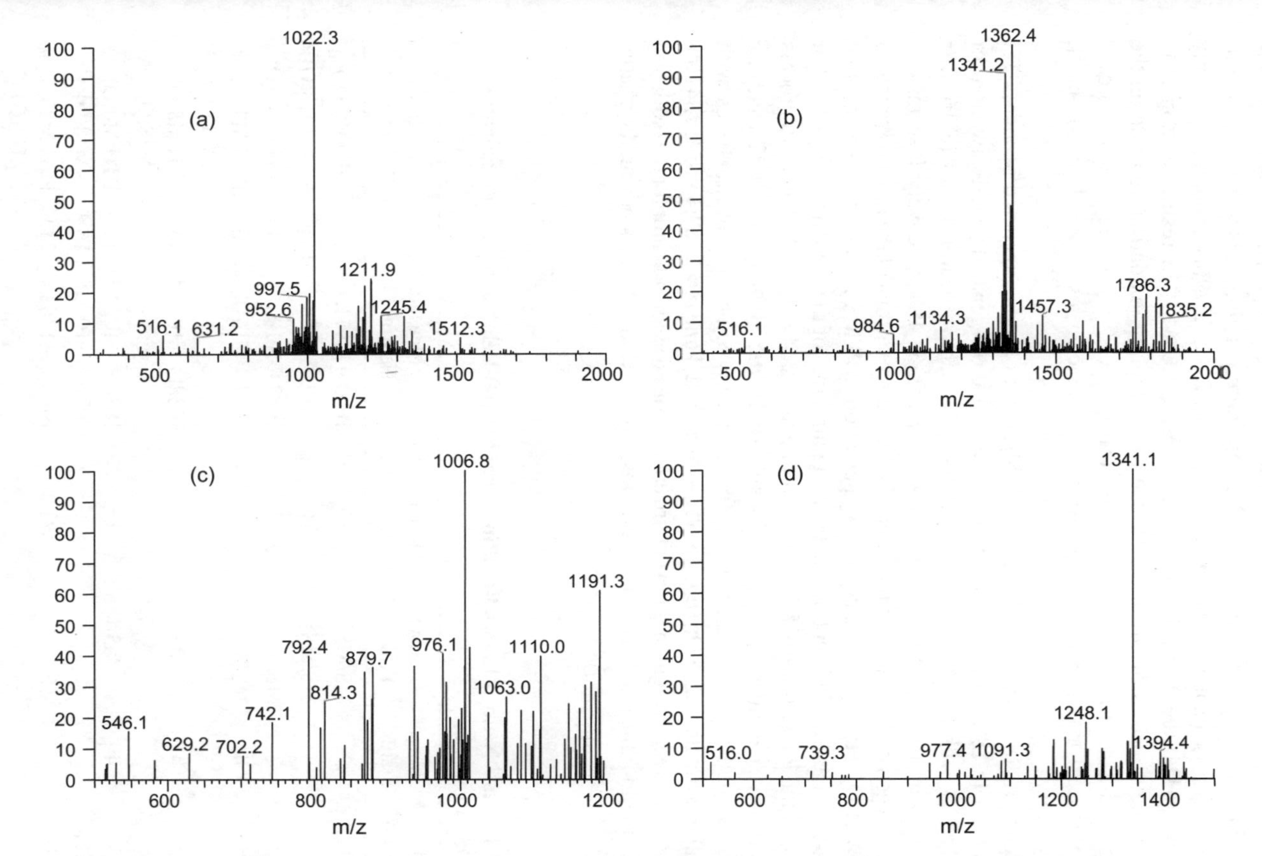

FIGURE 10.28 Mass spectra of a dioxygenated peptide, PTH Met8(O)–Met18(O), which is related to methionine oxidation: (a) MS/MS spectrum of *m/z* 1038.3 (+4); (b) MS/MS spectrum of *m/z* 1384.1 (+3); (c) MS/MS/MS spectrum of *m/z* 1038.3 → *m/z* 1022.4; (d) MS/MS/MS spectrum of *m/z* 1384.1 → *m/z* 1362.4.

loss of 32 (*m/z* 1003.9 – *m/z* 971.7), 50% of the MW 64 of sulfinic acid. For the ion of *m/z* 1082.8 with Z = 3 that is related to fragment 9–34 [Met18(O)], its neutral loss is 21.4 (*m/z* 1082.8 – *m/z* 1061.4), which represents 33% of the MW of sulfinic acid. The ion of *m/z* 812.2 for fragment 9–34 [Met18(O)] has four charges (Z = 4); its neutral loss is 15.8 (*m/z* 812.2 – *m/z* 796.4), representing only 25% of 64 Da, the MW of sulfinic acid. These LC-MS/MS results demonstrate that the neutral loss of sulfinic acid due to methionine oxidation depends on the charge status of the parent ions. The neutral losses of 64, 32, 21.3, and 16 could be observed for the same peptide with different charges of Z = 1, 2, 3, and 4, respectively.

Figure 10.28 shows the LC-MS/MS/MS spectra of a dioxygenated peptide, PTH Met8(O)–Met18(O). In the MS/MS studies of this dioxygenated peptide, two ions of *m/z* 1038.3 with Z = 4 and *m/z* 1384.1 with Z = 3 have neutral losses of 16 (*m/z* 1038.3 – *m/z* 1022.3) and 21.9 (*m/z* 1384.1 – *m/z* 1362.4), respectively. It seems that the di-oxygenated peptide has the same MS/MS pattern as that of the monooxygenated PTH Met(O). However, further ion fragmentation (MS/MS/MS) studies indicate that the di-oxygenated PTH peptide can have a neutral loss of the second sulfinic acid. For example, the ion of *m/z* 1038.3 contains four charges (Z = 4); its MS/MS/MS process generates two neutral losses of approximately 16 Da (*m/z* 1038.3 – *m/z* 1022.4 and *m/z* 1022.4 – *m/z* 1006.8). For the ion *m/z* 1084.1 with Z = 3, it has two neutral losses, with each approximately 21.3 Da (*m/z* 1384.1 – *m/z* 1362.4 and *m/z* 1362.4 – *m/z* 1341.1). The observed MS^3 fragmentation pattern could be used as an indicator for the oxidation of two methionine residues. The neutral loss observed in MS^n depends on the charge

TABLE 10.5 Neutral Losses Resulting from Met(O)-Related Peptide Fragments in LC-MS^n Processes

Parent Ion (*m/z*)	Charge Status	Collision Process	Anticipated Neutral Loss	Measured Neutral Loss	Fragment Assignment
1003.9	2	MS/MS	32	1003.9 – 971.7 = 32.2	1-18(M8(O))
812.2	4	MS/MS	16	812.2 – 796.4 = 15.8	9-34 (M18(O))
1082.8	3	MS/MS	21.3	1082.8 – 1061.4 = 21.4	9-34 (M18(O))
1034.4	4	MS/MS	16	1034.4 – 1018.4 = 16.0	PTH M8(O) or PTH M18(O)
1038.3	4	MS/MS	16	1038.3 – 1022.4 = 15.9	PTH M8(O) M18(O)
1038.3 → 1022.4	4	MS/MS/MS	16	1038.3 – 1022.4 = 15.9 1022.4 – 1006.8 = 16.4	PTH M8(O) M18(O)
1384.1	3	MS/MS	21.3	1384.1 – 1362.4 = 21.9	PTH M8(O) M18(O)
1384.1 → 1362.4	3	MS/MS/MS	21.3	1384.1 – 1362.4 = 21.7 1362.4 – 1341.1 = 21.3	PTH M8(O) M18(O)

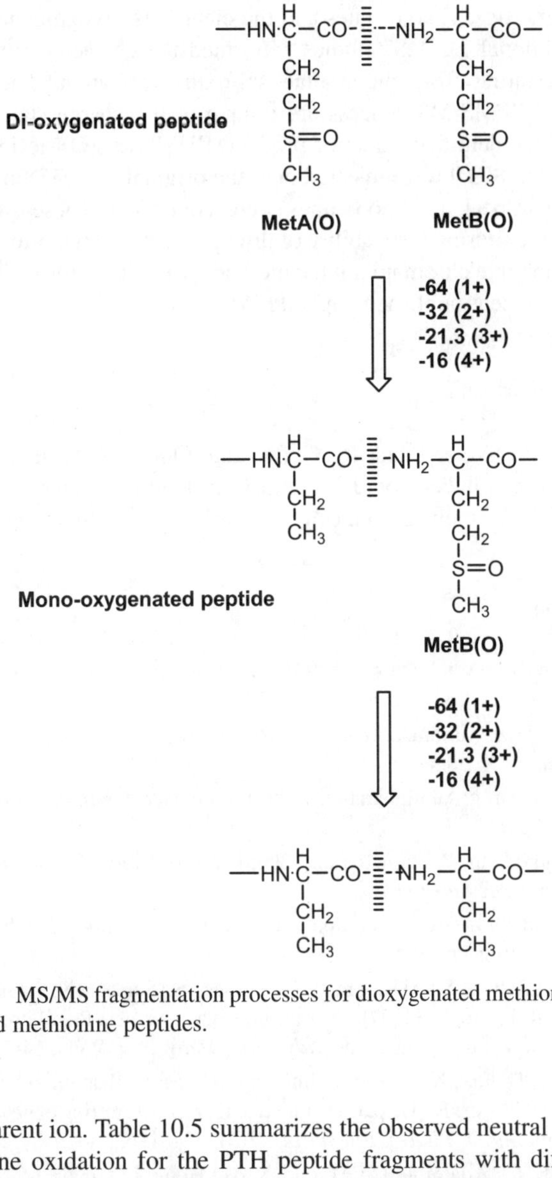

Di-oxygenated peptide

MetA(O) MetB(O)

-64 (1+)
-32 (2+)
-21.3 (3+)
-16 (4+)

Mono-oxygenated peptide

MetB(O)

-64 (1+)
-32 (2+)
-21.3 (3+)
-16 (4+)

FIGURE 10.29 MS/MS fragmentation processes for dioxygenated methionine peptides and monooxygenated methionine peptides.

status of the parent ion. Table 10.5 summarizes the observed neutral losses resulting from methionine oxidation for the PTH peptide fragments with different charges. The MS/MS/MS fragmentation processes for peptide with different oxidized methionine residues are illustrated in Figure 10.29. These MS^n studies positively confirm the oxidation of methionine in PTH1-34 to methionine sulfoxide.

The approach in using high-resolution and tandem mass spectrometry in conjunction with CNBr digestion was effective in identifying oxidation products of a PTH1-34 formulation drug. All anticipated CNBr digested peptide fragments, including both oxidized and nonoxidized peptide fragments, were positively identified for methionine sulfoxide formation at positions 8 and 18. TOF-MS capability facilitated the observation of multiple charge states and the clear mass assignments of the peptide fragments. Additional LC/MS^n studies confirmed the characteristic neutral loss of sulfinic acid resulting from methionine sulfoxide formation. The neutral losses detected in the MS/MS/MS process are unique to the charge status of the parent peptide ions. Three identified oxidation products [PTH Met8(O)Met18(O), PTH Met8 (O), and PTH Met18(O)] were injected onto the original non-MS-compatible HPLC method, and their peak retention times were confirmed. These strategies can be generalized for monitoring the stability of drug products, which often require the use of non-MS-compatible chromatographic methods. Detailed information on this study can be seen in a recent published paper. [135]

ACKNOWLEDGMENT

The authors would like to thank R. Vivilecchia, Qin Ji, W. Wang, D. Drinkwater, F. Harmon, K. Toscano, P. Sutton, J. Valente, R. LoBrutto, J. Pickett, J. Pepper, F. Lin, and A. Kuemmell for their assistance and contribution to this manuscript.

REFERENCES

1. US Food and Drug Administration (2003), *Guidance for Industry, Q3A Impurities in New Drug Substances.*
2. US Food and Drug Administration (2006), *Guidance for Industry, Q3B Impurities in New Drug Products.*
3. US Food and Drug Administration (1997), *Guidance for Industry, Q3C Impurities: Residual Solvents.*
4. US Food and Drug Administration (2005), *Draft Guidance for Industry, ANDAs: Impurities in Drug Substances.*
5. US Food and Drug Administration (2005), *Draft Guidance for Industry, ANDAs: Impurities in Drug Products.*
6. Basak, A. K.; Raw, A. S.; Al Hakim, A. H.; Furness, S.; Samaan, N. I.; Gill, D. S.; Patel, H. B.; Powers, R. F.; Yu, L. (2007), Pharmaceutical impurities: Regulatory perspective for abbreviated new drug applications, *Adv. Drug Deliv. Rev. 59*(1), 64–72.
7. Grimm, W.; Thomas,K., (1995), International harmonization of stability-tests for pharmaceuticals—the ICH tripartite guideline for stability testing of new drug substances and products, *Eur. J. Pharm. Biopharm. 41*(3), 194–196.
8. Waterman, K. C.; Adami, R. C.; Alsante, K. M.; Hong, J.; Landis, M. S.; Lombardo, F.; Roberts, C. J. (2002), Stabilization of pharmaceuticals to oxidative degradation, *Pharm. Devel. Technol. 7*(1), 1–32.
9. Waterman, K. C.; Adami, R. C.; Alsante, K. M.; Antipas, A. S.; Arenson, D. R.; Carrier, R.; Hong, J. Y.; Landis, M. S.; Lombardo, F.; Shah, J. C.; Shalaev, E.; Smith, S. W.; Wang, H., (2002), Hydrolysis in pharmaceutical formulations, *Pharm. Devel. Technol. 7*(2), 113–146.
10. Puz, M. J.; Johnson, B. A.; Murphy, B. J., (2005), Use of the antioxidant BHT in asymmetric membrane tablet coatings to stabilize the core to the acid catalyzed peroxide oxidation of a thioether drug, *Pharm. Devel. Technol. 10*(1), 115–125.

11. Baertschi S.; Jansen P. (2005), Stress testing: A predictive tool, in Baertschi, S. ed., *Pharmaceutical Stress Testing*, CRC Press, Boca Raton, FL, pp. 13–50.
12. Jansen P.; Smith K.; Baertschi S. (2005), Stress testing: Analytical considerations, in Baertschi S. ed., *Pharmaceutical Stress Testing*, CRC Press, Boca Raton, FL, pp. 141–172.
13. Klick S.; Muijselaar P.; Waterval J.; Eichinger T.; Korn C.; Gerding T.; Debets A.; Sanger-van de Groemd C.; Beld C.; Somsen G.; DeJong G. (2005), Stress testing for drug substances and drug products, *Pharm. Technol.*, (Feb.), 48–60.
14. Aman, W.; Thoma, K. (2003), ICH guideline for photostability testing: Aspects and directions for use, *Pharmazie 58*(12), 877–880.
15. Chow, S. C.; Shao, J. (2007), Stability analysis for drugs with multiple active ingredients, *Statist. Med. 26*(7), 1512–1517.
16. Giron, D.; Mutz, M.; Garnier, S. (2004), Solid-state of pharmaceutical compounds - Impact of the ICH Q6 guideline on industrial development, *J. Thermal Anal. Calorim. 77*(2), 709–747.
17. Grimm, W. (1996), Stability testing of clinical samples, *Drug Devel. Industr. Pharm. 22* (9–10), 851–871.
18. Pack, D. J. (1998), The ICH guideline for stability testing—what matrixing does it permit? *Drug Inform. J. 32*(2), 561–567.
19. Baertschi S.; Alsante K. (2005), Stress testing: The chemistry of drug degradation, in Baertschi, S. ed., *Pharmaceutical Stress Testing*, CRC Press, Boca Raton, FL, pp. 51–140.
20. Aksoy, B.; Kucukguzel, I.; Rollas, S. (2007), Development and validation of a stability-indicating HPLC method for determination of ciprofloxacin hydrochloride and its related compounds in film-coated tablets, *Chromatographia 66*, S57–S63.
21. Baboota, S.; Faiyaz, S.; Ahuja, A.; Ali, J.; Shafiq, S.; Ahmad, S. (2007), Development and validation of a stability-indicating HPLC method for analysis of celecoxib (CXB) in bulk drug and microemulsion formulations, *Acta Chromatogr. 18*, 116–129.
22. Bakshi, M.; Singh, S. (2004), HPLC and LC-MS studies on stress degradation behaviour of tinidazole and development of a validated specific stability-indicating HPLC assay method, *J. Pharm. Biomed. Anal. 34*(1), 11–18.
23. Bansal, G.; Singh, M.; Jindal, K. C.; Singh, S. (2008), Ultraviolet-photodiode array and high-performance liquid chromatographic/mass spectrometric studies on forced degradation behavior of glibenclamide and development of a validated stability-indicating method, *J. Assoc. Official Anal. Chem. 91*(4), 709–719.
24. Bhardwaj, S. P.; Singh, S. (2008), Study of forced degradation behavior of enalapril maleate by LC and LC-MS and development of a validated stability-indicating assay method, *J. Pharm. Biomed. Anal. 46*(1), 113–120.
25. Khedr, A.; Sheha, M. (2008), Stress degradation studies on betahistine and development of a validated stability-indicating assay method. *J. Chromatogr. B (Anal. Technol. Biomed. Life Sci.) 869*(1–2), 111–117.
26. Malesuik, M. D.; Cardoso, S. G. A.; Steppe, M. (2008), Development of a validated stability-indicating LC method for nitazoxanide in pharmaceutical formulations, *Chromatographia 67*(1–2), 131–136.
27. Nogueira, D. R.; D'Avila, F. B.; Rolim, C. M. B.; Dalmora, S. L. (2007), Development and validation of a stability-indicating LC method for the determination of rupatadine in pharmaceutical formulations, *Chromatographia 66*(11–12), 915–919.

28. Rao, D. V. S.; Radhakrishnanand, P. (2008), Stress degradation studies on dutasteride and development of a stability-indicating HPLC assay method for bulk drug and pharmaceutical dosage form, *Chromatographia 67*(9–10), 841–845.
29. Tagliari, M. P.; Stulzer, H. K.; Murakami, F. S.; Kuminek, G.; Valente, B.; Oliveira, P. R.; Silva, M. A. S. (2008), Development and validation of a stability-indicating LC method to quantify hydrochlorothiazide in oral suspension for pediatric use, *Chromatographia 67* (7–8), 647–652.
30. Crowley, P.; Martini, L. (2001), Drug-excipient interactions, *Pharm. Technol. Eur. 34*, 15–26.
31. Freed, A. L.; Strohmeyer, H. E.; Mahjour, M.; Sadineni, V.; Reid, D. L.; Kingsmill, C. A. (2008), pH control of nucleophilic/electrophilic oxidation, *Int. J. Pharm. 357*(1–2), 180–188.
32. Li, S. H.; Patapoff, T. W.; Overcashier, D.; Hsu, C.; Nguyen, T. H.; Borchardt, R. T. (1996), Effects of reducing sugars on the chemical stability of human relaxin in the lyophilized state, *J. Pharm. Sci. 85*(8), 873–877.
33. Ren, S.; Park, M. J.; Sah, H.; Lee, B. J. (2008), Effect of pharmaceutical excipients on aqueous stability of rabeprazole sodium, *Int. J. Pharm. 350*(1–2), 197–204.
34. Schildcrout, S. A.; Risley, D. S.; Kleemann, R. L. (1993), Drug-excipient interactions of seproxetine maleate hemi-hydrate–isothermal stress methods, *Drug Devel. Industr. Pharm. 19*(10), 1113–1130.
35. Alsante, K. M.; Ando, A.; Brown, R.; Ensing, J.; Hatajik, T. D.; Kong, W.; Tsuda, Y. (2007), The role of degradant profiling in active pharmaceutical ingredients and drug products, *Adv. Drug Deliv. Rev. 59*(1), 29–37.
36. Wu Y. (2000), The use of liquid chromatography-mass spectrometry for the identification of drug degradation products in pharmaceutical formulations, *Biomed. Chromatogr. 14*, 384–396.
37. Hovorka S.; Schoneich C. (2001), Oxidative degradation of pharmaceuticals: Theory, mechanisms and inhibition, *J. Pharm. Sci. 90*, 253–269.
38. Bruni, G.; Amici, L.; Berbenni, V.; Marini, A.; Orlandi, A. (2002), Drug-excipient compatibility studies—search of interaction indicators, *J. Thermal Anal. Calorim. 68*(2), 561–573.
39. Huang, W. X.; Desai, A.; Tang, Q.; Yang, R.; Vivilecchia, R. V.; Joshi, Y. (2006), Elimination of metformin-croscarmellose sodium interaction by competition, *Int. J. Pharm. 311*(1–2), 33–39.
40. Qiu, F. H.; Cobice, D.; Pennino, S.; Becher, M.; Norwood, D. L. (2008), Identification of drug meglumine interaction products using LC/MS and forced degradation studies, *J. Liquid Chromatogr. Related Technol. 31*(15), 2331–2336.
41. Sarisuta, N.; Lawanprasert, P.; Puttipipatkhachorn, S.; Srikummoon, K. (2006), The influence of drug-excipient and drug-polymer interactions on butt adhesive strength of ranitidine hydrochloride film-coated tablets, *Drug Devel. Industr. Pharm. 32*(4), 463–471.
42. Stulzer, H. K.; Rodrigues, P. O.; Cardoso, T. M.; Matos, J. S. R.; Silva, M. A. S. (2008), Compatibility studies between captopril and pharmaceutical excipients used in tablets formulations, *J. Thermal Anal. Calorim. 91*(1), 323–328.
43. Vandooren, A. A. (1983), Design for drug-excipient interaction studies, *Drug Devel. Industr. Pharm. 9*(1–2), 43–55.

44. Byrn, S. R.; Xu, W.; Newman, A. W. (2001), Chemical reactivity in solid-state pharmaceuticals: Formulation implications, *Adv. Drug Deliv. Rev. 48*(1), 115–136.
45. Chen, J. G.; Markovitz, D. A.; Yang, A. Y.; Rabel, S. R.; Pang, J.; Dolinsky, O.; Wu, L. S.; Alasandro, M. (2005), Degradation of a fluoropyridinyl drug in capsule formulation: Degradant identification, proposed degradation mechanism, and formulation optimization *Pharm. Devel. Technol. 5*(4), 561–570.
46. Dubost, D. C.; Kaufman, M. J.; Zimmerman, J. A.; Bogusky, M. J.; Coddington, A. B.; Pitzenberger, S. M. (1996), Characterization of a solid state reaction product from a lyophilized formulation of a cyclic heptapeptide. A novel example of an excipient-induced oxidation *Pharm. Res. 13*(12), 1811–1814.
47. Qiu, F. H.; Norwood, D. L. (2007), Identification of pharmaceutical impurities, *J. Liquid Chromatogr. Related Technol. 30*(5–8), 877–935.
48. Murakami, T.; Fukutsu, N.; Kondo, J.; Kawasaki, T.; Kusu, F. (2008), Application of liquid chromatography-two-dimensional nuclear magnetic resonance spectroscopy using pre-concentration column trapping and liquid chromatography-mass spectrometry for the identification of degradation products in stressed commercial amlodipine maleate tablets, *J. Chromatogr. A 1181*(1–2), 67–76.
49. Degenhardt, A.; Winterhalter, P. (2001), Isolation and purification of isoflavones from soy flour by high-speed countercurrent chromatography, *Eur. Food Res. Technol. 213*(4–5), 277–280.
50. Dhingra, O. D.; Jham, G. N.; Barcelos, R. C.; Mendonca, F. A.; Ghiviriga, I. (2007), Isolation and identification of the principal fungitoxic component of turmeric essential oil, *J. Essential Oil Res. 19*(4), 387–391.
51. Galanopoulou, O.; Rozou, S.; Antoniadou-Vyza, E. (2008), HPLC analysis, isolation and identification of a new degradation product in carvedilol tablets *J. Pharm. Biomed. Anal. 48*(1), 70–77.
52. Hiriyanna, S. G.; Basavaiah, K. (2008), Isolation and characterization of process related impurities in anastrozole active pharmaceutical ingredient, *J. Brazil. Chem. Soc. 19*(3), 397–404.
53. Hoshi, H.; Yagi, Y.; Iijima, H.; Matsunaga, K.; Ishihara, Y.; Yasuhara, T. (2005), Isolation and characterization of a novel immunomodulatory alpha-glucan-protein complex from the mycelium of Tricholoma matsutake in basidiomycetes, *J. Agric. Food Chem. 53*(23), 8948–8956.
54. Ling, Z. Q.; Xie, B. J.; Yang, E. L. (2005), Isolation, characterization, and determination of antioxidative activity of oligomeric procyanidins from the seedpod of Nelumbo nucifera Gaertn *J. Agric. Food Chem. 53*(7), 2441–2445.
55. Zhou, X.; Peng, J. Y.; Fan, G. R.; Wu, Y. T. (2005), Isolation and purification of flavonoid glycosides from Trollius ledebouri using high-speed counter-current chromatography by stepwise increasing the flow-rate of the mobile phase, *J. Chromatogr. A 1092*(2), 216–221.
56. Clarkson, C.; Staerk, D.; Hansen, S. H.; Jaroszewski, J. W. (2005), Hyphenation of solid-phase extraction with liquid chromatography and nuclear magnetic resonance: Application of HPLC-DAD-SPE-NMR to identification of constituents of Kanahia laniflora, *Anal. Chem. 77*(11), 3547–3553.
57. Lambert, M.; Staerk, D.; Hansen, S. H.; Sairafianpour, M.; Jaroszewski, J. W. (2005), Rapid extract dereplication using HPLC-SPE-NMR: Analysis of isoflavonoids from Smirnowia iranica, *J. Nat. Products 68*(10), 1500–1509.

58. Shertzer, H. G.; Tabor, M.W. (1985), Peroxide removal from organic solvents and vegetable oils, *J. Environ. Sci. Health, Part A A20*, 845–855.

59. Guengerich, C. P.; Schug, K. (1978), Reaction of nitrosylpentaammineruthenium(ii) with aromatic-aldehydes in aqueous base—example of catalyzed metal-ion hydrolysis, *Inorg. Chem. 17*(10), 2819–2821.

60. Kimura, M.; Muto, T. (1979), Metal-ion catalyzed oxidation of steroids. 8. Reaction of cholesteryl acetate with tert-butyl hydroperoxide in the presence of tris-(acetylacetonato) iron(iii), *Chem. Pharm. Bull. 27*(1), 109–112.

61. Nishigaichi, Y.; Takuwa, A.; Iihama, K.; Yoshida, N. (1991), Metal-ion-catalyzed fluxionality of pentadienyltins and its application to the diels-alder reaction, *Chem. Lett. 20*(4), 693–696.

62. Jakel, D.; Keck, M. (2000), Purity of excipients, *Excip. Toxic. Safety 103*, 21–58.

63. Rowe, R. C; Sheskey, P. J; Owen, S. C. eds. (2006), *Handbook of Pharmaceutical Excipients*, 5th ed. , American Pharmaceutical Ass., Washington, DC.

64. Hartauer, K. J.; Arbuthnot, G. N.; Baertschi, S. W.; Johnson, R. A.; Luke, W. D.; Pearson, N. G.; Rickard, E. C.; Tingle, C. A.; Tsang, P. K. S.; Wiens, R. E. (2005), Influence of peroxide impurities in povidone and crospovidone on the stability of raloxifene hydrochloride in tablets: Identification and control of an oxidative degradation product, *Pharm. Devel. Technol. 5*(3), 303–310.

65. Wasylaschuk, W.; Harmon, P.; Wagner, G.; Harman, A.; Templeton, A.; Xu, H.; Reed, R. (2007), Evaluation of hydroperoxides in common pharmaceutical excipients, *J. Pharm. Sc. 96*(1), 106–116.

66. Tallon, M. A.; Malawer, E. G.; Machnicki, N. I.; Brush P.J.; Wu, C.S.; Cullen, J. P. (2007), The effect of crosslinker structure upon the rate of hydroperoxide formation in dried, crosslinked poly(vinylpyrrolidone), *J. Appl. Polym. Sci. 107*, 2776–2785.

67. Nassar, M. N.; Nesarikar, V. N.; Lozano, R.; Parker, W. L.; Huang, Y. D.; Palaniswamy, V.; Xu, W. W.; Khaselev, N. (2004), Influence of formaldehyde impurity in polysorbate 80 and PEG-300 on the stability of a parenteral formulation of BMS-204352: Identification and control of the degradation product, *Pharm. Devel. Technol. 9*(2), 189–195.

68. Frontini, R.; Mielck J. B. (1995), Formation of formaldehyde in polyethyleheglycol and in poloxamer under stress conditions, *Int. J. Pharm. 114*, 121–123.

69. Johnson, D. M.; Taylor, W. F. (1984), Dagradation of fenprostalene in polyethylene glycol 400 solution, *J. Pharm. Sci. 73*, 1414–1417.

70. Bergh, M.; Magnusson, K.; Nilsson, L.; Karlberg, A. (1998), Formation of formaldehyde and peroxides by air oxidation of high purity polyoxyethylene surfactants, *Contact Dermatitis 39*, 14–20.

71. Waterman, K. C.; Adami, R. C.; Hong, J. Y. (2003) Impurities in drug product, in Ahuja, S.; Alsante, K. M. eds., *Handbook of Isolation and Characterization of Impurities in Pharmaceuticals*, Academic Press/Elsevier Science, San Diego, pp. 75–88.

72. Digenis, G. A.; Thomas, B.; Shah, V. P. (1994), Cross-linking of gelatin capsules and its relevance to their in vitro-in vivo performance, *J. Pharm. Sci. 83*, 915–921.

73. Pan, C.; Harmon, F.; Toscano, K.; Liu, F.; Vivilecchia, R. (2008), Strategy for identification of leachables in packaged pharmaceutical liquid formulations, *J. Pharm. Biomed. Anal. 46*(3), 520–527.

74. Jenke, D. (2002), Extractable/leachable substances from plastic materials used as pharmaceutical product containers/devices, *PDA J. Pharm. Sci. Technol. 56*(6), 332–371.

75. Jenke, D., (2003), Organic extractables from packaging materials: Chromatographic methods used for identification and quantification, *J. Liquid Chromatogr. Related Technol. 26*(15), 2449–2464.

76. Jenke, D. R. (2005), Linking extractables and leachables in container/closure applications, *PDA J. Pharm. Sci. Technol. 59*(4), 265–281.

77. Rogalewicz, R.; Batko, K.; Voelkel, A. (2006), Identification of organic extractables from commercial resin-modified glass-ionomers using HPLC-MS, *J. Environ. Monitor. 8*(7), 750–758.

78. Mawhinney, D. B.; Stanelle, R. D.; Hamelin, E. I.; Kobelski, R. J. (2007), Enhancing the response of alkyl methylphosphonic acids in negative electrospray ionization liquid chromatography tandem mass spectrometry by post-column addition of organic solvents, *J. Am. Soc. Mass Spectrom. 18*, 1821–1826.

79. Carabias-Martinez, R.; Rodriguez-Gonzalo, E.; Revilla-Ruiz, P. (2004), Determination of weakly acidic endocrine-disrupting compounds by liquid chromatography-mass spectrometry with post-column base addition, *J. Chromatogr. A 1056*(1–2), 131–138.

80. Cheng, C.; Tsai, H. R.; Chang, K. C. (2006), On-line cut-off technique and organic modifier addition aided signal enhancement for trace analysis of carbohydrates in cellulase hydrolysate by ion exclusion chromatography-electrospray ionization mass spectrometry, *J. Chromatogr. A 1119*(1–2), 188–196.

81. Voyksner, R. D. (1997), Combining liquid chromatography with electrospray mass spectrometry, in Cole, R. B.ed., *Electrospray Ionization Mass Spectrometry*, Wiley, New York, 1997, Part III, pp. 321–343.

82. Collin, F.; Khoury, H.; Rousselot, D. B.; Therond, P.; Legrand, A.; Jore, D.; Albert, M. G. (2004), Liquid chromatographic/electrospray ionization mass spectrometric identification of the oxidation end-products of metformin in aqueous solution, *J. Mass Spectrom. 39*, 890–902.

83. Novak, T. J.; Helmy, R.; Santos, I. (2005), Liquid chromatography-mass spectrometry using the hydrogen/deuterium exchange reaction as a tool for impurity identification in pharmaceutical process development, *J. Chromatogr. B 825*, 161–168.

84. Chen, G. D.; Khusid, A.; Daaro, I.; Irish, P.; Pramanik, B. N., (2007), Structural identification of trace level enol tautomer impurity by on-line hydrogen/deuterium exchange HR-LC/MS in a LTQ-Orbitrap hybrid mass spectrometer, *J. Mass Spectrom. 42*(7), 967–970.

85. Li, Y. S.; Williams, T. D.; Topp, E. M. (2008), Effects of excipients on protein conformation in lyophilized solids by hydrogen/deuterium exchange mass spectrometry, *Pharm. Res. 25*(2), 259–267.

86. Tsutsui, Y.; Wintrode, P. L. (2007), Hydrogen/deuterium, exchange-mass spectrometry: A powerful tool for probing protein structure, dynamics and interactions, *Curr. Med. Chem. 14*(22), 2344–2358.

87. Liu D.; Wu L.; Sun M.; MacGregor P. (2007), On-line H/D exchange LC-MS strategy for structural elucidation of pharmaceutical impurities, *J. Pharm. Biomed. Anal. 44*, 320–329.

88. Iwasa, K.; Takahashi, T.; Nishiyama, Y.; Moriyasu, M.; Sugiura, M.; Takeuchi, A.; Tode, C.; Tokuda, H.; Takeda, K. (2008), Online structural elucidation of alkaloids and other

constituents in crude extracts and cultured cells of Nandina domestica by combination of LC-MS/MS, LC-NMR, and LC-CD analyses, *J. Nat. Products*, *71*(8), 1376–1385.

89. Kang, S. W.; Kim, C. Y.; Jung, S. H.; Um, B. H. (2008), The Rapid Identification of Isoflavonoids from *Belamcanda chinensis* by LC-NMR and LC-MS, *Chem. Pharm. Bull.* *56*(10) 1452–1454.
90. Kim, M. C.; Kang, S. W.; Kim, S. J.; Um, B. H. (2008), The rapid analysis and identification of prenylflavonoids in Sophora flavescens roots by on flow LC-NMR/MS, *Planta Medica 74*(9), PC71.
91. Feng, W. Q.; Liu, H. Y.; Chen, G. D.; Malchow, R.; Bennett, F.; Lin, E.; Pramanik, B.; Chan, T. M. (2001), Structural characterization of the oxidative degradation products of an antifungal agent SCH 56592 by LC-NMR and LC-MS, *J. Pharm. Biomed. Anal.* *25* (3–4), 545–557.
92. Li, N.; Yang, J.; Qin, F.; Li, F. M.; Gong, P. (2007), Isolation and identification of a major impurity in a new bulk drug candidate by preparative LC, ESI-MSn, LC-MS-MS, and NMR *J. Chromatogr. Sci. 45*(1), 45–49.
93. Novak, P.; Tepes, P.; Fistric, I.; Bratos, I.; Gabelica, V. (2006), The application of LC-NMR and LC-MS for the separation and rapid structure elucidation of an unknown impurity in 5-aminosalicylic acid, *J. Pharm. Biomed. Anal. 40*(5), 1268–1272.
94. Pan, C. K.; Liu, F.; Ji, Q.; Wang, W.; Drinkwater, D.; Vivilecchia, R. (2006), The use of LC/MS, GC/MS, and LC/NMR hyphenated techniques to identify a drug degradation product in pharmaceutical development, *J. Pharm. Biomed. Anal. 40*(3), 581–590.
95. Sieber, M.; Wagner, S.; Rached, E.; Amberg, A.; Mally, A.; Dekant, W. (2008), A combined GC-MS, LC-MS and NMR metabonomics approach for early detection of ochratoxin a nephrotoxicity, *Naunyn-Schmied. Arch. Pharmacol. 377*, 74–74.
96. Stulten, D.; Lamshoft, M.; Zuhlke, S.; Spiteller, M. (2008), Isolation and characterization of a new human urinary metabolite of diclofenac applying LC-NMR-MS and high--resolution mass analyses, *J. Pharm. Biomed. Anal. 47*(2), 371–376.
97. Tatsis, E. C.; Boeren, S.; Exarchou, V.; Troganis, A. N.; Vervoort, J.; Gerothanassis, I. P. (2007), Identification of the major constituents of Hypericum perforatum by LC/SPE/NMR and/or LC/MS, *Phytochemistry 68*(3), 383–393.
98. Wu, L. M.; Hong, T. Y.; Vogt, F. G. (2007), Structural analysis of photo-degradation in thiazole-containing compounds by LC-MS/MS and NMR, *J. Pharm. Biomed. Anal. 44*(3), 763–772.
99. Yang, X. Z.; Yang, Y. P.; Tang, C. P.; Ke, C. Q.; Ye, Y. (2007), Rapid identification of bibenzyls of Stemona sessilifolia using hyphenated LC-UV-NMR and LGMS methods, *Chem. Res. Chinese Univ. 23*(1), 48–51.
100. Alsante, K. M.; Friedmann, R. C.; Hatajik, L. L.; Sharp, T. R.; Snyder, K.D.; Szczesny, E. J. (2000) in Ahuja, S. ed., *Handbook of Modern Pharmaceutical Analysis*, Academic Press/Elsevier Science, San Diego.
101. Lee, M. S.; Kerns, E. H., (1999) LC/MS applications in drug development, *Mass Spectrom. Rev. 18*(3–4), 187–279.
102. Alsante, K. M.; Boutros, P.; Couturier, M. A.; Friedmann, R. C.; Harwood, J. W.; Horan, G. J.; Jensen, A. J.; Liu, Q. C.; Lohr, L. L.; Morris, R.; Raggon, J. W.; Reid, G.; et al. (2004), Pharmaceutical impurity identification: A case study using a multidisciplinary approach, *J. Pharm. Sci. 93*(9), 2296–2309.

103. Ermer, J. (1998), The use of hyphenated LC-MS technique for characterisation of impurity profiles during drug development, *J. Pharm. Biomed. Anal. 18*(4–5), 707–714.

104. Ermer, J.; Vogel, M. (2000), Applications of hyphenated LC-MS techniques in pharmaceutical analysis, *Biomed. Chromatogr. 14*(6), 373–383.

105. Chen, G. D.; Pramanik, B. N.; Liu, Y. H.; Mirza, U. A. (2007), Applications of LC/MS in structure identifications of small molecules and proteins in drug discovery, *J. Mass Spectrom. 42*(3), 279–287.

106. Sandvoss, M.; Bardsley, B.; Beck, T. L.; Lee-Smith, E.; North, S. E.; Moore, P. J.; Edwards, A. J.; Smith, R. J. (2005), HPLC-SPE-NMR in pharmaceutical development: Capabilities and applications, *Magn. Resonance Chem. 43*(9), 762–770.

107. Murakami, T.; Fukutsu, N.; Kondo, J.; Kawasaki, T.; Kusu, F. (2008), Application of liquid chromatography-two-dimensional nuclear magnetic resonance spectroscopy using pre-concentration column trapping and liquid chromatography-mass spectrometry for the identification of degradation products in stressed commercial amlodipine maleate tablets, *J. Chromatogr. A 1181*(1–2), 67–76.

108. Baertschi, S. W.; Brunner, H.; Bunnell, C. A.; Cooke, G. G.; Diseroad, B.; Dorman, D. E.; Jansen, P. J.; Kemp, C. A. J.; Maple, S. R.; McCune, K. A.; Speakman, J. L. (2008), Isolation, identification, and synthesis of two oxidative degradation products of olanzapine (LY170053) in solid oral formulations, *J. Pharm. Sci. 97*(2), 883–892.

109. Baertschi, S. W. (2006), Analytical methodologies for discovering and profiling degradation-related impurities, *Trends Anal. Chem. 25*(8), 758–767.

110. Blanchard, A.; Lee, C.; Nickerson, B.; Lohr, L. L.; Jensen, A. J.; Alsante, K. M.; Sharp, T. R.; Santafianos, D. P.; Morris, R.; Snyder, K. D. (2004), Identification of low-level degradants from low dose tablets, *J. Pharm. Biomed. Anal. 36*(2), 265–275.

111. Fujinari, E. M. (1998), Chemiluminescent nitrogen detectors (CLND) for GC, SimDis, SFC, HPLC and SEC applications, *Instrum. Meth. Food and Beverage Anal. 39*, 376–378.

112. Letot, E.; Koch, G.; Falchetto, R.; Bovermann, G.; Oberer, L.; Roth, H. J. (2005), Quality control in combinatorial chemistry: Determinations of amounts and comparison of the "purity" of LC-MS-purified samples by NMR, LC-UV and CLND *J. Combin. Chem. 7*(3), 364–371.

113. Liang, X. Z.; Patel, H.; Young, L.; Shah, P.; Raglione, T. (2005), Practical application of implementing the equimolarity principle of LC-CLND in pharmaceutical analysis, *Abstracts of Papers of Am. Chemical Society*, Abstract 230, pp. U281–U281.

114. Seeling, A.; Lehmann, J. (2006), NO-donors, part X [1]: Investigations on the stability of pentaerythrityl tetranitrate (PETN) by HPLC-chemoluminescence-N-detection (CLND) versus UV-detection in HPLC, *J. Pharm. Biomed. Anal. 40*(5), 1131–1136.

115. Chiarotti, M.; Marsili, R.; Moreda-Pineiro, A. (2002), Gas chromatographic-mass spectrometric analysis of residual solvent trapped into illicit cocaine exhibits using head-space solid-phase microextraction, *J. Chromatogr. B (Anal. Technol. Biomed. Life Sci.) 772*(2), 249–256.

116. Colomb, A.; Yassaa, N.; Williams, J.; Peeken, I.; Lochte, K. (2008), Screening volatile organic compounds (VOCs) emissions from five marine phytoplankton species by head space gas chromatography/mass spectrometry (HS-GC/MS), *J. Environ. Monitor. 10*(3), 325–330.

117. Sakurai, K.; Sugaya, N.; Nakagawa, T.; Saito, H.; Uchiyama, T.; Fujimoto, Y.; Takahashi, K. (2007), Simultaneous analysis of residual 4-alkylphenols in synthetic resin products for drug and food use using head-space gas chromatography-mass spectrometry (HS-GC/MS), *J. Health Sci. 53*(3), 263–270.

118. Sugaya, N.; Nakagawa, T.; Sakurai, K.; Morita, M.; Onodera, S. (2001), Analysis of aldehydes in water by head space-GC/MS, *J. Health Sci. 47*(1), 21–27.

119. Uematsu, Y.; Suzuki, K.; Iida, K.; Hirata, K.; Ueta, T.; Kamata, K. (2002), Determination of low levels of methanol and ethanol in licorice extract by large volume injection head-space GC, *J. Food Hygien. Soc. Jpn. 43*(5), 295–300.

120. Fliszar, K.; Wiggins, J. M.; Pignoli, C. M.; Martin, G. P.; Li, Z. (2004), Analysis of organic volatile impurities in pharmaceutical excipients by static headspace capillary gas chromatography, *J. Chromatogr. A 1027*(1–2), 83–91.

121. Degenhardt, A.; Winterhalter, P. (2001), Isolation and purification of isoflavones from soy flour by high-speed countercurrent chromatography, *Eur. Food Res. Technol. 213*(4–5), 277–280.

122. Dhingra, O. D.; Jham, G. N.; Barcelos, R. C.; Mendonca, F. A.; Ghiviriga, I. (2007), Isolation and identification of the principal fungitoxic component of turmeric essential oil, *J. Essential Oil Res. 19*(4), 387–391.

123. Galanopoulou, O.; Rozou, S.; Antoniadou-Vyza, E. (2008), HPLC analysis, isolation and identification of a new degradation product in carvedilol tablets, *J. Pharm. Biomed. Anal. 48*(1), 70–77.

124. Hiriyanna, S. G.; Basavaiah, K. (2008), Isolation and characterization of process related impurities in anastrozole active pharmaceutical ingredient, *J. Brazil. Chem. Soc. 19*(3), 397–404.

125. Hoshi, H.; Yagi, Y.; Iijima, H.; Matsunaga, K.; Ishihara, Y.; Yasuhara, T. (2005), Isolation and characterization of a novel immunomodulatory alpha-glucan-protein complex from the mycelium of Tricholoma matsutake in basidiomycetes, *J. Agric. Food Chem. 53*(23), 8948–8956.

126. Ling, Z. Q.; Xie, B. J.; Yang, E. L. (2005), Isolation, characterization, and determination of antioxidative activity of oligomeric procyanidins from the seedpod of Nelumbo nucifera Gaertn, *J. Agric. Food Chem. 53*(7), 2441–2445.

127. Bates, M.; Bruno, P.; Caputi, M.; Caselli, M.; de Gennaro, G.; Tutino, M. (2008), Analysis of polycyclic aromatic hydrocarbons (PAHs) in airborne particles by direct sample introduction thermal desorption GC/MS, *Atmosph. Environ. 42*(24), 6144–6151.

128. Crifasi, J. A.; Bruder, M. F.; Long, C. W.; Janssen, K. (2006), Performance evaluation of thermal desorption system (TDS) for detection of basic drugs in forensic samples by GC-MS, *J. Anal. Toxicol. 30*(8), 581–592.

129. Cummins, W.; Duggan, P.; McLoughlin, P. (2008), Thermal desorption characterisation of molecularly imprinted polymers. Part I: A novel study using direct-probe GC-MS analysis *Anal. Bioanal. Chem. 391*(4), 1237–1244.

130. Holland, N.; Duggan, P.; Owens, E.; Cummins, W.; Frisby, J.; Hughes, H.; McLoughlin, P. (2008), Thermal desorption characterisation of molecularly imprinted polymers. Part II: Use of direct probe GC-MS analysis to study crosslinking effects *Anal. Bioanal. Chem. 391*(4), 1245–1253.

131. Pan, C. K.; Liu, F.; Sutton, P.; Vivilecchia, R. (2005), Use of thermal desorption GC/MS to study weight loss in thermogravimetric analysis of di-acid salts, *Thermochim. Acta 435*(1), 11–17.

132. Nabuchi, Y.; Fujiwara, E.; Ueno, K.; Kuboniwa, H.; Asoh, Y.; Ushio, H. (1995), Oxidation of recombinant human parathyroid hormone: Effect of oxidized position on the biological activity, *Pharm. Res. 12*(12), 2049–2052.

133. Nabuchi, Y.; Fujiwara, E.; Kuboniwa, H.; Asoh, Y.; Ushio, H. (1998), Kinetic study of methionine oxidation in human parathyroid hormone, *Anal. Chim. Acta 365*(1–3), 301–307.

134. Gross, E.; Witkop, B. (1961), Selective cleavage of methionyl peptide bonds in ribonuclease with cyanogen bromide, *J. Am. Chem. Soc. 83*(6), 1510–1511.

135. Pan, C.; Valente, J.; LoBrutto, R.; Pickett, J.; Motto, M. (2010), Combined application of high resolution and tandem mass spectrometers to characterize methionine oxidation in a parathyroid hormone formulation, *J. Pharm. Sci, 99*(3), 1169–1179.

CHAPTER 11

Investigation of Degradation Products and Extractables in Developing Topical OTC (Over the Counter) and NCE (New Chemical Entity) Consumer Healthcare Medication Products

FA ZHANG

Analytical Development, Johnson & Johnson Consumer and Personal Products Worldwide, Skillman, NJ 08558

11.1 INTRODUCTION

Topical medication is applied to body surfaces such as skin, hair, or mucous membranes, including the vagina, eyes, and other surfaces. Topical drug products are usually available in semisolid forms such as ointments, cream, gel, lotion, and foam, to facilitate application. The active ingredient is absorbed transdermally. To maintain the desired physical, chemical, and pharmacological characteristics, topical medication products usually contain complicated ingredients and need special formulation manufacturing processes [1–6]. Topical Medicare products also should to be kept at ambient conditions for convenient usage. All of the aforementioned factors make it a challenging task to maintain the stability of topical drugs such as to reduce active ingredient degradation or interaction with formulation components. Topical drugs are also prone to contamination such as leachables or extractables from packages. Similar to other types of drugs, including oral drugs, topical drugs also contain the categories of new-drug application (NDA) or over-the-counter (OTC) drug. In new-chemical-entity (NCE) development for NDA, understanding the degradation chemistry of the active ingredient becomes critical to accelerate

Characterization of Impurities and Degradants Using Mass Spectrometry, First Edition.
Edited by Birendra N. Pramanik, Mike S. Lee, and Guodong Chen.

active ingredient structural modification, analytical methodology establishment, formulation optimization, packaging selection, and metabolism/toxicity assessment. For OTC or other marketed topical drugs, continuous monitoring of degradation phenomena of the active ingredients or their interactions with formulation ingredients or leachables/extractables contamination are still vital to ensure quality of the products and minimize toxicological risks.

High performance liquid chromatography–mass spectrometry has been developed as an excellent technique to fulfill the types of tasks described here because of its high sensitivity and rich structural information with its versatile ionization (ESI—electrospray ionization, APCI—atmospheric-pressure chemical ionization, APPI—atmospheric photoionization, etc.) and detection (MS/MS to the end, high accuracy, high resolution, etc.) capabilities [7–10]. However, in many cases, HPLC-MS may still not be able to provide absolute structural confirmation because of its inherent limitation. For example, it is always challenging to use HPLC-MS to distinguish isomers. It becomes necessary and effective to strategically use a combination of HPLC-MS with other analytical technologies, such as HPLC-UV (ultraviolet and visible light absorbance), GC-MS (gas chromatography–mass spectrometry), HPLC-NMR (nuclear magnetic resonance spectrometry), offline NMR, chemical derivatization, and classic organic synthesis for structural elucidation. In addition to structural identification, further steps toward understanding the degradation mechanism and root cause of impurities or contamination are always needed and vital to provide insights into the related drug development issues and as a guideline for preventive actions.

In this chapter, several selected examples are illustrated regarding degradation product identification, degradation pathway investigation, drug–formulation ingredient interaction, and extractables studies for NEC evaluation and marketed topical OTC product assessment.

11.2 OXIDATIVELY INDUCED COUPLING OF MICONAZOLE NITRATE WITH BUTYLATED HYDROXYTOLUENE IN A TOPICAL OINTMENT

Miconazole nitrate (**1**), 1-(2-(2,4-dichlorobenzyloxy)-2-(2,4-dichlorophenyl)ethyl) imidazolium nitrate, is an imidazole antifungal agent that has shown fungistatic activity against a number of pathogenic fungi and yeasts [11–20]. Compound **1** can be administered by intravenous infusion in the treatment of severe systemic fungal infections, including candidiasis, coccidioidomycosis, cryptococcosis, paracoccidioidomycosis, and infections due to *Pseudeliescheria boydii.* Compound **1** can also be applied topically for dermatophytic and vaginal infections or infectious vaginitis [21–24]. In the development of a new topical ointment formulation containing **1** as the active ingredient and 2,6-di-*tert*-butyl-4-methylphenol (BHT) as an antioxidant in the petrolatum vehicle, a trace-level degradation product was observed. Studies using HPLC coupled with electrospray ionization mass spectrometry (HPLC-ESI-MS) revealed that it is a 1 : 1 adduct of miconazole with BHT, 1-(3,5-di-*tert*-butyl-4-hydroxy-benzyl)-3-[2-(2,4-dichlorobenzyloxy)-2-(2,4-dichloro-phenyl)-ethyl]-

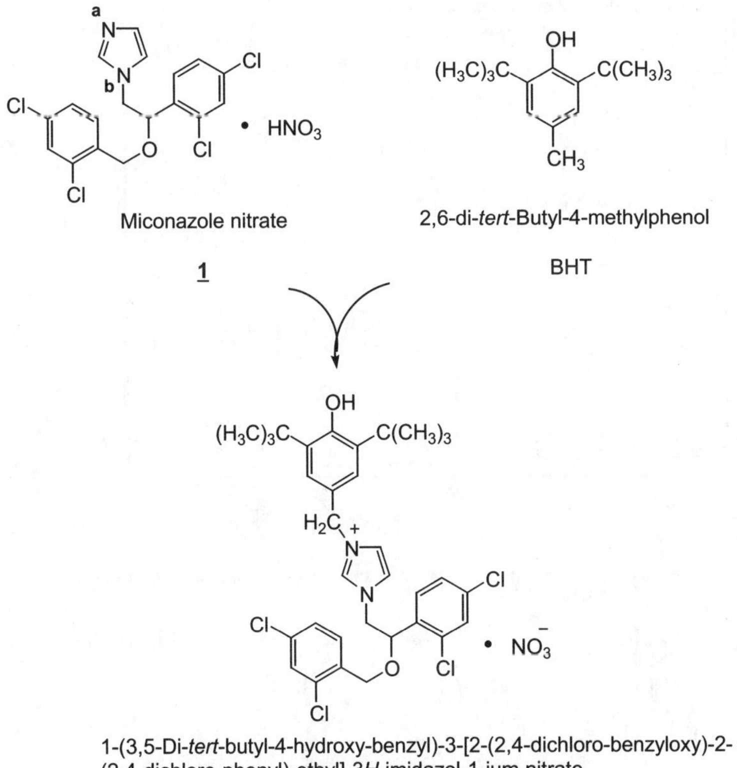

Miconazole nitrate

1

2,6-di-*tert*-Butyl-4-methylphenol

BHT

1-(3,5-Di-*tert*-butyl-4-hydroxy-benzyl)-3-[2-(2,4-dichloro-benzyloxy)-2-(2,4-dichloro-phenyl)-ethyl]-3*H*-imidazol-1-ium nitrate

SCHEME 11.1 Interaction of miconazole nitrate with BHT.

3*H*-imidazol-1-ium nitrate (**2**) (Scheme 11.1) [25]. Its absolute structure and formation mechanism were confirmed by synthetic approaches.

11.2.1 HPLC-MS Screening

The topical ointment under investigation consists of miconazole nitrate (**1**) (0.25%) as the active ingredient and BHT ($<=$ 20 ppm) as an antioxidant in the petrolatum vehicle. The typical HPLC-UV chromatogram of the ointment after one year of storage at 40°C/75% RH is presented in Figure 11.1a. Miconazole nitrate (**1**) eluted at 44 min, BHT eluted later than 50 min, and the degradation product (**2**) eluted at 45.4 min. The level of **2** is approximately 0.5% of the miconazole nitrate (peak area %). The remaining peaks were due to the complicated formulation ingredients. Under the same analytical conditions, the unaged ointment and a placebo containing all the ingredients except for miconazole nitrate after storage at 40°C/75% RH for one year did not contain peak **2**, indicating that **2** is related to miconazole nitrate. The HPLC-ESI mass spectrum of peak **2** presented in Figure 11.1b, exhibits a predominant ion cluster at *m*/*z* 633, 635, and 637 with an abundance ratio of 7 : 10 : 5. This is consistent

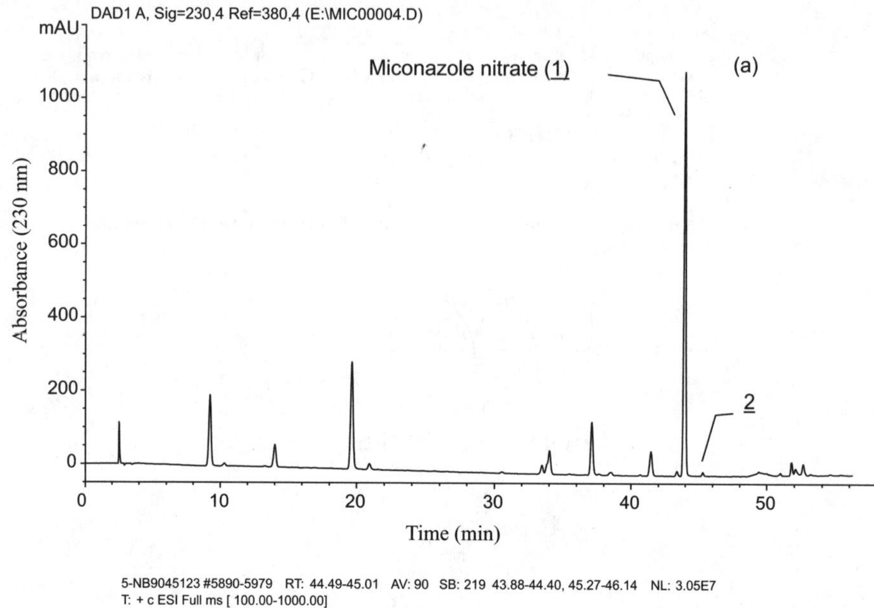

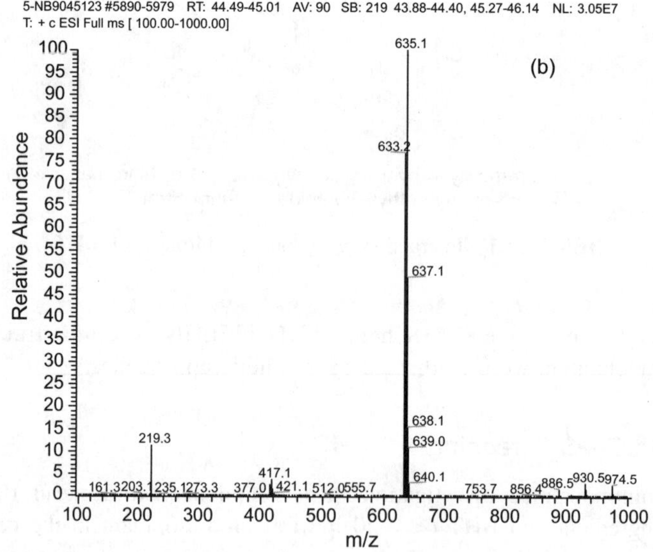

FIGURE 11.1 (a) HPLC-UV chromatogram of the topical ointment after one year at 40°C/75% RH; (b) ESI-MS spectrum of peak **2**; (c) ESI-MS spectrum of peak **2** at elevated CID level (20%) (reproduced with permission from Ref. 25).

with the theoretical isotope ratio of an ion having the weight of 633 with four chlorine atoms, indicating that the four chlorine atoms of miconazole may remain intact. This observed ion cluster also indicates that compound **2** might be a 1 : 1 adduct between miconazole and BHT since miconazole and BHT have molecular weights of 414 and

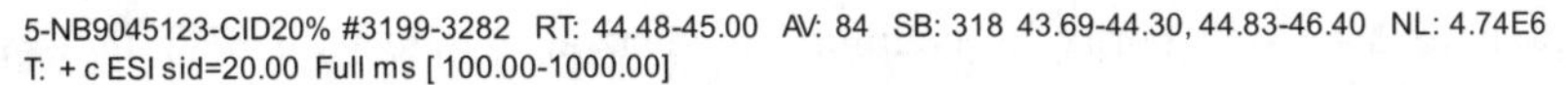

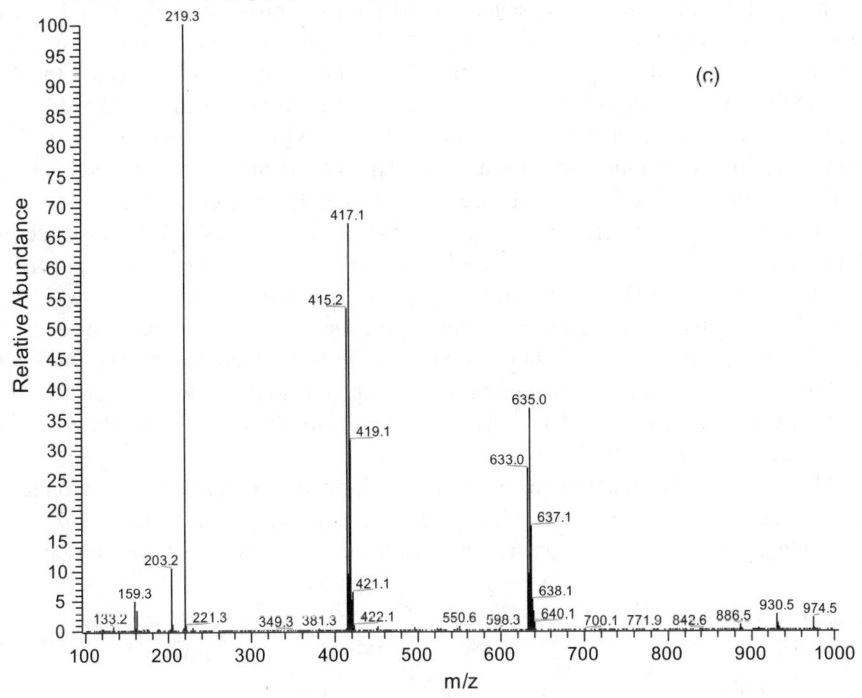

FIGURE 11.1 (*Continued*)

220, respectively. In addition, the ESI-MS spectrum of peak **2** was also obtained at an elevated level of source collision-induced dissociation (CID) energy (20%). As expected, more fragmentation of the ion cluster at *m/z* 633 occurred to generate the predominant ions at *m/z* 415 and 219 (Figure 11.1c). The ions at *m/z* 415 and 219 clearly indicated the presence of miconazole and BHT, respectively. The ion at *m/z* 415 is actually an ion cluster having predominant ions at *m/z* 415, 417, and 419 with an abundance ratio of 7 : 10 : 5, indicating the presence of four chlorine atoms. The structure of **2** was proposed to be a salt with the structure of 1-(3,5-di-*tert*-butyl-4-hydroxybenzyl)-3-[2-(2,4-dichlorobenzyloxy)-2-(2,4-dichlorophenyl)-ethyl]-3*H*-imidazol-1-ium nitrate. The positive counter ion of **2** was observed directly by the mass spectrometer to give the ion at *m/z* 633 in Figure 11.1b,c.

11.2.2 Organic Synthesis

In order to confirm the structural assignment of **2** and to prepare an adequate amount of **2** to be used as reference material to quantify **2** in the drug product, it became necessary to obtain pure solid **2**. It is not practical to isolate **2** directly from the ointment because of its low level and the interference from the complicated formulation matrix. In this report, a synthetic method was successfully established to prepare **2** in high purity by coupling miconazole nitrate with BHT in DMSO (dimethylsulfoxide) in the presence of potassium nitrosodisulfonate and TEA (triethylamine). The reaction was conducted at ambient temperature, and the synthesized **2** was isolated using a semipreparative HPLC method and subjected to MS and NMR analyses. The structure of the synthesized **2** was identified as 1-(3,5-di-*tert*-butyl-4-hydroxy-benzyl)-3-[2-(2,4-dichloro-benzyloxy)-2-(2,4-dichloro-phenyl)-ethyl]-3*H*-imidazol-1-ium trifluoroacetate. This is a trifluoroacetate since it was isolated using an HPLC mobile phase containing trifluoroacetic acid.

For mass spectrometric analysis, a solution of the synthesized **2** was directly infused into the ESI source of a quadruple ion trap mass spectrometer in the positive-ion detection mode. The synthesized **2** exhibits a positive ion of *m/z* 633 with the typical ion cluster for four chlorine atoms to give the most abundant isotope peaks at *m/z* 633, 635, and 637 with an abundance ratio of 7 : 10 : 5. Increasing the source CID energy level resulted in enhancement of the fragment ions at *m/z* 219 and the ion cluster, with the most abundant ions occurring at *m/z* 415, 417, and 419 with an abundance ratio of 7 : 10 : 5.

MS^n ($n = 1,2,3,4$) experiments were also performed using the quadruple ion trap mass spectrometer, and the results obtained are presented in Figure 11.2. For simplicity, only the ion at *m/z* 633 was isolated at the first step for subsequent

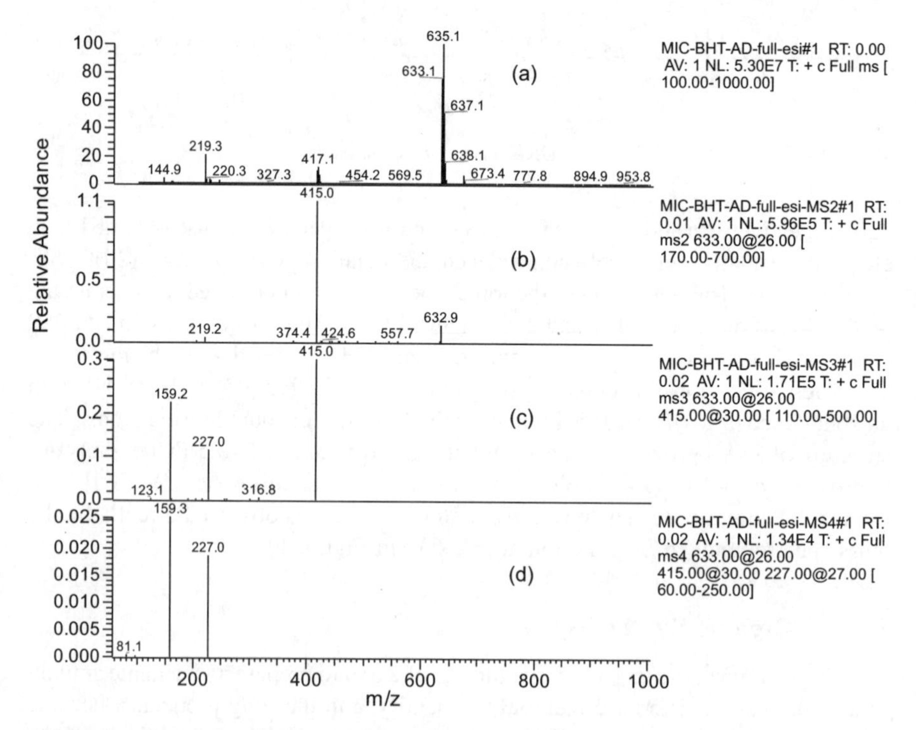

FIGURE 11.2 ESI-MS^n spectra of the synthesized **2**: (a) full-scan MS^1; (b) Full scan MS^2, 633>; (c) full-scan MS^3, 633 > 415>; (d) full-scan MS^4, 633 > 415 > 227> (reproduced with permission from Ref. 25).

fragmentation. All four chlorine atoms in the ion at *m/z* 633 had the same mass of 35. As a consequence, the chlorine atoms in the fragment ions in Figure 11.2b–d must also have the same mass of 35. Figure 11.2 indicates that the ion at *m/z* 633 generates ions at *m/z* 219 and 415. The ion at *m/z* 415 subsequently produces the ion at *m/z* 227. The ion at *m/z* 227 generates the ion at *m/z* 159. The proposed MS fragmentation pathway is presented in Scheme 11.2. ^{1}H, ^{13}C, ^{19}F, ^{1}H$-^{1}$H homonuclear chemical shift correlation spectroscopy (COSY), ^{1}H$-^{13}$C heteronuclear chemical shift correlation

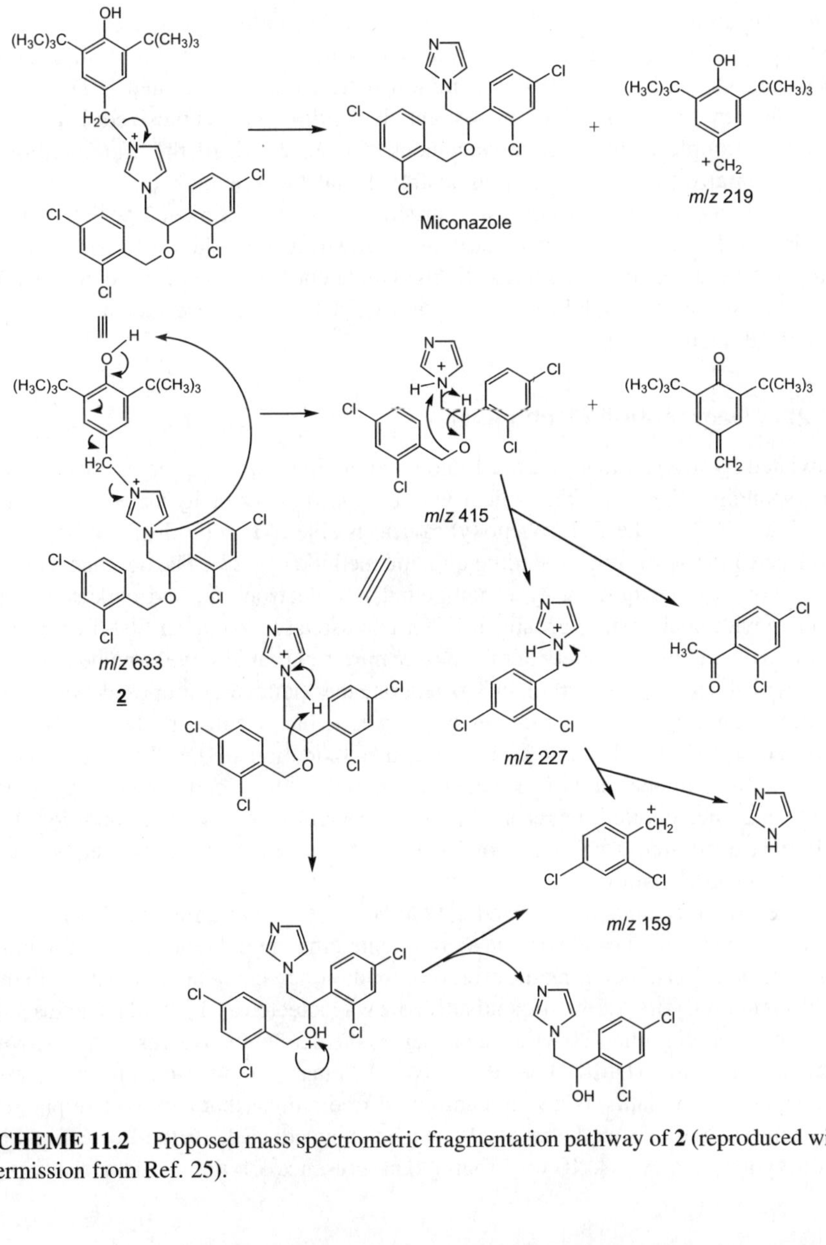

SCHEME 11.2 Proposed mass spectrometric fragmentation pathway of **2** (reproduced with permission from Ref. 25).

heteronuclear multi bond coherence (^{1}H, ^{13}C-HMBC and ^{1}H, ^{15}N-HMBC), and distortionless enhancement by polarization transfer (DEPT) NMR measurements were performed to elucidate the structure of the synthesized **2**.

To confirm the structural assignment of **2** formed in the topical ointment, the HPLC-UV and HPLC-MS characteristics of **2** detected in the ointment and that of the synthesized **2** were compared. HPLC-UV analysis revealed that compound **2** formed in the ointment and the synthesized compound **2** eluted at the same retention time. They also exhibited the same UV absorption pattern using diode array detection. In addition, the adduct **2** formed in the ointment and the synthesized **2** also gave identical ESI mass spectra with the same isotope distribution at *m/z* 633, 635, and 637 and identical fragmentation patterns at the elevated CID level as described in Figure 11.1b,c. As expected, spiking the synthesized **2** into the ointment sample resulted in enhancement of peak **2**. All of the aforementioned experimental evidence clearly demonstrated that the miconazole-BHT adduct **2** detected in the ointment has the same structure as the synthesized **2**, which, in turn, confirmed the assignment of its structure. However, the adduct **2** formed in the ointment has a negative counterion (NO_3^-) different from that of the synthesized **2** which was isolated by HPLC (counterion $CF_3CO_2^-$) because the mobile phase contained trifluoroacetic acid.

11.2.3 Degradation Mechanism

Butylated hydroxytoluene is a well-known antioxidant and can be oxidized to the corresponding phenoxyl free radical via electron transfer or hydrogen abstraction reactions [26,27]. The BHT phenoxyl radical is able to rapidly disproportionate to produce BHT and the corresponding quinone methide (**3**) [27]. The quinone methide **3**, similar to other quinone-type compounds, is electrophilic and expected to be reactive with nucleophilic groups. It is not unreasonable to expect that the nitrogen atoms in the imidazole moiety of miconazole nitrate (**1**) could serve as nucleophiles to react with the methylene carbon in **3** to generate the adduct **2**. A proposed oxidatively induced coupling reaction mechanism between miconazole nitrate and BHT is depicted in Scheme 11.3. It is expected that oxidation of BHT is necessary to generate the quinone methide **3**, which can couple with miconazole nitrate, which serves as a nucleophile. The reaction must also be promoted by basic conditions that will release the imidazole moiety of **1** from a protonated cation to the free base and increase its nucleophilicity.

To test the mechanism described in Scheme 11.3, the coupling reaction between miconazole nitrate (**1**) and BHT was investigated in a model reaction system under various conditions, including the effects of oxidation reagent, base, and acid. In this model reaction, potassium nitrosodisulfonate was selected as the oxidant, and acetic acid and triethylamine (TEA) were chosen as the acid and base, respectively, with dimethylsulfoxide (DMSO) as the solvent. Potassium nitrosodisulfonate, also known as *Fremy's salt*, is a stable aminoxyl free radical that can oxidize phenols to phenoxyl radicals and that itself can be reduced back to the corresponding hydroxylamine [28,29]. It was found that presence of the oxidant potassium

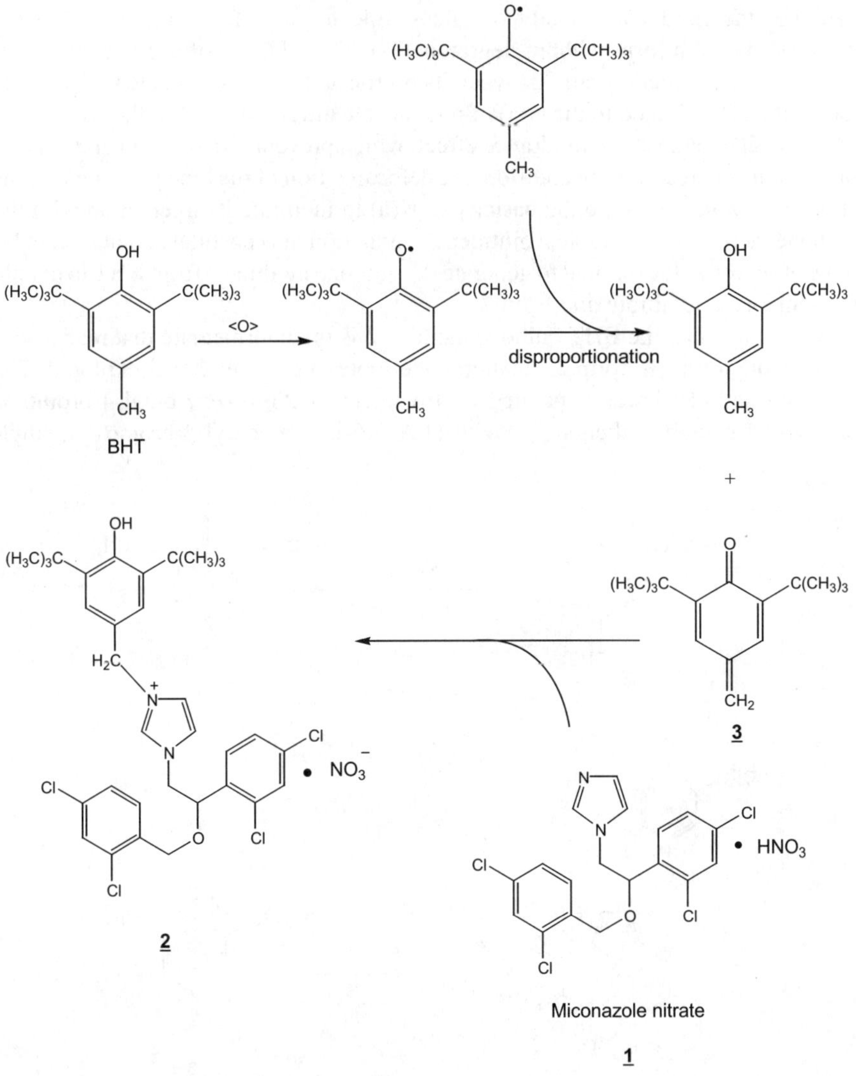

SCHEME 11.3 Proposed reaction mechanism between miconazole nitrate and BHT (reproduced with permission from Ref. 25).

nitrosodisulfonate is necessary for the formation of **2**. This is consistent with our proposal that BHT needs to be oxidized first to generate the quinone methide **3**, which then reacts with miconazole **1**. It was also found that addition of TEA considerably increases the yield of **2** due to deprotonation of the imidazole moiety in miconazole nitrate to increase its electron density and nucleophilicity. This is also the optimized condition for the synthesis of **2**. Interestingly, addition of acetic acid did not suppress the formation of **2**. This suggests that in the nonaqueous reaction medium, the imidazole moiety of miconazole nitrate exists completely in the protonated cation form without deprotonation. The addition of acetic acid has no effect on the population ratio between its protonated and deprotonated forms. The fact that BHT is linked to the N(a) of miconazole nitrate, instead of the N(b) atom, might be due to the steric hindrance effect, which prevents N(b) from participating in the coupling reaction. In addition, the delocalization of the lone pair of electrons at N(b) may also increase the basicity of N(a) to facilitate its interaction with the quinone methide. In the topical ointment formulation, it is postulated that molecular oxygen serves as the oxidant to generate the quinone methide **3** from BHT to couple with miconazole nitrate and form **2**.

To confirm that the BHT quinone methide **3** is the intermediate that reacts with miconazole nitrate to form **2**, an alternative route to generate **3** was explored. The quinone methide **3** was generated by treatment of 2,6-di-*tert*-butyl-4-bromo-4-methyl-2,5-cyclohexadienone (**4**) with TEA. 2,6-Di-*tert*-butyl-4-bromo-4-methyl-

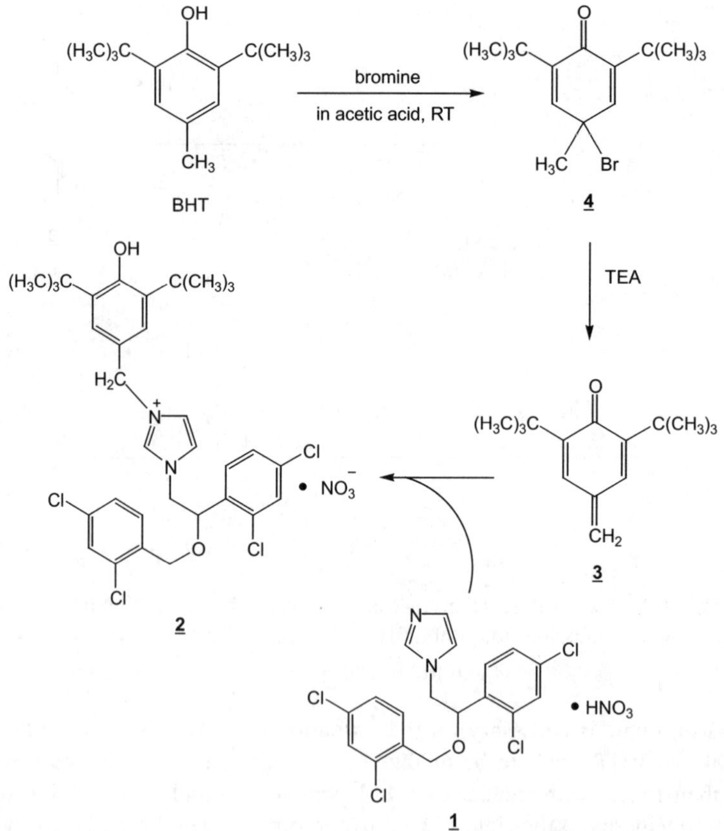

SCHEME 11.4 Interaction of miconazole nitrate (**1**) with quinone methide (**3**) generated from TEA treatment of 2,6-di-*tert*-butyl-4-bromo-4-methyl-2,5-cylclohexdienone (**4**). (reproduced with permission from Ref. 25).

2,5-cyclohexadienone (**4**) was synthesized by treatment of BHT with bromine in acetic acid and water based on a modified method from that reported by Coppinger and Campbell [30]. Compound **4** formed in the reaction precipitated from the solution shortly after the addition of bromine at ambient conditions and was collected by filtration and then purified by crystallization from petroleum ether to give yellow, needle-like crystals. The quinone methide **3** can be generated by treating the synthesized **4** with TEA [31–33]. It was indeed found that **2** did form by incubation of miconazole nitrate (**1**) with **4** in the presence of TEA. The route for synthesis of **4**, generation of **3** by treating **4** with TEA, and the subsequent reaction of **1** with the generated **3** is presented in Scheme 11.4. This scheme can also serve as an alternative synthetic route for adduct **2**.

11.3 EXTRACTABLES FROM RUBBER CLOSURES OF A PREFILLED SEMISOLID DRUG APPLICATOR

Extractables or leachables from packages including rubber closures into drug products can be detrimental since many of the materials extracted from closures are active chemicals that could be toxic, be pyrogenic, or even affect the stability of the active ingredients or interfere with the assays [34–39]. Regulatory authorities have increased their scrutiny of drug delivery products or medical devices that come into direct or indirect contact with rubber/plastic materials, inks, and adhesives. Drug manufacturers are expected to investigate the possibility of compounds leaching into drug products or the human body and determine any effects on the quality and/or safety of the products. Drug manufacturers are also expected to show due diligence in characterizing and identifying major leachables or extractables and evaluating the potential toxicity of those compounds [40–44]. In the event that a known toxic compound is found, it will be necessary to monitor its level with a validated analytical method. Since excessive quantities of accelerators, activators, and other additives are generally used to obtain complete vulcanization of rubber components along with antioxidants and fillers, various quantities of the unreacted components that remain in the rubber stock or their reaction products formed during the rubber manufacturing process may leach into the drug product [45]. It becomes a challenging task to identify the individual leachables or extractables in the drug product since their levels are usually low, if present. In this study, the possible extractables from the rubber closures that have been used for a selected applicator containing semisolid topical drug products were investigated [46]. The purpose of this work was to extract the rubber closures under exaggerated conditions and identify the major extractables that will be used as reference materials to monitor their possible presence in the drug products to ensure their quality. In this study, five extractables were structurally identified: 4-(1,1-dimethylpropyl)phenol (**5**), sulfur (**6**), 2,6-di-*tert*-butyl-[1,4]benzoquinone (**7**), furan-2-yl-(5-hydroxymethyl-furan-2-yl)-methanol (**8**), and 2-bromo-4-(1,1-dimethylpropyl)phenol (**9**) using HPLC, GC, MS, organic synthesis, and comparison with authentic compounds.

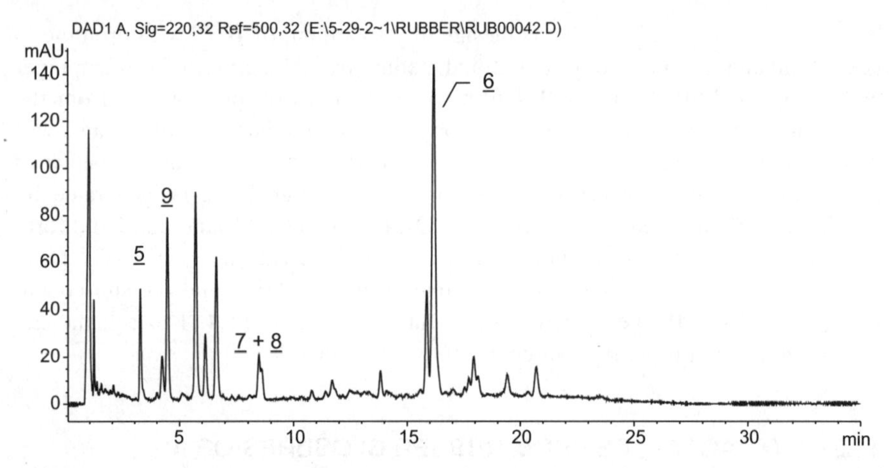

FIGURE 11.3 HPLC chromatograms of the rubber closures extract at 220 nm (reproduced with permission from Ref. 46).

11.3.1 Isolation of the Extractables

An exaggerated condition was employed to extract the possible extractables from the rubber closures using acetonitrile refluxed for 8 h. The acetonitrile solution (i.e., the rubber closures extract) was then used for the structural identification investigation. The analytical HPLC chromatogram of the rubber closures extract is presented in Figure 11.3, which indicates that the extractables are quite complicated. This report will focus only on the structural identification of extractables **5–9** as assigned in Figure 11.3. Extractables **5–9** were isolated using a semipreparative HPLC method. The fractions containing **5–9** were collected separately for the structural identification investigation. The original attempts to use HPLC-MS with electrospray (ESI) and atmospheric-pressure chemical ionization (APCI) to identify the extractables were unsuccessful because of the unsatisfactory ionization efficiency and lack of spectral library. An alternative approach was applied through offline coupling the HPLC separation and GC-MS with electron impact (EI) and chemical ionization (CI) sources. The HPLC fractions containing **5–9** were injected into a GC-MS system for structural identification.

11.3.2 Structural Identification of Extractables 5 and 6

Extractable **5** was identified as 4-(1,1-dimethylpropyl)phenol by interpreting its mass spectrum and by performing a search to known mass spectra libraries (NIST). The EI mass spectrum of **5** exhibited ions at *m/z* 164 (M), 149 ($M-CH_3$), 135 ($M-C_2H_5$), 119 ($M-C_2H_5-CH_4$), and 107 ($M-C_2H_5-C_2H_4$). In addition, the structure of **5** was also confirmed by its practically identical EI mass spectrum, GC retention time, UV spectrum, and HPLC retention time compared with those of an authentic compound under identical experimental conditions.

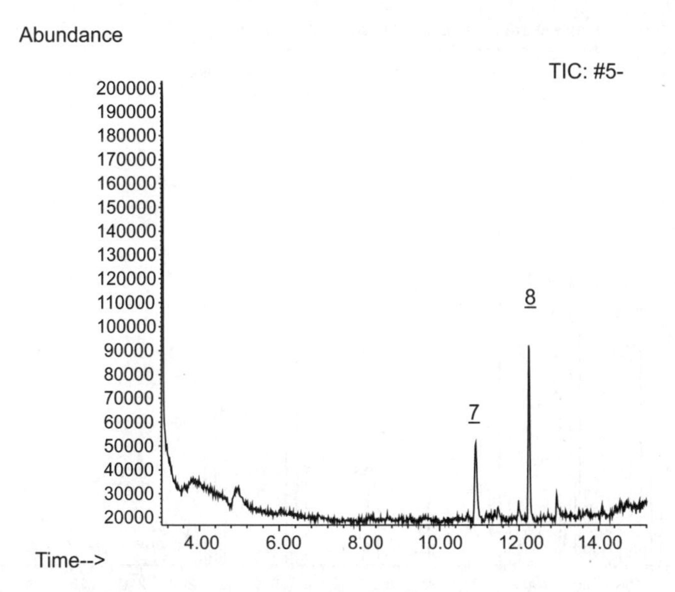

FIGURE 11.4 Full-Scan total-ion chromatogram of the collected fraction containing **7** and **8** from semipreparative HPLC isolation of the rubber closures extract (reproduced with permission from Ref. 46).

Extractable **6** was identified as sulfur by comparing its HPLC retention time and UV spectrum with that of an authentic compound under the identical experimental conditions. As expected, **6** did not exhibit decent mass spectrum under HPLC-MS (ESI and APCI) or GC-MS (EI and CI) conditions.

11.3.3 Structural Identification of Extractables 7 and 8

Extractables **7** and **8** were initially isolated as a mixture using the semipreparative HPLC method. To separate them from each other, the collected fraction containing **7** and **8** was introduced into the GC-MS systems with EI and CI capabilities. The GC chromatogram of **7** and **8** are presented in Figure 11.4. Extractable **7** was identified as 2,6-di-*tert*-butyl-[1,4] benzoquinone by interpreting its mass spectrum and by performing a search to known mass spectra libraries. The EI mass spectrum of **7** exhibited ions at *m/z* 220 (M), 205 ($M-CH_3$), 192 (M−CO), 177 ($M-CH_3-CO$), 163 [$M-(CH_3)_3C$], 149 ($M-CH_3-2CO$), 135 [$M-(CH_3)_3C-CO$], 107 [$M-(CH_3)_3C-2CO$], 121 [$M-CH_3-CO-(CH_3)_2CCH_2$], 91 [$M-CH_3-2CO-(CH_3)_3CH$], and 77 [$M-(CH_3)_3C-CO-(CH_3)_3CH$]. The structure of **7** was also confirmed by its practically identical EI mass spectrum and HPLC retention time compared with those of an authentic compound under identical experimental conditions.

Extractable **8** was tentatively proposed to be furan-2-yl-(5-hydroxymethyl-furan-2-yl)-methanol, based on interpretation of the mass spectrum. The CI mass spectra of **8** are presented in Figure 11.5a, which reveals that **8** has a protonated

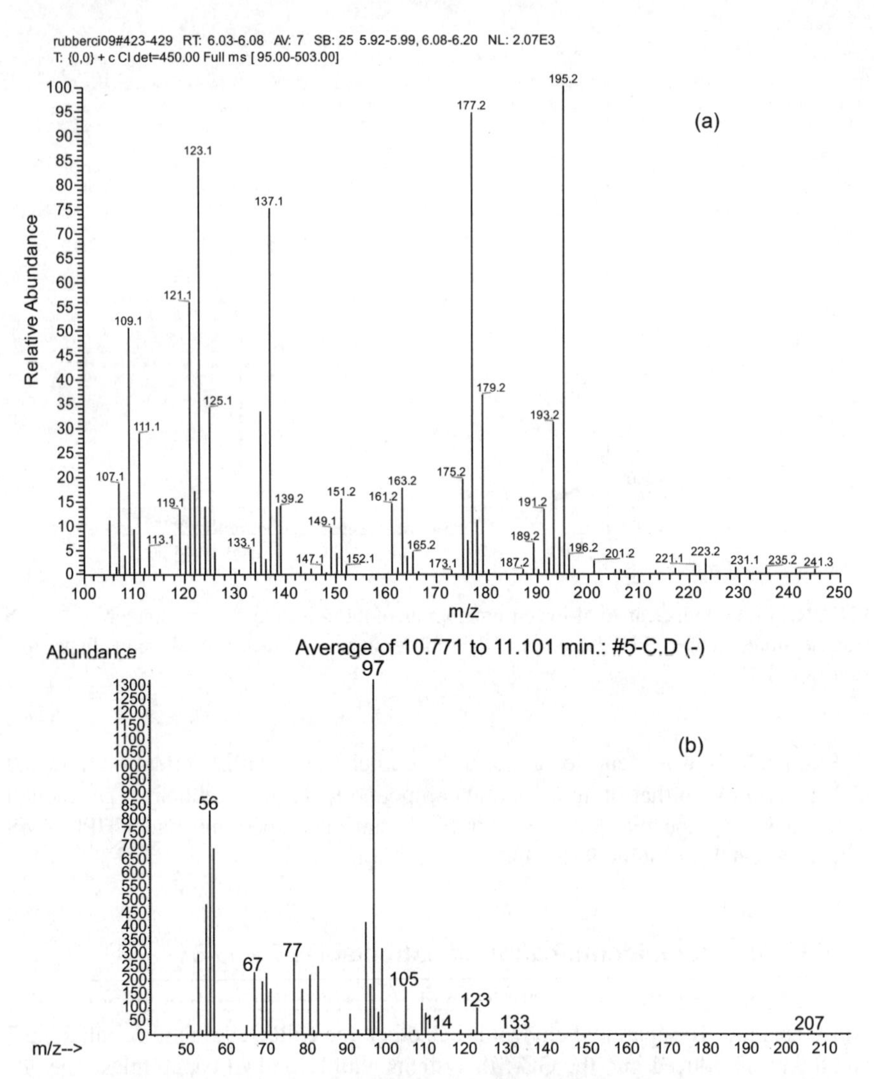

FIGURE 11.5 Mass spectrum of **8** in (a) chemical ionization (CI) mode and (b) electron impact ionization (EI) modes (reproduced with permission from Ref. 46).

molecular ion at *m/z* 195 with a fragment ion at *m/z* 177, indicating the propensity for **8** to lose a water molecule. Additional fragmentation ions at *m/z* 137, 123, and 109 were observed as well, which were possibly due to the neutral loss of 40, 54, and 68 units from the ion at *m/z* 177. The EI mass spectrometric study (Figure 11.5b) shows that 8 exhibits a predominant peak at *m/z* 97 and lacks the molecular ion peak at *m/z* 194, indicating that a facile mass spectrometric (MS) fragmentation occurred in the EI ionization source. The proposed MS fragmentation mechanism of **8** is presented in Scheme 11.5.

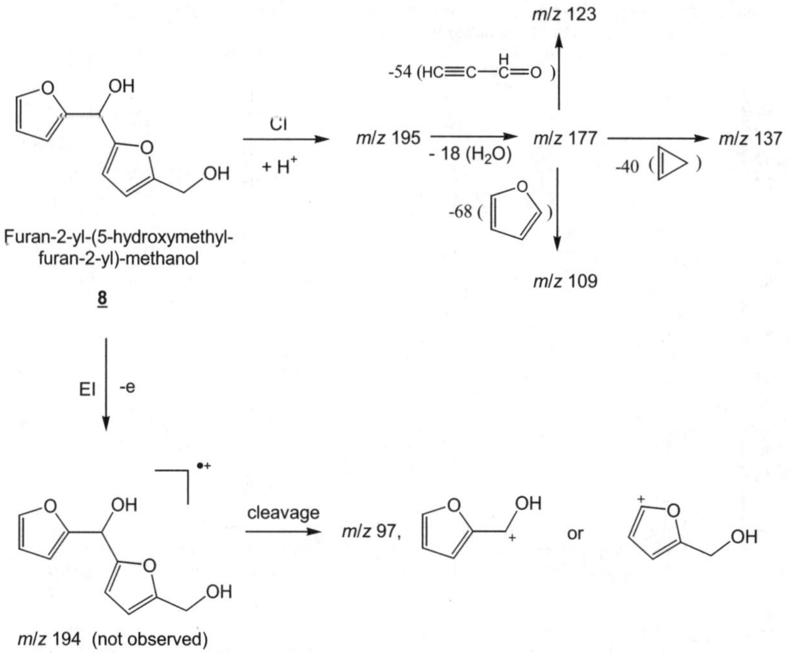

SCHEME 11.5 Proposed mass spectrometric fragmentation pathway of **8** (reproduced with permission from Ref. 46).

11.3.4 Structural Identification of Extractable 9

Extractable **9** was determined to be 2-bromo-4-(1,1-dimethyl-propyl)-phenol. The collected fraction containing **9** from semipreparative HPLC was introduced into the GC-MS system. The CI and EI mass spectra of **9** are presented in Figure 11.6. The CI mass spectrum of **9** exhibits a dominant peak cluster at *m/z* 243, which is responsible for the protonated molecular ion containing one bromine atom. The EI-MS spectrum of **9** shows a positive molecular ion at *m/z* 242 with a typical ion cluster containing one pair of bromine isotopes with masses of 79 and 81 to give the isotope peaks at *m/z* 242 and 244 at nearly equal abundance. The proposed MS fragmentation pattern is presented in Scheme 11.6.

To confirm the structure of extractable **9**, authentic 2-bromo-4-(1,1-dimethyl-propyl)phenol was synthesized chemically since it was not commercially available then. 2-Bromo-4-(1,1-dimethylpropyl)phenol was synthesized by bromination of 4-(1,1-dimethylpropyl)phenol (**5**) in acetic acid and then purified by the semipreparative HPLC method. The synthetic method (Scheme 11.7) was modified based on the basis of a report by Romadane and Chizhikova [47]. The structure of the synthesized 2-bromo-4-(1,1-dimethylpropyl)phenol was confirmed by EI-MS and NMR, including ^{1}H, ^{13}C, and 2D analyses.

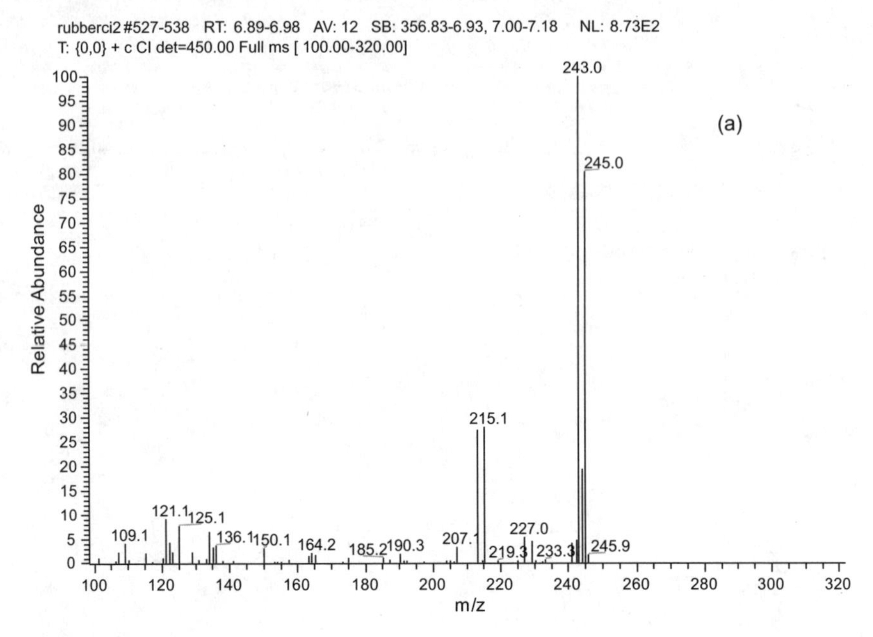

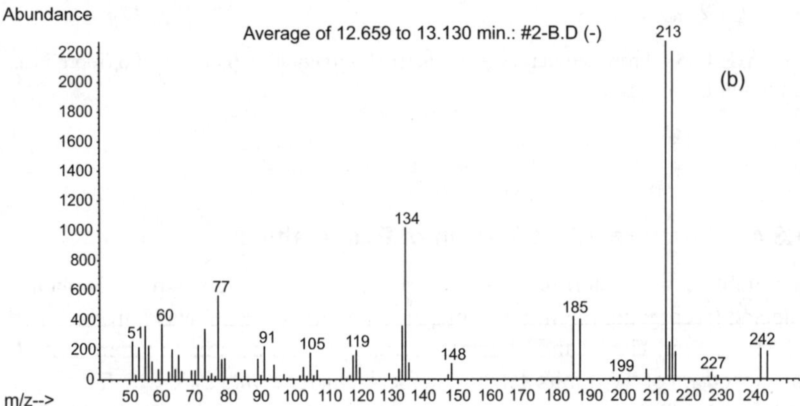

FIGURE 11.6 Mass spectra of **9** in (a) chemical ionization (CI) mode and (b) electron impact ionization (EI) modes (reproduced with permission from Ref. 46).

11.4 NEW DEGRADATION PRODUCTS AND PATHWAYS OF VITAMIN D AND ITS ANALOGS

The chemistry and biochemistry of vitamin D_3 (cholecalciferol) have been extensively studied for over half a century for the great diversity of its chemistry and, especially, its important roles in calcium regulation, immunological regulation, and inducing cancer cell differentiation [48–59]. Over 30 natural metabolites of vitamin D_3 have been identified from human beings and animals and many more synthetic

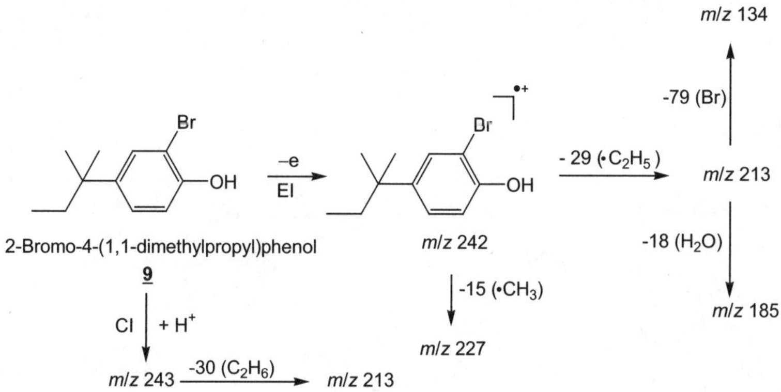

SCHEME 11.6 Proposed mass spectrometric fragmentation pathway of **9** (reproduced with permission from Ref. 46).

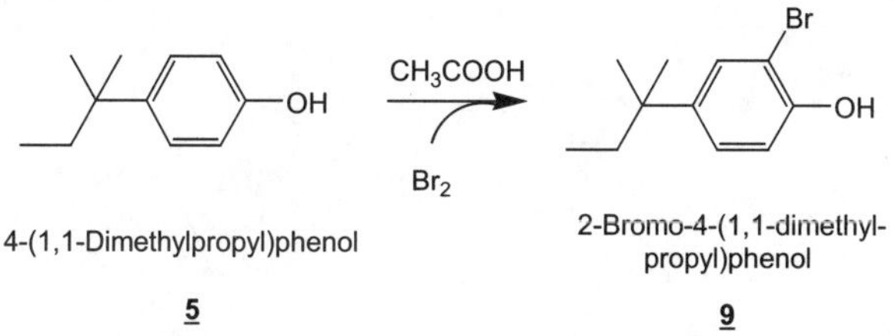

SCHEME 11.7 Synthesis of **9** (reproduced with permission from Ref. 46).

analogs. The chief natural source of vitamin D_3 is provitamin D_3, which is photolyzed by light in the skin to previtamin D_3, which is then spontaneously isomerized to vitamin D_3 (Scheme 11.8). Regarding the degradation chemistry, it is known that thermal equilibrium exists in solution between vitamin D_3 and previtamin D_3 at room temperature and intermediate temperature (20–80°C), while at higher temperatures (100–180°C), vitamin D_3 is transferred irreversibly to pyrocholecalciferol and isopyrocholecalciferol [53,59–62]. Previtamin D_3 could also be converted to tachysterol or isotachysterol catalyzed by iodine or acids [53–68]. Acidic conditions also promote transformation of tachysterol to isotachysterol. Vitamin D_3 can also be oxidized to form various epoxides [53,59,63–68]. Light or iodine could cause reversible conversion between vitamin D_3 and 5,6-*trans*-vitamin D_3 (Scheme 11.8).

Our investigation demonstrated that thermal degradation products of vitamin D_3 in solution is very sensitive to acidity of the solvent [73]. Heating vitamin D_3 in dimethylsufoxide (DMSO) at 140°C under nitrogen in the dark did not produce previtamin D_3, pyrocholecalciferol, or isopyrocholecalciferol, but instead gave isovitamin D_3 and isotachysterol. We also found that isotachysterol, the acid-catalyzed isomerization product of vitamin D_3, is very labile in air even in the dark; we isolated and identified seven autoxidation products [74,75].

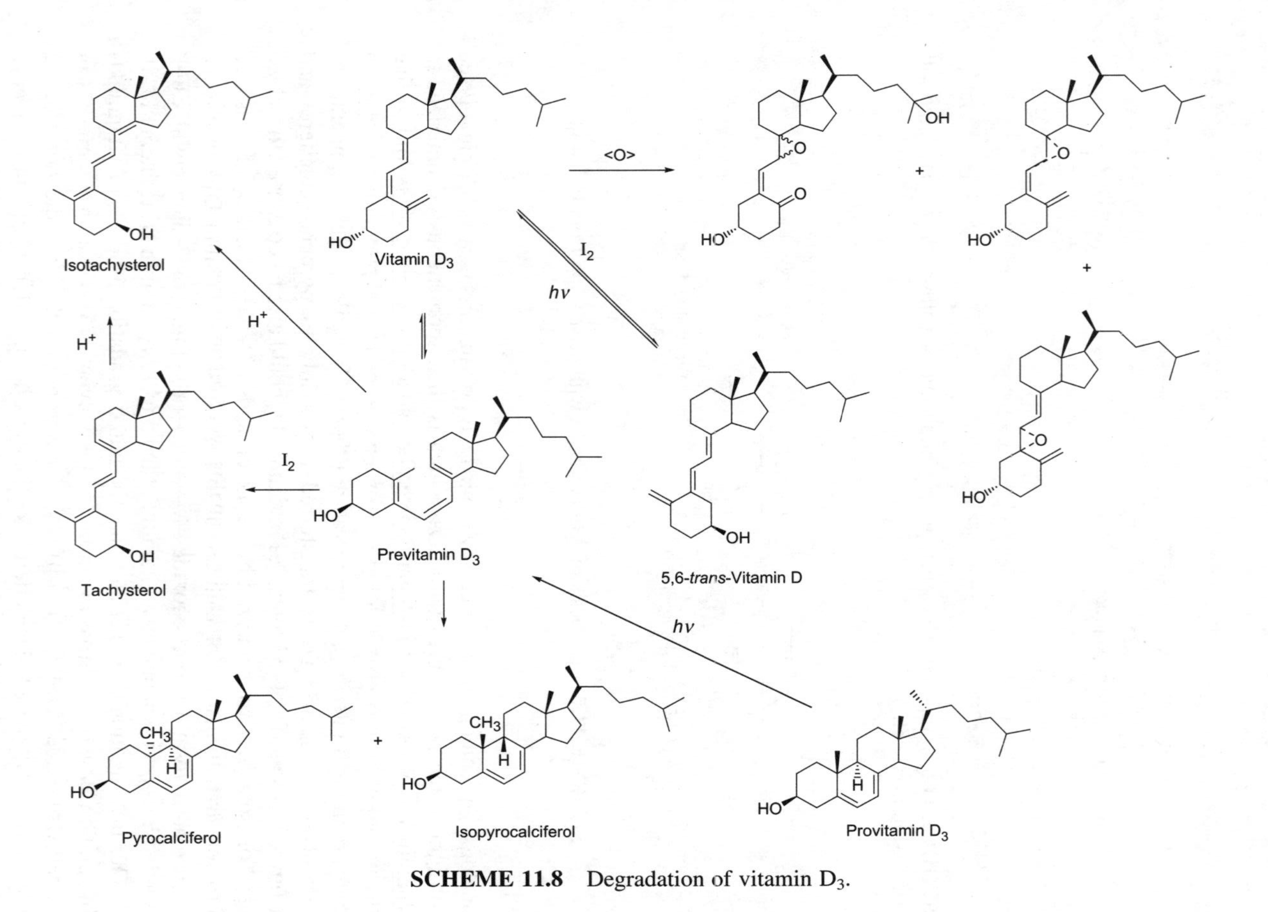

SCHEME 11.8 Degradation of vitamin D_3.

Vitamin D_3 itself is a prohormone that requires two in vivo sequential hydroxylations before manifesting optimal physiological activity [53,59,76]. The first occurs in the liver to produce 25-hydroxyvitamin D_3, which is the major circulating metabolite of vitamin D_3. The second hydroxylation occurs in the kidney and is controlled directly or indirectly by blood calcium and phosphate levels. The product, 1,25-dihydroxyvitamin D_3, is the hormonally active form of vitamin D_3 and functions as a classical steroid hormone to induce its physiological effects, intestinal calcium absorption, and bone calcium mobilization via genomic mechanism. 1,25-dihydroxyvitamin D_3 also promotes normal cell differentiation and proliferation and also evokes a variety of biological responses through nongenomic mechanism. It is well established that the 1-hydroxyl group is necessary for the biological activities of 1,25-dihydroxyvitamin D_3 [77]. The elucidation of the metabolism of vitamin D has led to a change in emphasis from study of the parent vitamin D itself to study of its metabolites. This also led to a renewed interest in vitamin D research with regard to its potential use as a therapeutic agent, such as in the treatment of cancer and skin disorders such as psoriasis [78–80]. The use of 1,25-dihydroxyvitamin D for the treatment of malignancy resulted in toxic hypercalcemia and stimulated research to find analogs with high potency in inducing cell differentiation but with low calcitropic effects [81,82]. In fact, many hydroxylated derivatives of vitamin D have been synthesized to study the structure–activity relationships and to search for compounds showing an increase in specificity or overall activity for potential therapeutic and cosmetic applications. Some of the hydroxylated derivatives are already in development or being used therapeutically [83]. As an example, more than 10 analogs of 1-hydroxylvitamin D have already been approved by governmental agencies or are currently under development by various industrial and/or university research groups for the therapeutic indications of psoriasis, osteoporosis, hypocalcemia, hypoparathyroidism, leukemia, cancer, and immune diseases [84]. Ecalcidene (1-[(1α,3β,5*Z*,7*E*,20*S*)-1,3-dihydroxy-24-oxo-9,10-secochola-5,7,10(19)-trien-24-yl]piperidine), as a new 1-hydroxylvitamin D analog, was introduced by Hesse et al. and has the potential to serve as a cell modulator [85]. Understanding the degradation properties of ecalcidene is necessary for the development of its potential pharmaceutical applications. We studied the thermal degradation, acid-induced degradation, and iodine-induced degradation of ecalcidene [86]. The degradation products were identified using the combination of HPLC-UV, HPLC-MS, HPLC-NMR, and chemical derivatization. The degradation mechanisms were proposed. Some new chemical features regarding the reactions of this secosteroid 1-hydroxyvitamin D analog were reported for the first time.

As illustrated in this section, degradation of vitamin D_3 and its analogs forms various isomers. HPLC-MS alone is not adequate to reveal their absolute structures. Accordingly, a strategic combination of HPLC-MS, NMR, including HPLC-NMR, chemical derivatization, and so on becomes necessary.

11.4.1 Thermal Isomerization of Vitamin D_3 in DMSO

As indicated in Scheme 11.8, the thermal reversible conversion between vitamin D_3 and previtamin D_3 is a well-known process. The previtamin D_3 formed could further

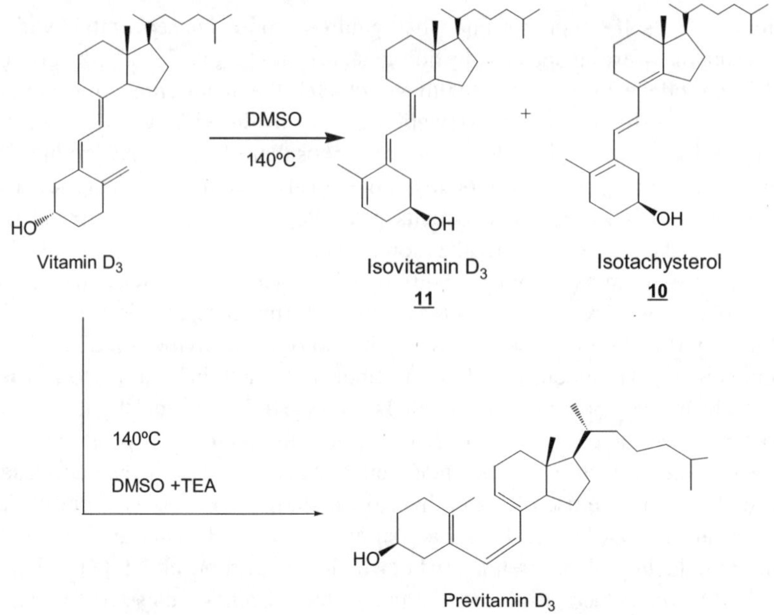

SCHEME 11.9 Thermal degradation of vitamin D_3 in DMSO.

generate two products through cyclization: pyrocholecalciferol and isopyrocholecalciferol. However, we found that incubating vitamin D_3 in DMSO at 140°C under nitrogen in the dark produced two different products (**10** and **11**) [73]. The M + 1 peaks in HPLC-FT-ESI-MS of **10** and **11** corresponded to the same molecular formula $C_{27}H_{44}O$, namely, isomers of vitamin D_3. Their UV spectra exhibited strong absorption at 287 and 288 nm respectively, indicating the presence of an all-*trans*-triene chromophore in the two molecules. MS or MS/MS of **10**, **11**, vitamin D_3, and previtamin D_3 all exhibited similar patterns, and were not able to distinguish between these compounds. Additional NMR studies including ^1H, ^{13}C, COSY, and nuclear overhauser enhancement spectroscopy (NOESY) revealed that **10** and **11** have structures of isotachysterol (**10**) and isovitamin D_3 (**11**) (Scheme 11.9).

It was reported previously that isotachysterol (**10**) was formed by acid-catalyzed isomerization of vitamin D_3 [64], and **10** and **11** were also detected on an aerosil surface loaded with vitamin D_3 [87]. Therefore, it was considered probable that the formation of **10** and **11** in the present case was due to the weak acidity of solvent DMSO. Indeed, addition of couple of drops of triethylamine (TEA) to the reaction system produced previtamin D_3 as the principal product rather than **10** and **11**. Further incubation of previtamin D_3 at elevated temperature produced pyrocholecalciferol and isopyrocholecalciferol as reported [60–62].

11.4.2 Autoxidation of Isotachysterol

It is well known that vitamin D_3 is relatively stable in the air at ambient temperature, while its acid-catalyzed isomerization product, isotachysterol (**10**), is very labile in air, even in the dark [63–68]. However, no effort was made previously to identify the complex autoxidation products of isotachysterol. We report here the isolation and identification of the principal autoxidation products of isotachysterol, including (5*R*)-5,10-epoxy-9,10-secocholesta-6,8(14)-dien-3β-ol (**12a**), (5*S*)-5,10-epoxy-9,10-secocholesta-6,8(14)-dien-3β-ol (**12b**), (10*R*)-9,10-secocholesta-5,7,14-trien-3β,10-diol (**13a**), (10*S*)-9,10-secocholesta-5,7,14-trien-3β,10-diol (**13b**), (7*R*,10*R*)-7,10-epoxy-9,10-secocholesta-5,8(14)-dien-3β-ol (**14**), 5,10-epidioxy-isotachysterol (**15**), and 3,10-epoxy-5-oxo-5,10-seco-9,10-secocholesta-6,8(14)-dien-10-ol (**16**) (Scheme 11.10) [74,75]. The formation of these products is discussed in terms of free-radical peroxidation chemistry.

Isotachysterol (**10**) was prepared by HCl-catalyzed isomerization of vitamin D_3 in methanol [64]. The pale yellow oil of **10** was placed in a small beaker at ambient temperature in the dark. Isotachysterol **10** was found to oxidize rapidly to a very complex mixture as monitored using HPLC-UV and HPLC-FT-ESI-MS. Seven major oxidation products were focused and isolated for structural investigation.

The HPLC-FT-ESI-MS of **12a**, **12b**, **13a**, **13b**, and **14** gave molecular ion peaks of 401.3413, 401.3422, 401.3416, 401.3411, and 401.3411, respectively, corresponding to the same molecular formula with one more oxygen than **10** ($C_{27}H_{44}O_2 + H$ requires 401.3420), indicating that they are all isomeric compounds. In addition to MS, it was necessary to use other techniques such as NMR to establish their absolute structures, and even UV could also give some very valuable structural information.

The UV spectra of **12a** and **12b** were almost identical, showing a band at 248 nm that is characteristic of conjugated double bonds. Comparison of their ^{1}H and ^{13}C NMR spectra with those of vitamin D_3 and its metabolites [88] and with those of isotachysterol [89] clearly demonstrates that **12a** and **12b** are 5,10-epoxides of **10** since the remarkable changes on ^{13}C chemical shifts are observable only for 5- and 10-Cs (from double-bond carbons to epoxy carbons) and on ^{13}C and ^{1}H chemical shifts for 19-Me, and to a less extent, for 4-C. The coupling constants of 3-H are 8.0, 8.0, 4.5, and 4.5 Hz for **12a**, and 9.6, 9.6, 4.7, and 4.7, Hz for **12b**, respectively, demonstrating that the 3-H is axial in both **12a** and **12b**. The NOESY spectrum of **12a** showed clear cross-peaks between 1α-H, 3α-H, and 19-CH_3 and between 6-H, 1α-H, and 19-CH_3 as indicated by a molecular mechanics (MM2)-optimized model (Figure 11.7), indicating that the epoxy ring and the 3-hydroxyl are located on the same side of the molecule. On the other hand, clear NOESY correlations between 6-, 4β-, 2β-,1β-Hs, and 19-CH_3 of **12b** (Figure 11.7) demonstrates that the epoxy ring and the 3-hydroxyl are at the opposite sides of the molecule. In addition, we also demonstrated that epoxidation of isotachysterol with anhydrous *tert*-butylhydroperoxide (TBHP) in benzene in the presence of VO(acac)$_2$ (0.01 equiv) gave **12a** as the sole epoxy product (yield 45%). It is well known that epoxidation of homoallylic alcohols with TBHP/VO(acac)$_2$ produces stereospecifically *syn*-epoxy alcohols [90–93]. Therefore, **12a** and **12b** are assigned as (5*R*)-5,10-epoxy-9,10-secocholesta-6,8(14)-dien-3β-ol (5β,10-epoxy-isotachysterol) and (5*S*)-5,10-epoxy-9,10-secocholesta-6,8(14)-dien-3β-ol (5α,10-epoxyisotachysterol), respectively.

Both **13a** and **13b** exhibited an UV absorption maximum at 278 nm, suggesting the existence of a conjugated triene chromophore. Comparison of their ^{1}H and ^{13}C NMR

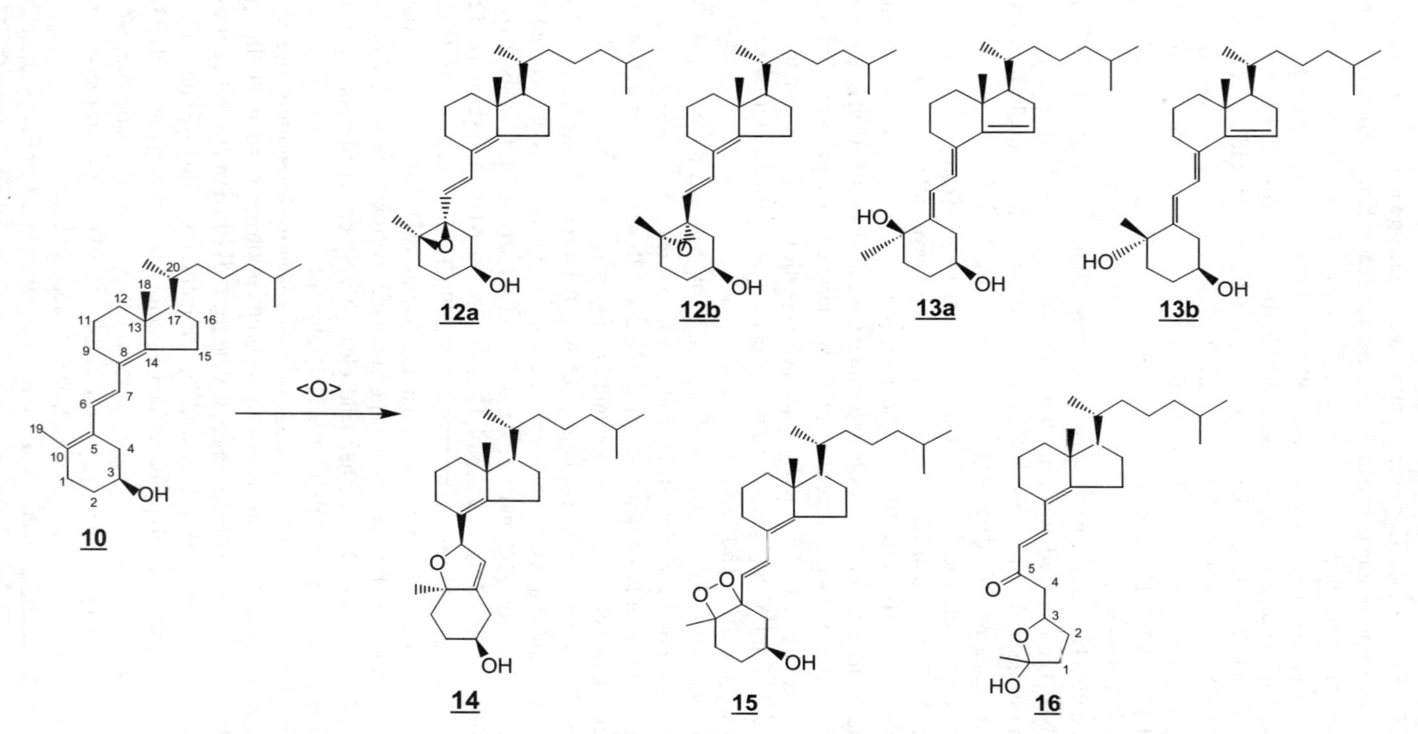

SCHEME 11.10 Autoxidation of isotachysterol (**10**) (reproduced with permission from Ref. 74).

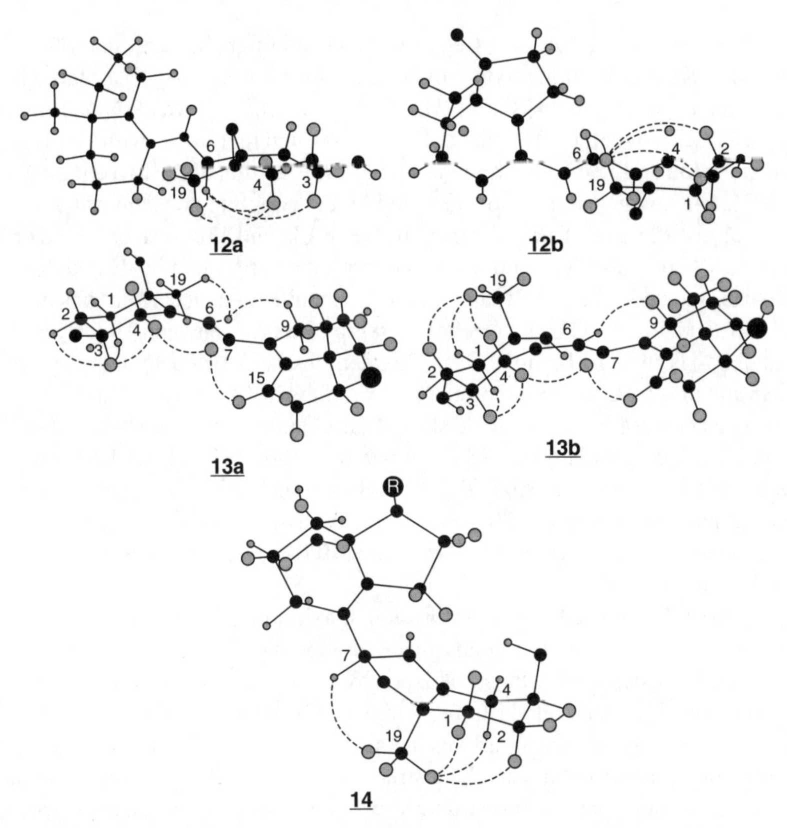

FIGURE 11.7 Principal NOE correlations of **12a**, **12b**, **13a**, **13b**, and **14** presented with ball-and-stick representations of the MM2-optimized structures (reproduced with permission from Ref. 74).

spectra with those of isotachysterol [89] showed remarkable differences on ^{13}C chemical shifts of 5-, 10- and 15-Cs, and to a less extent, on 16- and 19-Cs, which suggests that the 5, 6, 8-triene structure in **10** might change to a 5,7,14-triene system in both **13a** and **13b**. The A-ring structure of **13a** and **13b** was confirmed by their gCOSY spectra, which exhibited spin coupling network between the hydroxymethine proton 3-H and 4α-, 4β-, 2α-, 2β-, 1α- and 1β-Hs, and by their HMBC spectra, which showed correlations between 19-CH_3 and 1-, 5- and 10-Cs. The structure of the C ring was confirmed by their gCOSY spectra, which showed correlations between the allylic 9β-H and 9α-, 11α-, 11β-, 12α-, and 12β-Hs, and by their HMBC spectra, which show correlations between the olefinic 7-H and 8-, 9-, and 14-Cs. The structure of the D ring was confirmed by their gCOSY (gradient-selective COSY) spectra, which showed correlations of the olefinic 15-H with 16α- and 16β-Hs, and 17-H with 16α-H, together with the HMBC correlation of 18-CH_3 with 13- and 14-Cs. The structure of the *seco*-B ring was confirmed by their HMBC spectra, which showed correlations between the olefinic 6-H and 5-, 7-, 8-, and 10-Cs. The coupling constants of 3-H (8.0, 8.0, 4.4, and 4.4 Hz for **13a**, and 9.2, 9.2, 4.6, and 4.6 Hz for **13b**, respectively) suggest that the 3-H is axial and that the A ring of **13a** and **13b** might be partitioned between a 30/70 and 24/76 equilibrium mixture of chair conformers favoring an a chair with the 3β-OH equatorially oriented [88,94]. The nuclear Overhauser effect (NOE) enhancement was observed for the 7-H with 4α-H and 15-H, and for the 6-H with 9β-H and 19-CH_3 in both **13a** and **13b**, indicating the triene configuration of the two compounds to be (5*E*,7*E*,14*E*), which were also supported by the coupling constant of the olefinic protons ($J_{6,7} = 12.0$ Hz). The differences between **13a** and **13b** were observed only in their NOESY spectra, which showed clear correlations between 3α-H and 4α-, 2α-, and 1α-Hs, and between 19-CH_3 and 6-H in **13a**, while no such correlations occurred in **13b**. Instead, clear NOESY correlations were observed between 19-CH_3 and 4β-, 2β-, and 1β-Hs in **13b** (Figure 11.7). This demonstrates that **13a** and **13b** are 10-epimers and that 19-CH_3 is equatorial and α-oriented in **13a**, but axial and β-oriented in **13b**. The comparatively downfield shift of the chemical shifts of 2β-H and 4β-H (*d* 1.74 and 2.55, respectively) in **13a** compared to those of **13b** (*d* 1.51 and 2.04, respectively) also indicated that the 10-OH is axial and β-oriented in **13a** and equatorial and α-oriented in **13b**. Therefore, **13a** and **13b** were assigned as (10*R*)-9,10-secocholesta-5,7,14-trien-3β,10-diol and (10*S*)-9,10-secocholesta-5,7,14-trien-3β,10-diol, respectively.

Compound **14** showed a UV absorption maximum of 206 nm, indicating the absence of conjugated double bonds in the compound. The comparison of its ^1H and ^{13}C NMR spectra with those of isotachysterol (**10**) [89] showed remarkable differences on ^{13}C chemical shifts of 7- and 10-Cs, from olefinic carbons in **10** to oxygen-connecting quaternary carbons in **14**, suggesting that **14** is a 7,10-epoxide of **10** containing a dihydofuran ring. The structure of **14** was fully assigned by its 2D NMR spectroscopy. The A-ring structure was confirmed by its gCOSY spectrum (Figure 11.8a), in which the hydroxymethine proton 3α-H (*d* 4.06) correlated with 2α-, 2β-, 4α-, and 4β-Hs, and the 2β-H correlated with 1α- and 1β-Hs, and by its gHMBC spectrum (Figure 11.8b), which shows correlations between 19-CH_3 and 1-, 5-, and 10-Cs. The dihydrofuran ring was supported by the HMBC correlations of the olefinic 6-H with 4-, 5-, 7-, and 10-Cs. The C- and D-ring structures were confirmed by the H,H-COSY correlations of 9β-, 9α-, 11α-, 11β -, 12α-, and 12β-Hs; of 15α-, 15β-, 16α-, 16β-, and 17α-Hs; and by the HMBC correlations between 18-CH_3 and 12-, 13-, and 14-Cs. The connection between the tetrahydrofuran ring and the C ring was supported by the HMBC correlations between the oxygen-connecting 7-H and 6-, 8-, and 14-Cs. The NOESY 1D spectrum showed clear correlations between 19-CH_3 and 1α-, 2α-, and 4α-Hs, indicating that the 19-CH_3 is axial and the 3-H is equatorial, which is consistent with the coupling constant of 3-H (3.2, 3.2, 2.4, and 2.4 Hz). The NOESY 1D spectrum also exhibited a correlation between the 19-CH_3 and 7-H, demonstrating that they are located on the same side of the dihydrofuran ring. Thus **14** was assigned as (7*R*,10*R*)-7,10-epoxy-9,10-secocholesta-5,8(14)-dien-3β-ol.

Compound **15** showed that the molecular ion peak of 439.3211 corresponds to a molecule with two more oxygens than **10** ($C_{27}H_{44}O_3{}^{+}Na$ requires 439.3188). The UV absorption maximum at 296 nm suggested the presence of an extended conjugated system. However, **15** was unstable and gradually converted to a new compound **16** during the process of semipreparative HPLC separation. Compound **16** gave a

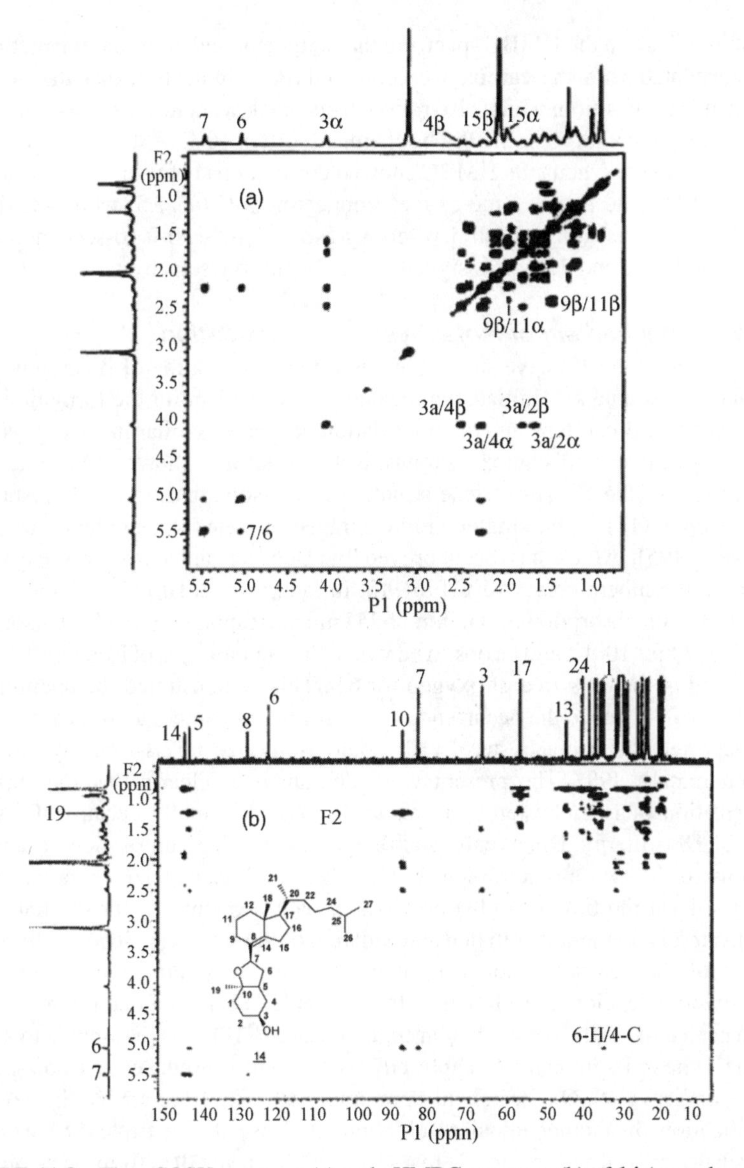

FIGURE 11.8 The gCOSY spectrum (a) and gHMBC spectrum (b) of **14** (reproduced with permission from Ref. 74).

molecular ion peak of 417.3373 corresponding to a same molecular formula of **15** ($C_{27}H_{44}O_3{}^{+}H$ requires 417.3369). The UV spectrum of **16** showed a strong absorption maximum at 305 nm, suggesting the existence of an extended conjugated system. The ^{13}C NMR spectrum revealed the presence of –C=O, –OCH– and –O–C–O– moieties. Comparison of its ^{13}C NMR chemical shifts with those of **10** demonstrated that, besides the A-ring carbons as well as 7- and 14-Cs, other chemical shifts are almost identical. In the HMBC spectrum the olefinic 6- and 7-Hs and the methylenic 4-Hs correlated with the carbonyl carbon (*d* 198.7), indicating that the –C=O is located at the 5 position, which also rationalizes the downfield shift of 4-, 7-, and 14-Cs in comparison with those of **10**. The chemical shift of 10-C (*d* 108.2) suggested that it bonded to two oxygens. Its HMBC showed correlations between the 10-C and 19-CH_3 and 2-Hs, and between the oxygen connecting 3-C (*d* 77.5) with 2- and 4-Hs. Therefore, **16** was assigned as 3,10-epoxy-5-oxo-5,10-seco-9,10-secocholesta- 6, 8(14)-dien-10-ol, and **15** was assigned as 5,10-epidioxyisotachysterol.

11.4.2.1 Mechanism of Isotachysterol Autoxidation

Mordi and Walton [95] have studied in detail the autoxidation of β-carotene in the dark and proposed a self-initiated autocatalytic mechanism for the formation of the 5,6- epoxide of β-carotene and other oxidation products. Similar mechanisms might also be applicable to this autoxidation of isotachysterol as shown in Scheme 11.11. In other words, the all-*trans*-triene structure in **10** isomerizes to the corresponding 6,7-*cis*-isomer (**17**) via the singlet biradical transition state (**18**), similar to the case of β-carotene [95], which has been proved by Doering and coworkers capable of occurring at temperatures <40°C [96–98]. In fact, a small HPLC peak close to the peak of **10** with absorption maximum at 253 nm corresponding to 6,7-*cis*-isotachysterol (**17**) [74,99,100] could be observed when a hexane solution of isotachysterol (**10**) was placed in the dark free of oxygen for 6 h. This demonstrated the unambiguous formation of **17**, hence the occurrence of the *trans*/*cis*-isomerization process. It had been reported that isotachysterol (**10**) could isomerize to *cis*-isotachysterol (**17**) photochemically [99]. The present work demonstrates clearly that the *trans/cis*-isomerization of isotachysterol can also take place thermally because the singlet biradical **18** is thermodynamically stabilized by delocalization of the two unpaired electrons to the two allylic moieties. This thermal *trans/cis*-isomerization of isotachysterol via the biradical (**18**) provides a ready explanation for the liability of isotachysterol and the self-initiated autoxidation of the substrate. Specifically, during twisting of the central carbon–carbon bond of isotachysterol, the unpaired spin density would develop in each half of the molecule, reaching a maximum (one free spin in each half) in the perpendicular transition state (**18**). It is reasonable to assume that the unpaired spin can be "captured" by oxygen to produce a carbon–peroxyl triplet biradical (**19**). Oxygen should preferably attack C10 to enable the extensive delocalization of another unpaired electron. Because it is a triplet, **19** would be relatively long-lived and able to add to a second molecule of **10** to form a new biradical **20**, again at 10-C. Obviously, **20** can be subject to the well-precedented intramolecular homolytic substitution (S_Hi) [101–105], producing the 5,10-epoxides **12** and the 7,10-epoxide **14** by 5,10- and 7,10-ring closure, respectively, of the intermediate alkoxyl biradical **21**. Compound **13** was also possibly derived from the alkoxyl biradical **21** by consecutive 1,5-sigmatropic rearrangement of the allylic 15-H to 6-C and 1,4-sigmatropic rearrangement of the 6-H to 10-O. On the other hand, the peroxyl biradical **19** may collapse to the thermally unstable dioxetane **15**, which would easily be subject to peroxide scission to produce the dicarbonyl intermediate followed by acetalation,

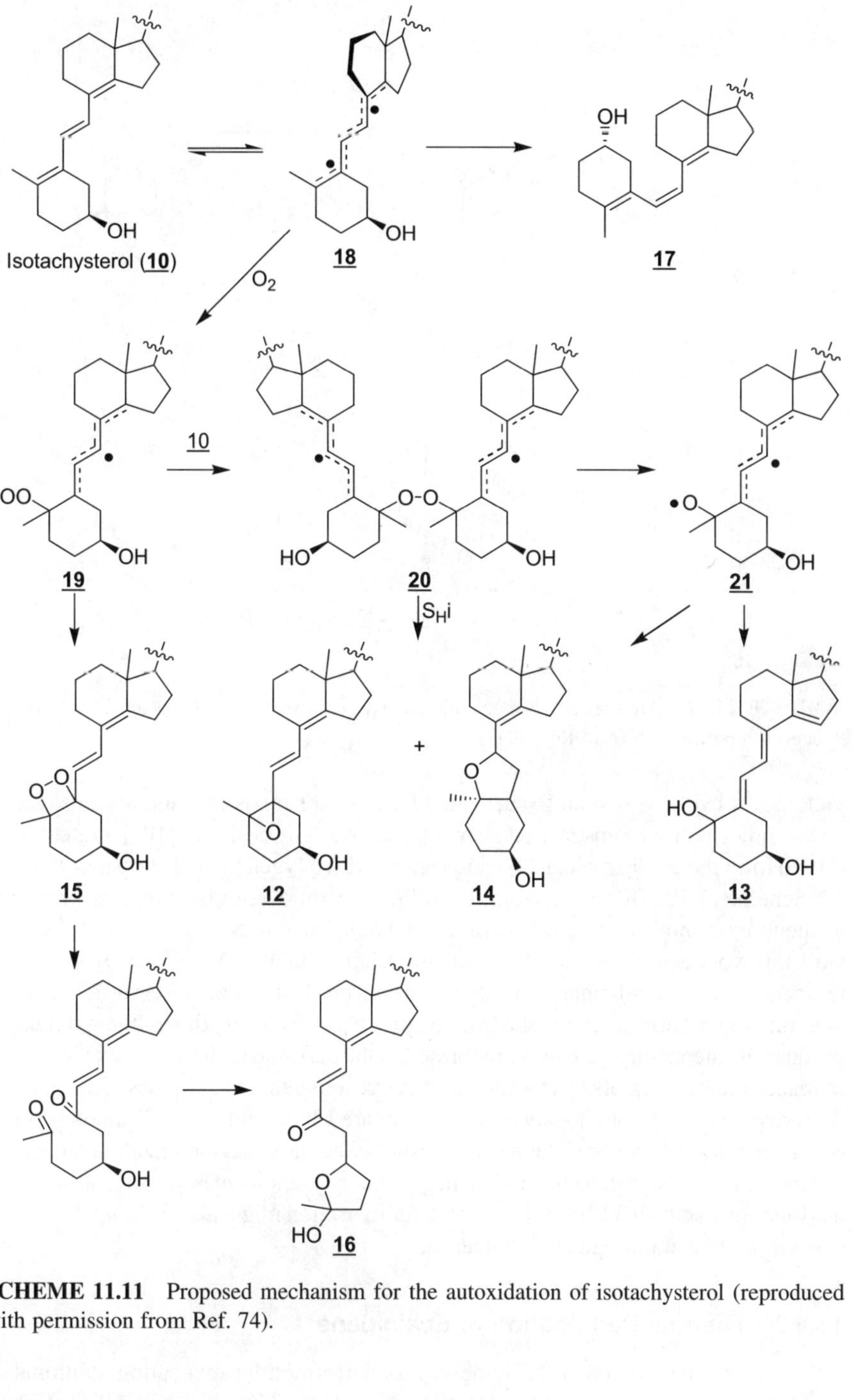

SCHEME 11.11 Proposed mechanism for the autoxidation of isotachysterol (reproduced with permission from Ref. 74).

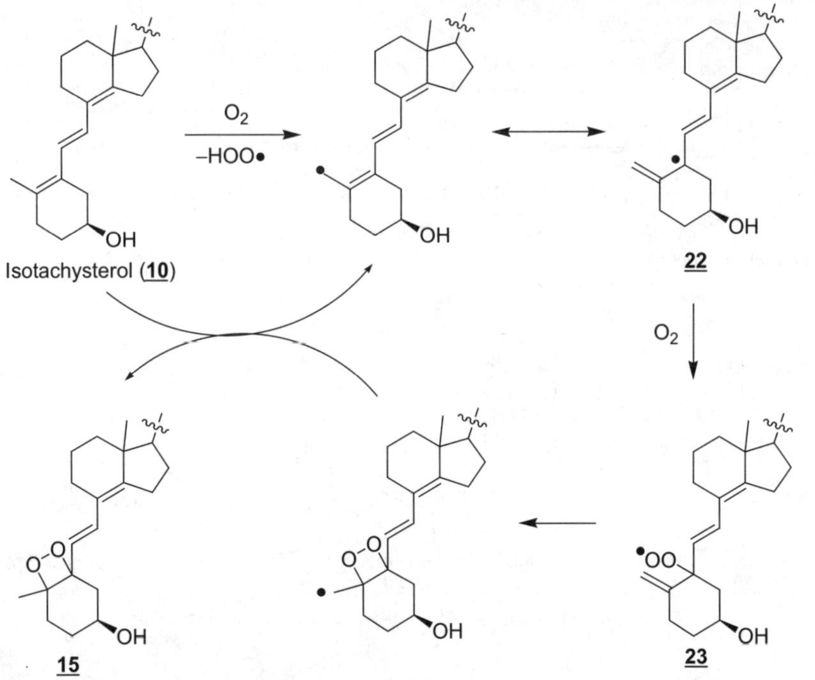

Isotachysterol (10) · 22 · 23 · 15 · O_2 · $-HOO\bullet$

SCHEME 11.12 An alternative mechanism for the autoxidation of isotachysterol (reproduced with permission from Ref. 74).

yielding the cyclic semiketal **16** (Scheme 11.12). Another possible initiation step might be the direct hydrogen abstraction by oxygen from allylic positions [106], preferably at C19 to form the allylic radical **22**, which reacts with oxygen to form the peroxy radical **23** (Scheme 11.12). Because it is similar to **19**, it can follow similar follow-up processes as mentioned above to give the products as exemplified in Scheme 11.12. In conclusion, this work demonstrates that despite the relative stability of vitamin D_3 at ambient temperatures, its acid-catalyzed isomerization product, isotachysterol, is liable to autoxidation to form a variety of oxidation products. The formation of these oxidation products is interesting since they are formed in the dark and in the absence of any other oxidants and/or initiators apart from atmospheric oxygen. Other oxides of vitamin D_3 derivatives reported previously were all prepared by chemical and photochemical oxidations [69–72]. Since isotachysterol is the acid-catalyzed isomerization product of vitamin D_3 and also can be formed in the presence of acidic vitamins such as ascorbic acid and folic acid [107], a similar autoxidation reaction might also take place in living systems and have biological significance.

11.4.3 Thermal Degradation of Ecalcidene

The thermal degradation of ecalcidene was performed by incubating solutions of ecalcidene in dimethylsulfoxide (DMSO) for 0.5 h at 25°C, 100°C, 140°C, 150°C,

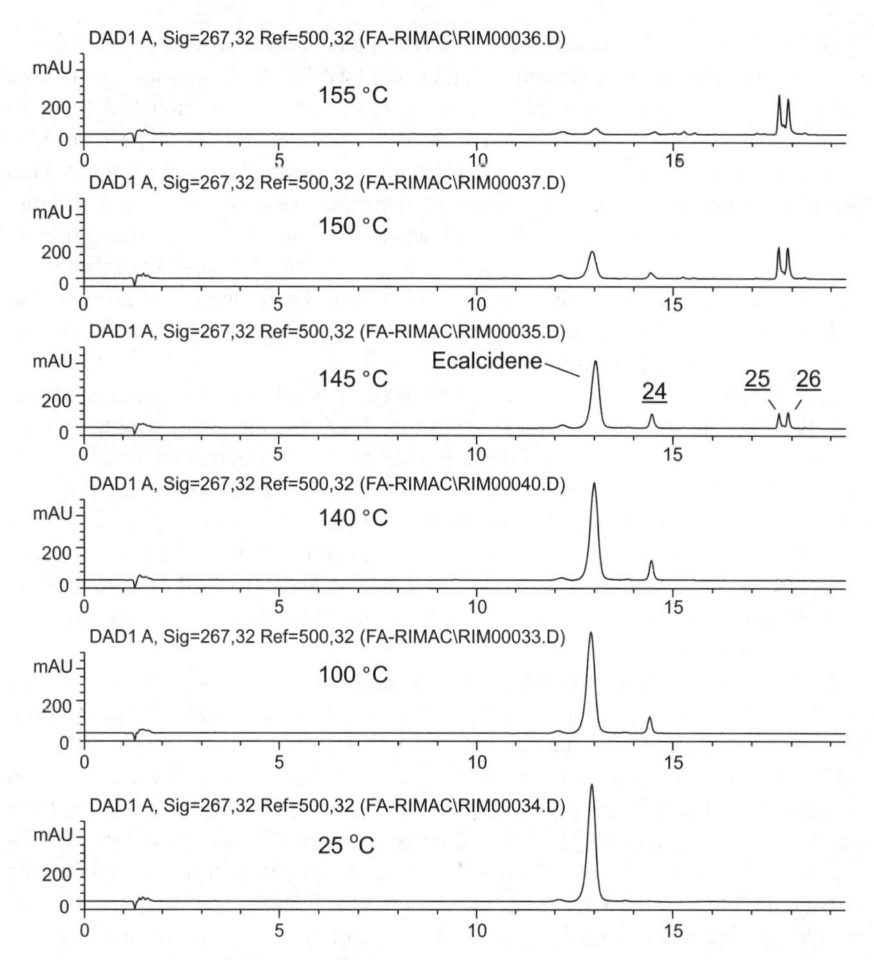

FIGURE 11.9 HPLC-UV chromatograms of ecalcidene in DMSO for 0.5 h (reproduced with permission from Ref. 86).

and 155°C [86]. The HPLC-UV chromatograms of the reaction solutions are shown in Figure 11.9. At temperatures of 140°C or lower, compound **24** was observed as the major degradation product. At temperatures higher than 140°C, two additional major degradation products, **25** and **26**, were generated. Figure 11.9 indicates that the yields of **24**–**26** were very dependent on temperature. HPLC-APCI mass spectrometric studies revealed that, as expected, ecalcidene exhibits ions at *m/z* 456 $[M + H]^+$, 438 $[M + H - H_2O]^+$, and 420 $[M + H - 2H_2O]^+$, which correspond to the protonated molecular ion and the ions losing one or two water molecules. The UV spectrum of ecalcidene shows absorption at λ_{max} 267 nm, due to the 5,7-diene system, This is consistent with the well-known fact that the 5,7-diene nucleus in vitamin D_2 or D_3 has a similar UV absorption at λ_{max} 265 nm [108]. The APCI mass spectrum for degradation product **24** exhibits ions at 456 $[M + H]^+$, 438 $[M + H - H_2O]^+$, and

420 $[M + H - 2H_2O]^+$ similar to those of ecalcidene, indicating that **24** is an isomer of ecalcidene. To obtain the structural details of **24**, HPLC-NMR measurements were performed. By comparing the ^1H NMR spectrum of **24** with solution NMR data of previtamin D_3 [109], it is proposed that **24** is a previtamin D_3–type isomer of ecalcidene. The UV spectrum of **24** exhibited a λ_{max} of 263 nm, which has a 4 nm blueshift compared with the absorption of the parent ecalcidene λ_{max} of 267 nm. Similar phenomena were also observed when comparing the UV absorption of previtamin D_3 to that of vitamin D_3. Previtamin D_3 and vitamin D_3 exhibit UV absorptions at 260 and 265 nm, respectively [110]. The structure of **24** was also confirmed by its unfavorable reactivity toward maleic anhydride, which will be discussed in following sections.

Further, HPLC-NMR experiments indicated that **24** was not stable and can partially revert to the parent ecalcidene in the HPLC mobile phase. Compound **24** was separated by HPLC and trapped in the NMR probe at room temperature followed by periodic ^1H NMR measurements. The ^1H NMR spectrum of **24** after 4 days of separation clearly indicates the appearance of the resonances associated with ecalcidene [i.e., the resonances at 6.28 (d, $J = 11.2$ Hz, H6), 5.96 (d, $J = 11.1$ Hz, H7), 5.19 (s, H19), and 4.81 (s, H19)). In contrast, the ^1H NMR spectrum of **24** obtained immediately after HPLC separation did not show the presence of ecalcidene at significant levels.

The APCI mass spectra of both **25** and **26** exhibit ions at *m/z* 438 $[M + H]^+$ and 420 $[M + H - H_2O]^+$. Compound **25** exhibits UV absorption peaks at λ_{max} 300 nm, 314 nm, and 328 nm. Compound **26** exhibits UV absorption peaks at λ_{max} 302, 316, and 332 nm. The fact that the UV and MS spectra of **25** is quite similar to those of **26** suggests that **25** and **26** are a pair of dehydrated isomers of ecalcidene. It has been reported that 3-acetoxyergosta-3,5,7,22-tetraene (Scheme 11.13) exhibits UV absorption at λ_{max} 303, 316, and 331 nm [111]. These three UV absorption peaks must be contributed solely by the 3,5,7-triene moiety since the isolated 3-acetoxy group or the isolated double bond at position 22 or the rest of the molecule should have no contribution to the UV spectrum at a wavelength higher than 250 nm, according to the well-known Woodward–Fieser rules [112]. The fact that the UV absorption pattern of

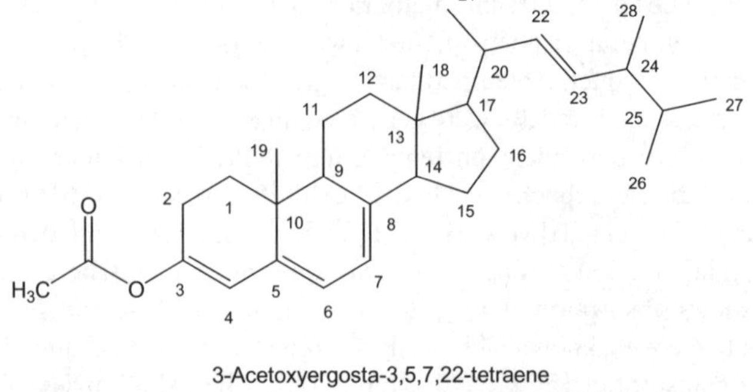

3-Acetoxyergosta-3,5,7,22-tetraene

SCHEME 11.13 Structure of 3-acetoxyergosta-3,5,7,22-tetraene.

3-acetoxyergosta-3,5,7,22-tetraene is quite similar to the UV spectra of **25** and **26** suggests that both **25** and **26** may contain a similar 3,5,7-triene moiety. Accordingly, **25** and **26** respectively are proposed to be the dehydrated pyrocalciferol analog and the dehydrated iso-pyrocalciferol analog of ecalcidene. Additional work is still needed to distinguish the absolute structures of **25** and **26**. The proposed structures and generation mechanism of **24**–**26** are shown in Scheme 11.14.

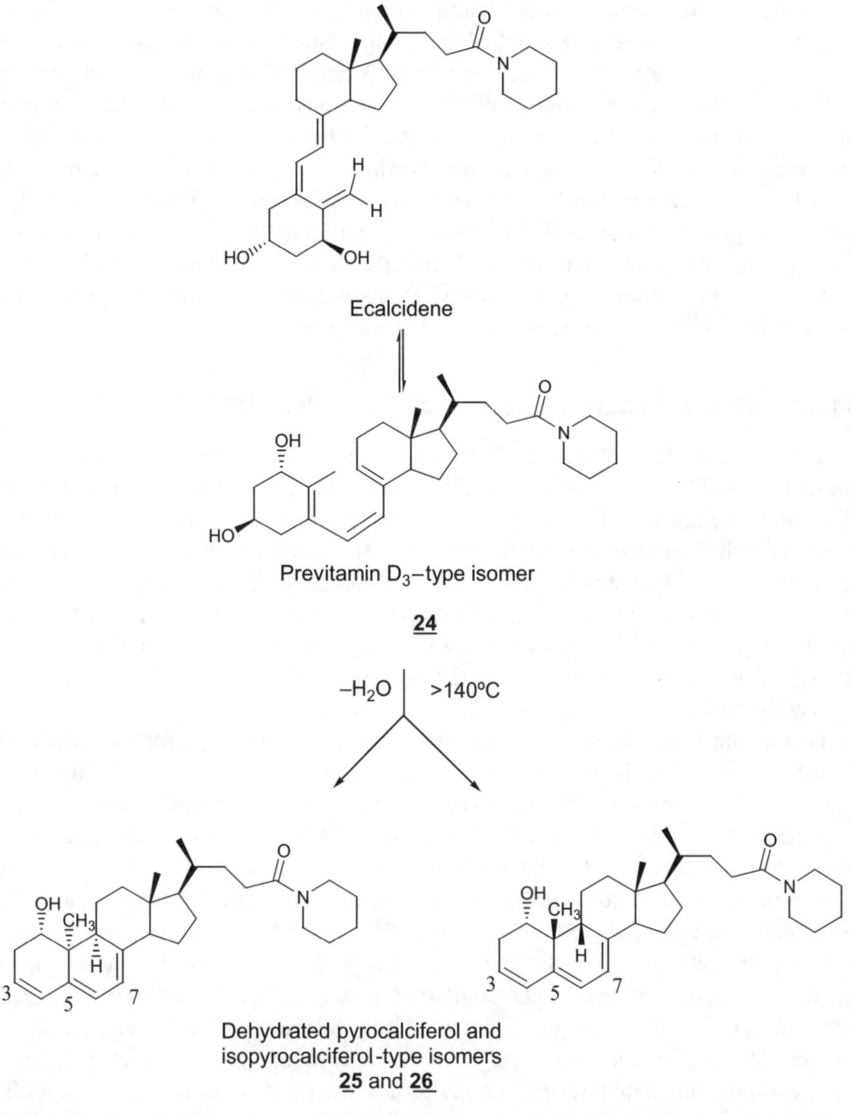

Ecalcidene

Previtamin D_3–type isomer

24

$-H_2O$ >140°C

Dehydrated pyrocalciferol and isopyrocalciferol-type isomers

25 and 26

SCHEME 11.14 Proposed thermal degradation mechanism of ecalcidene (reproduced with permission from Ref. 86).

Ecalcidene is first reversibly converted to the corresponding previtamin D–type isomer **24**, which then generates **25** and **26** by cyclization and dehydration on the B ring at elevated temperatures. As shown in Figure 11.9, increasing the temperature resulted in increase of peaks **25** and **26** along with decrease of the ecalcidene peak. Peak **24** increased first and then decreased, as would be expected with the mechanism depicted in Scheme 11.14, where **24** behaves as an intermediate generated from ecalcidene and then further reacts to produce the dehydrated pyrocalciferol analog **25** and the dehydrated isopyrocalciferol analog **26** at elevated temperature.

It is well known that in solution vitamin D_3 can reversibly convert to previtamin D_3, which can subsequently be rearranged to form pyrocalciferol and isopyrocalciferol at elevated temperatures by cyclization of the B ring [60–62] without dehydration (Scheme 11.8). It is not unreasonable to expect that ecalcidene, a close analog of vitamin D_3, has a similar thermal degradation pathway but is accompanied by an additional dehydration process to form a dehydrated pyrocalciferol-type isomer and a dehydrated isopyrocalciferol-type isomer, namely, **25** and **26** (Scheme 11.14). The occurrence of dehydration with **24**, instead of previtamin D_3, at elevated temperature, may be due to the structural difference between ecalcidene and vitamin D_3. It is likely that the electron-withdrawing 1-hydroxyl group of **24** may facilitate the dehydration of the 3-hydroxyl group to form a favorable triene system.

11.4.4 Acid-Induced Degradation of Ecalcidene [86]

The HPLC-UV chromatograms of ecalcidene in 1 : 1 mixture of acetonitrile and phosphate buffer with varying pH values after 5 days at room temperature are illustrated in Figure 11.10. At pH $\geq$7, only the previtamin D_3–type isomer **24** was observed to be the major degradation product. At pH < 5, a new degradation product (**27**) was formed. Also at pH 1, more minor degradation products eluting at 17–19 min were observed, and the peak at $t_R \sim 1$ min was caused by the buffer. An APCI mass spectrum of **27** exhibits ions at *m/z* 438 $[M + H - H_2O]^+$ and 420 $[M + H - 2H_2O]^+$. The molecular weight of **27** was determined to be 455 by ESI-MS, indicating that **27** is a isomer of the parent ecalcedene.

Compound **27** was then isolated through semipreparative HPLC followed by NMR investigations. It was identified to be an isomer of ecalcidene with C1–C9 hydroxyl migration. The structure of **27** and a postulated generation mechanism are presented in Scheme 11.15. The absolute configuration at C9 remains to be determined. As postulated in Scheme 11.15, the previtamin D_3–type isomer **24** might be generated first from ecalcidene and subsequently converted to the tachysterol-type isomer **28**. Acid facilitates the dehydration on C1 of **28** to yield the cation **29** to which a nucleophilic addition of H_2O may occur to produce **27**. The C9-hydroxylated tachysterol-type structure of **27** suggested that a tachysterol-type intermediate (**28**) may exist in this acid-induced degradation reaction. Alternatively, the dehydration of C1 and addition of H_2O to C9 may occur in a concerted manner. Figure 11.10 shows that ecalcidene gives only the previtamin D_3–type isomer **24** as the major product and that its level remains the same in pH $\geq$7 solutions. This indicated that only thermal isomerization of ecalcidene occurred under neutral or

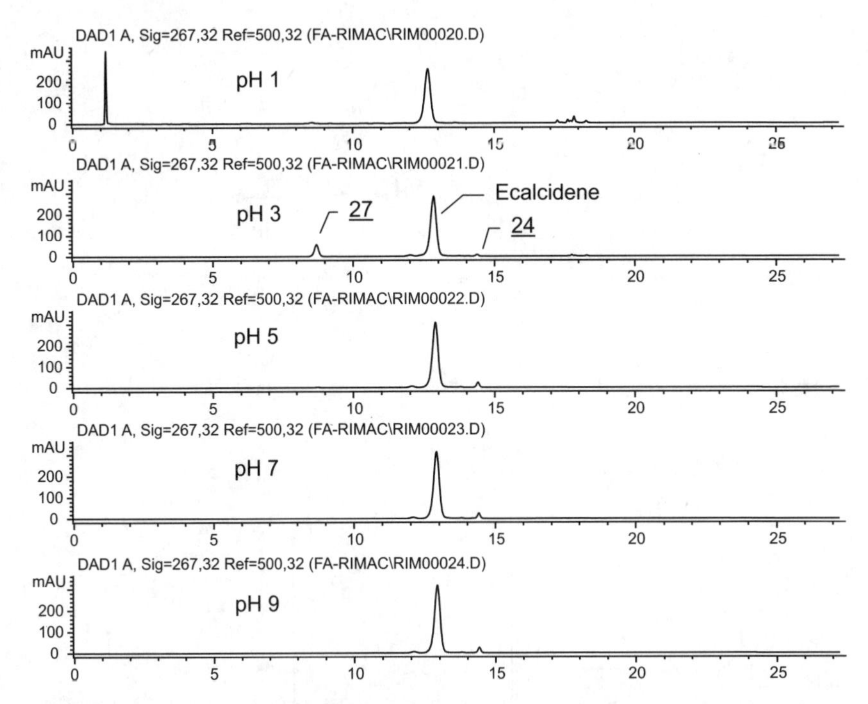

FIGURE 11.10 HPLC-UV chromatograms of ecalcidene at different pHs after 5 days at room temperature (reproduced with permission from Ref. 86).

basic conditions. When the pH of the solution was decreased to below <5, the level of **24** decreased and **27** was observed, indicating that **24** further reacted in acidic media and served as the precursor of **27**. At pH 3, **27** became the major degradation product. Further decreasing the pH to 1 resulted in the disappearance of **27** and **24** along with the formation of other products that eluted within 17–19 min, suggesting that additional reactions of **27** occurred in more acidic solutions. This observation is in agreement with the proposed mechanism that **24** is an intermediate for the formation of **27** as depicted in Scheme 11.15.

The acid-induced degradation of vitamin D_3 is an important aspect of vitamin D chemistry. However, there are only very limited reports regarding this process. It was reported that vitamin D_3, previtamin D_3, and tachysterol can be converted to isotachysterol in various acidic conditions such as HCl, BF_3, and H_3PO_3 [63–68]. It was reported that 25-hydroxyvitamin D_3 or 24,25-dihydroxyvitamin D_3 could also be transformed into the corresponding isotachysterols by treatment with hydrochloric acid [67]. However, when 1-hydroxy vitamin D_3 analogs were treated with acidic reagents, they were either completely destroyed or yielded a number of unknown products [68]. To the best of our knowledge, there were been no prior reports of the degradation products or the reaction mechanisms, and this was the first report of a degradation product that was isolated and structurally identified from the acid-induced degradation of a 1-hydroxlated vitamin D_3 analog. The established

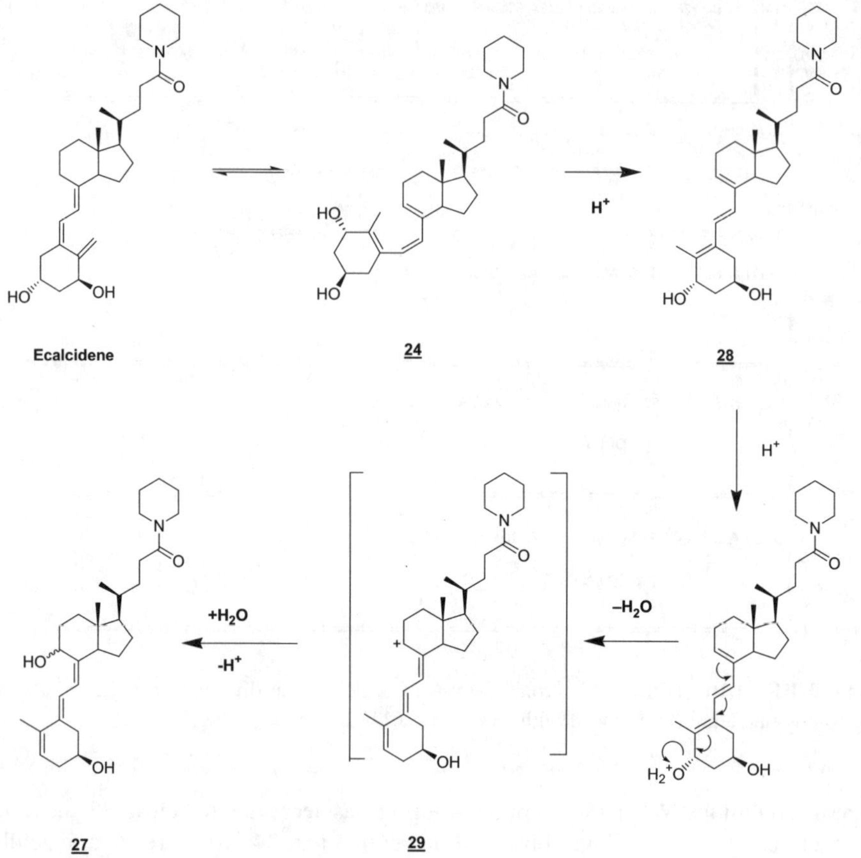

SCHEME 11.15 Proposed acid-induced degradation mechanism of ecalcidene (reproduced with permission from Ref. 86).

C9-hydroxylated tachysterol-type structure of **27** can be considered as a water-trapped tachysterol-type intermediate generated during the acid-induced degradation. Scheme 11.15 could serve as a new reaction model for vitamin D_3, and it also suggests that acid-induced formation of isotachysterol from vitamin D_3 may sequentially proceed through the intermediates of previtamin D_3 and tachysterol.

11.4.5 Iodine-Induced Degradation of Ecalcidene

Iodine is an important element in biological systems. It plays a central role in thyroid physiology, as it is both a major constituent of thyroid hormones and a regulator of thyroid gland function [113]. Iodine disorders induce biological and/or clinical expressions of thyroid dysfunction, and in some cases can disclose preexisting thyroid abnormalities. Investigation of the interaction between iodine and ecalcidene may shed light on the metabolism and toxicity of ecalcidene for its potential therapeutic application. Similar to vitamin D_3, ecalcidene has a *cis* configuration in terms of its triene system. In this section, the iodine induced degradation of ecalcidene and its previtamin D_3–type isomer **24** is summarized [86].

11.4.5.1 cis/trans-Isomerization of Ecalcidene

The iodine-induced degradation of ecalcidene was performed by addition of iodine (25 μg/mL) into a solution of ecalcidene (0.21 mg/mL) in DMSO. The reaction solution was incubated at room temperature for more than 30 min and then monitored using HPLC-UV and HPLC-APCI-MS. The HPLC-UV chromatogram (Figure 11.11 a) of the solution of ecalcidene without iodine exhibits only the ecaldidene peak. The addition of iodine to the ecalcidene solution resulted in the formation of the new product **30** (Figure 11.11 b). The APCI mass spectrum of **30** gave a predominant ion at *m/z* 456 $[M + H]^+$ indicating that **30** is an isomer of ecalcidene. The HPLC-^{1}H NMR investigation on **30** was performed. Compound **30** was identified to be the *epi–trans*-isomer of ecalcidene by comparing the HPLC- ^{1}H NMR data of **30** with NMR data of *epi–trans*-vitamin D_3 [114].

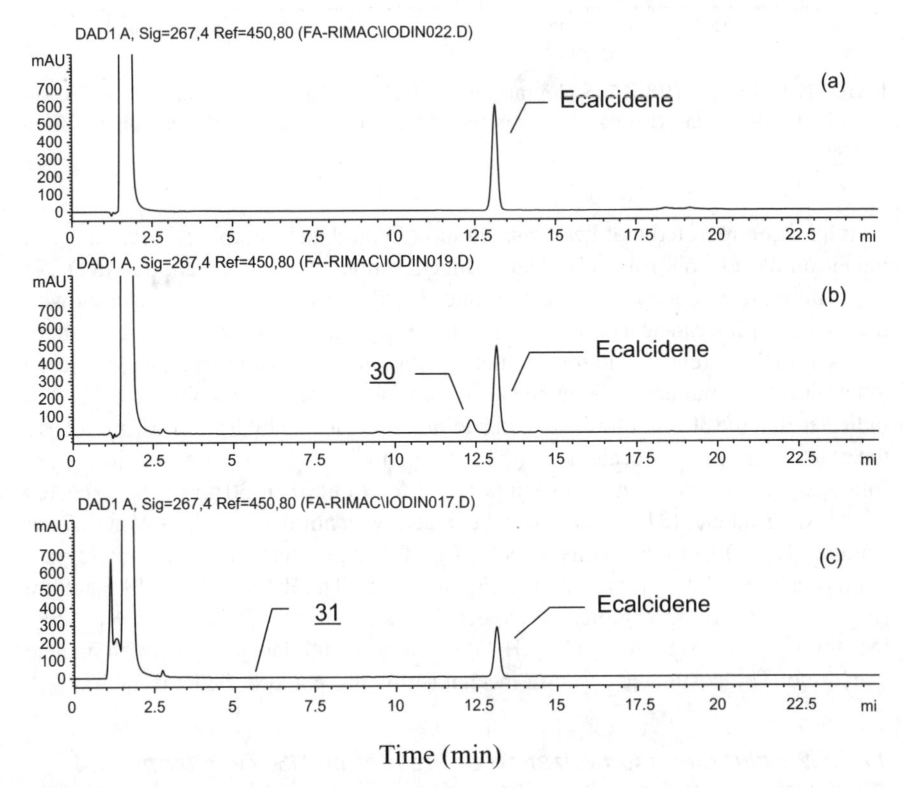

FIGURE 11.11 HPLC-UV (267 nm) chromatograms of (a) ecalcidene, (b) ecalcidene + iodine, and (c) ecalcidene + iodine + maleic anhydride (reproduced with permission from Ref. 86).

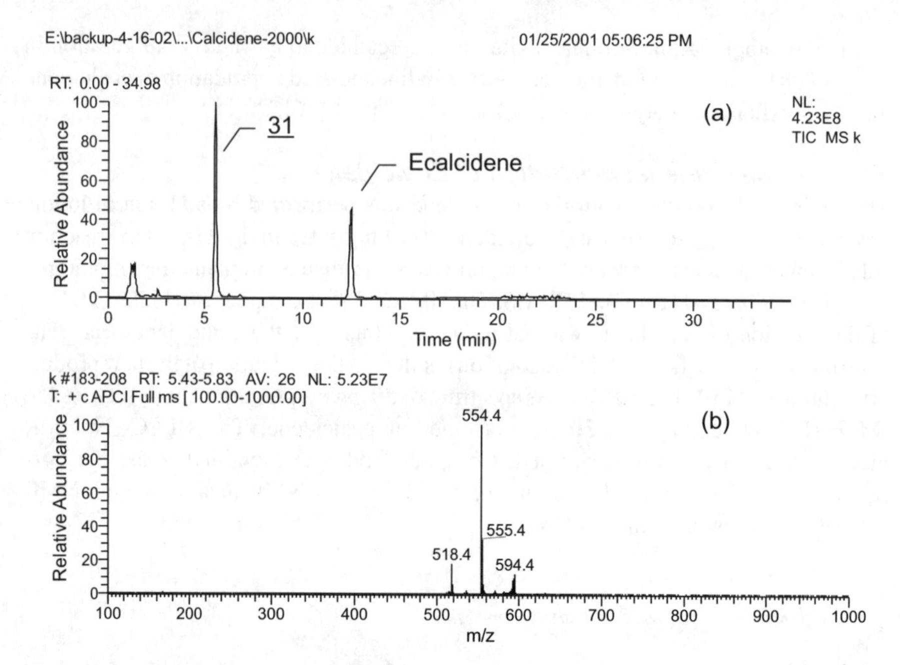

FIGURE 11.12 (a) HPLC-MS chromatogram (TIC) of ecalcidene + iodine + maleic anhydride in DMSO and (b) mass spectrum of adduct 31 (reproduced with permission from Ref. 86).

It has been reported that *epi–trans*-vitamin D_3 and tachysterol react rapidly with maleic anhydride to form Diels–Alder adducts, while vitamin D_3 (*cis*-vitamin D_3) or previtamin D_3 have very slow reaction rates [115]. Accordingly, derivatization with maleic anhydride can be used to distinguish the *epi–trans*-vitamin D_3 and tachysterol forms from the parent *cis*-vitamin D_3 and previtamin D_3 forms. This strategy was used to confirm the structure of **30** as the *epi–trans*-isomer of ecalcidene. Figure 11.11c indicates that the **30** was eliminated on addition of maleic anhydride (1.37 mg/mL) to the reaction mixture of ecalcidene (0.21 mg/mL) with iodine (25 μg/mL) in DMSO followed by incubation at room temperature for more than 30 min. The expected Diels–Alder adduct (**31**) of **30** with maleic acid was not detected by UV at 267 nm (Figure 11.11 c), but was readily detected by APCI-MS. The total-ion chromatogram of this reaction solution is presented in Figure 11.12a. The HPLC-APCI-MS spectrum (Figure 11.12b) of **31** exhibits the expected ions at *m/z* 594 $[M + CH_3CN]^+$, 554 $[M + H]^+$, and 518 $[M + H - 2H_2O]^+$ for a 1 : 1 adduct of **30** with maleic anhydride. The postulated reaction mechanism is presented in Scheme 11.16.

11.4.5.2 cis/trans-Isomerization of Previtamin D3–Type Isomer 24

The previtamin D_3–type isomer **24** was generated by thermal degradation of ecalcidene. A solution of ecalcidene (0.21 mg/mL) in DMSO was maintained at 140°C for 30 min. The HPLC-UV chromatogram of this reaction solution, presented

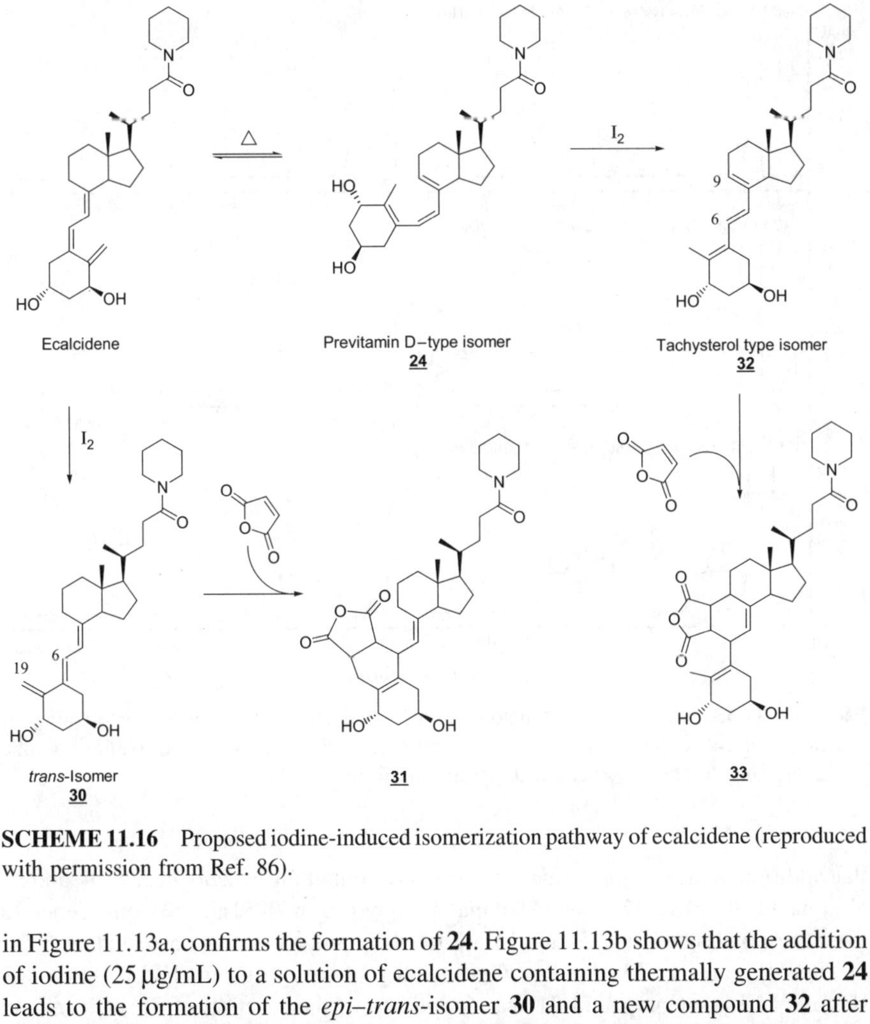

SCHEME 11.16 Proposed iodine-induced isomerization pathway of ecalcidene (reproduced with permission from Ref. 86).

in Figure 11.13a, confirms the formation of **24**. Figure 11.13b shows that the addition of iodine (25 μg/mL) to a solution of ecalcidene containing thermally generated **24** leads to the formation of the *epi–trans*-isomer **30** and a new compound **32** after incubation at room temperature for more than 30 min. Compound **32** must be produced from **24** since ecalcidene generates **30** only in presence of iodine. The structure of **32** is proposed to be a tachysterol-type isomer of ecalcidene, that is, a *trans*-isomer of **24**. The UV spectrum of **32** exhibits an absorption peak at 281 nm with shoulders at 270 and 296 nm. This is similar to the UV of tachysterol, which is also a *trans*-isomer of previtamin D_3, with an absorption peak at 281 nm and shoulders at 272 and 289 nm [110]. The mass spectrum of **32** shows ions at *m/z* 456 $[M + H]^+$, 438 $[M + H - H_2O]^+$, and 420 $[M + H - 2H_2O]^+$, revealing that 32 is an isomer of **24** and ecalcidene. The structure of **32** and a proposed formation pathway are presented in Scheme 11.16. De Vries et al. reported that besides *epi–trans*-vitamin D_3, tachysterol also has a high reactivity toward maleic anhydride [115]. The structure of **32**, a tachysterol-type isomer, was confirmed by using the derivatization with maleic anhydride, as was used to confirm the structure of **30**. Figure 11.13c indicates

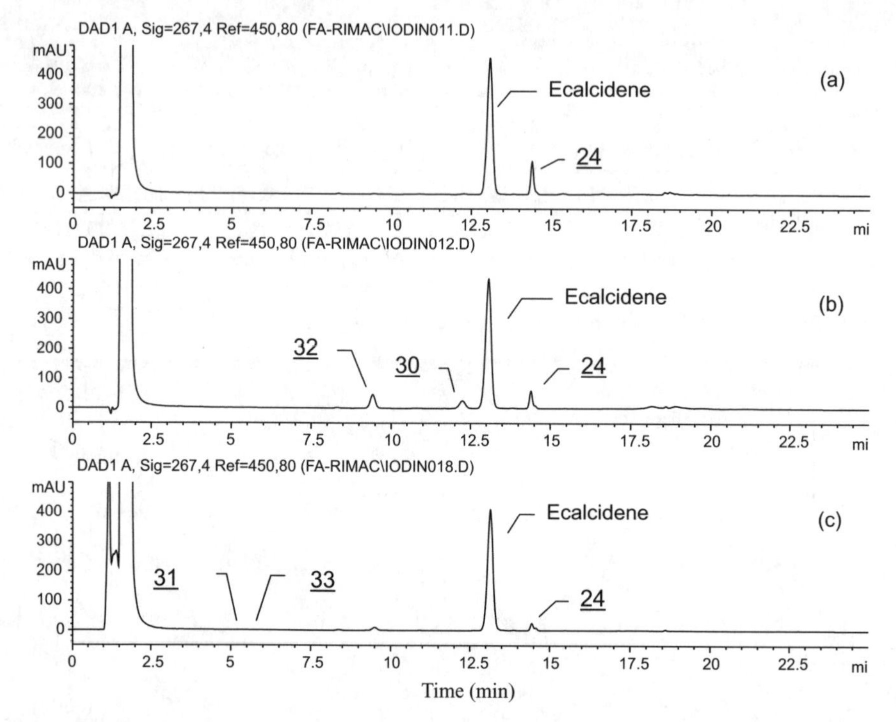

FIGURE 11.13 HPLC-UV chromatograms of (a) ecalcidene at 140°C for 30 min, (b) ecalcidene at 140°C for 30 min + iodine, and (c) ecalcidene at 140°C for 30 min + iodine + maleic anhydride (reproduced with permission from Ref. 86).

that addition of maleic anhydride to the mixture containing ecalcidene, **30**, **24**, and **32** eliminated **30** and **32**. The reaction of maleic anhydride with **30** and **32** would generate the corresponding adducts **31** and **33**. Although **31** and **33** were not detected by UV at 267 nm (Figure 11.13c), they were readily detected using HPLC-APCI-MS (Figure 11.14a). Compound **31** was formed from the reaction of **30** with maleic anhydride, and accordingly **33** must be formed from the reaction of **32** with maleic anhydride. The mass spectra (Figure 11.14b,c) of **31** and **33** exhibit ions at *m/z* 594 ($M^+ + CH_3CN$), 554 (MH^+), and 518 ($MH^+ - 2H_2O$), which are consistent with a 1 : 1 adduct of **30** or **32** with maleic anhydride. The postulated structures of **31** and **33** and formation pathway are presented in Scheme 11.16. Since the Diels–Alder reaction requires that the conjugated diene system in the dienophile be in an easily accessible cisoid conformation [116], maleic anhydride added to the C6 and C19 positions of **30** to form **31**, but added to C6 and C9 of **32** to form **33**. By similar reasoning, ecalcidene, vitamin D_3, previtamin D_3, and the previtamin D_3–type isomer **24** have low reactivity toward maleic anhydride. That ecalcidene, which has a *cis* configuration, and its corresponding previtamin D_3–type isomer **24** can be converted by iodine to *epi–trans*-ecalcidene **30** and the tachysterol-type isomer **32** is consistent with the well-known fact that vitamin D_3, which also has a *cis* configuration, and

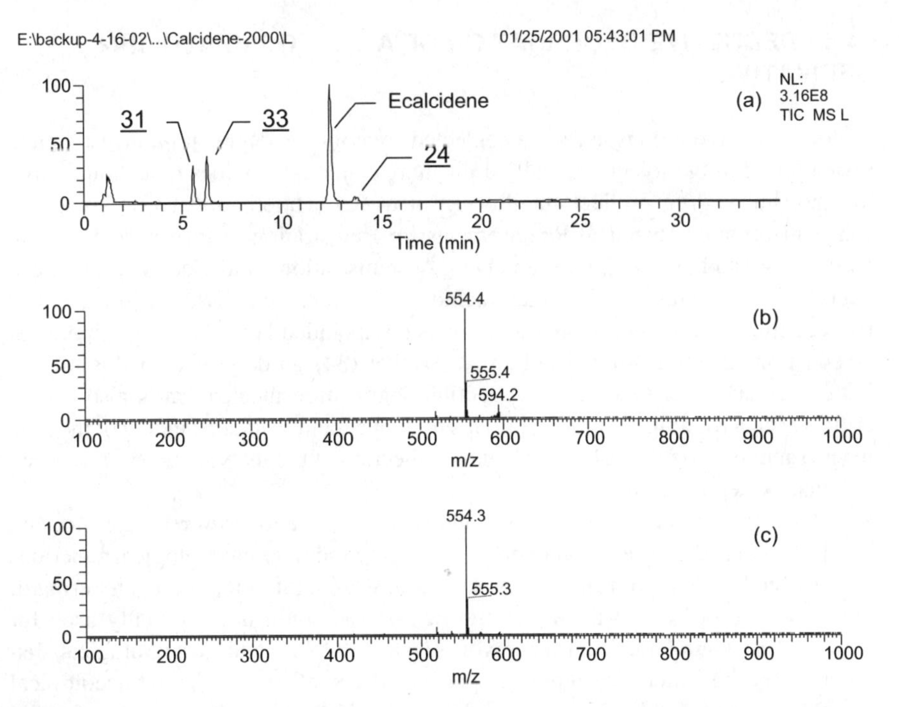

FIGURE 11.14 HPLC-MS chromatogram (a) of ecalcidene in DMSO at 140°C for 30 min + iodine + maleic anhydride and APCI mass spectra of **31** (b) and **33** (c) (reproduced with permission from Ref. 86).

previtamin D_3 can be converted by iodine to the corresponding *epi–trans*-vitamin D_3 and tachysterol [63].

As a summary of the thermal, acid induced, and iodine induced degradation of ecalcidene, a 1-hydroxylated analog of vitamin D_3, the following discovery was achieved. Like vitamin D_3, ecalcidene has a reversible isomerization to the corresponding previtamin D_3 type isomer **24**, which can subsequently be transformed into the pyro and isopyro isomers at elevated temperatures via cyclization of the B ring. However, this was accompanied by an unexpected dehydration of the 3-hydroxyl group. The presence of the 1-hydroxyl moiety may have been responsible for this since the 3-hydroxyl group in vitamin D remains intact during a similar thermal transformation [60–62]. In aqueous acidic media, ecalcidene underwent a novel C1–C9 hydroxyl migration. That this occurred, possibly via a tachysterol-type intermediate, would be the first observation of this new type of reaction so far for 1-hydroxy vitamin D analogs. Ecalcidene and its corresponding previtamin D–type isomer can undergo *cis/trans* conversion by interaction with iodine, which is similar to the behavior of vitamin D_3 [19]. A combination of various techniques, including online HPLC-UV, HPLC-MS, HPLC-NMR, maleic anhydride chemical derivatization, and sample isolation followed by offline spectrometric analyses, were successfully applied to determine the degradation products and mechanisms.

11.5 REDUCTIVE DEGRADATION OF A 1,2,4-THIADIAZOLIUM DERIVATIVE

Reductive degradation is usually a neglected area of drug degradation investigation. Oxidation of drug molecule usually draw more concern than does reduction in drug metabolism or drug product stability. Even in ICH (International Conference on Harmonization of Technical Requirements for Registration of Pharmaceuticals for Human Use) and FDA (US Food and Drug Administration) guidelines on drug forced degradation or stress testing that include oxidation, acid–base hydrolysis, and photodegradation, reduction investigation is not included [117–119]. Our finding of a reduction of a 1,2,4-thiadiazolium derivative (**34**) as described in this section indicated that understanding the reduction degradation mechanism is also truly a critical aspect for drug development. This finding benefited understanding of the drug in vitro and in vivo metabolism and guided selection of bioanalytical marker for drug pharmacokinetic evaluation.

1,2,4-Thiadiazolium derivatives could serve as melanocortin receptor modulators [120,121]. Because melanocortin receptors modulate physiological functions such as feeding behavior, nerve regeneration, drug addiction, and even dermatological conditions, 1,2,4-thiadiazolium derivatives could be potentially used for treatment of metabolic, central nervous system (CNS), and dermatological disorders. Despite numerous reports regarding the synthesis and pharmacological investigations on 1,2,4-thiadiazolium derivatives [120–125], there are very limited reports on the chemical properties of this type of compound. In fact, pharmaceutical properties, including chemical properties of new chemical entities (NCEs), are one of the major factors that impact their potential for development and attrition rate in the pharmaceutical development processes. Understanding the chemistry of NCEs become critical to facilitate those processes. Kurzer et al. [126] reported the isomerization of 2-aryl-5-arylamino-3-arylimino-Δ^4-1,2,4-thiodiazolines to 2-guanidinobenzothiazoles. Other authors [121] found that 2-(4-methylphenyl)-3-phenyl-5-(4-methoxyphenylamino)-[1,2,4] thiadiazolium bromide could be reduced to 1-(4-methoxyphenyl)-3-[N-(4-methylphenyl)benzimidoyl]-2-thiourea by dithioerythritol in dichloromethane. It was also reported [127] that 2,3-bis-(2-methoxy-phenyl)-5-phenylamino-[1,2,4]-thiadiazolium bromide (**34**), which is a 1,2,4-thiadiazolium derivative, could undergo a thermally promoted rearrangement reaction in solution to form the corresponding 2-amidinobenzothiazole (**35**). We found that **34** could be reduced to the corresponding imidoylthiourea (**36**) by 2-thioethanol and some biologically interesting reducing reagents, including glutathione, cysteine, and ascorbic acid [128]. The reduction also occurred through incubation of **34** with Sprague–Dawley rat and Yorkshire swine plasma. Both of the thermal arrangement and reduction are chemical reactions and nonenzymatic processes (Scheme 11.17).

The reaction of 2,3-bis(2-methoxy-phenyl)-5-phenylamino[1,2,4]-thiadiazolium bromide (**34**) with 2-thioethanol was also discovered to be a convenient method for synthesis of 1-[(2-methoxyphenyl)-(2-methoxyphenylimino)methyl]-3-phenylthiourea (**36**). 2-Thioethanol serves as a reducing reagent and reaction solvent. Compound

35 34 36

SCHEME 11.17 Thermoreductive degradation of **34**.

34 was dissolved in 2-thioethanol and incubated at room temperature for about 10 min followed by addition of water to generate precipitate **36** followed by HPLC-UV, ESI-MS, and NMR characterization. Compound **36** exhibits a positive ion at *m/z* 392 that is the protonated molecular ion. MS^n ($n=1,2,3$) (Figure 11.15) indicated that the ion at *m/z* 392 generates an ion at *m/z* 257, which subsequently generates the ion at *m/z* 240. The proposed MS fragmentation pathway is presented in Scheme 11.18. ^1H, ^{13}C, $^1H-^1H$ homonuclear chemical shift correlation (COSY), $^1H-^{13}C$

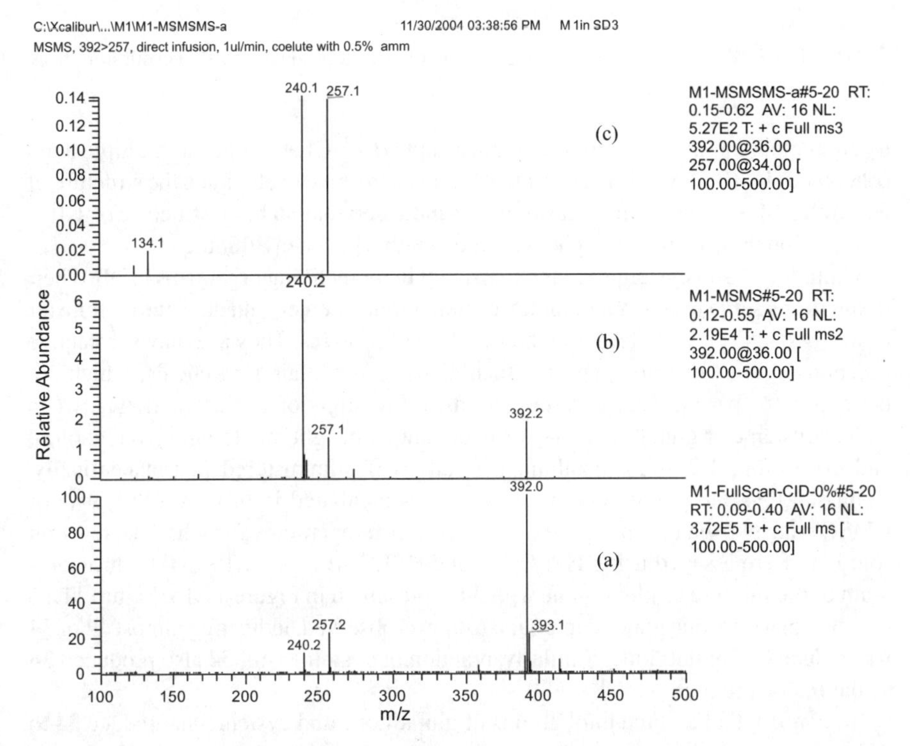

FIGURE 11.15 MS^n [$n=1$(a), 2(b), 3(c)] spectra of **36** (reproduced with permission from Ref. 128).

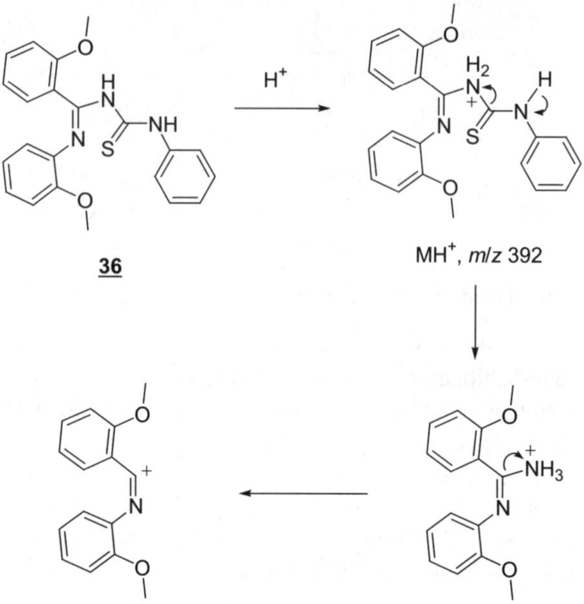

SCHEME 11.18 MS fragmentation pathway of **36** (reproduced with permission from Ref. 128).

heteronuclear multiple quantum coherence, and $^{1}H-^{13}C$ heteronuclear multiple bond coherence (HMBC) NMR measurements were performed to elucidate the structure of **36**. Both ESI-MS and NMR measurements indicated that **36** has a structure of 1-[(2-methoxy-phenyl)-(2-methoxyphenylimino)-methyl]-3-phenylthiourea.

Glutathione and cysteine occur extensively in many living organisms [129]. Their thiol groups serve as a metal complex center, a redox center, and an electron transfer center, and also play vital roles in the structure of proteins. They also have protective functions against oxidative species such as oxygen-containing radicals, which are believed to be partially cause certain diseases. Investigation of the reaction between **34** and cysteine or glutathione may provide unique insight into the in vivo fate of **34** and even other 1,2,4-thiadiazolium derivatives if administered pharmaceutically. Compound **34** and glutathione or cysteine was incubated in mixture (3 : 7, v/v) of DMSO and phosphate buffer (pH 7.4) at room temperature for 4 h. The reaction solutions were assayed using HPLC-UV and HPLC-MS. The HPLC-UV chromatogram of the mixture of glutathione with **34** is presented in Figure 11.15. Figure 11.16 revealed that only one major compound (**36**) was observed, indicating almost all of **34** was reduced by glutathione. Similarly, reaction of cysteine with **34** also produced **36** as the major product.

To confirm that it is the thiol groups of glutathione and cysteine that reduce **34** to **36**, the reaction solution described in Figure 11.16 was incubated with addition of maleimide, which is a well-known thiol-trapping reagent by forming 1 : 1 adduct [130]

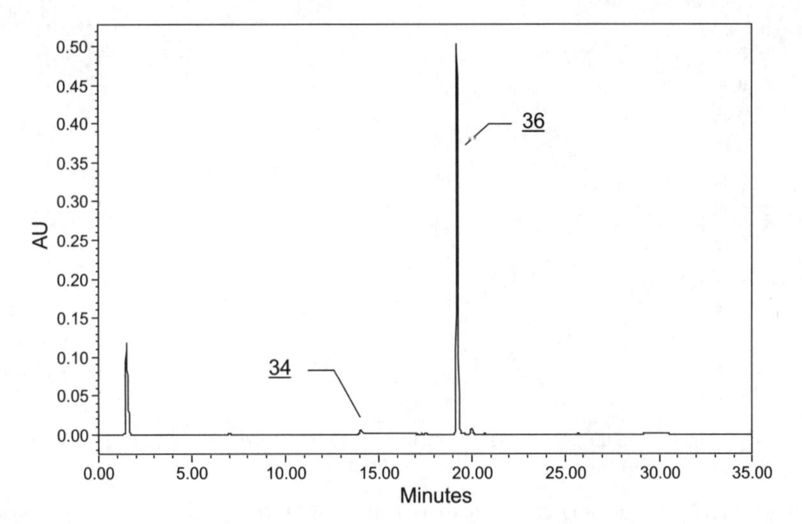

FIGURE 11.16 Chromatograms of reaction solutions of **1** with glutathione (reproduced with permission from Ref. 128).

as described in Scheme 11.19. As an example, the chromatogram of the reaction solution from incubation of **34** with glutathione in the presence of maleimide is presented in Figure 11.17. As expected, in contrast to Figure 11.16, Figure 11.17 revealed that the major peak observed is the starting compound **34**. The peak for

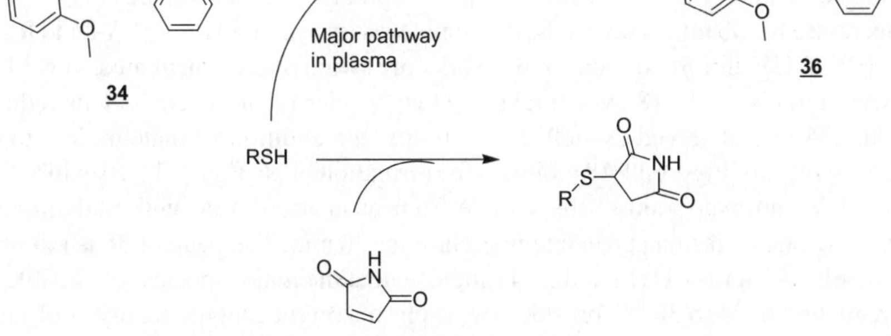

Ascorbic acid and other reducing species

Other products

Major pathway in plasma

34

36

RSH

R = residuals of glutathione, cysteine, peptides/proteins, and other thiol-containing species

SCHEME 11.19 Reduction of **34** (reproduced with permission from Ref. 128).

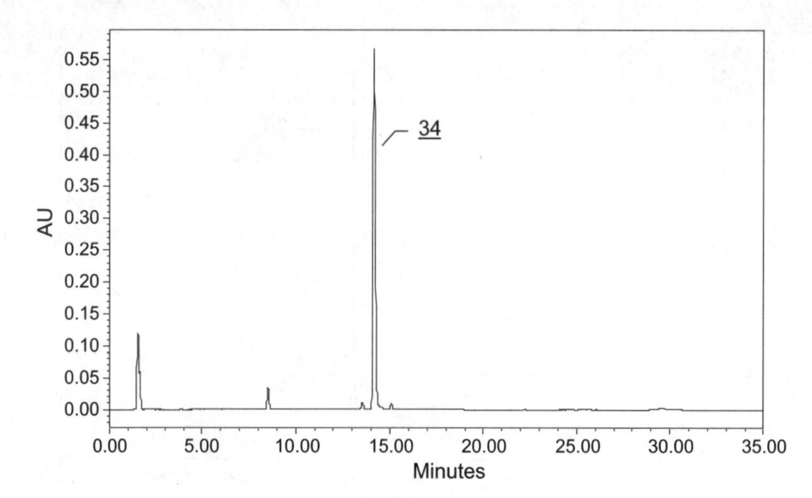

FIGURE 11.17 Chromatogram of reaction solutions of **34** with glutathione in the presence of maleimide (reproduced with permission from Ref. 128).

product **36** was not observed. Incubation of **34** with cysteine in the presence of maleimide resulted in similar phenomena. The aforementioned results indicated that the reactions between **34** and glutathione or cysteine were inhibited by trapping the thiol groups of glutathione or cysteine using maleimide.

Similar to glutathione and cysteine, ascorbic acid (vitamin C) is also one of the most important reducing agents in biochemistry and plays an important role in many biochemical processes [131]. Therefore, it is also important to understand the interaction of ascorbic acid with 1,2,4-thiadiazolium derivatives when they are used for therapeutic applications. Similar to glutathione or cysteine, ascorbic acid could also reduce **34** to **36**, but the reactions were slower and additional products were formed, which remain to be identified.

The metabolism of a drug molecule is a crucial part in its safety assessment. Investigating the stability of a drug molecule in plasma is one of the most common approaches in drug metabolism evaluation. As a preliminary investigation, **34** was incubated with Yorkshire swine plasma or Sprague–Dawley rat plasma at room temperature for 30 min. The resulting sample was assayed by HPLC-UV and HPLC-MS. HPLC-UV chromatograms of the Yorkshire swine plasma incubated with **34** are presented in Figure 11.18, which revealed that besides the unreacted **34**, its reduced product **36** was observed as well. Interestingly, the addition of maleimide into the plasma sample of Figure 11.18 inhibited the formation of **36**. Figure 11.19 exhibits the HPLC-UV chromatogram of the sample from incubation of **34** with maleimide in Yorkshire swine plasma at room temperature for 30 min. The peak of **36** is minimal. The results of Figures 11.18 and 11.19 indicate that the major species responsible for the reduction of **34** to **36** in Yorkshire swine plasma might contain mainly thiol, since maleimide can scavenge the thiol groups and prevent their reactions with **34**. For example, the thiol-containing species in plasma could be cysteine, glutathione, and other peptides or proteins. Similarly, incubation of **34** in Sprague–Dawley rat plasma

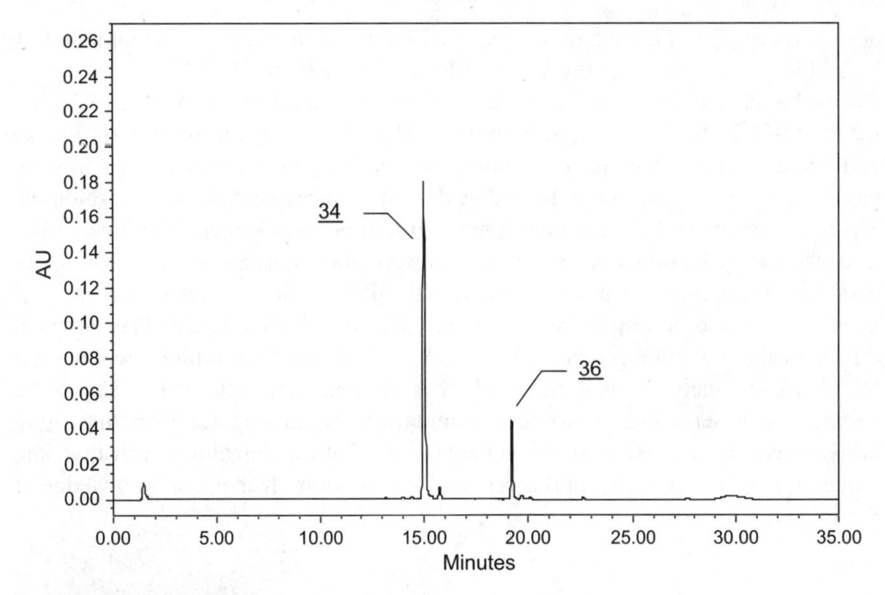

FIGURE 11.18 Chromatograms of Yorkshire swine plasma incubated with **34** (reproduced with permission from Ref. 128).

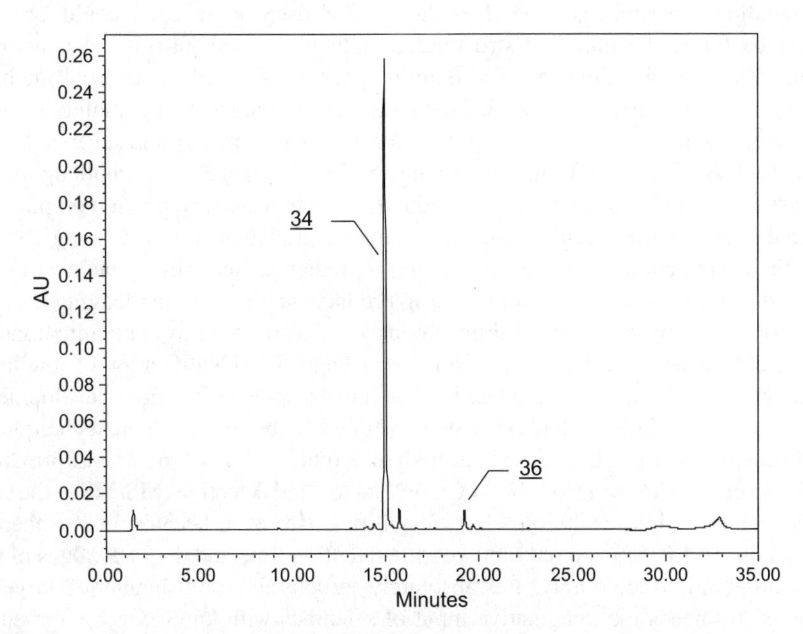

FIGURE 11.19 Chromatograms of Yorkshire swine plasma incubated with **34** and maleimide (reproduced with permission from Ref. 128).

also generated **36** as the major product, and the reaction could also be inhibited by maleimide. The reaction pathways are illustrated in Scheme 11.19.

On the basis of the aforementioned results, 2,3-bis-(2-methoxyphenyl)-5-phenylamino-[1,2,4]-thiadiazolium bromide (**34**), a 1,2,4-thiadiazolium derivative and melanocortin receptor modulator, is sensitive to biologically interesting reducing reagents and could be reduced to the corresponding imidoylthiourea, 1-[(2-methoxyphenyl)-(2-methoxy-phenylimino)methyl]-3-phenylthiourea (**36**). Because the reduction occurs via a nonenzymatic pathway, it is important to understand this reaction prior to traditional P450 studies in metabolism assessment. It will also be important to evaluate pharmacological and/or toxicological effects of the reduction products. In fact, **36** was also used as a biological marker for pharmacokinetic evaluation of **34**. It is obvious that antioxidants should be avoided as preservatives in product formulations containing 1,2,4-thiadiazolium derivatives. The new redox reaction between a 1,2,4-thiadiazolium derivative and reducing reagents observed in this report could also enrich chemical knowledge of the related compounds.

11.6 CONCLUSIONS

In the course of drug development, it is important to understand the intrinsic degradation chemistries of active ingredients in their pure forms or in formulations, including identification of the structures of degradation products and elucidating the degradation mechanisms. The degradation chemistry knowledge could provide guidance for lead compound structural modification, metabolism understanding, formulation optimization, storage condition/package selection, analytical method/sample handling procedure establishment, and drug product stability/safety evaluation to ensure high quality of drug products with desired efficacy and acceptable low toxicity. Degradation and impurity studies become more vital and challenging for semisolid topical pharmaceutical products, which usually contain complicated formulation ingredients, enhancing the potential degradation of the active ingredients and their interactions with the formulation ingredients. Accordingly, achieving the desired stability of semisolid topical drug products is always a challenging task. In addition, the semisolid topical drug products are also prone to contaminations of extractables/leachables from their primary package. The identification of low-level degradation products and impurities has become a routine task in drug development. Even though HPLC-MS has been demonstrated to be an excellent technique to perform this type of task because of its high sensitivity and rich structural information with its versatile ionization (ESI, APCI, APPI, etc.) and detection (MS/MS to the end, high accuracy, high resolution, etc.) capabilities. However, because of its inherent limitation, HPLC-MS can only tell you the retention time and the *m/z* values of the ions observed. Accordingly, the absolutely structural establishment of targeted species still needs the imaginative input of scientists with knowledge, experience, and other associated information. Our work described in this chapter demonstrated the power of a strategic combination of HPLC-MS with other new analytical technologies such as HPLC-UV, HPLC-NMR, GC-MS, offline NMR, and traditional semipreparative HPLC, chemical derivatization, and classic organic synthesis for structural identification. To be effective and successful to utilize the information from HPLC-MS for structural identification, it is never enough to emphasize the importance of sample cleanup or preparation, including the crucial HPLC separation since the new technology HPLC-MS is sensitive not only to the interested species but also to undesired interference which could prevent or mislead the identification of the targeted molecules. In addition to structural identification of degradation products and impurities, the investigator is always encouraged to reveal the reaction mechanisms with additional experimental approaches as needed. Understanding the degradation mechanism during the early development phase of drug development offers advantages in the evaluation of drug candidates and their subsequent performance during the later development phases. Such knowledge will provide insight into stability, metabolism, and toxicity and serve as a diagnostic tool to predict potential critical issues and to guide the proactive development of corrective measures before they occur. For marketed products, in addition to the value in safety and quality evaluation, the degradation mechanism information could also serve as a guideline for optimization of formulation, packaging, manufacturing, and storage condition. Our results mentioned in this chapter may not only benefit the pharmaceutical developments of the drug molecules but also enrich the knowledge of the chemistry of related compounds.

REFERENCES

1. Mahalingam, R.; Li, X.; Jasti, B. R. (2008), Semisolid dosages: Ointments, creams, and gels, in Gad, S. C., ed., *Pharmaceutical Manufacturing Handbook*, Wiley, Hoboken, NJ, pp. 267–312.
2. Schaefer, U. F.; Lippold, B. C.; Leopold, C. S. (2008), Formulation issues, in Roberts, M. S.; Walters, K. A., eds, *Dermal Absorption and Toxicity Assessment*, 2nd ed, Drugs and the Pharmaceutical Sciences, Vol.177, Informa Healthcare USA, Inc., New York, pp. 117–134.
3. Girod, S.; Rodriguez, F.; Grossiord, J. L. (2005), Formulation strategy for semi-solid forms (ointments, creams): Contribution of rheology, *STP Pharma Pratiques 15*(3), 255–263.
4. Smith, E. W.; Surber, C.; Maibach, H. I. (2005), Topical Dermatological Vehicles: A Holistic Approach, in Bronaugh, R. L.; Maibach, H. I., eds, *Percutaneous Absorption, Drugs-Cosmetics-Mechanisms-Methodology*, 4th ed., Drugs and the Pharmaceutical Sciences, Vol. 155, Taylor and Francis, New York, pp. 655–661.
5. Niazi, S. K. (2004), *Handbook of Pharmaceutical Manufacturing Formulations : Semisolid Products*, CRC Press, Boca Raton, FL.
6. Luedtke, E. G.; Feger, M.; Fuehrer, C.; Schneider, W.; Singh-Verma, S. B.; Stanzl, K. (1989), Guidelines for product development, manufacture, and quality assurance of cosmetic products. Part 2. Special product development (formulations and process development, *Parfuem. Kosme. 70*(1), 14–16, 18–26, 28.

7. Marshall, A. G.; Hendrickson, C. L. (2008), High-resolution mass spectrometers, *Annu. Rev. Anal. Chem. 1*, 579–599.
8. Ackermann, B. L.; Berna, M. J.; Eckstein, J. A.; Ott, L. W.; Chaudhary, A. K. (2008), Current applications of liquid chromatography/mass spectrometry in pharmaceutical discovery after a decade of innovation, *Annu. Rev. Anal. Chem. 1*, 357–396.
9. Lee, M., ed. (2005), *Integrated Strateges for Drug Discovery Using Mass Spectrometry*, Wiley, Hoboken, NJ.
10. Lee, M. (2002), *LC/MS Applications in Drug Development*, Wiley, New York.
11. Budavari, S. ed.(1955), *The Merck Index*, the 25th ed., Merck, Rahway, NJ, pp. 1055–1056.
12. Kauffman, C. A.; Carver, P. L. (1997), Use of azoles for systemic antifungal therapy, *Adv. Pharmacol. 39*, 143–189.
13. Kujala P.; Ruutu, P. (1991), Systemic antifungal drugs, *Duodecim; laaketieteellinen aikakauskirja 107*(10), 767–776.
14. Georgiev, V. S. (1992), Treatment and developmental therapeutics in aspergillosis. 2. Azoles and other antifungal drugs, *Respiration 59*(5), 303–313.
15. Ellepola, A. N.; Samaranayake, L. P. (2000), Oral candidal infections and antimycotics, *Crit. Rev. Oral Biol. Med. 11*(2), 172–198.
16. Silingardi, M.; Ghirarduzzi, A.; Tincani, E. : Iorio, A.; Iori, I. (2000), Miconazole oral gel potentiates warfarin anticoagulant activity, *Thromb. Haemost. 83*(5), 794–795.
17. Roberts, D. T. (1982), The current status of systemic antifungal agents, *Br. J. Dermatol. 106*(5), 597–602.
18. Heel, R. C.; Brogden, R. N.; Pakes, G. E.; Speight, T. M.; Avery, G. S. (1980), Miconazole: A preliminary review of its therapeutic efficacy in systemic fungal infections, *Drugs 19*(1), 7–30.
19. Janssen, P. A. J.; Van Bever, W. F. M. (1979), Miconazole, *Pharmacol. Biochem. Properties Drug Substances 2*, 333–354.
20. Stevens, D. A.; Levine, H. B.; Deresinski, S. C. (1976), Miconazole in coccidiodomycosis. II. Therapeutic and pharmacologic studies in man, *Am. J. Med. 60*(2), 191–202.
21. Bodey, G. P. (1988), Topical and systemic antifungal agents, *Med. Clin. North Am. 72*(3), 637–659.
22. Landers, D. V. (1988), The treatment of vaginitis: Trichomonas, yeast, and bacterial vaginosis, *Clin. Obstet. Gynecol. 31*(2), 473–479.
23. Sobel, J. D. (1999), Vulvovaginitis in healthy women, *Comprehen. Ther. 25*(6–7), 335–346.
24. Doering, P. L.; Santiago, T. M. (1990), Drugs for treatment of vulvovaginal candidiasis: Comparative efficacy of agents and regimens, *Ann. Pharmacother (DICP). 24*(11), 1078–1083.
25. Zhang, F., Nunes, M. (2004), Structure and generation mechanism of a novel degradation product formed by oxidatively induced coupling of miconazole nitrate with butylated hydroxytoluene in a topical ointment studied by HPLC-ESI-MS and organic synthesis, *J. Pharm. Sci. 93*(2), 300–309.
26. Zhang, N.; Kawakami, S.; Higaki, M.; Wee, V. T. (1997), New oxidation pathway of 3,5-di-tert-butyl-4-hydroxytoluene: an ionspray tandem mass spectrometric and gas chromatographic/mass spectrometric study, *J. Am. Oil. Chem. Soc. 74*(7), 781–786.

27. Bauer, R. H.; Coppinger, G. M. (1963), Chemistry of hindered phenols. Reactivity of 2,6-di-tert-butyl-4-methylphenoxyl, *Tetrahedron 19*(8), 1201–1206.

28. Zimmer, H.; Lankin, D. C.; Horgan, S. W. (1971), Oxidations with potassium nitrosodisulfonate (Fremy's radical). Teuber reaction, *Chem. Rev. 71*(2), 229–246.

29. Ishii, H. (1972), Oxidation with Fremy's salt, *Yuki Gosei Kagaku Kyokaishi 30*(11), 922–941.

30. Coppinger, G. M.; Campbell, T. W. (1953), Reaction between 2,6-di-tert-butyl-p-cresol and bromine, *J. Am. Chem. Soc. 75*, 734–736.

31. Omura, K. (1995), Antioxidant synergism between butylated hydroxyanisole and butylated hydroxytoluene, *J. Am. Oil Chem. Soc. 72*(12), 1565–1570.

32. Omura, K. (1992), Chemistry on the decay of the phenoxy radical from butylated hydroxytoluene, *J. Am. Oil Chem. Soc. 69*(5), 461–465.

33. Omura, K. (1992), Reinvestigation of the reaction of 2,6-di-tert-butylbenzoquinone methide and 2,6-di-tert-butylphenol, *J. Org. Chem. 57*, 306–312.

34. Lachman, L.; Urbanyi, T.; Weinstein, S. (1963), Stability of antibacterial preservatives in parenteral solutions. IV. Contribution of rubber closure composition on preservative loss, *J. Pharm. Sci. 52*, 244–249.

35. Royce, A.; Sykes, G. (1957), Losses of bacteriostats from injections in rubber-closed containers, *J. Pharm. Pharmacol. 9*, 814–822.

36. Weiner, S. (1955), The interference of rubber with the bacteriostatic action of thiomersalate, *J. Pharm. Pharmacol. 7*, 118–125.

37. Wing, W. T. (1956), An examination of rubber used as closures for containers of injectable solutions. III. Effect of the chemical composition of the rubber mix on phenol and chlorocresol absorption, *J. Pharm. Pharmacol. 8*, 738–744.

38. Berry, H. (1953), Pharmaceutical aspects of glass and rubber, *J. Pharm. Pharmacol. 5*, 1008–1017.

39. Christiansen, E. (1951), The influence of vulcanized rubber on the quality of distilled water, *Medd. Norsk Farm. Selskap. 13*, 121–130.

40. Lachman, L.; Pauli, W. A.; Sheth, P. B.; Pagliery, M. (1966), Lined and unlined rubber stoppers for multiple-dose vial solutions. II. Effect of Teflon lining on preservative sorption and leaching of extractives, *J. Pharm. Sci. 55*(9), 962–966.

41. Tiller, P. R.; El Fallah, Z.; Wilson, V.; Huysman, J.; Patel, D. (1997), Qualitative assessment of leachables using data-dependent liquid chromatography/mass spectrometry and liquid chromatography/tandem mass spectrometry, *Rapid Commun. Mass Spectrom. 11*(14), 1570–1574.

42. Gaind, V. S.; Jedrzejczak, K. (1993), HPLC determination of rubber septum contaminants in the iodinated intravenous contrast agent (sodium iothalamate), *J. Anal. Toxicol. 17*, 34–37.

43. Lattimer, R. P.; Harris, R. E.; Rhee, C. K.; Schulten, H. R. (1986), Identification of organic additives in rubber vulcanizates using mass spectrometry, *Anal. Chem. 58*(14), 3188–3195.

44. Paskiet, D. M. (1997), Strategy for determining extractables from rubber packaging materials in drug products, *FDA J. Pharm. Sci. Technol. 51*(6), 248–251.

45. Taylor, R.; Son, P. N. (1982), *Encylcopedia of Chemical Technology*, Interscience Publishers, New York, Vol. *20*, pp. 337–365.

46. Zhang, F.; Chang, A.; Karaisz, K.; Feng, R.; Cai, J. (2004), Structural identification of extractables from rubber closures used for pre-filled semisolid drug applicator by chromatography, mass spectrometry, and organic synthesis, *J. Pharm. Biomed. Anal. 34*, 841–849.

47. Romadane, I.; Chizhikova, V. P. (1971), Synthesis and structure of halo-substituted 4-alkylphenols, *Latv. PSR Zinat. Akad. Vestis, Kim. Ser. 5*, 563–567.

48. Feldman, D.; Glorieux, F. H.; Pike, J. W., eds. (1997), *Vitamin D*, Academic Press, San Diego.

49. Kumar, R., ed. (1984), *Vitamin D: Basic Clinical Aspects*, Martinus Nijhoff, Boston.

50. Havinga, E. (1973), Vitamin D, example and challenge, *Experientia 29*, 1181–1193.

51. Reischl, W.; Zbiral, E. (1985), Selective oxidation of the vitamin D3 triene system—the reaction of vitamin D3 with N-bromosuccinimide—a simple entry to C-19-substituted tachysterol and isotachysterol derivatives, *Liebigs Ann. Chem. 6*, 1210–1215.

52. Colston, K. W.; Mackay, A. G.; James, S. Y.; Binderup, L.; Chander, S.; Coombes, R. C. (1992), EB1089: A new vitamin D analog that inhibits the growth of breast cancer cells in vivo and in vitro, *Biochem. Pharmacol. 44*, 2273–2280.

53. Yeung, B.; Vouros, P. (1995), The role of mass spectrometry in vitamin D research, *Mass Spectrosc. Rev. 14*, 179–194.

54. Fujishima, T.; Konno, K.; Nakagawa, K.; Kurobe, M.; Okano, T.; Takayama, H. (2000), Efficient synthesis and biological evaluation of all A-ring diastereomers of 1α, 25-dihydroxyvitamin D_3 and its 20-epimer *Bioorg. Med. Chem. 8*, 123–134.

55. Norman, A. W.; Bouillon, R.; Thomasset, M., eds. (1991), *Vitamin D: Gene Regulation, Structure-Function Analysis and Clinical Application*, Walter de Gruyter, Berlin.

56. Norman A. W. (1979), *Vitamin D: The Calcium Homoeostatic Steroid Hormone*, Academic Press, New York.

57. De Luca, H. F.; Paaren, H. E.; Schnoes, H. K. (1979), Vitamin D and calcium metabolism, *Top. Curr. Chem. 83*, 1–65.

58. Reichel, H.; Koeffler, H. P.; Norman, A. W. (1989), The role of the vitamin D endocrine system in health and disease, *N. Engl. J. Med. 320*, 980–991.

59. Sebrell, W. H., Jr.; Harris, R. S., eds. (1968), *The Vitamins: Chemistry, Physiology, Pathology, Methods*, 2nd ed., Vol. 2, Academic Press, New York.

60. Pelc, B.; Marshall, D. H. (1978), Thermal transformation of cholecalciferol between 100–170°C, *Steroids 31*, 23–29.

61. Benmoussa, A.; Delaurent, C.; Lacout, J. L.; Loiseau, P. R.; Mikou, M. (1996), Determination of cholecalciferol and related substances by calcium phosphate hydroxy-apatite and calcium phosphate fluoroapatite high-performance liquid chromatography, *J. Chromatogr. A 731*, 153–160.

62. Jones, G.; Trafford, D. J. H.; Hollis, B. W.; Makin, H. L. J.; (1992), Vitamin D: cholecalciferol, ergocalciferol, and hydroxylated metabolites, in Nelis, H. J.; Lambert, W. E.; DeLeenheer, A. P., eds, *Modern Chromatographic Analysis of Vitamins*, 2nd ed., Chromatographic Science Series, Vol. 30, Marcel Dekker, New York, pp. 73–151.

63. Verloop, A.; Koevoet, A. L.; van Moorselaar, R.; Havinga, E. (1959), Studies on vitamin D and related compounds. IX. Remarks on the iodine-catalyzed isomerizations of vitamin D and related compounds, *Recueil Trav. Chim. Pays-Bas Belg. 78*, 1004–1014.

64. Agarwal, V. K. (1990), A new procedure for the isomerization of vitamin D and its metabolites, *J. Steroid Biochem. 35*, 149–150.

65. Verloop, A.; Corts, G. J. B.; Havinga, E. (1960), Studies on vitamin D and related compounds. X. Preparation and properties of cis-isotachysterol, *Recueil Trav. Chim. Pays-Bas Belg. 79*, 164–178.

66. Inhoffen, H. H.; Bruckner, K.; Grundel, R. (1954), The vitamin D series. I. Rearrangement of vitamin D2 to an isotachysterol and partial synthesis of isovitamin D2, *Chem. Berichte 87*, 1–13.

67. Seamark, D. A.; Trafford, D. J. H.; Makin, H. L. J. (1980), The estimation of vitamin D and some metabolites in human plasma by mass fragmentography, *Clin. Chim. Acta 106*(1), 51–62.

68. Coldwell, R. D.; Trafford, D. J.; Varley, M. J.; Makin, H. L.; Kirk, D. N. (1988), The measurement of vitamins D_2 and D_3 and seven major metabolites in a single sample of human plasma using gas chromatography/mass spectrometry, *Biomed. Environ. Mass Spectrom. 16*(1–12), 81–85.

69. DeLuca, H. F.; Schnoes, H. K.; Tanaka, Y.; Alper, J. B. (1980), *25-Hydroxycholecalciferol Derivatives,* US Patent 4,229,359.

70. Nakayama, K.; Yamada, S.; Takayama, H.; Nawata, Y.; Iitaka, Y. (1984), Studies of vitamin D oxidation 4. Regio- and stereoselective epoxidation of vitamin D, *J. Org. Chem. 49*, 1537–1539.

71. King, J. M.; Min, D. B. (1998), Riboflavin photosensitized singlet oxygen oxidation of vitamin D, *J. Food Sci. 63*, 31–34.

72. Yamada, S.; Nakayama, K.; Takayama, H.; Itai, A.; Iitaka, Y. (1983), Studies of vitamin D oxidation 3. Dye-sensitized photooxidation of vitamin D and chemical behavior of vitamin D 6,19-epidioxides, *J. Org. Chem. 48*, 3477–3483.

73. Jin. X. L.; Yang, X. P.; Yang, L.; Liu, Z. L.; Zhang, F. (2003), Thermal isomerization of Vitamin D_3 in dimethyl sulfoxide, *J. Chem. Res.(s)*, (11), 691–693.

74. Jin, X. L.; Yang, X. P.; Yang, L.; Liu, Z. L.; Zhang, F. (2004), Autoxidation of isotachysterol, *Tetrahedron 60*(12), 2881–2888.

75. Jin, X. L.; Yang, X. P.; Yang, L.; Liu, Z. L.; Zhang, F.; Wang, J. X.; Li, Y. (2003), Autoxidation of isotachysterol: Formation of new epoxides, *J. Chem. Res.(s)*, (8), 477–479.

76. Bouillon, R.; Okamura, W. H.; Norman, A. W. (1995), Structure-function relationships in the vitamin D endocrine system, *Endocr. Rev. 16*, 200–257.

77. Norman, A. W.; Zhou, J. Y.; Henry, H. L.; Uskokovic, M. R.; Koeffler, H. P. (1990), Structure-function studies on analogs of 1α,25-dihydroxyvitamin D3: differential effects on leukemic cell growth, differentiation, and intestinal calcium absorption, *Cancer Res. 50*, 6857–6864.

78. Munker, R.; Norman, A. W.; Koeffler, H. P. (1986), Vitamin D compounds. Effect on clonal proliferation and differentiation of human myeloid cells, *J. Clin. Invest. 78*, 424–430.

79. MacLaughlin, J. A.; Gange, W.; Taylor, D.; Smith, E.; Holick, M. F. (1985), Cultured psoriatic fibroblasts from involved and uninvolved sites have a partial but not absolute resistance to the proliferation-inhibition activity of 1,25-dihydroxyvitamin D_3, *Proc. Natl. Acad. Sci. USA 82*, 5409–5412.

80. Morimoto, S.; Onishi, T.; Imanaka, S.; Yukawa, H.; Kozuka, T.; Kitano, Y.; Yoshikawa, Y.; Kumahara, Y. (1986), Topical administration of 1,25-dihydroxyvitamin D3 for psoriasis: Report of five cases, *Calcif. Tissue Int. 38*, 119–122.

81. Koeffler, H. P.; Hirji, K.; Itri, L. (1985), 1,25-Dihydroxyvitamin D3: In vivo and in vitro effects on human preleukemic and leukemic cells, *Cancer Treat. Rep. 69*, 1399–1407.

82. Okamura, W. H.; Palenzuela, J. A.; Plumet, J.; Midland, M. M. (1992), Vitamin D: Structure-function analyses and the design of analogs, *J. Cell. Biochem. 49*(1), 10–18.

83. Jones G.; Makin, H. L. J. (2000), Vitamin Ds: Metabolites and analogs, in De Leenheer, A. P.; Lambert, W. E.; Van Bocxlaer, J. F., eds, *Modern Chromatographic Analysis of Vitamins*, 3rd ed., Vol. 84, Chromatographic Science Series, Marcel Dekker, New York, pp. 75–141.

84. Jones, G.; Strugnell, S. A.; DeLuca, H. A. (1998), Current understanding of the molecular actions of vitamin D, *Physiol. Rev. 78*, 1193–1231.

85. Hesse, R. H.; Reddy, G. S.; Setty, S. K. S. (1993), *Preparation of Vitamin D Amide Derivatives as Cell Modulators,* Int. Patent Application (PCT), p. 36.

86. Zhang, F.; Nunes, M.; Segmuller, B.; Dunphy, R. Hesse, R. H.; Setty, S. K. S. (2006), Degradation chemistry of a vitamin D analogue (Ecalcidene) investigated by HPLC-MS, HPLC-NMR and chemical derivatization *J. Pharm. Biomed. Anal. 40*(4), 850–863.

87. Dmitrenko, O. G.; Telbiz, G. M.; Kikteva, T. A.; Eremenko, A. M.; Chuiko, A. A. (1997), Stability and thermoconversion of vitamin D_3 on Aerosil surfaces, *Proc. Indian Acad. Sci. (Chem. Sci.) 109*(5), 333–337.

88. Mizhiriskii, M. D.; Konstantinovskii, L. E.; Vishkautsan, R. (1996), 2D NMR study of solution conformations and complete 1H and 13C chemical shifts. Assignments of vitamin D metabolites and analogs, *Tetrahedron 52*, 1239–1252.

89. Boomsma, F.; Jacobs, H. J. C.; Havinga, E.; Van der Gen, A. (1977), Studies on vitamin D and related compounds. Part XXVI. The overirradiation products of previtamin D and tachysterol: Toxisterols, *Recueil Trav. Chim. Pays-Bas 96*, 104–112.

90. Sharpless, K. B.; Michaelson, R. C. (1973), High stereo- and regioselectivities in the transition metal catalyzed epoxidations of olefinic alcohols by tert-butyl hydroperoxide, *J. Am. Chem. Soc. 95*(18), 6136–6137.

91. Sharpless, K. B.; Verhoeven, T. R. (1979), Metal-catalyzed, highly selective oxygenations of olefins and acetylenes with tert-butyl hydroperoxide. Practical considerations and mechanisms, *Aldrichim. Acta 12*(4), 63–74.

92. Mihelich, E. D. (1979), Vanadium-catalyzed epoxidations. I. A new selectivity pattern for acyclic allylic alcohols, *Tetrahedron Lett. 49*, 4729–4732.

93. Mihelich, E. D.; Daniels, K.; Eickhoff, D. J. (1981), Vanadium-catalyzed epoxidations. 2. Highly stereoselective epoxidations of acyclic homoallylic alcohols predicted by a detailed transition-state model, *J. Am. Chem. Soc. 103*(25), 7690–7692.

94. Okamura, W. H.; Hammond, M. L.; Rego, A.; Norman, A. W.; Wing, R. M. (1977), Studies on vitamin D (calciferol) and its analogs. 12. Structural and synthetic studies of 5,6-trans-vitamin D_3 and the stereoisomers of 10,19-dihydrovitamin D_3 including dihydrotachysterol, *J. Org. Chem. 42*(13), 2284–2291.

95. Mordi, R. C.; Walton, J. C.; Burton, G. W.; Hughes, L.; Ingold, K. U.; Lindsay, D. A.; Moffatt, D. J. (1993), Oxidative degradation of β-carotene and β-apo-8′-carotenal *Tetrahedron 49*(4), 911–928.

96. Doering, W. von E.; Kitagawa, T. (1991), Thermal cis-trans rearrangement of semirigid polyenes as a model for the anticarcinogen β-carotene: An all-trans-pentaene and an all-trans-heptaene, *J. Am. Chem. Soc. 113*(11), 4288–4297.

97. Doering, W. von E.; Birladeanu, L.; Cheng, X. H.; Kitagawa, T.; Sarma, K. (1991), Two factors in thermal cis-trans rearrangement of pentaenes: configuration in (E,E)-octahydro-2,2′(3H,3′H)-binaphthylidene and extensivity in 2,2′- and 3,3′-bicholestadienylidenes *J. Am. Chem. Soc.* *113*(12), 4558–4563.

98. Doering, W. von E.; Sarma, K. (1992), Stabilization energy of polyenyl radicals: All-trans-nonatetraenyl radical by thermal rearrangement of a semirigid {4-1-2} heptaene. Model for thermal lability of β-carotene *J. Am. Chem. Soc.* *114*(15), 6037–6043.

99. Verloop, A.; Corts, G. J. B.; Havinga, E. (1960), Studies on vitamin D and related compounds. X. Preparation and properties of cis-isotachysterol, *Recueil Trav. Chim. Pays-Bas Belg.* *79*, 164–178.

100. Onisko, S. L.; Schnoes, H. K.; Deiuca, H. F. (1978), Two new vitamin D isomers. Formation of (3*S*,10*R*)-(*Z*,*Z*)-9,10-secocholesta-5,7,14-trien-3-ol and its 10*S*-epimer from cis-isotachysterol via facile [1,7]sigmatropic rearrangements, *J. Org. Chem.* *43*(18), 3441–3444.

101. Mayo, F. R. (1958), The oxidation of unsaturated compounds. V. The effect of oxygen pressure on the oxidation of styrene, *J. Am. Chem. Soc.* *80*, 2465–2480.

102. Mayo, F. R.; Miller, A. A. (1958), The oxidation of unsaturated compounds. VI. The effect of oxygen pressure on the oxidation of α -methylstyrene, *J. Am. Chem. Soc.* *80*, 2480–2493.

103. Porter, N. A.; Cudd, M. A.; Miller, R. W.; McPhail, A. T. (1980), A fixed-geometry study of the SH2 reaction on the peroxide bond, *J. Am. Chem. Soc.* *102*(1), 414–416.

104. Porter, N. A.; Zuraw, P. J. (1984), Stereochemistry of hydroperoxide cyclization reactions, *J. Org. Chem.* *49*(8), 1345–1348.

105. Bourgeois, M. J.; Maillard, B.; Montaudon, E. (1986), Homolytic intramolecular displacements. 12. Decomposition of ethylenic peroxides: Effect of chain length *Tetrahedron* *42*(19), 5309–5320.

106. Howard, J. A. (1973), Homogeneous liquid-phase autoxidations, in Kochi, J. K. ed, *Free Radicals*, Wiley, New York, pp. 23–62.

107. Takahashi, T.; Yamamoto, R. (1969), Stability of vitamin D_2 powder preparations. VI. Influence of some vitamins and acids on the isomerization of vitamin D_2, *Yakugaku Zasshi* *89*(7), 938–942.

108. Napoli, J. L.; Koszewski, N. J.; Horst, R. L. (1986), Isolation and identification of vitamin D metabolites, *Meth. Enzymol.* *123*, 127–140.

109. Dauben, W. G.; Funhoff, D. J. H. (1988), NMR spectroscopic investigation of previtamin D_3: Total assignment of chemical shifts and conformational studies, *J. Org. Chem.* *53*(22), 5376–5379.

110. Hofsass, H.; Grant, A.; Alicino, N. J.; Greenbaum, S. B. (1976), High-pressure liquid chromatographic determination of vitamin D_3 in resins, oils, dry concentrates, and dry concentrates containing vitamin A, *J. Assoc. Official Anal. Chem.* *59*(2), 251–260.

111. Jones, G.; Schnoes, H. K.; DeLuca, H. F. (1975), Isolation and identification of 1,25-dihydroxyvitamin D_2, *Biochemistry* *14*(6), 1250–1256.

112. Scott, A. I. (1964), *Interpretation of UV Spectra of Natural Products*, Pergamon, New York.

113. Plantin-Carrenard, E.; Beaudeux, J. L.; Foglietti, M. J. (2000), Physiopathology of iodine: Current interest of its measurement in biological fluids, *Ann. Biol. Clin.* *58*(4), 395–403.

114. Berman, E.; Luz, Z.; Mazur, Y.; Sheves, M. (1977), Conformational analysis of vitamin D and analogs. 1. Carbon-13 and proton nuclear magnetic resonance study, *J. Org. Chem.*, *42*(21), 3325–3330.

115. De Vries, E. J.; Mulder, F. J.; Borsje, B. (1977), Analysis of fat-soluble vitamins. XV. Confirmation of isotachysterol in vitamin D concentrates, *J. Assoc. Official Anal. Chem.*, *60*(5), 989–992.

116. March, J. (1992), *Advanced Organic Chemistry*, 4th ed., Wiley, New York, pp. 839–852.

117. Reynolds, D. W.; Fachine, K. L.; Mullaney, J. F.; Alsante, K. M.; Hatajit, T. D.; Motto, M. G. (2002), Available guidance and best practices for conducting forced degradation studies, *Pharm. Technol.* *26*(2), 48–56.

118. Reynolds D. W. (2004), Forced degradation of pharmaceuticals, *Am. Pharm. Rev.* (May/June), *7*(3), 56–61.

119. Alsante, K. M.; Martin, L.; Baertschi, S. W. (2003), A stress testing benchmarking study, *Pharm. Technol.* *27*(2), 60–72.

120. Eisinger, M.; Fitzpatrick, L. J.; Lee, D. H.; Pan, K.; Plata-Salaman, C.; Reitz, A. B.; Smith-Swintosky, V. L.; Zhao, B. (2003), *Preparation of 1,2,4-Thiadiazolium Derivatives as Melanocortin Receptor Modulators,* Int. Patent Application (PCT), p. 75.

121. Pan, K.; Scott, M. K.; Lee, D. H. S.; Fitzpatrick, L. J.; Crooke, J. J.; Rivero, R. A.; Rosenthal, D. I.; Vaidya, A. H.; Zhao, B.; Reitz, A. B. (2003), 2,3-Diaryl-5-anilino[1,2,4] thiadiazoles as melanocortin MC4 receptor agonists and their effects on feeding behavior in rats, *Bioorg. Med. Chem.* *11*(2), 185–192.

122. Goerdeler, J.; Lőbach, W. (1979), Ring cleaving cycloadditions. VI. Reaction of 5-imino-δ 3-1,2,4-thiadiazolines with heterocumulenes (preparative aspects) *Chem., Beriohte.* *112* (2), 517–531.

123. Chetia, J. P.; Mazumder, S. N.; Mahahan, M. P. (1985), One-pot synthesis of 2-aryl-3-phenyl(benzyl)-5-phenylimino-δ 4-1,2,4-thiadiazolines using N-chlorosuccinimide *Synthesis.* *1985*, 83–84.

124. Kihara, Y.; Kabashima, S.; Uno, K.; Okawara, T.; Yamasaki, T.; Furukawa, M. (1990), Oxidative heterocyclization using diethyl azodicarboxylate *Synthesis* *1990*, 1020–1023.

125. Bohrisch, J.; Patzel, M.; Grubert, L.; Liebscher, J. (1993), Syntheses of S,N-heterocycles from N-thioacyllactamimines *Phosp. Sulf. Silicon Related Elem.* *84*(1–4), 253–256.

126. Kurzer, F.; Sanderson, P. M. (1960), Thiadiazoles. X. The synthesis and isomerization of 2-aryl-5-arylamino-3-arylimino-δ 4-1,2,4-thiadiazolines, *J. Chem. Soc.* *1960*, 3240–3249.

127. Pan, K.; Reitz, A. B. (2003), The synthesis of aminobenzothiazoles from 2,3-biaryl-5-anilino-δ 3-1,2,4-thiadiazolines, *Synth. Communi.* *33*(12), 2053–2060.

128. Zhang, F.; Estavillo, C.; Mohler, M.; Cai, R. (2008), Non-enzymatic reduction of a 1,2,4-thiadiazolium derivative, *Bioorg. Med. Chem. Lett.* *18*, 2172–2178.

129. Jocelyn, P. C. (1972), *Biochemistry of the SH Group*, Academic Press, London, New York.

130. Shimada, K.; Mitamura, K. (1994), Derivatization of thiol-containing compounds, *J. Chromatog. B (Biomed. Appl.)* *659*(1/2), 227–241.

131. Davies, M. B.; Austin, J.; Partridge, D. A. (1991), *Vitamin C: Its Chemistry and Biochemistry*, Royal Society of Chemistry, Cambridge, UK.

CHAPTER 12

Characterization of Impurities and Degradants in Protein Therapeutics by Mass Spectrometry

LI TAO, MICHAEL ACKERMAN, WEI WU, PEIRAN LIU, and REB RUSSELL
Bristol-Myers Squibb Co., 311 Pennington-Rocky Hill Road, Pennington, NJ 08534

12.1 INTRODUCTION TO THERAPEUTIC PROTEINS

Since human insulin was launched as the first protein therapeutic drug in 1982, more than 130 recombinant protein therapeutics have been approved by the FDA [1]. Protein therapeutics have continued to play a significant role in almost every field of medicine. In more recent years, the growth rate of the US market for biologics—therapeutics consisting of proteins and/or nucleic acids—in which protein therapeutics are the most prominent, has maintained double-digit growth [2]. While protein therapeutics are gaining momentum in becoming increasingly important in the marketplace, many challenges remain in discovering and developing new efficacious and safe products for the patients.

Protein-based drugs have the potential to be more specific toward therapeutic targets resulting in fewer side effects due to specific interactions among biomolecules. But the complex nature of biomolecules and the sophisticated processes of manufacturing may cause major issues in biologics' efficacy, clearance, and toxicity. Protein molecules consist of hundreds of amino acid residues. Any variation in primary, secondary, tertiary, or quaternary structure can cause heterogeneity in proteins, such that almost no protein therapeutic is composed of a single species. In addition, the degree of molecular heterogeneity is even further complicated by various modifications that occur during expression (e.g., posttranslational modification), processing, and storage (e.g., through degradation). Therefore, the characterization and analysis of impurities and degradants in therapeutic proteins present unique challenges that are different from those arising from small-molecule drugs.

Characterization of Impurities and Degradants Using Mass Spectrometry, First Edition.
Edited by Birendra N. Pramanik, Mike S. Lee, and Guodong Chen.

Even the concepts of impurities and degradants take on different meanings for protein therapeutics relative to classical small-molecule drugs.

Generally speaking, the molecular composition of protein therapeutics, excluding formulation components, can be classified into the following categories:

1. *Product-related substances*—molecular variants that have properties comparable to the desired form of the drug with regard to activity, efficacy, and safety
2. *Product-related impurities*—molecular variants that do not have properties comparable to the desired form with regard to activity, efficacy, and safety
3. *Process-related impurities*—molecules that are derived from the manufacturing process, or downstream processing

Therefore, degradation products of protein therapeutics, depending on their therapeutic properties regarding efficacy and safety, can be product-related substances or product-related impurities. Unlike small molecule drugs in which most covalent modifications result in change in a drug's activities, many modifications on protein therapeutics do not affect their intended activities. In addition, conformational changes in which all covalent bonds are intact, such as denaturation and aggregation, can have significant and detrimental effects on protein drugs. Moreover, many protein variants, especially those minor ones with unknown properties, cannot even be categorized, since it is scientifically impossible to isolate all variants and test their properties individually. The assurance in efficacy and safety of a protein drug is not achieved by exhaustive characterization of all components or variants; rather, it is by well-controlled processes that generate highly reproducible biomolecules and ultimately by the corresponding satisfactory results from clinical trials.

The manufacturing processes for therapeutic proteins are also longer and more sophisticated than those for small-molecule drugs. Impurities that are inadvertently introduced into the protein therapeutics during manufacturing need to be closely monitored and tightly controlled. Although some process-related impurities are benign in terms of toxicity, others may pose serious side effects if their contents reach certain levels. Among the latter ones, the most critical species are host cell proteins, host cell DNAs, endotoxins, microbials, and leachables. Their contents are critical attributes defining the quality of protein therapeutic drugs. In this chapter, the analysis for host cell proteins, host cell DNAs, and endotoxins will be briefly described.

12.2 RECENT ADVANCES IN MASS SPECTROMETRY

Chronologically, the emergence of recombinant proteins as therapeutic drugs is followed by two critical advances in macromolecular mass spectrometry (MS): matrix-assisted laser dissociation/ionization (MALDI) [3] and electrospray ionization (ESI) [4]. These two ionization techniques have revolutionized macromolecular MS in such a way that the landscape of protein characterization and analysis has changed dramatically since the 1980s [5–8]. Nowadays, protein mass spectrometry is used extensively in drug discovery [9–21] and drug development [22–26]. In the discovery stage, MS is an indispensible tool in ensuring that the protein targets have the correct identities, the intended constructs, the expected posttranslation modifications, and so on. In the development stage, the search for efficient manufacturing processes requires an in-depth understanding of how protein molecules behave in each manufacturing step to achieve optimum product quality and manufacturing efficiency. Moreover, MS is essential in the long-term stability studies of protein therapeutics, in which sensitive detection of small and site-specific changes need to be monitored.

In general, there are two levels to MS analysis of proteins: (1) the intact mass analysis by liquid chromatography coupled with mass spectrometry (LC-ESI-MS or LC-MS) or matrix-assisted laser dissociation/ionization coupled with time-of-flight mass spectrometry (MALDI-TOF) and (2) detailed analysis focusing on amino acid residues by tandem MS (LC-MS^n)—usually on peptides generated by enzymatic digestion of proteins. More recent advances even allow LC-MS/MS to be carried out directly on large proteins using high-end MS such as Fourier transform MS (FT-MS) [27,28]. Intact mass analysis by MS is a direct and the most accurate way of measuring molecular weights of biological molecules. It is also the most comprehensive and definitive way of assessing the molecular integrity of a protein. Currently, the mass accuracy of many ESI interfaced commercial mass spectrometers can achieve a mass accuracy of 0.01%, or ±5 Da for a 50-kDa protein. This mass accuracy allows detection of most chemical changes that may occur within a protein, provided that the primary sequence of the protein is known. Many modifications to a protein can be detected straightforwardly by MS. If the sites of modifications or variations need to be determined, LC-MS^n can be performed. Even though many modifications can readily be detected by intact mass analysis, a definitive conclusion regarding the nature and locations of the modifications still requires LC/MS^n analysis. After a therapeutic target or a protein therapeutic has been fully characterized, modifications and variations of protein molecules can be determined by peptide mapping methods based on reversed-phased high-performance liquid chromatography (RP-HPLC). RP-HPLC is also used in all ESI-based mass spectrometry systems as a separation, and more importantly, as a desalting step. This chapter focuses on the applications of MS in the analysis and characterization of impurities and degradants in proteins from a drug discovery–drug development perspective.

12.3 IMPURITIES

During the drug discovery stage, the presence of impurities in a therapeutic target is not a cause of concern since recombinant proteins as therapeutic targets are used mainly for assay development and high-throughput screenings (HTS), neither of which is sensitive to the presence of contaminants. In fact, recombinant proteins with purities in the range of 70–80% have been used frequently in HTS. Impurities become more of a concern when they interfere with the target's function or cleave (proteolize) the target at a significant rate.

In contrast to the requirements for drug discovery, during drug development the impurities and degradants in protein therapeutics need to be well characterized and tightly controlled, since their contents can directly affect the safety and quality of a drug. While impurities are mainly the residual species that are not completely removed from the drug, the degradants can stem from every step of the lifecycle of a protein drug, such as during expression, purification, formulation, packaging, and storage. The impurities are extremely complex in nature and may include host cell DNAs, host cell proteins, the expression media components, substances that are used in the downstream purification and formulation, and leachables from packaging material.

Since the identities and contents of these impurities are specific to each manufacturing and delivering process, analyses of these impurities are diverse and complicated and usually require multiple analytical methodologies that are based on chromatography, spectroscopy, and MS. MS is frequently used for identifying impurity bands on gels from sodium dodecylsulfate polyacrylamide gel electrophoresis (SDS-PAGE) and isoelectric focusing (IEF). The procedure involves in-gel digestion of the protein bands followed by LC-MS/MS analysis of the digested peptides. Since this procedure is well documented in many protocols of practical applications of MS [29–32], it will not be discussed in this chapter. Among the impurities, the levels of endotoxin, residual host cell DNA, and residual host cell proteins (HCPs) are among the most critical attributes in protein therapeutics. Their analyses are briefly described below.

12.3.1 Endotoxin

Endotoxins are lipopolysaccharides (LPSs) that exist in the outer membrane of gram-negtive bacteria. These toxins belong to the strongest elicitors of the mammalian immune system by inducing a series of cytokine responses. If an excessive amount of endotoxin is injected into an animal, it can induce massive inflammation that can lead to organ failure or even death [33,34]. Therefore, endotoxins are closely monitored during biopharmaceutical development and for final product release. Since 1983, limulus amebocyte lysate (LAL) assay has been accepted as the standard test method for endotoxin in pharmaceuticals, surgical implants, water, and food [35,36]. This assay is based on the coagulation cascade of the LAL that is extremely sensitive to the presence of LPS. The traditional LAL assay suffers from three drawbacks: (1) low specificity due to broad reactivity of LAL, (2) high variation of batch-to-batch lysate reactivity, and (3) the dwindling supply of LAL from horseshoe crabs. Consequently, new assays have been developed to overcome these drawbacks. One of them is based on recombinant factor C (rFC), an endotoxin-sensitive serine protease that initiates the coagulation cascade [37]. The sensitivities of LAL and rFC assays for endotoxins are in the pg/mL–fg/mL range.

12.3.2 Residual DNA

The levels of residual host cell DNA are tightly controlled in protein therapeutics also for safety reasons. Commercial kits targeting generic DNA exist with quantification limits in the low ng/mL range [38]. In order to meet the FDA's criteria for well-characterized biopharmaceuticals that suggest a 100 pg/dose-limit for residual DNA, methods with higher sensitivities are required. Traditionally, membrane hybridization assays utilizing radiolabeled DNA probes prepared from the host cell's genomic DNA have been used with low-picogram sensitivities. Nonradioactive methods have been developed also, such as the dot or slot blot hybridization technique using nonisotopic DNA probe and a immunoenzymatic detection system, which has been demonstrated to achieve a minimum detectable limit of 10 pg DNA/mg protein [39,40]. This method has been shown to be suitable for quality control purposes.

12.3.3 Residual HCP

During the early process development, impurities at levels of a few percentages in recombinant proteins can be detected by SDS-PAGE or gel isoelectric focusing. Their identities can be determined by *N*-terminal amino acid sequencing, Western blot, or in-gel digestion and LC-MS/MS analysis followed by a database search. Impurities at this stage are usually copurified HCP or degraded proteins of interest. Determining the identities of the impurity bands is critical for downstream processes, for which MS can provide the most definitive identifications. Western blot and *N*-terminal sequencing can only show whether the gel band contains the certain binding epitopes or the first ~15 residues at the *N* terminus. Both analyses can rarely indicate whether the backbone of the protein is intact or specify the location of the truncation, if there is any. But for MS, this information can readily be obtained by intact mass analysis on the liquid sample or by the proteomics approach on the gel band. Information obtained from the abovementioned analyses can direct optimization of the purification processes to remove HCP impurities most efficiently or to reevaluate construct design, expression system, and formulations.

During later stages of process development when production development has matured and a robust process has been established, a more process-specific and custom-tailored assay for HCP is needed. It has been reported that HCPs can cause immune responses in patients at levels as low as 100 ppm [41]. An upper limit of 1–100 ppm of residual HCPs has been quoted as a regulatory benchmark for protein therapeutics [42]. A working target for HCP content of 1–10 ppm is regularly achieved by leading pharmaceutical companies [42]. Process-specific HCP assays are mostly immunoassays—usually immunoligand assays (ILAs) or enzyme-linked immunosorbent assays (ELISA)—using polyclonal antibodies raised against host cell proteins [43–46].

12.4 DEGRADATION PRODUCTS

Degradation products in therapeutic proteins refer to modified molecules brought about over time through various degradation pathways, generated by physical degradation or chemical degradation during production and storage. Physical degradation changes the oligomeric state or conformation of proteins only through denaturation, aggregation, precipitation, and adsorption to surfaces. Degradants generated through physical degradation pathways are more efficiently analyzed by chromatography and spectroscopic techniques, even though MS-based techniques have been used successfully to probe protein conformational changes [47–49].

Physical degradation is rarely encountered in small-molecule drugs. But for proteins, because of their polymeric nature and ability to adopt various secondary, tertiary, and quaternary structures, their properties can be drastically different without change of chemical bonds [50,51]. Among all physical degradation pathways, denaturation and aggregation are two of the major changes that directly affect the functionality of proteins and ultimately the efficacy and safety of protein therapeutics. Consequently, protein denaturation and aggregation are among the most critical attributes of protein therapeutics. Denaturation and aggregation are characterized and analyzed by various techniques [52,53], such as size exclusion chromatography (SEC), field flow fractionation (FFF) [54], dynamic or static light scattering [55], and analytical ultracentrifugation (AUC) [56], etc. Since their analyses are outside the scope of this chapter, they will not be further discussed.

12.4.1 Chemical Degradation

Degradation products in protein therapeutics can also be generated through various chemical reactions that involve forming or breaking covalent bonds. Analysis of chemical degradants of proteins is a daunting task because of the multiple pathways of protein degradation, the minor differences among various forms of protein molecules, and the complex processes of protein production. The major degradation pathways for protein therapeutics include deamidation/isomerization, oxidation, and fragmentation. These degradation pathways are of major concern during protein therapeutics development [57]. It is essential to understand the mechanisms governing these chemical reactions in order to optimize manufacturing processes to reduce chemical degradation. Each of these degradation pathways is discussed in detail below.

12.4.1.1 Deamidation/Isomerization

Deamidation is a common chemical degradation pathway involving mainly asparagine (Asn) residues under physiological conditions, in which Asn residues lose NH_3 with a concurrent change transforming Asn into Asp and/or isoaspartic acid (isoAsp), thus changing charge states and possibly the conformations of proteins and peptides [58–61]. Deamidation may also occur at glutamine (Gln) residues, but at a much slower rate [62]. Based on calculated results of 1371 Asn residues on 126 representative human proteins, at neutral pH and 37°C, deamidation half-times are in the range of 1–500 days for Asn and 100–5000+ days for Gln [62].

Proteins were found to undergo deamidation in vivo in the late 1960s [63,64]. Since then, they have been generally viewed as an unintended and undesirable form of protein aging or damage. However, it has been proposed that deamidation plays a regulatory role such that it serves as a molecular timer of biological processes [65–67].

Endogenous IgGs isolated from human serum have been reported to possess 23% deamidation at one of the most susceptible sites [68]. Deamidation of antibodies affects activity, especially if the deamidation sites are located in the complementarity-determining region (CDR) binding loops [57,69–78]. For example, the deamidated counterparts of a human growth hormone releasing factor analog are 25–500 times less potent than the unmodified peptide [79]. Deamidation of Asn residue 30 in one of the light chains of a commercial recombinant monoclonal antibody reduced its potency to 70% [80]. The dysfunctional isoAsp variants can have a negative impact on the efficacy of a drug if it competes with normal molecules in binding to the therapeutic targets [81]. Examples of protein functions not affected by deamidation on some residues can also be found and probably underreported [82,83]. Nevertheless, in protein therapeutics, deamidation is treated as a degradation pathway that is closely controlled to ensure the safety, efficacy, and quality of drugs, since the affected protein molecules may lose activity or become immunogenic. Usually, the site most susceptible to deamidation is chosen as a marker and monitored during process development of a protein therapeutic.

Under acidic conditions, protein deamidation proceeds by direct hydrolysis of the Asn residue to yield only Asp residues [64], whereas under neutral or alkaline conditions, the deamidation mechanism involves cyclic imide intermediates that hydrolyze to yield both isoAsp and Asp residues, at approximately 3 : 1 ratio for peptides [63,64]. Some stable imide intermediates can even be isolated if they are purified under mild conditions [84,85]. Figure 12.1 shows a schematic of protein deamidation under neutral or alkaline conditions—the physiological condition. Racemization can occur during ring opening for the two endproducts [63]. In addition,

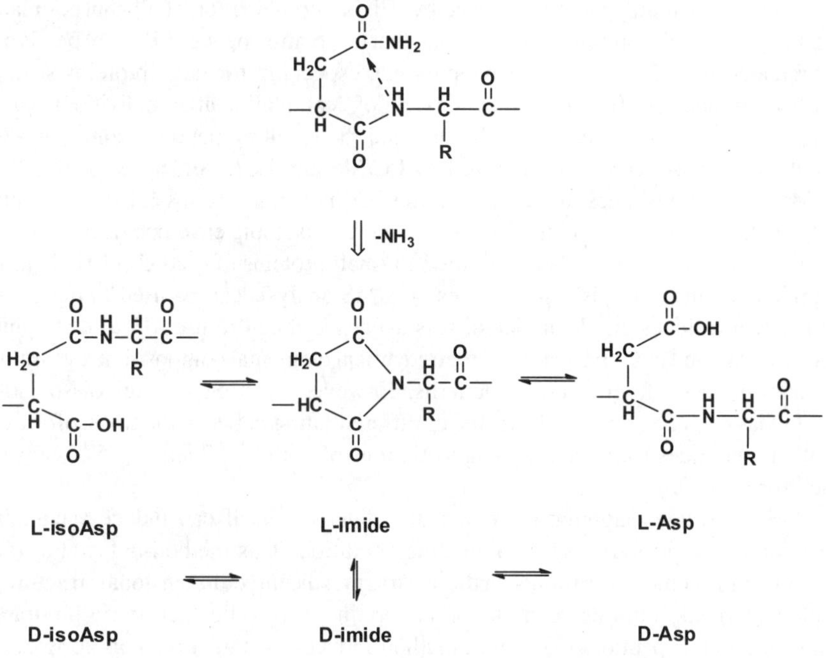

$-NH_3$

L-isoAsp L-imide L-Asp

D-isoAsp D-imide D-Asp

FIGURE 12.1 Deamidation through cyclic imide intermediate.

when deamidation converts an Asn to an isoAsp residue, the protein backbone is extended by a methylene group, causing termination of the Edman degradation reaction.

The cyclic imide has a mass shift of −17 Da from the native form, which potentially can be isolated if the intermediate is stable. The Asp/isoAsp product has a mass shift of + 1 Da compared with the original peptide/protein, which can be detected in peptides but not in large proteins by intact mass analysis. The protease Asp-N can be used to differentiate between the Asp-containing peptide and isoAsp-containing peptide, since Asp is recognized by Asp-N as a cleavage point but not the isoAsp. Because of the difference in charge and hydrophobicity, the deamidated peptides usually can be separated from the native form by various chromatography and electrophoretic techniques. In addition to the most widely used RP-HPLC-based methods, other separation techniques have been used, such as ion exchange chromatography (IEC) [86,87], hydrophobic interaction chromatography (HIC) [88,89], and isoelectric focusing [85].

In a typical RP-HPLC chromatogram, the unmodified peptide is flanked on both sides by deamidated peptides; the center peptide is 1 Da less than its deamidated counterparts on both sides. Alternatively, without separating deamidated species from the original form, the amount of isoaspartyl residue can be indirectly assessed by monitoring *S*-adenosylhomocysteine (SAH), a product of protein isoaspartylmethyltransferase (PIMT) when it transfers a methyl group from *S*-adenosylmethionine (SAM) onto the free α-carboxyl of the isoaspartyl residue [90,91]. Deamidation can also be monitored by isoelectric focusing (IEF) since deamidation causes pI changes. However, the banding pattern observed by IEF is a combination of all charge-related modifications to the protein molecule such that correlation between IEF gel bands and deamidated residues are not always possible, especially for large proteins such as antibody molecules. To determine the extent of deamidation at an individual Asn or Gln residue, proteins need to be digested and the resulting peptides containing the Asn/Gln residues need to be analyzed by LC/MS and LC/MS/MS.

More recent advances in high-resolution MS have made it possible to determine deamidation in intact proteins using a mass defect–isotopic envelope deconvolution method. However, this method is limited to small proteins of ~20 kDa [92–94]. If a peptide contains multiple Asn residues, MS/MS analysis can be used to determine which Asn residues are deamidated. It is usually difficult to use MS to distinguish between Asp and isoAsp because of their identical elemental composition and similar collision-induced fragmentation patterns. However, electron transfer dissociation (ETD) has been demonstrated capable of differentiating between the two isoforms in synthetic peptides using unique diagnostic ions of the $c + 57$ and $z - 57$ peaks for each form [95,96].

During peptide mapping, the enzymatic digestion itself can induce asparagine deamidation, especially under an alkaline condition. This method-induced artifact arises from structural constraints on the asparagine; the three-dimensional structure of protein sterically inhibits deamidation at certain sites, while such steric hindrance may not occur on peptides [97]. Since method-induced deamidation is proportional to incubation time, reducing digestion time and lowering buffer pH can significantly reduce this deamidation artifact. Alternatively, a timecourse study can be performed to determine the relationship between the observed deamidation and the incubation time. Then, a timecourse can be extrapolated to time zero to estimate the existing deamidation levels of the concerned residues prior to digestion. An extreme measure for differentiating Asn deamidation that occurred prior to and during sample preparation, the sample preparation can be carried out in ^{18}O – enriched water [98]. By this method, deamidation that occurs during sample preparation will result in a mass increase of 3 Da, instead of 1 Da when the sample is digested in regular solution. This molecular mass difference will be readily detected, thus allowing differentiation of method-induced artifacts from existing deamidation. However, this method is rarely feasible because of the high cost of ^{18}O-enriched water.

The susceptibility of Asn residues to deamidation depends on their primary sequences and local conformations. Protein deamidation is estimated to be influenced by approximately 60% from primary structure and 40% from conformation [62]. At the primary sequence level, rates of Asn deamidation in small peptides have been shown to correlate with the residue to the *C*-terminal side of the Asn residue. On model peptides, Gly, Ser, and His residues at the *C*-terminal side of Asn have been shown to promote Asn deamidation [99,100]. On the contrary, the residue on the *N*-terminal side of Asn does not appear to affect the deamidation rate [97,101]. On protein molecules, well-folded structure and flexibility constraints appear to inhibit deamidation [97,102,103]. For example, out of the 25 Asn residues on an IgG antibody molecule [97], only four Asn residues (SNG, ENN, LNG, and LNN) were observed to deamidate with slow rates at pH 7.5 and 37°C. The resistance to deamidation probably results from rigid constraints exerted from a well-folded protein structure. On the other hand, Asn located in the complementarity-determining regions (CDRs) undergo deamidation more rapidly, possibly because of their higher flexibility and increased solvent exposure. Interestingly, the folded structure of a protein may change the susceptibility of an Asn residue in both directions. In other words, the folded structure of proteins can render some Asn residues more susceptible to deamidation. It has been estimate that in about 6% deamidating Asn residues, the conformation actually accelerates deamidation [62]. In a more recent report, deamidation of a 22–amino acid peptide, and that of the Fc fragment of a human monoclonal antibody (Fc IgG1) that contained the 22mer peptide sequence, were studied [104]. On the peptide, out of three potential deamidation sites, only N382 was found to deamidate to form Asp382 and isoAsp382 with a ratio of ~1 : 4. But on the Fc protein, Asp387 was detected in addition to isoAsp 382 and small amounts of Asp382. The protein structure can also affect the susceptibility of Gln to deamidation. Although usually much slower than Asn deamidation, Gln deamidation has been observed in some protein molecules [105–107]

Deamidation is affected by pH, temperature, and formulation components of the protein, in which high pH and high temperature accelerate the reaction rate [89,107–109]. Little deamidation in proteins is observed when the pH is in the range of 3–5. However, this pH range may promote aggregation and fragmentation. At pH > 5, deamidation of Asn begins to occur. As a consequence, lower pH (pH 5–6) is often found to be the optimum pH range for protein products of which Asn deamidation is a major degradation pathway. For most proteins whose Asp isomerization is a major degradation pathway, the optimum pH range for formulation is usually in the range pH 6–8 [109]. The rate of Asn deamidation was significantly reduced in the presence of alcohol or glycerol, or in solutions with low-dielectric constants [110]. This reduction is presumably due to the decreased stabilization of the ionic intermediates formed during the cyclization step of the Asn deamidation pathway.

In addition to Asn deamidation, some Asp residues can undergo isomerization to form isoAsp and cyclic imide (Asu) if the latter are stable enough [60,62,63,111,112], as was reported in CDRs of the light chains of two recombinant monoclonal antibodies (MAbs) [89]. Isomerization of Asp was also observed in the loop region of recombinant human interleukin 11 under stressed conditions [111]. Some Asp isomerizations have been shown to result in the loss of activities or trigger immunogenicity [60,81,89]. The isomerization rate increases with higher temperature and lower pH. Similar to deamidation, the Asp isomerization rate depends on (1) the *C*-terminal side residue to Asp, (2) local conformational flexibility surrounding Asp, and (3) the extent of solvent exposure of the labile Asp residue [89]. Unlike the case in Asn deamidation, the decrease in Asp isomerization is not as pronounced with the addition of alcohol or glycerol. It was proposed that the destabilization of the intermediate may be offset by the increase of the reactivity of Asp [89].

12.4.1.2 Protein Fragmentation

12.4.1.2.1 Asx- and Glx-Related Fragmentation

Besides being involved in protein deamidation/isomerization, Asn and Gln residues are also involved in spontaneous peptide bond cleavage at their carboxy sides [63,83,113]. When deamidation proceeds through direct hydrolysis or cyclic imide formation, it usually exhibits activation barriers lower than that in peptide bond cleavage, and therefore is a more favorable pathway. But under certain circumstances, bond cleavage does occur. The fundamental distinction between the mechanisms leading to deamidation via succinimides and backbone cleavage was found to be the difference in nucleophilic entities involved in the cyclization process (backbone vs. sidechain amide nitrogen). If deamidation is prevented by protein three-dimensional structure, cleavage may become a competing pathway. In addition, peptide bond cleavage at Asn residues is more likely to take place after it has deamidated into Asp [114]. Peptide bond cleavages at Asp and Glu residues have been reported also in peptides and proteins, and the occurrence rate is higher than that in Asn and Gln [115–117].

The mechanism of Asp/Glu-related cleavages is closely related to that of deamidation process, in which it was proposed to involve nucleophilic attack of the ionized sidechain carboxylate on the protonated carbonyl carbon of the peptide bond to give a cyclic anhydride intermediate [116]. The cleavage rate is affected by the sidechain carboxylic acid group, [117,118]. Cleavage of Asn–Pro residues turned out to be the fastest where succinimide formation could not occur [119,120]. Such peptide bond cleavage can render proteins inactive. For instance, one of the modifications observed in lens proteins is the progressive, age-dependent cleavage of specific peptide bonds in bovine A-crystallin [121]. In fact, peptide bond cleavages are observed in most peptide deamidation reactions, but the fragmentation is usually much slower than deamidation. Asn peptide deamidation half-time ranges from about 1 to 400 days, and Asn cleavage rate ranges from about 200 to >10000 days [67].

12.4.1.2.2 Ab Fragmentation

Another type of fragmentation is observed in the hinge region of monoclonal antibodies [104,122–126]. This cleavage is not affected by protease inhibitors or EDTA, which inhibits protease activities, indicating its spontaneous nature. This cleavage was observed on some antibody molecules after long-term storage for an extended period of time at different temperatures, even at 5°C [73,127]. The site of cleavage is usually in the heavy-chain hinge region near the papain cleavage site, generating Fab and Fab + Fc fragments that can be detected by MALDI-TOF-MS [126]. Metal-mediated cleavage of the antibody molecules in the hinge region has also been reported [127].

Two mechanisms for antibody fragmentation in the hinge region have been proposed: β-elimination and direct hydrolysis [98,123]. β-Elimination is more pronounced at pH ≥ 7, which causes cleavage in between S/C in the SCDKTHTC region. β-Elimination of the disulfide bond leads to the formation of a dehydroalanine residue, which hydrolyzes to form an amide group at the newly formed *C* terminus and a pyruvyl group at the newly formed *N* terminus [123]. Direct hydrolysis is accelerated by acidic and basic pH. While cleavage was found in every peptide bond in SCDKTHTC of the hinge region, the major cleavage sites have been identified to be located in between S/C, C/D, D/K, and H/T [122]. It was also found that the major cleavage sites in the hinge region shift toward the *C* terminus when pH changes from 9 to 5. At pH 4, the major cleavage site shifted to the CH2 domain. In addition, oligosaccharides inhibit hinge region fragmentation only at pH 4. This shift was not observed from pH 9 to 5 [98].

12.4.1.3 Oxidation

Protein oxidation is defined as covalent modification on proteins caused by oxidizing reagents or radiation. The oxidizing reagents include reactive oxygen intermediates, peroxides, metals ions, HOCl/HOBr, activated phagocytes, mitochondria, oxidoreductases, and some drugs and their metabolites [128–138]. Protein oxidation can cause loss of function and activity as a result of backbone or side chain modification, dimerization, aggregation, unfolding, conformational changes, and other phenomena [137–142]. Different oxidation pathways may have different consequences on protein functions. For example, the formation of carbonyls and nonnative disulfide bonds are more disruptive to protein functions compared with methionine oxidation, which is confined more to individual residues [143–145].

Factors affecting rate of oxidation include temperature, pH, metal ions, and oxygen level. The dependence of oxidation on pH for various amino acid sidechains is different. For example, oxidation of His is rapid at neutral pH when uncharged but is quite slow at low pH when charged. At higher pH, Tyr is most reactive, while Trp and Met are the only amino acids readily oxidized below pH 4 [146]. It is worth mentioning that some of the common pharmaceutical excipients may contain peroxides which oxidize proteins: polyethylene glycol (PEG), polyvinylypyrrolidone (PVP), polysorbate 80 (P80), and hydroxypropyl cellulose (HPC) [147,148].

Structurally, protein oxidation can originate from the backbone or the sidechains. Backbone oxidation involves primarily hydrogen atom abstraction at the α-carbon, resulting in backbone fragmentation [128,131,135]. If the oxidation reaction originates from the sidechain of aliphatic residues, hydrogen abstraction tends to occur. The resulting products are usually heterogeneous since multiple C–H groups exist on most residues. Major products from aliphatic sidechain oxidations are peroxides, alcohols, and carbonyls [149,150]. The oxidation at aromatic sidechains usually results in addition rather than abstraction of hydrogen with heterogeneous products as well [151]. Among all amino acid residues, Cys and Met are the only ones whose oxidized states can be reversed, with the resulting damage repaired. Protein oxidation may also result in fragmentation. Backbone fragmentation commonly is induced by radicals, which takes place through formation of an α-carbon radical followed by subsequent formation of peroxyl radical in the presence of O_2. Little backbone fragmentation is observed in the absence of O_2 [131,134,137].

12.4.1.3.1 Cystine/Cysteine Oxidation

Up to three oxygen atoms can be inserted into the positions surrounding the sulfur atom in cystine or cysteine residues. The oxidation of Cys by nonradical oxidants form protein thiols, peroxides, disulfides, and oxyacids [152]. The thiol group of Cys (RSH) can be oxidized in three steps, to a sulfenic acid (RSOH), a sulfinic acid (RSO_2H), and a sulfonic acid (RSO_3H), which correspond to mass increases of $+16$, $+32$ and $+48$ Da, respectively [153,154]. The oxidation of free sulfhydryl groups can lead to the formation of intra- or intermolecular disulfide bonds. The oxidation of thiol groups by molecular oxygen can be accelerated by the presence of catalytic quantities of metal ions, such as iron and copper. In addition, the speed of oxidation of thiol groups can be greatly influenced by the neighboring residues [154]. Since the mercaptide ion is oxidized more easily than the undissociated thiol group, its oxidation rate increases with increasing pH. Several MS-based methods of quantitative analysis for thiol and disulfide groups in proteins have been described [155–157]. Furthermore, affinity probes have been developed that selectively react with oxidized Cys residues. One approach traps sulfenic acid as a stable thioether by oxidation with dimedone, generating a residue mass shift of 138 Da.

Most Cys residues of the antibodies or biologics are in the form of disulfide bonds. However, free sulfhydryl groups were detected in recombinant antibodies [157]. As a result of the more reactive nature of the free sulfhydryl group, Cys residues not involved in disulfide formation are usually avoided in various platforms of biologics construct design. On the other hand, the reactivity of the sulfhydryl group can be exploited as an anchoring point for specific conjugation. For example, protein constructs containing a single Cys residue have been PEGylated through a maleimide bond to reduce non-specific PEGylation through primary amines.

12.4.1.3.2 Methionine Oxidation

Methionine (Met) is one of the residues most susceptible to oxidation. It can even be oxidized by atmospheric oxygen to form methionine sulfoxide, which increases the mass by 16 Da. Under extreme conditions, Met can be oxidized to form sulfone [158]. This mass shift can be detected by LC-MS on most biologics, except for proteins with extensive glycosylation, where a deglycosylation pretreatment may be necessary to obtain a legible LC-MS spectrum. It has been shown that within a given protein, the susceptibility of Met residues toward oxidation is position-dependent. For example, in human growth hormone (hGH), Met^{170} was found to be completely resistant to oxidation by hydrogen peroxide [159]. In human chorionic somatomammotropin, Met64, Met16, and Met179 have markedly different reaction rates [159]. Oxidation also increases the polarity of the sidechain of Met, which allows separation of oxidized from nonoxidized peptides. For example, the Fc domain with oxidized Met residues elutes earlier relative to the nonoxidized molecules on a HIC column [160,161], antibodies with oxidized Met residues elute later on a weak cation-exchange column [162], while peptides containing oxidized Met usually elute earlier than the unmodified peptides on a RP column [73,160,162].

Not all oxidation of Met residues results in loss of protein function [146]. Some proteins lose activity after certain Met residues are oxidized [163–166]. For example, the monosulfoxide derivatives of pancreatic ribonuclease, chymotrypsin-A, and Kunitz trypsin inhibitor have all been shown to be active after being treated by hydrogen peroxide [167–169].

The oxidation of Met is reversible. Methionine sulfoxide oxidation of α-crystallin (Met138 of αA and Met68 of αB) resulted in loss of its chaperone activity, and subsequent treatment with protein methionine sulfoxide (PMSO) reductase A repaired its chaperone activity [170]. In many cases, such as parathyroid hormone, ribonuclease *S*-peptide, ribonuclease, and lysozyme, reduction of Met sulfoxide by thiols results in the recovery of nearly full biological activity [146]. The cell culture media components and formulation excipients can be used to protect proteins from oxidative damage. Oxidation of Met residues to methionine sulfoxide occurs frequently when cell culture media are switched from serum-containing to serum-free [171].

Characterization and quantitative analysis of oxidation plays an important role in biopharmaceutical development. Determination of oxidized Met in proteins is problematic for conventional amino acid analysis, since Met sulfoxide is converted to Met during acid hydrolysis. By MS-based methods, Met-containing peptides can be separated from those containing oxidized Met by ion-exchange chromatography, counter-current distribution, RP-HPLC, or affinity chromatography. LC-MS- and LC-MS/MS-based methods are then used to determine the extent and the locations of the oxidized Met residues. It has been demonstrated that the extent of oxidation of Met residues can be assessed by peptide mapping with MS detection with good linearity [$R^2 > 0.99$; RSD (relative standard of deviation) 4–9%] [172].

In addition to oxidation on Met and Cys, oxidation on other residues has been reported. Oxidation of His forms predominantly oxohistidine [158]. His is especially susceptible to metal-catalyzed oxidation because of its tendency to interact with transition metals [173]. Oxidation of Trp generates multiple products, such as 5-hydroxy-Trp, oxyindole alanine, kynurenine, and *N*-formylkynurenine. [158]. Oxidation of Trp residues of several monoclonal antibodies—all located on CDR regions of corresponding molecules—have also been reported [174–176].

12.4.2 Variants Caused by Posttranslational Modification

Posttranslational modification (PTMs) is an important step in protein biosynthesis, in which a translated protein is covalently modified by various functional groups. PTMs convey functional regulation, structural stability, and pharmacokinetic properties to proteins. The most prominent and most complex form of PTM is glycosylation, which occurs with proteins expressed mainly by mammalian cells [177,178]. Glycosylation can have a significant impact on many properties of protein therapeutics [177], such as structural stability [179,180], potency [181], efficacy [182], immunogenecity [183–185], and pharmacokinetics [186,187]. Since glycosylation of therapeutic proteins is too broad a topic for this chapter to cover [180,188–192], and the applications of MS in this field have been extensively reviewed in some recent publications [193–197], it will not be discussed further in this chapter.

The application of MS in characterization of PTMs has been discussed in several reviews [198–206]. In addition to characterization of protein glycosylation, MS has also been successfully used to control cell culture conditions to manipulate oligosaccharide profiles for glycoproteins in mammalian expression systems [207]. In bacterial fermentations, feeding strategies have been used to mitigate amino acid misincorporation or translation errors that result in unexpected protein sequences or unwanted modifications [208]. LC-MS has also been used to monitor chemical modifications following storage of formulated bulk solutions or lyophilized products [209].

While many PTMs occur intracellularly during protein expression, others result from protein production processes. Many PTMs in protein therapeutics are not relevant to protein function, and therefore constitute a source of product related variants. However, it is usually challenging to determine whether a modification affects the pharmaceutical properties of a protein, due to the technical difficulty in separating the modified molecules from the rest of the population. In other words, many of the minor species resulting from various PTMs cannot be categorized in the development of protein therapeutics. Instead, during the manufacturing of biologics, processes are well controlled in which all variants are monitored closely by sensitive and comprehensive techniques such as MS to ensure a high degree of reproducibility. Ultimately, the efficacy and safety of biologics are guaranteed by the well-controlled manufacturing processes and the corresponding clinical trials.

Although many types of modifications can occur with proteins, the nature of modifications correlates with the physiological conditions of the host cells and the specific processes in protein production. Some variants are generated during protein expression, such as glycosylation and gluconoylation. The latter was frequently observed in *Escherichia coli*–expressed recombinant proteins [210–212]. Other modifications are generated in the downstream manufacturing processes such as carbamylation, deamidation, and oxidation. Information about the expression system and the manufacturing processes can significantly facilitate identification of the modifications. In addition to deamidation, oxidation, and glycosylation, the common PTMs observed in therapeutic targets and proteins include phosphorylation, acetylation, methylation, *N*-terminal pyroglutamation, carbamylation, gluconoylation/phosphorgluconoylation, *S*-thiolation, and β-mercaptoethanol adducts. If molecular heterogeneity is not overwhelmingly complicated, a simple intact mass analysis can reveal the extent and nature of the PTMs. If the locations of PTMs are to be determined, LC-MS/MS analysis on enzymatically digested proteins is usually the method of choice. If the heterogeneity is overwhelmingly complicated for ESI-MS, then MALDI-TOF can be used to obtain an approximate mass spectrum at the expense of fine molecular mass profiles due to limited resolution of MALDI on larger molecules. Table 12.1 lists the common PTMs observed during the production of recombinant proteins as therapeutic drugs and therapeutic targets.

TABLE 12.1 Commonly Observed PTMs in Recombinant Proteins

Average Mass change (Da)	Modification
−34	Dehydroalanine
−18	Dehydration, pyroglutamate formation succinimide formation from Asp
−2	Disulfide formation
−1	Deamidation
2	Reduction of a S−S bond
14	Mythylation
16	Oxidation of Met
	Oxidation of His
	Oxidation of cysteine
22	Sodium adduct
28	Formylation
32	Oxidation of Met
42	Acetylation
43	Carbamylation
48	Oxidation of cysteine to form cysteic acid
76	β-Mercaptoethanol adduct
80	Sulfonation
	Phosphorylation
119	*S*-Cysteine
133	*S*-Homocystine adduct
178	Gluconoylation
226	Biotinylation
258	Phosphogluconoylation
305	*S*-Glutathione
356	4-Phosphopantetheine
685	Dephosphorylated coenzyme A

12.4.2.1 Case Study: Characterization of S-Thiolation on Secreted Proteins from E. coli

Protein *S*-thiolation is a posttranslational modification in which nonprotein thiols are linked to protein molecules through disulfide bonds. It has been shown that in vivo, S-thiolation is responsible for regulating cellular redox status, nitrogen oxide–mediated signal transduction, protein quaternary structure, and DNA binding. Although physiologically important for the cells, *S*-thiolation during recombinant therapeutic protein production is usually undesirable for the following reasons: (1) it has the potential effect of interfering with the biological functions of protein molecules due to steric hindrance, (2) it may disrupt the formation of native disulfide bonds that are essential for some protein functions, (3) *S*-thiolation on the active site cysteine by glutathione has been shown to cause significant change to some protein functions [213–215], and (4) *S*-thiolation on recombinant therapeutic proteins interferes with down-stream processes, such as the PEGylation reaction, which requires a free sulfhydryl group.

Even though *S*-thiolation can be removed from recombinant proteins by a reducing reagent such as dithiothreitol (DTT) or 2-mercaptoethanol, the reduction reaction is indiscriminate because desired disulfide bonds may also be reduced. Moreover, an additional reduction step results in increased manufacturing time and cost in recombinant therapeutic protein production. With more therapeutic proteins being manufactured using the secretory pathways in *E. coli*, there is a concomitant impetus to understand the mechanism of *S*-thiolation of secreted therapeutics in order to control and reduce heterogeneity. This work identified several small thiol-containing metabolites that frequently modify secreted recombinant proteins in one of the most commonly used *E. coli* strains, BL21 (DE3). On the basis of these results, approaches can be taken during upstream process fermentation to better understand and control against *S*-thiolation on secreted therapeutic proteins expressed by *E. coli*.

S-Thiolation is routinely detected in many recombinant proteins secreted from the *E. coli* BL21 (DE3) strain in our work. In this report, a model protein (MP) that carried most of the observed *S*-thiolations is used to demonstrate the characterization process through which the four most abundant thiol modifiers are identified. The MP is a proprietary recombinant protein currently being developed as a therapeutic drug. It has a molecular mass of 12255.8 Da and contains only one cysteine residue at the *C* terminus as the intended PEGylation site. The extent of *S*-thiolation on MP can be assessed by its mass spectra before and after reduction as shown in Figure 12.2. The nonreduced MP was quite heterogeneous with five major species, each one of them is marked by its extra mass compared to unmodified molecule; the reduced MP was more homogeneous with one major species also marked by its extra mass. From comparison of these two mass spectra, it is evident that the multiple species in nonreduced MP were mainly caused by disulfide linked small thiols. In addition, after thiols were removed from MP by reduction, the main component was still 177 Da larger than the unmodified molecule. This indicates that a nonreducible modification exists on MP.

The *C*-terminal cysteine on MP is predicted to be the *S*-thiolation site since it is the sole residue that contains the free-sulfhydryl group. This prediction was confirmed

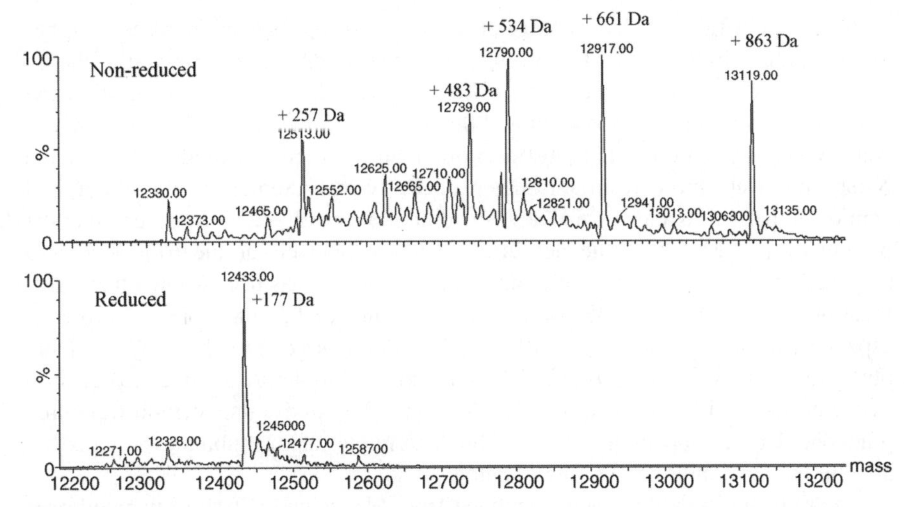

FIGURE 12.2 Deconvoluted mass spectra of the model protein before and after being reduced. All major mass peaks are marked by their mass increase relative to that of unmodified protein molecule. The modified protein molecules had been enriched by ion exchange chromatography such that the unmodified molecules were barely detected. (Reprinted from Ref. 219, with permission of Wiley-Blackwell.)

by trypsin digestion and LC-MS/MS analysis on nonreduced MP. The results showed that the *C*-terminal cysteine of MP was modified by four small molecules, resulting in mass increases of 305, 356, 483, and 685 Da, respectively (Table 12.2). In addition, these four small molecules can be cleaved from MP by reduction and be analyzed by LC-MS. Their masses were determined to be 307, 358, 485, and 687 Da, respectively. This result provided orthogonal confirmation that the modifiers were thiols with masses 2 Da larger than the mass increases that they brought to MP through *S*-thiolation.

TABLE 12.2 Mass Increases Caused by *S*-Thiolation on the Model Protein and Its *C*-Terminal Peptide[a]

Modification on Whole Protein (Da)	Modification on C-Terminal Cysteine (Da)	Mass Difference (Da)
257		
483	305	178
534	356	178
661	483	178
863	685	178

[a]These mass increases on protein molecules were determined from the five most abundant peaks in intact mass analysis. By tryptic digestion and LC-MS/MS analysis, these mass increases were attributed to modifications on the *C*-terminal cysteine and on the *N*-terminal amino acid.

Source: Reprinted from Ref. 219, with permission of Wiley-Blackwell.

As shown in Table 12.2, except for the one with +257 Da mass increase, all other modifications observed on MP correlate with one modification observed at the *C*-terminal cysteine, in which their mass difference is 178 Da. The trypsin digestion and LC-MS/MS analysis also demonstrated that the +177 and +257 Da modifications were at the *N* terminus of MP (data not shown). Furthermore, the modifications at the N-terminus were more accurately determined, by measurement of the *N*-terminal peptide, to be +178 and +258 Da, instead of +177 and +257 Da observed at the protein level. The less accurate measurement at protein level was due to the decreased accuracy associated with determining larger masses. It has been known that modifications of +178 and +258 Da at the *N* terminus of *E. coli*–expressed proteins, especially those expressed by the BL21 (DE3) strain, are due to gluconoylation and phosphogluconoylation [210–212]. The same modifications are assumed to occur on MP without further verification. Results from this study also demonstrate that gluoconoylation happens not only at the *N* terminus of recombinant proteins but also on the thiol modifier, as shown below.

The four thiol modifiers were removed from MP using DTT and separated on a reversed-phase column followed by LC-MS and LC-MS/MS analysis as described earlier. The thiol modifiers are denoted as TM, followed by their masses in daltons. For example, the thiol molecule that causes mass increase of 305 Da will be designated as TM307. The 2 Da difference between mass increase and modifier mass is due to the loss of two protons during disulfide formation. The identification process for each thiol is described below.

12.4.2.2 TM307

TM307 eluted as a single peak in the reversed-phase chromatogram. Additional studies showed that the difference in retention times of glutathione (GSH) and TM307 was within experimental error. The MS/MS spectrum of the protonated glutathione ion was identical to that of protonated TM307 ion (Figure 12.3). Thus, it can be concluded that TM307 is GSH. GSH is a modulator that keeps the cytoplasm of *E. coli* intact and functional in a reducing environment with concentrations of <10 mM [216,217]. GSH-labeled protein molecules are usually formed in the periplasm when the local environment is nonreducing. If GSH exists at a sufficiently high enough concentration in periplasm, it is possible that some expressed protein molecules react with GSH to form disulfide bonds. Another possibility is that the secreted protein molecules may react with GSH after being secreted into the media. However, since free GSH was not detected in the expression media, the second possibility is less likely.

12.4.2.3 TM485

TM485 eluted as two peaks in the reversed-phase chromatogram with retention times of 5.7 and 7.3 min, and at a peak area ratio of 2:1. These two species were designated as TM485_1 and TM485_2. The MS/MS spectra of the two protonated species were essentially the same (Figure 12.4). The MS/MS spectra of protonated TM485_1 and TM485_2 share strong similarities with those of $[GSH + H]^+$ (Figure 12.3). The fragment ions of $[TM485\_1 + H]^+$ and $[TM485\_2 + H]^+$ overlap with y_1 and y_2

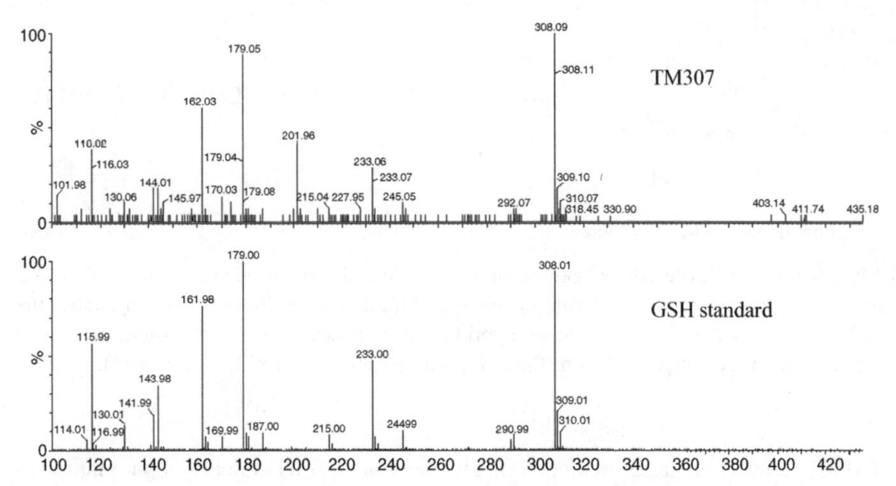

FIGURE 12.3 MS/MS spectra of TM307 and GSH standard (reduced form) showing the same fragmentation pattern (reprinted from Ref. 219, with permission of Wiley-Blackwell).

fragments of GSH. Overall, the MS/MS spectra of [TM485_1 + H]$^+$ and [TM485_2 + H]$^+$ were consistent with *N*-terminal-modified GSH with a mass increase of + 178 Da. Since the + 178 Da modification on the protein was assigned to gluconoylation, TM485_1 and TM485_2 were likewise identified as *N*-terminal gluconoylated GSH. The existence of chromatographically separated TM485_1 and

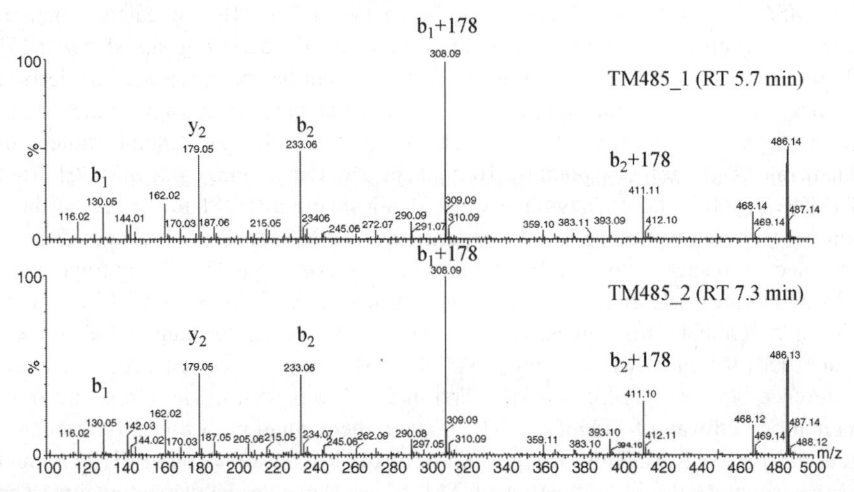

FIGURE 12.4 MS/MS spectra of the two modifiers TM485_1 and TM485_2, in which ions from *b*, *y*, and modified *b* series are shown; the virtually identical fragmentation patterns from two well separated modifiers indicate that they are diastereomers (reprinted from Ref. 219, with permission of Wiley-Blackwell).

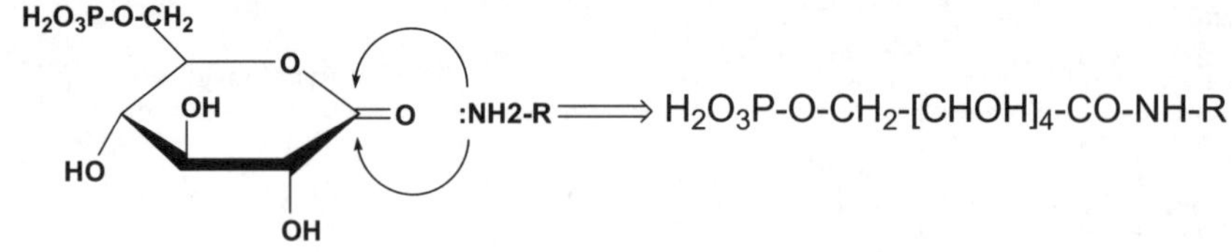

H_2O_3P-O-CH_2 ... :NH2-R ⟹ H_2O_3P-O-CH_2-[CHOH]$_4$-CO-NH-R

6-phosphoglucono-1,5-lactone

FIGURE 12.5 Diastereomers generated from nucleophilic addition to carbonyl group. The primary amine from GSH molecule experiences different accessibilities when it attacks the carbon atom from both sides of the sp^2 hybrid plane, thus resulting in two diastereomers with different quantities (reprinted from Ref. 219, with permission of Wiley-Blackwell).

TM485_2 itself is strong evidence of gluconoylated GSH, since gluconoylation of GSH generates two diastereomers. This is because the gluconoylation process involves a nucleophilic addition to the carbonyl group step in which the nitrogen of *N*-terminal primary amine attacks the sp^2 hybrid carbon of the 6-phosphoglucono-1,5-lactone from either side of the sp^2 plane [210], thus resulting in two diastereomers (Figure 12.5). Since the accessibilities from both sides are different, the ratio of the two isomers would not be 1 : 1. The observed ratio of 2 : 1 confirmed this prediction, and provided orthogonal data supporting the identification of TM485_1 and TM485_2 as diastereomers of gluconoylated GSH.

12.4.2.4 TM358 and TM687

Both TM358 and TM687 eluted as a single peak in the reversed-phase chromatogram. The MS/MS spectra of $[TM358 + H]^+$ and $[TM687 + H]^+$ revealed structural similarities between these two species since both contained a strong signal at *m/z* 261 (Figure 12.6). The existence of the *m/z* 261 fragment has been reported as derived from a series of coenzyme A–related compounds that have strong signals at *m/z* 261 upon fragmentation [218]. The *m/z* 261 fragment is the pantetheine moiety of coenzyme A–related compounds. To confirm that the fragments at *m/z* 261 from TM358, TM687, and coenzyme A were identical, the *m/z* 261 ion from the three sources was fragmented in an MS^3 experiment. The MS^3 spectra of the *m/z* 261 ion obtained from each source were compared as shown in Figure 12.7. The fragmentation patterns of *m/z* 261 generated from coenzyme A, TM358, and TM687 were virtually identical. This suggests TM358 and TM687 contain the pantetheine moiety. Since both the *m/z* 261 ion and $[TM358 + H]^+$ were singly charged, the mass difference between $[TM358 + H]^+$ and *m/z* 261 was 98 Da. Therefore, the fragmentation pathway of $[TM358 + H]^+$ favors the formation of *m/z* 261 through a neutral loss of 98 Da. This strongly indicates the loss of a phosphate (H_3PO_4) and further supports the identification of TM358 as phosphorylated pantetheine. The structure is shown in Figure 12.8.

For coenzyme A fragmentation (Figures 12.6 and 12.8), since the sum of 261 and 508 equals 769, the molecular ion *m/z* 768 appears to be fragmented into two

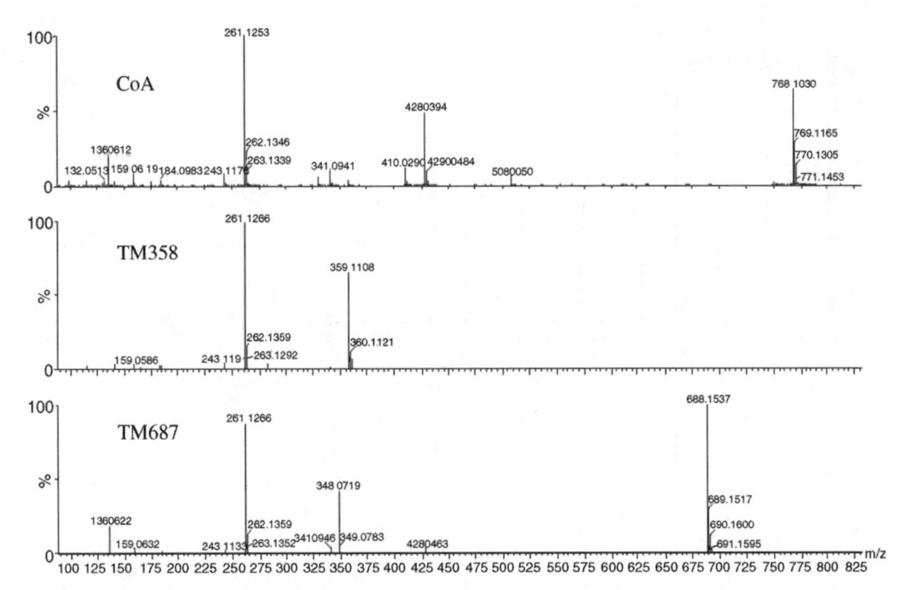

FIGURE 12.6 MS/MS spectra of coenzyme A, TM358, and TM687 indicate all three molecules fragment similarly such that the *m/z* 261 ion is generated (reprinted from Ref. 219, with permission of Wiley-Blackwell).

complementary pieces at *m/z* 261 and *m/z* 508, with the former and latter consisting of pantetheine and ATP portions of the coenzyme A molecule, respectively. There was a strong ion at *m/z* 428 that was 80 *m/z* unit below *m/z* 508, indicating a favored pathway for the formation of the ATP fragment followed by an additional loss of 80 Da (HPO_3 group) from the *m/z* 508 ion. The loss of the HPO_3 group was most likely on the ribose ring, otherwise a further loss of H_2O to generate the *m/z* 410 ion would have lacked a structural foundation. The ion at *m/z* 341 was likely formed when the phosphate group closest to pantetheine broke away from the other two phosphate groups. For $[TM687 + H]^+$, the fragmentation pattern was very similar to that of coenzyme A. Specifically, the ions at *m/z* 428 and *m/z* 261 appeared to be the two complementary fragments. The more likely pathway was loss of the HPO_3 group at *m/z* 348. Similarly, the *m/z* 341 ion was likely formed when the phosphate group closest to pantetheine was lost. Considering that the mass between coenzyme A and TM687 is 80 Da, and considering the very similar fragmentation patterns of these two thiols, we propose that TM687 is dephosphorylated coenzyme A, as shown in Figure 12.8.

Both dephosphorylated coenzyme A and phosphorylated pantetheine are intermediates in coenzyme A biosynthesis. It is not surprising that these two molecules are found as modifiers to the expressed protein molecules. However, it is interesting that coenzyme A itself was not observed as a modifier. There are two possibilities regarding the origin of these two modifiers. If intact coenzyme A is linked to the expressed protein through disulfide bond first, the observed two modifiers may have

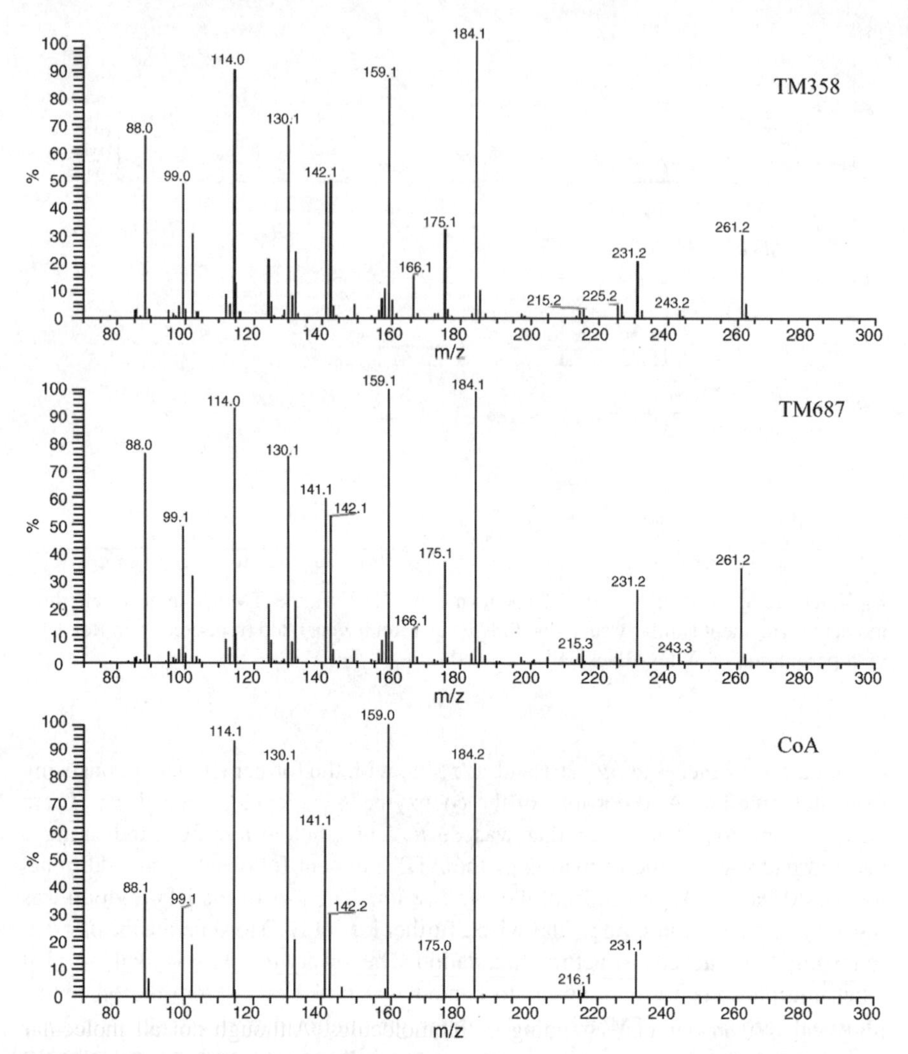

FIGURE 12.7 MS^3 spectra of *m/z* 261 ions from coenzyme A, TM358, and TM687, respectively; the three virtually identical MS^3 spectra indicate that all three molecules contain the same structural moiety (reprinted from Ref. 219, with permission of Wiley-Blackwell).

been generated through certain degradation pathways while coenzyme A is disulfide linked to the protein via a disulfide bond. The other possibility is that the two thiols exist at relatively high concentrations in the periplasm or in the expression media, where they react with expressed protein molecules in a nonreducing environment. Since phosphorylated pantetheine was detected in the expression media, the probability of the second is more likely.

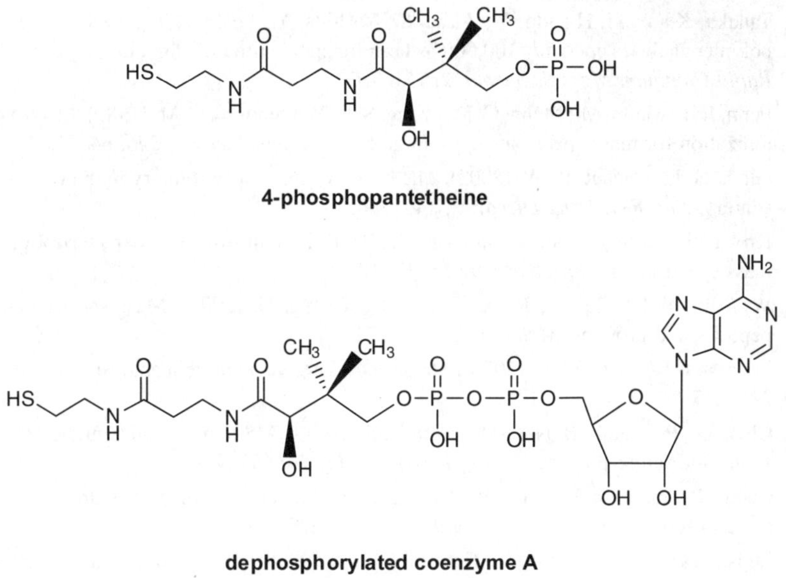

4-phosphopantetheine

dephosphorylated coenzyme A

FIGURE 12.8 Proposed structures for TM358 (4-phosphopantetheine) and TM687 (dephosphorylated coenzyme A) (reprinted from Ref. 219, with permission of Wiley-Blackwell).

12.5 CONCLUSIONS

The ultimate goal of drug development is to determine the safety and efficacy of drug candidates, and to ensure that drug candidates are produced by highly reproducible and well-controlled processes. The strengths of biologics and the associated challenges during drug development result from the complex biophysical properties of biomolecules. While the complex interaction between biomolecules allows specific inhibition or stimulation of therapeutic targets (thus with fewer side effects), it is quite challenging to characterize biomolecules extensively. Since the 1990s, MS has played a pivotal role in our understanding of biomolecules. Although not all molecular variants can be tested for safety and efficacy, since it is almost impossible to isolate all components, MS has made it possible to characterize and monitor these minor components to ensure that the biomolecules are produced reproducibly with regard to the amounts of variants. Through this approach, the correlation of complex biomolecules and the corresponding clinical trials can be firmly established.

REFERENCES

1. Leader, B.; Baca, Q. J.; Golan, D. E. (2008), Protein therapeutics: A summary and pharmacological classification, *Nat. Rev. Drug Discov.* *7*, 21–39.
2. Aggarwal, S. (2009), What's fueling the biotech engine—2008, *Nat. Biotechnol.* *27*, 987–993.

3. Tanaka, K.; Waki, H.; Ido, Y.; Akita, S.; Yoshida, Y.; Yoshida, T. (1988), Protein and polymer analyses up to m/z 100,000 by laser ionization time-of-flight mass spectrometry, *Rapid Commun. Mass Spectrom. 2*, 151–153.
4. Fenn, J. B.; Mann, M.; Meng, C. K.; Wong, S. F.; Whitehouse, C. M. (1989), Electrospray ionization for mass spectrometry of large biomolecules, *Science 246*, 64–71.
5. Glish, G. L.; Vachet, R. W. (2003), The basics of mass spectrometry in the twenty-first century, *Nat. Rev. Drug Discov. 2*, 140–150.
6. Kriwacki, R.; Reisdorph, N.; Siuzdak, G. (2004), Protein structure characterization with mass spectrometry, *Spectroscopy 18*, 37–47.
7. Wysocki, V. H.; Resing, K. A.; Zhang, Q.; Cheng, G. (2005), Mass spectrometry of peptides and proteins, *Methods 35*, 211–222.
8. Domon, B.; Aebersold, R. (2006), Mass spectrometry and protein analysis, *Science 312*, 212–217.
9. Chen, G.; Pramanik, B. N. (2009), Application of LC/MS to proteomics studies: Current status and future prospects, *Drug Discov. Today 14*, 465–471.
10. Chen, G., Pramanik, B. N. (2008), LC-MS for protein characterization: Current capabilities and future trends, *Expert Rev. Proteom. 5*, 435–444.
11. Janiszewski, J. S.; Liston, T. E.; Cole, M. J. (2008), Perspectives on bioanalytical mass spectrometry and automation in drug discovery. *Curr. Drug Metab. 9*, 986–994.
12. Annis, D. A.; Nickbarg, E.; Yang, X.; Ziebell, M. R.; Whitehurst, C. E. (2007), Affinity selection-mass spectrometry screening techniques for small molecule drug discovery, *Curr. Opin. Chem. Biol. 11*, 518–526.
13. Servais, A.-C.; Crommen, J.; Fillet, M. (2006), Capillary electrophoresis-mass spectrometry, an attractive tool for drug bioanalysis and biomarker discovery, *Electrophoresis 27*, 2616–2629.
14. Deng, G.; Sanyal, G. (2006), Applications of mass spectrometry in early stages of target based drug discovery, *J. Pharm. Biomed. Anal. 40*, 528–538.
15. Zhang, J.; McCombie, G.; Guenat, C.; Knochenmuss, R. (2005), FT-ICR mass spectrometry in the drug discovery process, *Drug Discov. Today 10*, 635–642.
16. Flook, P. K.; Yan, L.; Szalma, S. (2003), Target validation through high throughput proteomics analysis, *Drug Discov. Today (TARGETS) 2*, 217–223.
17. Valle, R. P. C.; Jendoubi, M. (2003), Antibody-based technologies for target discovery, *Curr. Opin. Drug Discov. Devel. 6*, 197–203.
18. Ryan, T. E.; Patterson, S. D. (2002), Proteomics: Drug target discovery on an industrial scale. *Trends Biotechnol. 20*, S45–51.
19. Weng, Z.; DeLisi, C. (2002), Protein therapeutics: Promises and challenges for the 21st century, *Trends Biotechnol. 20*, 29–35.
20. Veenstra, T. D. (2006), Proteomic approaches in drug discovery, *Drug Discov. Today (Technol.) 3*, 433–440.
21. Papac, D. I.; Shahrokh, Z. (2001), Mass spectrometry innovations in drug discovery and development, *Pharm. Res. 18*, 131–145.
22. John, H.; Walden, M.; Schäfer, S.; Genz, S.; Forssmann, W.-G. (2004), Analytical procedures for quantification of peptides in pharmaceutical research by liquid chromatography-mass spectrometry, *Anal. Bioanal. Chem. 378*, 883–897.

23. Ovsyannikova, I. G.; Johnson, K. L.; Bergen, H. R., III; Poland, G. A. (2007), Mass spectrometry and peptide-based vaccine development, *Clin. Pharmacol. Ther. 82*, 644–652.

24. Azad, N. S.; Rasool, N.; Annunziata, C. M.; Minasian, L.; Whiteley, G.; Kohn, E. C. (2006), Proteomics in clinical trials and practice: Present uses and future promise, *Molec. Cell. Proteom. 5*, 1819–1829.

25. Daugherty, A. L.; Mrsny, R. J. (2006), Formulation and delivery issues for monoclonal antibody therapeutics, *Adv. Drug Deliv. Rev. 58*, 686–706.

26. Baumann, A. (2006), Early development of therapeutic biologics—Pharmacokinetics, *Curr. Drug Metab. 7*, 15–21.

27. Hicks, L. M.; Mazur, M. T.; Miller, L. M.; Dorrestein, P. C.; Schnarr, N. A.; Khosla, C.; Kelleher, N. L. (2006), Investigating nonribosomal peptide and polyketide biosynthesis by direct detection of intermediates on >70 kDa polypeptides by using fourier-transform mass spectrometry, *ChemBioChem. 7*, 904–907.

28. Horn, D. M.; Zubarev, R. A.; McLafferty, F. W. (2000), Automated de novo sequencing of proteins by tandem high-resolution mass spectrometry, *Proc. Natl. Acad. Sci. USA 97*, 10313–10317.

29. Rosenfeld J.; Capdevielle, J.; Guillemot, J. C.; Ferrara, P. (1992), In-gel digestion of proteins for internal sequence analysis after one- or two-dimensional gel electrophoresis, *Anal. Biochem. 203*, 173–179.

30. Hellman, U.; Wernstedt, C.; Góñez, J.; Heldin, C. H. (1995), Improvement of an "In-Gel" digestion procedure for the micropreparation of internal protein fragments for amino acid sequencing, *Anal. Biochem. 224*, 451–455.

31. Shevchenko A., Wilm M., Vorm O., Mann M. (1996), Mass spectrometric sequencing of proteins from silver-stained polyacrylamide gels, *Anal. Chem. 68*, 850–858.

32. Aebersold, R.; Mann, M. (2003), Mass spectrometry-based proteomics, *Nature, 422*, 198–207.

33. Liebers, V.; Raulf-Heimsoth, M.; Brüning, T. (2008), Health effects due to endotoxin inhalation, *Arch. Toxicol. 82*, 203–210.

34. Liebers, V.; Brüning, T.; Raulf-Heimsoth, M. (2006), Occupational endotoxin-exposure and possible health effects on humans, *Am. J. Industr. Med. 49*, 474–491.

35. Bryans, T. D.; Braithwaite, C.; Broad, J.; Cooper, J. F.; Darnell, K. R.; Hitchins, V. M.; Karren, A. J.; Lee, P. S. (2004), Bacterial endotoxin testing: A report on the methods, background, data, and regulatory history of extraction recovery efficiency. *Biomed. Instrum. Technol. 38*, 73–78.

36. Brandenburg, K.; Howe, J.; Gutsman, T.; Garidel, P. (2009), The expression of endotoxic activity in the Limulus test as compared to cytokine production in immune cells, *Curr. Med. Chem. 16*, 2653–2660.

37. Ding, J. L.; Ho, B. (2001), A new era in pyrogen testing, *Trends Biotechnol. 19*, 277–281.

38. Bolger, R.; Lenoch, F.; Allen, E.; Meiklejohn, B.; Burke, T. (1997), Fluorescent dye assay for detection of DNA in recombinant protein products, *BioTechniques 23*, 532–537.

39. Riggin, A.; Luu, V. T.; Lobdell, J. K.; Wind, M. K. (1997), A non-isotopic probe-hybridization assay for residual DNA in biopharmaceuticals, *J. Pharm. Biomed. Anal. 16*, 561–572.

40. Ji, X.; Lee, K.; DiPaolo, B. (2002), High-sensitivity hybridization assay for quantitation of residual E. coli DNA, *BioTechniques 32*, 1162–1167.
41. Konrad, M. (1989), Immunogenecity of proteins administered to humans for therapeutic purposes, *Trends Biotechnol. 7*, 175–179.
42. Eaton, L. C. (1995), Host cell contaminant protein assay development for recombinant biopharmaceuticals, *J. Chromatogr. A 705*, 105–114.
43. Ghobrial, I. A.; Wong, D. T.; Sharma, B. G. (1997), An immuno-ligand assay for the detection and quantitation of contaminating proteins in recombinant human erythropoietin (r-HuEPO), *Pharm. Technol. 21*, 48–56.
44. Chen, A. B.; Championsmith, A. A.; Blanchard, J.; Gorrell, J.; Niepelt, B. A.; Federici, M. M.; Formento, J.; Sinicropi, D. V. (1992), Quantitation of E. coli protein impurities in recombinant human interferon-γ, *Appl. Biochem. Biotechnol. 36*, 137–152.
45. Whitmire, M. L.; Eaton, L. C. (1997), An immunoligand assay for quantitation of process specific Escherichia coli host cell contaminant proteins in a recombinant bovine somatotropin, *J. Immunoassay 18*, 49–65.
46. Dagouassat, N.; Haeuw, J.-F.; Robillard, V.; Damien, F.; Libon, C.; Corvaïa, N.; Lawny, F.; Nguyen, T. N.; Bonnefoy, J.-Y.; Beck, A. (2001), Development of a quantitative assay for residual host cell proteins in a recombinant subunit vaccine against human respiratory syncytial virus, *J. Immunol. Meth. 251*, 151–159.
47. Fowler, J. D.; Brown, J. A.; Kvaratskhelia, M.; Suo, Z. (2009), Probing conformational changes of human DNA polymerase λ using mass spectrometry-based protein footprinting. *J. Molec. Biol. 390*, 368–379.
48. Kvaratskhelia, M.; Miller, J. T.; Budihas, S. R.; Pannell, L. K.; Le Grice, S. F. J. (2002), Identification of specific HIV-1 reverse transcriptase contacts to the viral RNA:tRNA complex by mass spectrometry and a primary amine selective reagent, *Proc. Nat. Acad. Sci. USA 99*, 15988–15993.
49. Wood, T. D.; Guan, Z.; Borders, C. L., Jr.; Chen, L. H.; Kenyon, G. L.; Mclafferty, F. W. (1998), Creatine kinase: Essential arginine residues at the nucleotide binding site identified by chemical modification and high-resolution tandem mass spectrometry, *Proc. Nat. Acad. Sci. USA 95*, 3362–3365.
50. Philo, J. S.; Arakawa, T. (2009), Mechanisms of protein aggregation, *Curr. Pharm. Biotechnol. 10*, 348–351.
51. Chi, E. Y.; Krishnan, S.; Randolph, T. W.; Carpenter, J. F. (2003), Physical stability of proteins in aqueous solution: Mechanism and driving forces in nonnative protein aggregation, *Pharm. Res. 20*, 1325–1336.
52. Philo, J. S. (2009), A critical review of methods for size characterization of non-particulate protein aggregates, *Curr. Pharm. Biotechnol. 10*, 359–372.
53. Philo, J. S. (2006), Is any measurement method optimal for all aggregate sizes and types? *AAPS (Am. Assoc. Pharmceutical. Scientists) J. 8*, 564–571.
54. Cao, S.; Pollastrini, J.; Jiang, Y. (2009), Separation and characterization of protein aggregates and particles by field flow fractionation, *Curr. Pharm. Biotechnol. 10*, 382–390.
55. Kendrick, B. S.; Kerwin, B. A.; Chang, B. S.; Philo, J. S. (2001), Online size-exclusion high-performance liquid chromatography light scattering and differential refractometry methods to determine degree of polymer conjugation to proteins and protein-protein or protein-ligand association states, *Anal. Biochem. 299*, 136–146.

56. Philo, J. S. (2001), Ultracentrifugation, *BioPharm. 14*, 52–54.
57. Harris, R. J.; Kabakoff, B.; Macchi, F. D.; Shen, F. J.; Kwong, M.; Andya, J. D.; Shire, S. J.; Bjork, N.; Totpal, K.; Chen, A. B. (2001), Identification of multiple sources of charge heterogeneity in a recombinant antibody, *J. Chromatogr. B (Biomed. Sci. Appl.) 752*, 233–245.
58. Di Donato, A.; Ciardiello, M. A.; De Nigris, M.; Piccoli, R.; Mazzarella, L.; D'Alessio, G. (1993), Selective deamidation of ribonuclease A. Isolation and characterization of the resulting isoaspartyl and aspartyl derivatives, *J. Biol. Chem. 268*, 4745–4751.
59. Gupta, R.; Srivastava, O. P. (2004), Deamidation affects structural and functional properties of human αA-crystallin and its oligomerization with a B-crystallin, *J. Biol. Chem. 279*, 44258–44269.
60. Cacia, J.; Keck, R.; Presta, L. G.; Frenz, J. (1996), Isomerization of an aspartic acid residue in the complementarity-determining regions of a recombinant antibody to human IgE: Identification and effect on binding affinity, *Biochemistry 35*, 1897–1903.
61. Harris, R. J.; Kabakoff, B.; Macchi, F. D.; Shen, F. J.; Kwong, M.; Andya, J. D.; Shire, S. J.; Bjork, N.; Totpal, K.; Chen, A. B. (2001), Identification of multiple sources of charge heterogeneity in a recombinant antibody, *J. Chromatogr. B (Biomed. Sci. Appl.) 752*, 233–245.
62. Paborji, M.; Pochopin, N. L.; Coppola, W. P.; Bogardus, J. B. (1994), Chemical and physical stability of chimeric L6, a mouse-human monoclonal antibody, *Pharm. Res. 11*, 764–771.
63. Geiger, T.; Clarke, S. (1987), Deamidation, isomerization, and racemization at asparaginyl and aspartyl residues in peptides. Succinimide-linked reactions that contribute to protein degradation, *J. Biol. Chem.*, *262*, 785–794.
64. Kroon, D. J.; Baldwin-Ferro, A.; Lalan, P. (1992), Identification of sites of degradation in a therapeutic monoclonal antibody by peptide mapping, *Pharm. Res.*, *9*, 1386–1393.
65. Hsu, Y.-R.; Chang, W.-C.; Mendiaz, E. A.; Hara, S.; Chow, D. T.; Mann, M. B.; Langley, K. E.; Lu, H. S. (1998), Selective deamidation of recombinant human stem cell factor during in vitro aging: Isolation and characterization of the aspartyl and isoaspartyl homodimers and heterodimers, *Biochemistry 37*, 2251–2262.
66. Tsai, P. K.; Bruner, M. W.; Irwin, J. I.; Ip, C. C. Y.; Oliver, C. N.; Nelson, R. W.; Volkin, D. B.; Middaugh, C. R. (1993), Origin of the isoelectric heterogeneity of monoclonal immunoglobulin h1B4, *Pharm. Res.*, *10*, 1580–1586.
67. Noguchi, S.; Miyawaki, K.; Satow, Y. (1998), Succinimide and isoaspartate residues in the crystal structures of hen egg-white lysozyme complexed with tri-N-acetylchitotriose, *J. Molec. Biol. 278*, 231–238.
68. Huang, L.; Lu, J.; Wroblewski, V. J.; Beals, J. M.; Riggin, R. M. (2005), In vivo deamidation characterization of monoclonal antibody by LC/MS/MS, *Anal. Chem. 77*, 1432–1439.
69. Paul, S. R.; Bennett, F.; Calvetti, J. A.; Kelleher, K.; Wood, C. R.; O'Hara, R. M., Jr.; Leary, A. C.; Sibley, B.; Clark, S. C.; Williams, D. A.; Yang, Y.-C. (1990), Molecular cloning of a cDNA encoding interleukin 11, a stromal cell-derived lymphopoietic and hematopoietic cytokine, *Proc. Nat. Acad. Sci. USA 87*, 7512–7516.
70. Tyler-Cross, R.; Schirch, V. (1991), Effects of amino acid sequence, buffers, and ionic strength on the rate and mechanism of deamidation of asparagine residues in small peptides, *J. Biol. Chem.*, *266*, 22549–22556.

71. Solstad, T.; Carvalho, R. N.; Andersen, O. A.; Waidelich, D.; Flatmark, T. (2003), Deamidation of labile asparagine residues in the autoregulatory sequence of human phenylalanine hydroxylase: Structural and functional implications, *Eur. J. Biochem. 270*, 929–938.
72. Paborji, M.; Pochopin, N. L.; Coppola, W. P.; Bogardus, J. B. (1994), Chemical and physical stability of chimeric L6, a mouse-human monoclonal antibody, *Pharm. Res. 11*, 764–771.
73. Rao, P. E.; Kroon, D. J. (1993), Orthoclone OKT3. Chemical mechanisms and functional effects of degradation of a therapeutic monoclonal antibody, *Pharm. Biotechnol. 5*, 135–158.
74. Liu, Y. D.; van Enk, J. Z.; Flynn, G. C. (2009), Human antibody Fc deamidation in vivo, *Biologicals 37*, 313–322.
75. Hsu, Y.-R.; Chang, W.-C.; Mendiaz, E. A.; Hara, S.; Chow, D. T.; Mann, M. B.; Langley, K. E.; Lu, H. S. (1998), Selective deamidation of recombinant human stem cell factor during in vitro aging: Isolation and characterization of the aspartyl and isoaspartyl homodimers and heterodimers, *Biochemistry 37*, 2251–2262.
76. Tsai, P. K.; Bruner, M. W.; Irwin, J. I.; Ip, C. C. Y.; Oliver, C. N.; Nelson, R. W.; Volkin, D. B.; Middaugh, C. R. (1993), Origin of the isoelectric heterogeneity of monoclonal immunoglobulin h1B4, *Pharm. Res. 10*, 1580–1586.
77. Yan, B.; Steen, S.; Hambly, D.; Valliere-Douglass, J.; Vanden Bos, T.; Smallwood, S.; Yates, Z.; Arroll, T.; Han, Y.; Gadgil, H.; Latypov, R. F.; Wallace, A.; et al. (2009), Succinimide formation at Asn 55 in the complementarity determining region of a recombinant monoclonal antibody IgG1 heavy chain, *J. Pharm. Sci. 98*, 3509–3521.
78. Vlasak, J.; Bussat, M. C.; Wang, S.; Wagner-Rousset, E.; Schaefer, M.; Klinguer-Hamour, C.; Kirchmeier, M.; Corvaïa, N.; Ionescu, R.; Beck, A. (2009), Identification and characterization of asparagine deamidation in the light chain CDR1 of a humanized IgG1 antibody, *Anal. Biochem. 392*, 145–154.
79. Friedman, A. R.; Ichhpurani, A. K.; Brown, D. M.; Hillman, R. M.; Krabill, L. F.; Martin, R. A.; Zurcher-Neely, H. A.; Guido, D. M. (1991), Degradation of growth hormone releasing factor analogs in neutral aqueous solution is related to deamidation of asparagine residues. Replacement of asparagine residues by serine stabilizers, *Int. J. Peptide Protein Res. 37*, 14–20.
80. Harris, R. J.; Kabakoff, B.; Macchi, F. D.; Shen, F. J.; Kwong, M.; Andya, J. D.; Shire, S. J.; Bjork, N.; Totpal, K.; Chen, A. B. (2001), Identification of multiple sources of charge heterogeneity in a recombinant antibody, *J. Chromatogr. B (Biomed. Sci. Appl.) 752*, 233–245.
81. Aswad, D. W.; Paranandi, M. V.; Schurter, B. T. (2000), Isoaspartate in peptides and proteins: Formation, significance, and analysis, *J. Pharm. Biomed. Anal. 21*, 1129–1136.
82. Paul, S. R.; Bennett, F.; Calvetti, J. A.; Kelleher, K.; Wood, C. R.; O'Hara, R. M., Jr.; Leary, A. C.; Sibley, B.; Clark, S. C.; Williams, D. A.; Yang, Y.-C. (1990), Molecular cloning of a cDNA encoding interleukin 11, a stromal cell-derived lymphopoietic and hematopoietic cytokine, *Proc. Nat. Acad. Sci. USA*, *87*, 7512–7516.
83. Tyler-Cross, R.; Schirch, V. (1991), Effects of amino acid sequence, buffers, and ionic strength on the rate and mechanism of deamidation of asparagine residues in small peptides, *J. Biol. Chem.*, *266*, 22549–22556.

84. Huang, H. Z.; Nichols, A.; Liu, D. (2009), Direct identification and quantification of aspartyl succinimide in an IgG2 mAb by RapiGest assisted digestion, *Anal. Chem.*, *81*, 1686–1692.
85. Bischoff, R.; Lepage, P.; Jaquinod, M.; Cauet, G.; Acker-Klein, M.; Clesse, D.; Laporte, M.; Bayol, A.; Van Dorsselaer, A.; Roitsch, C. (1993), Sequence-specific deamidation: Isolation and biochemical characterization of succinimide intermediates of recombinant hirudin, *Biochemistry 32*, 725–734.
86. Gotte, G.; Libonati, M.; Laurents, D. V. (2003), Glycosylation and specific deamidation of ribonuclease B affect the formation of three-dimensional domain-swapped oligomers. *J. Biol. Chem.*, *278*, 46241–46251.
87. Zhang, W.; Czupryn, M. J. (2003), Analysis of isoaspartate in a recombinant monoclonal antibody and its charge isoforms, *J. Pharm. Biomed. Anal. 30*, 1479–1490.
88. Lindner, H.; Sarg, B.; Grunicke, H.; Helliger, W. (1999), Age-dependent deamidation of H1°histones in chromatin of mammalian tissues, *J. Cancer Res. Clin. Oncol. 125*, 182–186.
89. Wakankar, A. A.; Borchardt, R. T.; Eigenbrot, C.; Shia, S.; Wang, Y. J.; Shire, S. J.; Liu, J. L. (2007), Aspartate isomerization in the complementarity-determining regions of two closely related monoclonal antibodies, *Biochemistry 46*, 1534–1544.
90. Johnson, B. A.; Aswad, D. W. (1991), Optimal conditions for the use of protein L-isoaspartyl methyltransferase in assessing the isoaspartate content of peptides and proteins, *Anal. Biochem. 192*, 384–391.
91. Johnson, B. A.; Aswad, D. W. (1993), Kinetic properties of bovine brain protein L-isoaspartyl methyltransferase determined using a synthetic isoaspartyl peptide substrate, *Neurochem. Res. 18*, 87–94.
92. Schmid, D. G.; Von der Mulbe, F.; Fleckenstein, B.; Weinschenk, T.; Jung, G. (2001), Broadband detection electrospray ionization fourier transform ion cyclotron resonance mass spectrometry to reveal enzymatically and chemically induced deamidation reactions within peptides, *Anal. Chem. 73*, 6008–6013.
93. Zabrouskov, V.; Han, X.; Welker, E.; Zhai, H.; Lin, C.; Van Wijk, K. J.; Scheraga, H. A.; McLafferty, F. W. (2006), Stepwise deamidation of ribonuclease A at five sites determined by top down mass spectrometry, *Biochemistry 45*, 987–992.
94. Robinson, N. E.; Zabrouskov, V.; Zhang, J.; Lampi, K. J.; Robinson, A. B. (2006), Measurement of deamidation of intact proteins by isotopic envelope and mass defect with ion cyclotron resonance Fourier transform mass spectrometry, *Rapid Commun. Mass Spectrom. 20*, 3535–3541.
95. O'Connor, P. B.; Cournoyer, J. J.; Pitteri, S. J.; Chrisman, P. A.; McLuckey, S. A. (2006), Differentiation of aspartic and isoaspartic acids using electron transfer dissociation, *J. Am. Soc. Mass Spectrom. 17*, 15–19.
96. Cournoyer, J. J.; Pittman, J. L.; Ivleva, V. B.; Fallows, E.; Waskell, L.; Costello, C. E.; O'Connor, P. B. (2005), Deamidation: Differentiation of aspartyl from isoaspartyl products in peptides by electron capture dissociation, *Protein Sci. 14*, 452–463.
97. Chelius, D.; Render, D. S.; Bondarenko, P. V. (2005), Identification and characterization of deamidation sites in the conserved regions of human immunoglobulin gamma antibodies, *Anal. Chem. 77*, 6004–6011.
98. Gaza-Bulseco, G.; Liu, H. (2008), Fragmentation of a recombinant monoclonal antibody at various pH, *Pharm. Res. 25*, 1881–1890.

99. Brennan, T. V.; Clarke, S. (1995), Effect of adjacent histidine and cysteine residues on the spontaneous degradation of asparaginyl- and aspartyl-containing peptides, *Int. J. Peptide Protein Res. 45*, 547–553.
100. Robinson, N. E.; Robinson, Z. W.; Robinson, B. R.; Robinson, A. L.; Robinson, J. A.; Robinson, M. L.; Robinson, A. B. (2004), Structure-dependent nonenzymatic deamidation of glutaminyl and asparaginyl pentapeptides, *J. Peptide Res. 63*, 426–436.
101. Patel, K.; Borchardt, R. T. (1990), Chemical pathways of peptide degradation. III. effect of primary sequence on the pathways of deamidation of asparaginyl residues in hexapeptides, *Pharm. Res. 7*, 787–793.
102. Kosky, A. A.; Razzaq, U. O.; Treuheit, M. J.; Brems, D. N. (1999), The effects of alpha-helix on the stability of Asn residues: Deamidation rates in peptides of varying helicity, *Protein Sci. 8*, 2519–2523.
103. Wearne, S. J.; Creighton, T. E. (1989), Effect of protein conformation on rate of deamidation: Ribonuclease A, *Proteins Struct. Funct. Genet. 5*, 8–12.
104. Sinha, S.; Zhang, L.; Duan, S.; Williams, T. D.; Vlasak, J.; Ionescu, R.; Topp, E. M. (2009), Effect of protein structure on deamidation rate in the Fc fragment of an IgG1 monoclonal antibody, *Protein Sci. 18*, 1573–1584.
105. Liu, H.; Gaza-Bulseco, G.; Sun, J. (2006), Characterization of the stability of a fully human monoclonal IgG after prolonged incubation at elevated temperature, *J. Chromatogr. B (Anal. Technol. Biomed. Life Sci.) 837*, 35–43.
106. Liu, H.; Gaza-Bulseco, G.; Chumsae, C. (2008), Glutamine deamidation of a recombinant monoclonal antibody, *Rapid Commun. Mass Spectrom. 22*, 4081–4088.
107. Wright, H. T. (1991), Nonenzymatic deamidation of asparaginyl and glutaminyl residues in proteins, *Crit. Rev. Biochem. Molec. Biol. 26*, 1–52.
108. Oliyai, C.; Borchardt, R. T. (1993), Chemical pathways of peptide degradation. IV. Pathways, kinetics, and mechanism of degradation of an aspartyl residue in a model hexapeptide, *Pharm. Res. 10*, 95–102.
109. Wakankar, A. A.; Borchardt, R. T. (2006), Formulation considerations for proteins susceptible to asparagine deamidation and aspartate isomerization, *J. Pharm. Sci. 95*, 2321–2336.
110. Brennan, T. V.; Clarke, S. (1993), Spontaneous degradation of polypeptides at aspartyl and asparaginyl residues: Effects of the solvent dielectric, *Protein Sci. 2*, 331–338.
111. Zhang, W.; Czupryn, M. J.; Boyle, P. T., Jr.; Amari, J. (2002), Characterization of asparagine deamidation and aspartate isomerization in recombinant human interleukin-11, *Pharm. Res. 19*, 1223–1231.
112. Clarke, S. (1987), Propensity for spontaneous succinimide formation from aspartyl and asparaginyl residues in cellular proteins, *Int. J. Peptide Protein Res. 30*, 808–821.
113. Violand, B. N.; Schlittler, M. R.; Toren, P. C.; Siegel, N. R. (1990), Formation of isoaspartate 99 in bovine and porcine somatotropins, *J. Protein Chem. 9*, 109–117.
114. Catak, S.; Monard, G.; Aviyente, V.; Ruiz-López, M. F. (2008), Computational study on nonenzymatic peptide bond cleavage at asparagine and apartic acid, *J. Phys. Chem. A 112*, 8752–8761.
115. Joshi, A. B.; Rus, E.; Kirsch, L. E. (2000), The degradation pathways of glucagon in acidic solutions, *Int. J. Pharm. 203*, 115–125.

116. Joshi, A. B.; Sawai, M.; Kearney, W. R.; Kirsch, L. E. (2005), Studies on the mechanism of aspartic acid cleavage and glutamine deamidation in the acidic degradation of glucagon, *J. Pharm. Sci. 94*, 1912–1927.

117. Pisskiewicz, D.; Michael L.; Smith, E. L. (1970), Anomalous cleavage of aspartyl-proline peptide bonds during amino acid sequence determination, *Biochem. Biophys. Res. Commun. 40*, 1173–1178.

118. Gerschler, J. J.; Wier, K. A.; Hansen, D. E. (2007), Amide bond cleavage: Acceleration due to a 1,3-diaxial interaction with a carboxylic acid, *J. Org. Chem. 72*, 654–657.

119. Landon, M. (1977), Cleavage at aspartyl-prolyl bonds, *Meth. Enzymol. 47*, 145–149.

120. Marcus, F. (1985), Preferential cleavage at aspartyl-prolyl peptide bonds in dilute acid, *Int. J. Peptide Protein Res. 25*, 542–546.

121. Voorter, C. E. M.; De Haard-Hoekman, W. A.; Van Den Oetelaar, P. J. M.; Bloemendal, H.; De Jong, W. W. (1988), Spontaneous peptide bond cleavage in aging α-crystallin through a succinimide intermediate, *J. Biol. Chem. 263*, 19020–19023.

122. Cordoba, A. J.; Shyong, B.-J.; Breen, D.; Harris, R. J. (2005), Non-enzymatic hinge region fragmentation of antibodies in solution, *J. Chromatogr. B (Anal. Technol. Biomed. Life Sci.) 818*, 115–121.

123. Cohen, S. L.; Price, C.; Vlasak, J. (2007), β-Elimination and peptide bond hydrolysis: Two distinct mechanisms of human IgG1 hinge fragmentation upon storage, *J. Am. Chem. Soc. 129*, 6976–6977.

124. Liu, H.; Gaza-Bulseco, G.; Lundell, E. (2008), Assessment of antibody fragmentation by reversed-phase liquid chromatography and mass spectrometry, *J. Chromatogr. B (Anal. Technol. Biomed. Life Sci.) 876*, 13–23.

125. Jiskoot, W.; Beuvery, E. C.; de Koning, A. M.; Herron, J. N.; Crommelin, D. J. A. (1990), *Pharm. Res. 7*, 1234–1241.

126. Alexander, A. J.; Hughes, D. E. (1995), Monitoring of IgG antibody thermal stability by micellar electrokinetic capillary chromatography and matrix-assisted laser desorption/ionization mass spectrometry, *Anal Chem. 67*, 3626–3632.

127. Smith, M. A.; Easton, M.; Everett, P.; Lewis, G.; Payne, M.; Riveros-Moreno, V.; Allen, G. (1996), Specific cleavage of immunoglobulin G by copper ions, *Int. J. Peptide Protein Res. 48*, 48–55.

128. Davies, K. J. (1987), Protein damage and degradation by oxygen radicals. I. General aspects, *J. Biol. Chem., 262*, 9895–9901.

129. Davies, K. J.; Delsignore, M. E.; Lin, S. W. (1987), Protein damage and degradation by oxygen radicals. II. Modification of amino acids, *J. Biol. Chem., 262*, 9902–9907.

130. Dean, R. T.; Fu, S.; Stocker, R.; Davies, M. J. (1997), Biochemistry and pathology of radical-mediated protein oxidation, *Biochem. J. 324*, 1–18.

131. Garrison, W. M. (1987), Reaction mechanisms in the radiolysis of peptides, polypeptides, and proteins, *Chem. Rev. 87*, 381–398.

132. Cecarini, V.; Gee, J.; Fioretti, E.; Amici, M.; Angeletti, M.; Eleuteri, A. M.; Keller, J. N. (2007), Protein oxidation and cellular homeostasis: Emphasis on metabolism, *Biochim. Biophys. Acta (Molec. Cell Res.) 1773*, 93–104.

133. Grune, T.; Klotz, L.-O.; Gieche, J.; Rudeck, M.; Sies, H. (2001), Protein oxidation and proteolysis by the nonradical oxidants singlet oxygen or peroxynitrite, *Free Radical Biol. Med. 30*, 1243–1253.

134. Hawkins, C. L.; Pattison, D. I.; Davies, M. J. (2003), Hypochlorite-induced oxidation of amino acids, peptides and proteins, *Amino Acids 25*, 259–274.
135. Stadtman, E. R. (1990), Metal ion-catalyzed oxidation of proteins: Biochemical mechanism and biological consequences, *Free Radical Biol. Med. 9*, 315–325.
136. Stadtman, E. R.; Levine, R. L. (2000), Protein oxidation, *Ann. NY Acad. Sci. 899*, 191–208.
137. Stadtman, E. R.; Levine, R. L. (2003), Free radical-mediated oxidation of free amino acids and amino acid residues in proteins, *Amino Acids 25*, 207–218.
138. Shacter, E. (2000), Protein oxidative damage, *Meth. Enzymol. 319*, 428–436.
139. Berlett, B. S.; Stadtman, E. R. (1997), Protein oxidation in aging, disease, and oxidative stress, *J. Biol. Chem. 272*, 20313–20316.
140. Stadtman, E. R.; Berlett, B. S. (1998), Reactive oxygen-mediated protein oxidation in aging and disease, *Drug Metab. Rev. 30*, 225–243.
141. Stadtman, E. R.; Oliver, C. N. (1991), Metal-catalyzed oxidation of proteins: Physiological consequences, *J. Biol. Chem. 266*, 2005–2008.
142. Steinberg, D.; Parthasarathy, S.; Carew, T. E.; Khoo, J. C.; Witztum, J. L. (1989), Beyond cholesterol: Modifications of low-density lipoprotein that increase its atherogenicity, *N. Engl. J. Med. 320*, 915–924.
143. Levine, R. L.; Berlett, B. S.; Moskovitz, J.; Mosoni, L.; Stadtman, E. R. (1999), Methionine residues may protect proteins from critical oxidative damage, *Mech. Ageing Devel. 107*, 323–332.
144. Levine, R. L.; Wehr, N.; Williams, J. A.; Stadtman, E. R.; Shacter, E. (2000), Determination of carbonyl groups in oxidized proteins, *Meth. Molec. Biol. 99*, 15–24.
145. Stadtman, E. R.; Moskovitz, J.; Levine, R. L. (2003), Oxidation of methionine residues of proteins: Biological consequences, *Antioxidants Redox Signal. 5*, 577–582.
146. Manning, M. C.; Patel, K.; Borchardt, R. T. (1989), Stability of protein pharmaceuticals, *Pharm. Res. 6*, 903–917.
147. Wasylaschuk, W. R.; Harmon, P. A.; Wagner, G.; Harman, A. B.; Templeton, A. C.; Xu, H.; Reed, R. A. (2007), Evaluation of hydroperoxides in common pharmaceutical excipients, *J. Pharm. Sci. 96*, 106–116.
148. Huang, T.; Garceau, M. E.; Gao, P. (2003), Liquid chromatographic determination of residual hydrogen peroxide in pharmaceutical excipients using platinum and wired enzyme electrodes, *J. Pharm. Biomed. Anal. 31*, 1203–1210.
149. Hawkins, C. L.; Davies, M. J. (1998), EPR studies on the selectivity of hydroxyl radical attack on amino acids and peptides, *J. Chem. Soc. Perkin Trans. 2*, 2617–2622.
150. Hawkins, C. L.; Davies, M. J. (2001), Generation and propagation of radical reactions on proteins, *Biochim. Biophys. Acta (Bioenergetics) 1504*, 196–219.
151. Davies, M. J.; Fu, S.; Wang, H.; Dean, R. T. (1999), Stable markers of oxidant damage to proteins and their application in the study of human disease, *Free Radical Biol. Med. 27*, 1151–1163.
152. Eaton, P. (2006), Protein thiol oxidation in health and disease: Techniques for measuring disulfides and related modifications in complex protein mixtures, *Free Radical Biol. Med. 40*, 1889–1899.
153. Reddie, K. G.; Carroll, K. S. (2008), Expanding the functional diversity of proteins through cysteine oxidation, *Curr. Opin. Chem. Biol. 12*, 746–754.

154. Krishnamurthy, R.; Manning, M. C. (2002), The stability factor: Importance in formulation development, *Curr. Pharm. Biotechnol. 3*, 361–371.

155. Shetty, V.; Spellman, D. S.; Neubert, T. A. (2007), Characterization by tandem mass spectrometry of stable cysteine sulfenic acid in a cysteine switch peptide of matrix metalloproteinases, *J. Am. Soc. Mass Spectrom. 18*, 1544–1551.

156. Wu, J.; Watson, J. T. (1997), A novel methodology for assignment of disulfide bond pairings in proteins, *Protein Sci. 6*, 391–398.

157. Xiang, T.; Chumsae, C.; Liu, H. (2009), Localization and quantitation of free sulfhydryl in recombinant monoclonal antibodies by differential labeling with ^{12}C and ^{13}C iodoacetic acid and LC-MS analysis, *Anal. Chem. 81*, 8101–8108.

158. Ji, J. A.; Zhang, B.; Cheng, W.; Wang, Y. J. (2009), Methionine, tryptophan, and histidine oxidation in a model protein, PTH: Mechanisms and stabilization, *J. Pharm. Sci. 98*, 4485–4500.

159. Teh, L. C.; Murphy, L. J.; Huq, N. L.; Surus, A. S.; Friesen, H. G.; Lazarus, L.; Chapman, G. E. (1987), Methionine oxidation in human growth hormone and human chorionic somatomammotropin. Effects on receptor binding and biological activities, *J. Biol. Chem. 262*, 6472–6477.

160. Shen, J. F.; Kwong, Y. M.; Keck, G. R.; Harris, J. R. (1996), The application of tert-Butylhydroperoxide oxidization to study sites of potential methionine oxidization in a recombinant antibody, in Marshak, D. R., ed., *Techniques in Protein Chemistry*, Academic Press, New York, Vol. VII, pp. 275–284.

161. Lam, X. M.; Yang, J. Y.; Clcland, J. L. (1997), Antioxidants for prevention of methionine oxidation in recombinant monoclonal antibody HER2, *J. Pharm. Sci. 86*, 1250–1255.

162. Chumsae, C.; Gaza-Bulseco, G.; Sun, J.; Liu, H. (2007), Comparison of methionine oxidation in thermal stability and chemically stressed samples of a fully human monoclonal antibody, *J. Chromatogr. B (Anal. Technol. Biomed. Life Sci.) 850*, 285–294.

163. Hsu, Y.-R.; Narhi, L. O.; Spahr, C.; Langley, K. E.; Lu, H. S. (1996), In vitro methionine oxidation of Escherichia coli-derived human stem cell factor: Effects on the molecular structure, biological activity, and dimerization, *Protein Sci. 5*, 1165–1173.

164. Labrenz, S. R.; Calmann, M. A.; Heavner, G. A.; Tolman, G. (2008), The oxidation of methionine-54 of epoetinum alfa does not affect molecular structure or stability, but does decrease biological activity, *PDA J. Pharm. Sci. Technol. 62*, 211–223.

165. Caldwell, P.; Luk, D. C.; Weissbach, H.; Brot, N. (1978), Oxidation of the methionine residues of Escherichia coli ribosomal protein L12 decreases the protein's biological activity, *Proc. Nat. Acad. Sci. USA 75*, 5349–5352.

166. Shechter, Y.; Burstein, Y.; Gertler, A. (1977), Effect of oxidation of methionine residues in chicken ovoinhibitor on its inhibitory activities against trypsin, chymotrypsin, and elastase, *Biochemistry 16*, 992–997.

167. Neumann, N. P.; Moore, S.; Stein, W. H. (1962), Modification of the methionine residues in ribonuclease, *Biochemistry 1*, 68–75.

168. Kassell, B. (1964), The basic trypsin inhibitor of bovine pancreas. II. Alteration of the methionine residue, *Biochemistry 3*, 152–155.

169. Van Patten, S. M.; Hanson, E.; Bernasconi, R.; Zhang, K.; Manavalan, P.; Cole, E. S.; McPherson, J. M.; Edmunds, T. (1999), Oxidation of methionine residues in antithrombin: Effects on biological activity and heparin binding, *J. Biol. Chem. 274*, 10268–10276.

170. Brennan, L. A.; Lee, W.; Giblin, F. J.; David, L. L.; Kantorow, M. (2009), Methionine sulfoxide reductase A (MsrA) restores α-crystallin chaperone activity lost upon methionine oxidation, *Biochim. Biophys. Acta (General Subjects) 1790*, 1665–1672.

171. Jenkins, N. (2007), Modification of therapeutic proteins: Challenges and prospects, *Cytotechnology 53*, 121–125.

172. Houde, D.; Kauppinen, P.; Mhatre, R.; Lyubarskaya, Y. (2006), Determination of protein oxidation by mass spectrometry and method transfer to quality control, *J. Chromatogr. A 1123*, 189–198.

173. Schöneich, C. (2000), Mechanisms of metal-catalyzed oxidation of histidine to 2-oxo-histidine in peptides and proteins, *J. Pharm. Biomed. Anal. 21*, 1093–1097.

174. Matamoros Fernandez, L. E.; Kalume, D. E.; Calvo, L.; Fernandez Mallo, M.; Vallin, A.; Roepstorff, P. (2001), Characterization of a recombinant monoclonal antibody by mass spectrometry combined with liquid chromatography, *J Chromatogr. B (Biomed. Sci. Appl.) 752*, 247–261.

175. Yang, J.; Wang, S.; Liu, J.; Raghani, A. (2007), Determination of tryptophan oxidation of monoclonal antibody by reversed phase high performance liquid chromatography, *J. Chromatogr. A 1156*, 174–182.

176. Wei, Z.; Feng, J.; Lin, H. Y.; Mullapudi, S.; Bishop, E.; Tous, G. I.; Casas-Finet, J.; Hakki, F.; Strouse, R.; Schenerman, M. A. (2007), Identification of a single tryptophanresidue as critical for binding activity in a humanized monoclonal antibody against respiratory syncytial virus, *Anal. Chem. 79*, 2797–2805.

177. Hashimoto, K.; Goto, S.; Kawano, S.; Aoki-Kinoshita, K. F.; Ueda, N.; Hamajima, M.; Kawasaki, T.; Kanehisa, M. (2006), KEGG as a glycome informatics resource, *Glycobiology 16*, 63R–70R.

178. Hossler, P.; Khattak, S. F.; Li, Z. J. (2009), Optimal and consistent protein glycosylation in mammalian cell culture, *Glycobiology 19*, 936–949.

179. Lis, H.; Sharon, N. (1993), Protein glycosylation. Structural and functional aspects, *Eur. J. Biochem. 218*, 1–27.

180. Sola, R. J.; Griebenow, K. (2009), Effects of glycosylate on the stability of protein pharmaceuticals, *J. Pharm. Sci. 98*, 1223–1245.

181. Coloma, M. J.; Trinh, R. K.; Martinez, A. R.; Morrison, S. L. (1999), Position effects of variable region carbohydrate on the affinity and in vivo behavior of an anti-(1 → 6) dextran antibody, *J. Immunol. 162*, 2162–2170.

182. Sethuraman, N.; Stadheim, T. A. (2006), Challenges in therapeutic glycoprotein production, *Curr. Opin. Biotechnol. 17*, 341–346.

183. Noguchi, A.; Mukuria, C. J.; Suzuki, E.; Naiki, M. (1995), Immunogenicity of N-glycolylneuraminic acid-containing carbohydrate chains of recombinant human erythropoietin expressed in Chinese hamster ovary cells, *J. Biochem. 117*, 59–62.

184. Jefferis, R. (2005), Glycosylation of recombinant antibody therapeutics, *Biotechnol, Progress 21*, 11–16.

185. Gala, F. A.; Morrison, S. L. (2004), V region carbohydrate and antibody expression, *J. Immunol. 172*, 5489–5494.

186. Delorme, E.; Lorenzini, T.; Giffin, J.; Martin, F.; Jacobsen, F.; Boone, T.; Elliott, S. (1992), Role of glycosylation on the secretion and biological activity of erythropoietin, *Biochemistry 31*, 9871–9876.

187. Walsh, G.; Jefferis, R. (2006), Post-translational modifications in the context of therapeutic proteins, *Nat. Biotechnol. 24*, 1241–1252.

188. Stallforth, P.; Lepenies, B.; Adibekian, A.; Seeberger, P. H. (2009), Carbohydrates: A frontier in medicinal chemistry, *J. Med. Chem. 52*, 5561–5577.

189. Brooks, S. A. (2009), Strategies for analysis of the glycosylation of proteins: Current status and future perspectives, *Molec. Biotechnol. 43*, 76–88.

190. Beck, A.; Wagner-Rousset, E.; Bussat, M.-C.; Lokteff, M.; Klinguer-Hamour, C.; Haeuw, J.-F.; Goetsch, L.; Wurch, T.; van Dorsselaer, A.; Corvaïa, N. (2008), Trends in glycosylation, glycoanalysis and glycoengineering of therapeutic antibodies and Fc-fusion proteins, *Curr. Pharm. Biotechnol. 9*, 482–501.

191. Kamoda, S.; Kakehi, K. (2008), Evaluation of glycosylation for quality assurance of antibody pharmaceuticals by capillary electrophoresis, *Electrophoresis 29*, 3595–3604.

192. Ko, K.; Ahn, M.-H.; Song, M.; Choo, Y.-K.; Kim, H. S.; Ko, K.; Joung, H. (2008), Glycoengineering of biotherapeutic proteins in plants, *Molec. Cells 25*, 494–503.

193. Amon, S.; Zamfir, A. D.; Rizzi, A. (2008), Glycosylation analysis of glycoproteins and proteoglycans using capillary electrophoresis-mass spectrometry strategies, *Electrophoresis 29*, 2485–2507.

194. Budnik, B. A.; Lee, R. S.; Steen, J. A. J. (2006), Global methods for protein glycosylation analysis by mass spectrometry, *Biochim. Biophys. Acta (Proteins Proteom.) 1764*, 1870–1880.

195. Morelle, W.; Canis, K.; Chirat, F.; Faid, V.; Michalski, J. C. (2006), The use of mass spectrometry for the proteomic analysis of glycosylation, *Proteomics 6*, 3993–4015.

196. Wuhrer, M.; Deelder, A. M.; Hokke, C. H. (2005), Protein glycosylation analysis by liquid chromatography-mass spectrometry, *J. Chromatogr. B (Anal. Technol. Biomed. Life Sci.) 825*, 124–133.

197. Harvey, D. J. (2005), Proteomic analysis of glycosylation: Structural determination of N- and O-linked glycans by mass spectrometry, *Expert Rev. Proteom. 2*, 87–101.

198. Amoresano, A.; Carpentieri, A.; Giangrande, C.; Palmese, A.; Chiappetta, G.; Marino, G.; Pucci, P. (2009), Technical advances in proteomics mass spectrometry: Identification of post-translational modifications, *Clin. Chem. Lab. Med. 47*, 647–665.

199. Sinha, S.; Pipes, G.; Topp, E. M.; Bondarenko, P. V.; Treuheit, M. J.; Gadgil, H. S. (2008), Comparison of LC and LC/MS methods for quantifying N-glycosylation in recombinant IgGs, *J. Am. Soc. Mass Spectrom. 19*, 1643–1654.

200. Witze, E. S.; Old, W. M.; Resing, K. A.; Ahn, N. G. (2007), Mapping protein post-translational modifications with mass spectrometry, *Nat. Meth. 4*, 798–806.

201. Su, X.; Ren, C.; Freitas, M. A. (2007), Mass spectrometry-based strategies for characterization of histones and their post-translational modifications, *Expert Rev. Proteom. 4*, 211–225.

202. Garcia, B. A.; Shabanowitz, J.; Hunt, D. F. (2007), Characterization of histones and their post-translational modifications by mass spectrometry, *Curr. Opin. Chem. Biol. 11*, 66–73.

203. Salzano, A. M.; Crescenzi, M. (2005), Mass spectrometry for protein identification and the study of post translational modifications, *Ann. Istit. Super. Sanita 41*, 443–450.

204. Jensen, O. N. (2004), Modification-specific proteomics: Characterization of post-translational modifications by mass spectrometry, *Curr. Opin. Chem. Biol. 8*, 33–41.

205. Küster, B.; Mann, M. (1998), Identifying proteins and post-translational modifications by mass spectrometry, *Curr. Opin. Struct. Biol. 8*, 393–400.

206. Barnes, C. S.; Lim, A. (2007), Applications of mass spectrometry for the structural characterization of recombinant protein pharmaceuticals, *Mass Spectrom. Rev. 26*, 370–388.

207. Yuk, I. H. Y.; Wang, D. I. C. (2002), Changes in the overall extent of protein glycosylation by Chinese hamster ovary cells over the course of batch culture, *Biotechnol. Appl. Biochem. 36*, 133–140.

208. Bogosian, G.; Violand, B. N.; Dorward-King, E. J.; Workman, W. E.; Jung, P. E.; Kane, J. F. (1989), Biosynthesis and incorporation into protein of norleucine by Escherichia coli, *J. Biol. Chem. 264*, 531–539.

209. Eng, M.; Ling, V.; Briggs, J. A.; Souza, K.; Canova-Davis, E.; Powell, M. F.; De Young, L. R. (1997), Formulation development and primary degradation pathways for recombinant human nerve growth factor, *Anal. Chem. 69*, 4184–4190.

210. Geoghegan, K. F.; Dixon, H. B. F.; Rosner, P. J.; Hoth, L. R.; Lanzetti, A. J.; Borzilleri, K. A.; Marr, E. S.; Pezzullo, L. H.; Martin, L. B.; Lemotte, P. K.; McColl, A. S.; Kamath, A. V.; Stroh, J. G. (1999), Spontaneous α-N-6-phosphogluconoylation of a "His tag" in Escherichia coli: The cause of extra mass of 258 or 178 Da in fusion proteins, *Anal. Biochem. 267*, 169–184.

211. Du, P.; Loulakis, P.; Luo, C.; Mistry, A.; Simons, S. P.; LeMotte, P. K.; Rajamohan, F.; Rafidi, K.; Coleman, K. G.; Geoghegan, K. F.; Xie, Z. (2005), Phosphorylation of serine residues in histidine-tag sequences attached to recombinant protein kinases: A cause of heterogeneity in mass and complications in function, *Protein Express. Purif. 44*, 121–129.

212. Aon, J. C.; Caimi, R. J.; Taylor, A. H.; Lu, Q.; Oluboyede, F.; Dally, J.; Kessler, M. D.; Kerrigan, J. J.; Lewis, T. S.; Wysocki, L. A.; Patel, P. S. (2008), Suppressing posttranslational gluconoylation of heterologous proteins by metabolic engineering of Escherichia coli, *Appl. Environ. Microbiol. 74*, 950–958.

213. Ward, N. E.; Stewart, J. R.; Ioannides, C. G.; O'Brian, C. A. (2000), Oxidant-induced S-glutathiolation inactivates protein kinase C-a (PKC-a): A potential mechanism of PKC isozyme regulation, *Biochemistry 39*, 10319–10329.

214. Melchers, J.; Dirdjaja, N.; Ruppert, T.; Krauth-Siegel, R. L. (2007), Glutathionylation of trypanosomal thiol redox proteins, *J. Biol. Chem. 282*, 8678–8694.

215. Chu, F.; Ward, N. E.; O'Brian, C. A. (2003), PKC isozyme S-cysteinylation by cystine stimulates the pro-apoptotic isozyme PKCd and inactivates the oncogenic isozyme PKCe, *Carcinogenesis 24*, 317–325.

216. Fahey, R. C.; Brown, W. C.; Adams, W. B.; Worsham, M. B. (1978), Occurrence of glutathione in bacteria, *J. Bacteriol. 133*, 1126–1129.

217. McLaggan, D.; Logan, T. M.; Lynn, D. G.; Epstein, W. (1990), Involvement of γ-glutamyl peptides in osmoadaptation of Escherichia coli, *J. Bacteriol. 172*, 3631–3636.

218. Gan-Schreier, H.; Okun, J. G.; Kohlmueller, D.; Langhans, C.-D.; Peters, V.; Ten Brink, H. J.; Verhoeven, N. M.; Jakobs, C.; Voelkl, A.; Hoffmann, G. F. (2005), Measurement of bile acid CoA esters by high-performance liquid chromatography-electrospray ionisation tandem mass spectrometry (HPLC-ESI-MS/MS), *J. Mass Spectrom. 40*, 882–889.

219. Liu, P.; Tarnowski, M. A.; O'Mara, B. W.; Wu, W.; Zhang, H.; Tamura, J. K.; Ackerman, M. S.; Tao, L.; Grace, M. J.; Russell, R. J. (2009), *Rapid Commun. Mass Spectrom. 23*, 3343–3349.

CHAPTER 13

Identification and Quantification of Degradants and Impurities in Antibodies

DAVID M. HAMBLY and HIMANSHU S. GADGIL
Amgen, 1201 Amgen Court West, Seattle, WA 98119

13.1 INTRODUCTION TO ANTIBODIES AND PROTEIN DRUGS

In the search for effective human therapeutics, protein drugs are widely recognized as an essential modality. As of 2007, the FDA had approved 54 biologics [1,2] and of these, 26 were monoclonal antibodies. It is estimated that in 2004, upward of 200 different monoclonal antibodies, hereafter called antibodies or MAbs, were in development worldwide [3]. As such, antibodies make up a large subset of the marketed and planned biological therapeutics and are the focus of this chapter.

13.1.1 Antibody Classification and Subtypes

In healthy individuals, antibodies produced by the immune system have high binding efficiencies to non-self-protein sequences, called the *antigen*. Antibodies come in five classes that activate different components of the immune system against the antigen. The immunoglobulin (Ig) IgD, IgE, and IgG antibodies have similar primary sequence and three-dimensional structure, while IgA is a dimer and IgM is a pentamer of the same basic immunoglobulin structure. Approximately 85% of immunoglobulin in serum consists of IgG antibodies [4]. Of the 26 marketed antibodies by the end of 2008, 18 are IgG subtypes and at least two more comprise IgG subdomains [5]. IgG antibodies are produced in subtypes 1–4, each with its own distinct functional purpose. While the sequences are nearly identical, aside from the antigen binding site, the disulfide bonding structure varies significantly as shown elsewhere [6].

Characterization of Impurities and Degradants Using Mass Spectrometry, First Edition.
Edited by Birendra N. Pramanik, Mike S. Lee, and Guodong Chen.

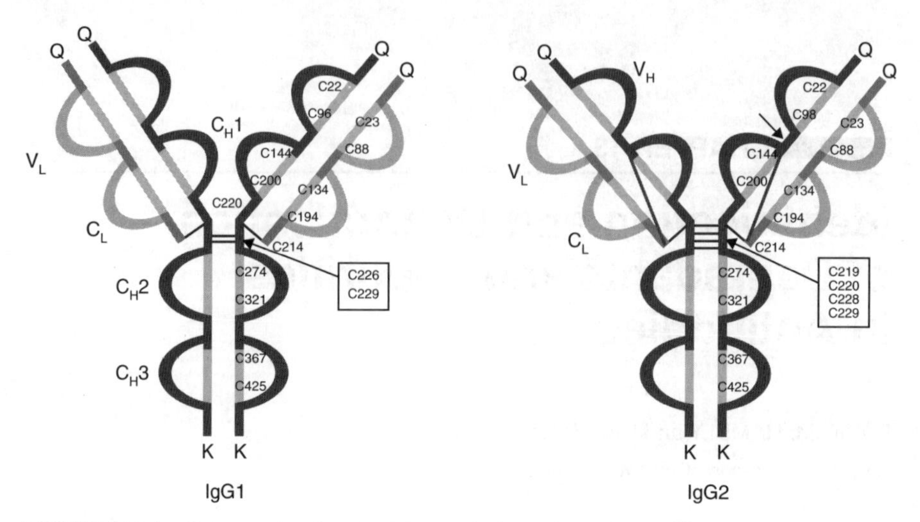

FIGURE 13.1 Scheme showing the EU numbering of cysteines and the varying disulfide bonding structure of the IgG1 and IgG2 subtypes.

Figure 13.1 demonstrates the disulfide bonding difference in the hinge and LC-HC connectivity. In vivo, IgG1 antibodies bind to the antigen and the Fc-γ receptor, which leads to antibody-dependent cell-mediated cytotoxicity (ADCC), a cell-killing immune response [7]. IgG1 antibodies can also initiate complement-dependent cytotoxicity (CDC) [8]. IgG 2 subtype antibodies have very weak activation signals for ADCC [9], but cannot activate CDC [10]. Careful selection of the IgG2 antibody epitope will serve to inhibit antigen binding to native receptors, making the therapeutic MAb a potent inhibitor of the antigen's in vivo function. The IgG3 subtype is a highly effective activator of ADCC and the most effective activator of CDC, while IgG4 shows variable ADCC activity but no initiation of the CDC pathway [11].

13.1.2 Antibody Structure

Antibody structure has been well characterized since the 1970s. In the next two sections, antibody types, basic structure, and nomenclature will be covered with minimal additional references [12–14]. The basic three-dimensional structure of an antibody can be visualized as a Y in Figure 13.1. The top of each arm is the antigen-binding site, such that each antibody can bind up to two ligands. IgA (dimers) and IgM (pentamers) can bind 4 and 10 ligands, respectively. An IgG antibody weighs ~150 kDa and consists of two heavy chains weighing about 50 kDa each, and two light chains weighing about 25 kDa each. Each prong of the Y consists of the *N*-terminal ends of the heavy chain and the light chain. Each light chain is anchored to a heavy chain with a disulfide bond. The two heavy chains are linked by disulfide bonds just below the middle of the *Y*, and the remainder of both heavy chains form the bottom of the Y structure. Papain cleavage of antibodies reported in 1959 resulted in two fractions: Fab and Fc [15]. The Fc fraction was crystallizable, hence Fc, but did not bind the antigen. The Fab fraction was antigen-binding, hence Fab, but was not readily crystallized. The papain cleavage site is located *C*-terminal to the light-chain (LC) disulfide anchor position, but *N*-terminal to the heavy-chain (HC) disulfide anchor position, and results in cleavage of the Y just above the middle of the antibody. This process forms two Fabs and one Fc per antibody. Each component weighs ~50 kDa, since the *N*-terminal half of the 50 kDa HC is bound to a 25 kDa LC in the antigen-binding Fab region, and the *C*-terminal half of the 50 kDa HC is covalently linked to the *C*-terminal half of the other HC [12].

13.1.3 Antibody-Domain Structure

The heavy chain is divided into four domains: the variable domain (V_H) and three constant domains: C_H1, C_H2 and C_H3. The smaller light chain is divided into a variable domain (V_L) and a constant domain (C_L). The heavy-chain constant region varies slightly between the different subtypes of IgG molecules. For example, IgG1 molecules have the sequence KTHTCPPCPAP between the C_H1 and C_H2 domains, where the heavy chains form two interchain disulfide bonds holding both heavy chains together. IgG2 molecules have the sequence KCCVECPPCPAP, where the heavy chains form up to four interchain disulfide bonds holding the two heavy chains together. It is interesting to note that, irrespective of subtype, the two C_H3 domains from each HC are bound by strong non-covalent interactions effectively holding the bottom of the Y together at two anchor points [16,17].

The light chain is bound to the heavy chain by a disulfide bond *N*-terminal to the hinge region. The V_L and V_H regions interact noncovalently and align to form the antigen-binding site, while the C_H1 and C_L regions interact noncovalently and covalently through a single disulfide bond. A unique element of the IgG4 subtype of antibodies is that in vivo, IgG4 light chains and heavy chains can be swapped between IgG4 antibodies, effectively generating antibodies with affinity for two different antigens [18].

The variable region of both the heavy chain (V_H) and the light chain (V_L) have three regions of high sequence divergence called *complementarity-determining regions* (CDRs) 1, 2, and 3. Regardless of IgG molecule, the specific CDR is always bracketed by nearly identical framework residues [19]. The V_H and V_L CDRs form the antigen binding pocket and determine the antigen specificity and binding affinity of the antibody. This is of critical importance since degradations in these residues have a strong likelihood of impacting antigen–antibody interactions.

13.1.4 Recombinant Antibody Production

It is highly relevant to consider the production processes used to produce recombinant antibodies, as the conditions employed during production seek to recover maximum yield with minimal impurities. These steps, however, can cause some amount of covalent degradations, some of which may copurify with the antibody.

Generally, recombinant antibodies are stably transfected into mammalian cell lines to ensure proper folding and glycosylation [3]. Chinese hamster ovary (CHO) or mouse myeloma (NS0) cell lines are commonly used to secrete MAb into the growth media. Amgen [20], Genentech [21–23], Merck [24], and Wyeth [25] (among many others) have adopted similar strategies for clinical and commercial-scale MAb purification, which are outlined below. Generally speaking, most processes begin with a centrifugation step to remove cells and other debris. Final removal of the cellular debris may be carried out using depth filtration, a porous medium that traps particulates. The MAb is then purified from the collected growth media using protein A (PrA) affinity capture chromatography at neutral pH. After the column is washed, the MAb is eluted using pH values ranging from 3 to 4 depending on the molecule. Given the low pH of the solution, this is an ideal place to perform a viral inactivation step. Following viral inactivation, the pH is neutralized. At this point, the virus-inactivated pool is frequently "polished" using cation exchange chromatography to bind the MAb and wash away contaminants, or conversely using anion exchange chromatography to bind contaminants while the MAb passes through. Other polishing options include hydrophobic interaction chromatography or hydroxyapatite-based chromatography. Whichever order or number of steps may be used, the polishing step is frequently followed by a viral filter and then ultrafiltration/diafiltration to exchange the product into the desired formulation buffer. The drug substance is then vialed to form the final drug product. These steps are optimized to maximize robustness, yield, and purity for clinical and commercial-scale processes. To evaluate the process and any potential changes, a number of assays are performed to track various contaminants such as host cell protein, PrA clearance, DNA, sterility, and MAb aggregates. Further tests confirming the potency, concentration, and identity are also performed. At this point, however, antibody characterization using mass spectrometry is an invaluable tool to confirm the process consistency, as any changes in the type or extent of covalent degradations can provide information for process development. Also, analysis of formulation samples under accelerated and real-time conditions can guide formulation development and selection of the final formulation conditions. In the following sections, methods used to characterize covalent degradations of monoclonal antibodies will be detailed with a focus on characterizing chromatographically resolved species.

13.1.5 Methods for Characterizing Antibody Degradation and Impurity

Conceptually, antibody purity can be evaluated by analyzing (1) the intact antibody [26,27], (2) an antibody after limited cleavage such as hinge cleavage with Lys-C [28,29], or (3) reduced and denatured antibody [30] since disulfide bonds anchor the light and heavy chains into the canonical immunoglobulin structure. A fourth method can also be used to investigate the Fab or Fc portions of the heavy chain by starting with Lys-C-cleaved antibody and then performing a reduction under denaturing conditions such that the Fc is reduced into two identical HC *C*-terminal pieces (Fc/2) and the Fab is reduced into a light chain and Fd, which is the *N*-terminal half of the heavy chain containing the variable region and first constant domain. The fifth method in the analytical chemist's arsenal is the peptide map. While the prior methods can deliver detailed and quantitative information on site-specific modifications of a protein's primary structure, and occasionally tertiary structure, high-resolution peptide maps can deliver quantitative information on multiple degradations from a single sample analysis. Peptide mapping is typically performed on fully reduced, alkylated, and digested antibody followed by analysis using reversed-phase high-performance liquid chromatography separation with online mass spectrometric detection (RP-HPLC-MS). Peptide mapping is an essential tool for degradant and impurity identification because each peptide is separated according to its interaction with the solid-phase support, often a C8 or C18 alkyl group. Therefore slight changes in the chemical structure of a peptide result in a change in the solid-phase partitioning, and hence a unique retention time for the degraded peptide. Since the degradants can be separated by HPLC, and quantified by UV detection, the power of mass spectrometry can be fully leveraged to identify the resolved degradation products. Accurate mass detection on a variety of high-end mass spectrometers enables peptide and degradant identification without the need for more complicated tandem mass spectrometry (MS/MS), but MS/MS is often required to determine the site of degradation.

13.2 OVERVIEW OF DEGRADATIONS AND IMPURITIES IN PROTEIN DRUGS AND ANTIBODIES

Protein degradations and impurities, thoroughly reviewed elsewhere [31], can impact any of the four structural levels of a protein fold. The primary structure can be altered at the genetic level giving rise to a mutant, or chemically modified to generate a degraded protein. Misfolding of the secondary or tertiary structure produces a protein that is chemically identical but structurally different. Finally, errors in the quaternary structure of the protein caused by aggregation, misfolding, clipping, or a different number of light or heavy chains can result in chemically unique and/or structurally distinct proteins. It must be remembered that a chemical modification may result in a change to the protein activity [32], bioavailability [33], or conformation [34], and a misfolded protein may have a different propensity for chemical degradation [35]. Thus the protein chemical sequence and the protein's structural fold are critical elements of the protein's function. However, protein function is not as simple as binding to the intended target. A change in the protein can have an impact on biological effect if the half-life is altered, the binding affinity is altered, an effector function is disturbed, or an immune response is raised against the protein therapeutic. While this chapter does not discuss these details, they should be considered when evaluating the potential safety issue of any degradation or impurity.

13.2.1 Chemical Degradations and Impurities

13.2.1.1 Methionine Oxidation

Perhaps the most widely known chemical modification is oxidation of methionine residues. The methionine sidechain contains a sulfur residue at the epsilon position that is susceptible to oxidation [36]. The oxidation of methionine may be a defense against protein oxidative damage [37–39]. In antibodies, two constant region methionines found in IgG1 and IgG2 molecules are known to be readily oxidized [40]. Oxidation of the methionine in the C_H2–C_H3 interface [Met358 by enzyme unit (EU) numbering [19]) region of IgG1 antibodies disrupts protein A binding [41]. This region is also key for Fcγ receptor binding, which triggers the antibody-dependent complement cascade [42], complement-dependent cytotoxicity, and phagocytosis [43]; thus methionine oxidation should be carefully monitored during process and pharmaceutical development. Methionine oxidation is of considerable interest in protein drugs, especially when the methionine residue is located near the antigen-binding site.

13.2.1.2 Disulfide Bonds or Reduced Cysteine

Another very important degradation or impurity relates to another sulfur-containing amino acid: cysteine. Cysteine is a unique amino acid in that it is used catalytically in some enzymes, such as cysteine proteases, or as a structural component in maintaining protein structure. Disulfide bonds are predominantly found in secreted proteins since the extracellular environment is oxidizing [44]. The IgG1 [45] and IgG4 [46] antibody classes have 16 disulfide bonds, although in a slightly different arrangement, while the IgG2 [6] class has the same structure as that of an IgG4, but with two extra hinge disulfide bonds for a total of 18. IgG3 has the canonical IgG1 sequence, followed by three repeats of a Cys–Pro-rich sequence for a total of 11 disulfide bonds [47,48]. These bonds maintain the structure of the subdomain, or anchor the LC to the HC, or anchor HC-HC. The presence of free cysteine residues instead of disulfide bonds [49,50] is considered an impurity in freshly prepared material, and these could cause degradation via disulfide bond rearrangement on processing or storage. More recent work has shown that the IgG2 antibodies have different disulfide bonding isoforms that form in vitro [6,51] and in vivo [52] which may have an effect on specific activity. Unpaired cysteine residues in antibodies may be disulfide bonded during production to free cysteine [28] or other free sulfur molecules, and this may need to be removed for full activity [53].

13.2.1.3 Deamidation of Asparagine and Glutamine

Asparagine [54–56] and glutamine [57] sidechains can be deamidated via acid- or base-catalyzed hydrolysis or formation of a succinimide cyclic intermediate [58–62]. Asparagine reacts with the $N+1$ amide backbone nitrogen to form a cyclic succinimide. Hydrolysis of this succinimide releases either the sidechain carbonyl or the backbone carbonyl, resulting in either aspartate (loss of amine from the sidechain) or isoaspartate (a rearranged backbone). As a result of entropic factors, glutamine rarely forms the succinimide in the polypeptide chain [63–65], but when glutamine is located at the *N* terminus, it readily cyclizes with the *N*-terminal primary amine, forming pyroQ [66]. The pyroQ form can be hydrolyzed to form an *N*-terminal glutamate. When glutamine in the polypeptide backbone does form the succinimide [62], hydrolysis results in either glutamate or isoglutamate (a rearranged backbone).

13.2.1.4 Isomerization of Aspartic Acid and Glutamic Acid

Aspartate [67,68] and glutamate residues [66] can also cyclize and hydrolyze to form isoaspartate and isoglutamate residues, respectively [58,69]. This reaction occurs in a manner identical to the deamidation reaction described above. Similarly, glutamate residues at the *N* terminus readily undergo succinimide formation, while glutamate residues positioned elsewhere in the protein backbone do not readily form the succinimide intermediate, and hence are rarely observed in the isoglutamate form. Aspartate does readily form the succinimide intermediate in a pH-dependent manner [70], and hydrolysis of the intermediate results in the formation of isoaspartate, or a return to the aspartate form. IsoD, as isoaspartate is commonly called, effectively moves the protein backbone onto the sidechain, and thus changes the stereochemistry of the α-carbon, lengthening the protein backbone by one carbon atom and decreasing the length of the aspartate sidechain. Functionally, IsoD has the same pK_a as does aspartate but may disrupt α-helical structure [71]. While the rate of formation is highly dependent on pH, it is also strongly dependent on the presence of a small residue at the N + 1 position [33]. Thus DG and NG are sequences that are highly susceptible to the formation of isoD residues and should be monitored carefully if they are present in a CDR.

13.2.1.5 Amide Backbone Hydrolysis Reactions

Backbone degradants can arise from protease contamination, light stress [72], kinetic effects [73], or metal-induced hydroxyl radicals [74,75]. The aspartate (D) proline (P) sequence [76] is also susceptible to acid-catalyzed hydrolysis of the backbone [77,78] and is a source of clipping in antibodies. This reaction is initiated by the aspartic acid sidechain forming an anhydride with the backbone carbonyl of the Asp sidechain [79]. This reactive species is hydrolyzed to an ester, from which the final cleavage is performed by ester hydrolysis. The cleavage reaction is dependent on both pH and the structure of the protein around the DP site. DP cleavage is not the only source of amide bond hydrolysis reactions in protein drugs [73]. Copurification of other proteases or metals can lead to backbone hydrolysis during processing or storage, and each hypothesis can be tested using protease inhibitors or the addition of a suitable metal chelator to inhibit backbone cleavage.

13.2.1.6 Glycation of Lysine Residues

Lysine residues have a sidechain primary amine that is reactive with reducing sugars such as glucose [80–82]. Reducing sugars react with the primary amine through Schiff base formation, leading to glycated lysine [27,83]. The glycated epsilon amine has been modified from primary to secondary, which may result in altered amine pK_a and differential ability to form hydrogen bonds. Furthermore, the glycated residue can decay into advanced glycation end products (AGEs) of differing mass and properties [84,85]. Certain AGEs are capable of crosslinking with other primary amines, leading to crosslinked proteins [86,87]. These AGEs have been implicated in various age-related diseases and have been observed in diabetic patients [88].

During protein production, cells require glucose in the media, and therefore the secreted antibody is at risk of glycation [89]. The risk of glycation is low during the purification steps since reducing sugars are seldom used. Typically, nonreducing sugars are used for frozen and liquid formulations; however, sucrose can degrade with time, giving rise to fructose and glucose, both of which are reducing sugars. Under accelerated storage conditions, sucrose degradation does cause antibody glycation [27,83,90]. Careful monitoring of glycation in liquid formulations indicates that normal storage conditions at 2–8°C does not result in additional glycation during storage [27,83,90].

13.2.1.7 C-Terminal Lysine Variants

The putative sequence of antibodies includes a leader sequence that targets the protein to the endoplasmic reticulum and Golgi bodies for processing and finally excretion to the extracellular space [91]. Leader sequences may be improperly processed by the Chinese hamster ovary (CHO) cells and have been observed at low levels. The CHO cells that are commonly used to express antibodies contain carboxy peptidases that remove the *C*-terminal lysine residue, leaving glycine as the terminal residue [92,93]. A second *C*-terminal variant was observed where the $N-1$ residue, a glycine, is partially removed by an enzymatic reaction, resulting in an α-amidated proline at the *C* terminus [94]. Both the lysine variant and amidated *C* terminus have different ionic charges from the typical *C*-terminal carboxylic acid [94]. In the case of the lysine, an extra positive charge is present at the terminus of the heavy chain and one less negative charge is observed in the α-amidated proline variant since the carboxylic acid is effectively removed. Since the antibody contains two heavy chains, these variants may be observed once or twice per antibody, based on the frequency of occurrence on a single chain. Thus, if the + lysine variant is observed on 10% of the heavy-chain sequences, it would be expected that 1% (0.10×0.10) of antibodies would contain a double-lysine variant.

13.2.1.8 Carbohydrate Variants

Human antibodies are expressed with a carbohydrate on a conserved asparagine in the C_H2 domain [95,96]. The biantennary structure of the carbohydrate may be modified with either one or two galactose residues, and it may have a fucose. When considering that there are two heavy chains, this heterogeneity gives rise to a complicated glycosylation pattern on intact antibodies that may contain 0–4 galactose residues [26]. The enzymatically driven glycosylation of the antibody passes through a mannose 5 (Man5) structure, which is rebuilt to the expected final carbohydrate, the biantennary sugar [97,98]. Processing errors result in some antibody with Man5 carbohydrates in the final product [99].

There are numerous other carbohydrate modifications, which are not covered in this chapter. Included among these are triantennary and tetraantennary structures, *O*-glycosylation on serine or threonine residues [100], and sialic acid addition to the carbohydrates [101]. Many of these carbohydrate modifications can be analyzed using fluorescence techniques, which have greater sensitivity [102].

13.3 METHODS USED TO IDENTIFY AND QUANTITATE DEGRADATIONS AND IMPURITIES

There is no perfect method for identifying and quantitating all degradations and impurities with one single assay. Generally, a combination of approaches must be used to fully evaluate a protein drug. This section details multiple methods that are useful for characterizing protein drugs, but these should not be considered comprehensive. As technology improves, for example, with small-particle-size beads for liquid chromatography, higher-sensitivity mass spectrometers, or new reversed-phase matrices, it should be expected that new methods will be developed that are more sensitive, are more specific, and can detect new modifications. As this book focuses on mass spectrometry, the methods detailed here will all use mass spectrometry for identification of protein degradations or impurities. However, other methods are very useful for monitoring; examples are charge state variants [94], higher-order oligomers [103], free cysteine residues [50], *N*-terminal sequences, protein clips, and structural isoforms [6,51]. To illustrate the current methods used to monitor degradations, an IgG1 antibody raised against streptavidin was incubated for 4 weeks at 37°C at either pH 5.2 or pH 8.0. The low-pH condition favors isomerization reactions, while the high-pH conditions favor deamidation reactions. The degraded antibody was then subjected to whole-mass analysis to demonstrate carbohydrate heterogeneity and lysine variants, limited digestion, and/or reduced analysis to demonstrate resolution of some site-specific degradations and peptide mapping to demonstrate a detailed analysis of covalent degradations.

13.3.1 Whole-Protein Mass Analysis Methods

13.3.1.1 Carbohydrate Variation

A very simple method used to analyze carbohydrate variation uses a 4-μg protein load on a 1 × 50-mm diphenyl column from Varian to desalt the protein (see method 1 in the Appendix at the end of this chapter) [104]. The outlet is directed into an electrospray TOF instrument to collect *m/z* data, which are then deconvoluted to mass data. Detection at 214 nm enables high-sensitivity detection of all eluting species; however, to ensure that the carbohydrate heterogeneity is evaluated using identical deconvolution conditions, the MS should be summed across the whole elution prior to deconvolution. Note in Figure 13.2 that the postpeak decreases with increasing incubation time. This is caused by formation of pyroQ at the *N* terminus resulting in a mass loss of ~17 Da × 2 = 34 Da (Figure 13.3) since both free amines at the termini are lost at the same rate. Deconvolution of the mass spectrum results in a mass chromatogram of the antibody carbohydrate heterogeneity (Figure 13.3). When using a resolving power greater than 2000, the various galactose carbohydrates should be readily resolved.

The heterogeneity observed in the whole-protein mass spectrum arises from two carbohydrate moieties, one on each HC anchored in the C_H2 domain. Since the carbohydrate can have the core structure (G_0) with fucose (G_0F), + 1 Gal (G_1F), or

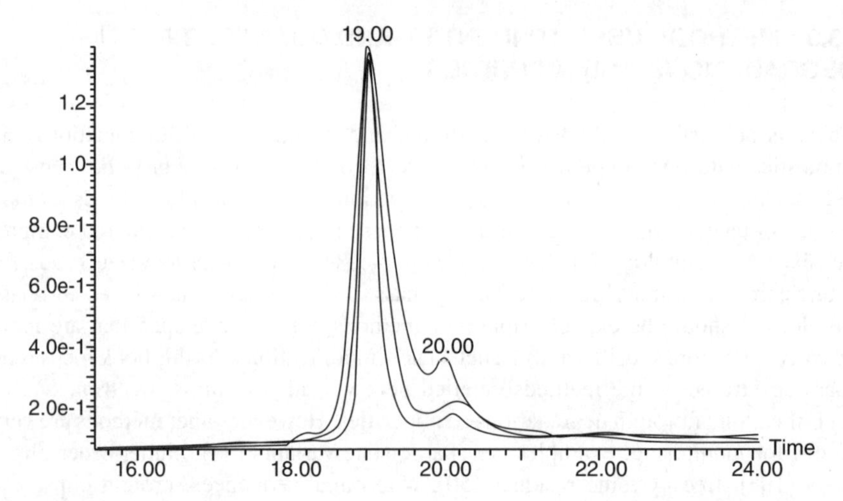

FIGURE 13.2 UV overlays of the elution profile of a control intact antibody overlaid with the same antibody incubated at pH 8 and 37°C for 2 weeks and 1 month, enabling covalent degradations to be identified. The light trace is the control, and the peak at 20 min decreases with increasing incubation length.

+ 2 Gal (G_2F), the combination of two HC in the antibody results in numerous antibody masses: G_0F/G_0F or G_1F/G_0F or G_1F/G_1F, G_2F/G_0F or G_2F/G_1F or G_2F/G_2F, where each peak is + 162 Da heavier than the previous peak.

The peak height is proportional to the abundance, but it is not quantitative, and a weighted average of the hexose distribution can be calculated [27]. The hexose

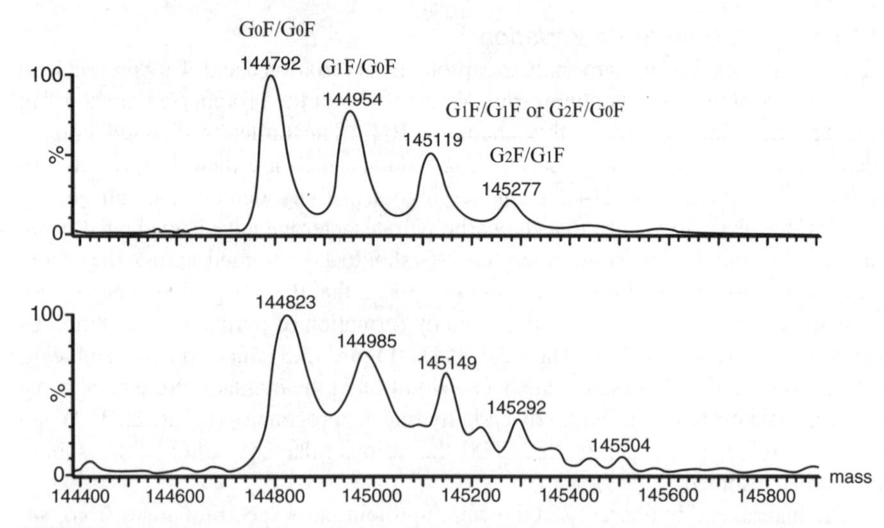

FIGURE 13.3 Deconvolution of intact antibody MS showing the carbohydrate structure and a second deconvolution of the peak at 20 min showing the + 34-Da species corresponding to Q at two *N* termini.

distribution can be compared across multiple samples such as different lot preparations, to evaluate the effect of various starting reagent variables on the process consistency.

13.3.1.2 Detection of Lysine C-terminal Variants and Glycated Lysine

Pretreatment of the protein sample with 500 units of PNGase F for 2 h in 10 mM phosphate buffer at pH 7 removes the *N*-linked carbohydrate structure. By performing the identical experiment as detailed above (method 1 in the Appendix), we note that the deconvoluted data from an antibody may have a few small peaks at higher mass compared to the main analyte peak. These peaks correspond to mass additions of approximately 128 Da (lysine addition at the *C* terminus) and 162 Da (glucose addition to a primary amine). While the mass separation between the IgG + lysine and IgG + glucose species is 34 Da, the isotope envelope of an intact IgG is about 25 Da wide. Therefore, even at infinite resolving power, deconvolution may be insufficient to determine the relative abundance for each of these species. However, the presence of these impurities can be determined from this simple and rapid assay, enabling an intelligent search for modifications in peptide-mapping data. If the protein has no glycation, the resulting deconvoluted data are sufficient for determining the percent relative abundance of the remaining lysine impurity directly from the percent relative signal of the deconvoluted intensity measurements (Figure 13.4).

13.3.1.3 Detection of Disulfide Bond Variants in IgG2 Antibodies

IgG2 antibodies have at least two disulfide bond populations, which can be assayed using method 2 in the Appendix. The normal disulfide bond structure links the LC to the HC between V_H and the C_H1 domain. However, an alternate population is

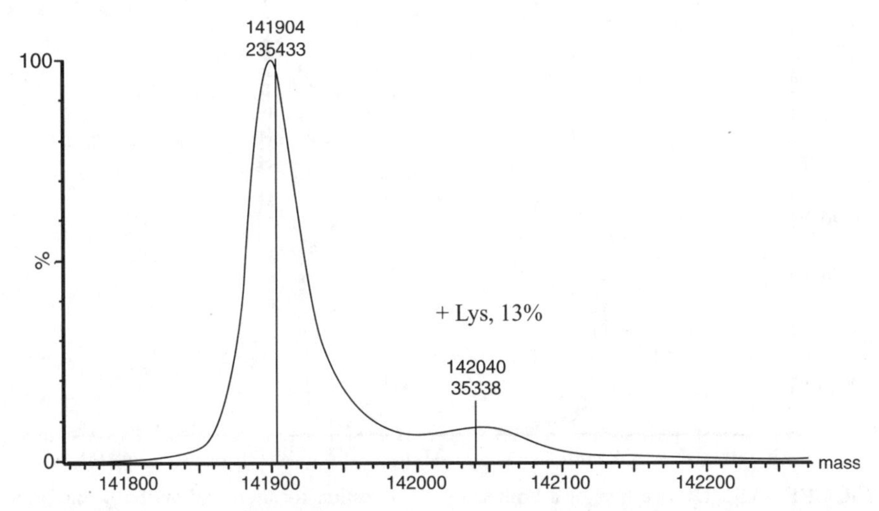

FIGURE 13.4 Deconvoluted MS spectrum of PNGase F–treated intact antibody where the lysine variant at, ∼ + 128 Da, accounts for 13% of the total antibody.

experimentally observed where the LC is linked to the HC in the hinge region. This linkage results in the C_H1 cysteine disulfide bonded to the hinge. Purification of these variants demonstrated that they may have different binding affinities [51], and interconvert in vivo [52].

13.3.2 Methods for Evaluating the Mass of Protein Fragments

13.3.2.1 Limited Digestion Method for Antibodies

The limited digestion method is capable of detecting methionine oxidation, formation of *N*-terminal succinimides, other succinimides, and free disulfide bonds, and may also be used to detect other degradations. For IgG1 antibodies, this method is most effectively performed using a Lys-C digestion. IgG2 antibodies require the use of papain or pepsin, although the ragged ends left by these enzymes reduces the utility of this method for degradant detection. For this reason, this section focuses on the use of limited Lys-C digestion for the analysis of IgG1 antibodies.

Lys-C is an enzyme that cleaves the amide bond *C*-terminal to lysine residues. While Lys-C is used as a protease for peptide mapping, low levels of this enzyme, for brief periods of time, cleave IgG1 antibodies in the hinge [28], on the *C*-terminal side of lysine 222 (EU numbering [14,19]). This cleavage results in the formation of the Fc and Fab portions as described in Section 13.1.2. While each portion weighs about 50 kDa, the portions readily resolve by hydrophobicity and UV signal intensity as two Fab fragments are produced for each Fc that is formed (Figure 13.5).

To perform this experiment, the antibody undergoes limited digestion with Lys-C as detailed in method 3 in the Appendix [28]. LC analysis is performed using method 4

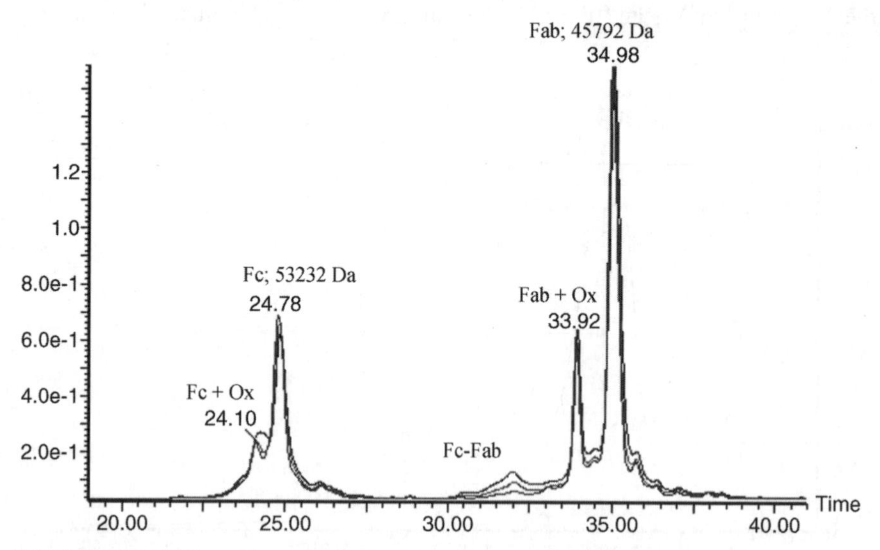

FIGURE 13.5 UV overlays of a limited Lys-C digestion for a control antibody and after incubation at 37°C for 2 weeks and 1 month at pH 8. The Fc-Fab and Fc-Ox peaks grow in intensity with increasing incubation length.

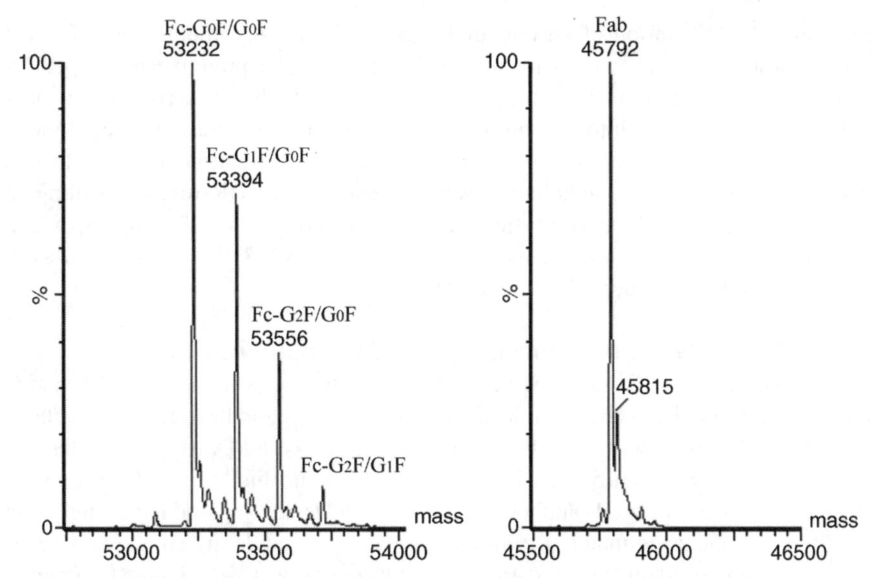

FIGURE 13.6 Deconvoluted MS spectrum of a limited Lys-C digested IgG1 showing the Fc with carbohydrate heterogeneity and the Fab with a small sodium adduct peak.

in the Appendix [104]. Deconvolution of the most abundant UV signal results in a mass spectrum corresponding to ~50 kDa, which should be the mass of the Fab fragment, while the lower abundance UV signal should have a mass corresponding to the 50 kDa Fc fragment, and containing significant heterogeneity from the carbohydrates (Figure 13.6). As in the whole-protein deconvoluted mass spectrum, the heterogeneity arises from a carbohydrate moiety anchored to the N297 (EU numbering) in the C_H2 domain. Since the carbohydrate can have the core structure (G_0) with fucose (G_0F), + 1 Gal (G_1F), or + 2 Gal (G_2F), the combination of two HCs in the Fc fragment results in multiple Fc fragment peaks: G_0F/G_0F or G_1F/G_0F or G_1F/G_1F, G_2F/G_0F or G_2F/G_1F or G_2F/G_2F, where each peak is + 162 Da heavier than the previous peak.

The UV spectrum for the limited Lys-C digest is capable of resolving multiple degradants. Oxidations are typically observed at lower retention times for both Fc and Fab fragments. For proteins with an *N*-terminal glutamate, the succinimide form is observed at later retention times, as well as additional protein succinimides and free-cysteine pairs. This method has also been used successfully to observe and quantitate cysteinylation [28] of cysteine and isomerization of certain aspartate residues. One such degradation is the hinge aspartate at position 221 (Figure 13.5). This residue is located N − 1 to the Lys-C cleaved K222 in all IgG 1 antibodies. When this residue is isomerized, K222 develops proteolytic resistance and only one arm is cleaved under the reaction conditions. This results in a Fab-Fc fragment that has a mass of ~100 kDa and contains the carbohydrate heterogeneity. This method was used to rapidly quantitate hinge aspartate isomerization and determine the rate and Arrhenius constant for the hinge aspartate isomerization. The method was used to detect and quantitate cysteinylation of an unpaired cysteine [28]. This impurity was formed during protein production when the free cysteine on the protein was capped by disulfide bond formation with a free-cysteine amino acid. This gives rise to a net mass gain of 119 Da, taking into account the fact that the residual mass of a cysteine is 121 Da and subtracting 2 Da from the disulfide bond formation reaction. In one documented case, the cysteinylation was present in the CDR region resulting in decreased antigen binding. The cysteinylation was removed by a refolding step added to the purification process, which increased the bioactivity [53]. Evidently, this was an important impurity to track for this particular antibody.

13.3.2.2 Limited and Reduced Method for Antibodies

This method is useful for tracking degradations to specific portions of the HC and is useful for further characterization work, as top–down and middle–down MS methods can be employed to sequence the *N* terminus of the ~25-kDa fragments that are obtained [105]. Essentially, this method uses the same limited Lys-C method as described above for IgG1 antibodies and adds a denaturing–reducing step. The digested and quenched material from the limited digestion described above is denatured by adding solid guanidine to a 7 M final concentration. (*Note:* Ignore the volume increase on guanidine addition.) To this solution, add 1 μL of TCEP (100 mg/mL in water) and incubate at 37°C for 30 min. This ensures complete reduction of all inter- and intrachain disulfide bonds (Figure 13.7). The Fab fragment is converted to LC and Fd [*N*-terminal half of the HC (V_H and C_H1)], while the Fc fragment is converted into Fc/2, consisting of two chains, each containing the same *C*-terminal half of the HC (C_H2 and C_H3) (Figure 13.8). This Fc/2 fragment will have some heterogeneity from the carbohydrate anchored in the C_H2 region, although it will be simpler than that

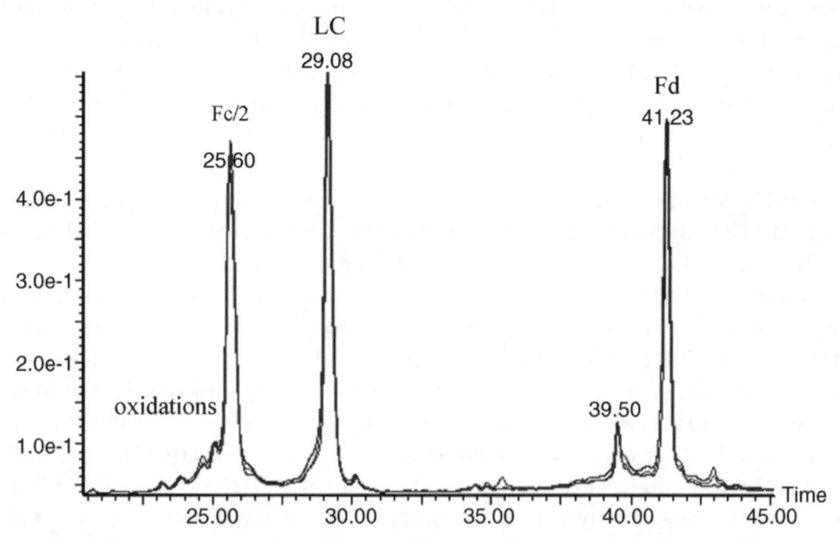

FIGURE 13.7 UV overlay of Lys-C limited digested and reduced control antibody and after incubation at 37°C for 2 weeks and 1 month at pH 8.

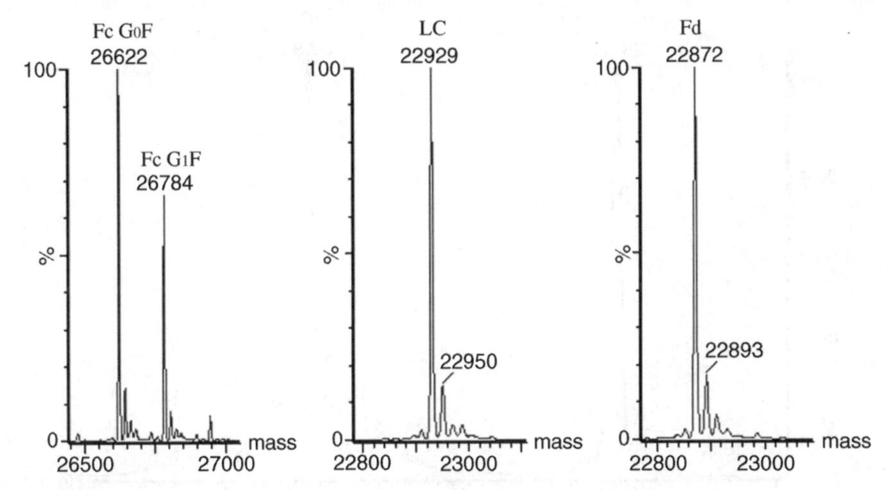

FIGURE 13.8 Deconvoluted MS spectrum of a limited Lys-C digested and reduced antibody showing, in order of retention time, Fc/2, LC, and Fd.

observed with the intact antibody since the single carbohydrate consists of the core structure (G_0), typically with focus (G_0F), +1 Gal (G_1F), or +2 Gal (G_2F)—hence only three main carbohydrate peaks are expected for the Fc/2 compared to five or more for the intact antibody, which has two carbohydrates, one on each HC [106].

The limited reduced method is capable of resolving methionine oxidations, *N*-terminal succinimide, succinimide within a chain, and was the method used to initially identify the hinge aspartate isomerization (observed as an intact HC). It is important to note that no cysteine heterogeneity is observed with this method, as all disulfide bonds have been completely reduced.

13.3.2.3 Reduced Protein Mass Analysis

It is also important to consider a simple reduced mass analysis for any protein drug. This confirms the LC and HC masses and demonstrates that there are no unanticipated backbone cleavages. It is also useful for evaluating methionine oxidation, *N*-terminal succinimide formation, and formation of other succinimides and may be applicable for other degradations. To reduce a MAb, incubate 5 mM DTT in the presence of 7.5 M GndHCl at pH 7.75 for 45 min at 45°C (method 5 in Appendix) [90]. Analysis using method 4, with detection at 214 nm, enables high-sensitivity detection of all eluting species (Figure 13.9), and the MS for the eluting peaks should be summed to get a complete mass spectrum for the eluting antibody fragments. Deconvolution of the lower abundance UV signal results in a mass spectrum corresponding to ~25 kDa, which is the mass of the LC, while the higher-abundance UV signal should have a mass corresponding to the 50 kDa HC, and containing the carbohydrate heterogeneity (Figure 13.10).

After reduction of the antibody, the LC and HC can then be alkylated to cap the cysteine residues (method 6 in Appendix) [90]. The alkylation step is important as it prevents disulfide bond formation during the alkylation or digestion steps, and

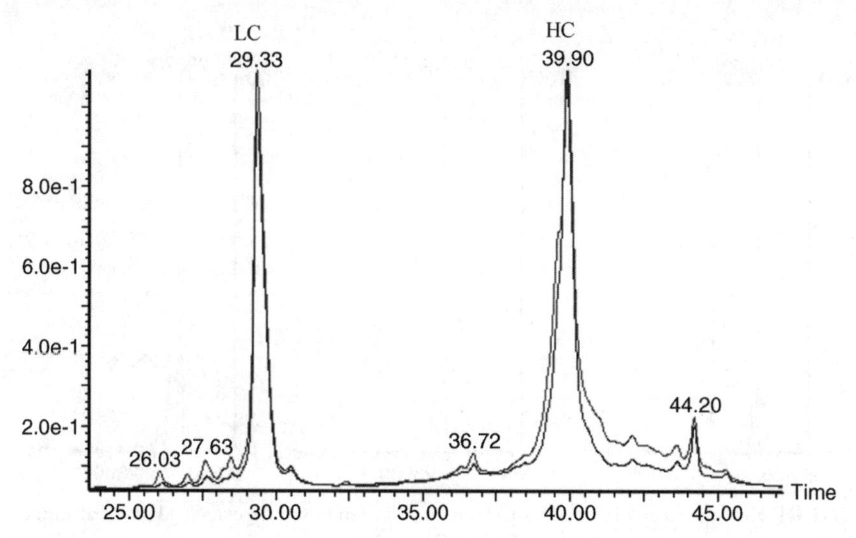

FIGURE 13.9 UV overlay of a reduced antibody showing the control sample and after incubation at 37°C for 1 month at pH 8 with increased oxidation on the light chain.

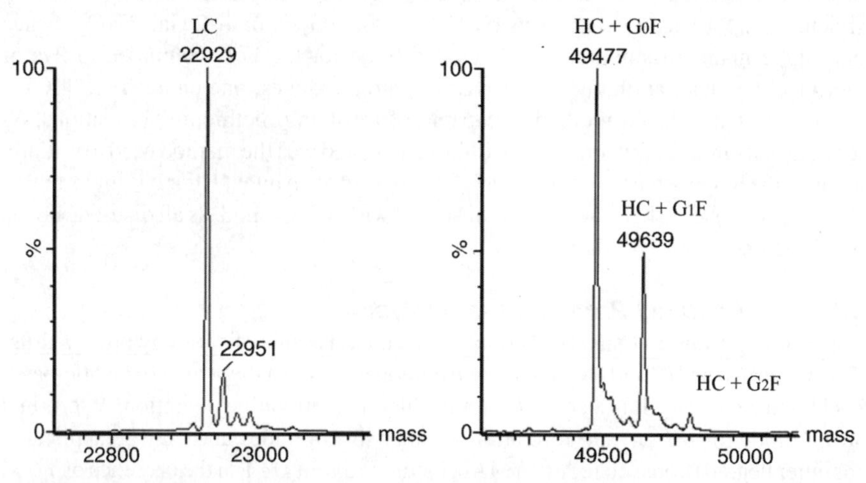

FIGURE 13.10 Deconvoluted MS spectrum of reduced antibody showing, in order of retention time, LC and HC manifesting carbohydrate heterogeneity.

peptides containing free cysteines tend to have a poor MS/MS spectrum, reducing the opportunity to identify the site of specific degradations during peptide mapping. For these reasons, it is important to ensure that the alkylation reaction is carried to completion at this step. The chromatographic separation of the LC and HC uses the same diphenyl method detailed several times already. A rapid screen can be carried out using a 1 × 50-mm column (method 1 in Appendix), while a higher-resolving-power assay can be used with a 2.1 × 150-mm separation (method 4 in Appendix) as shown in Figure 13.11. The deconvoluted MS data are shown in Figure 13.12.

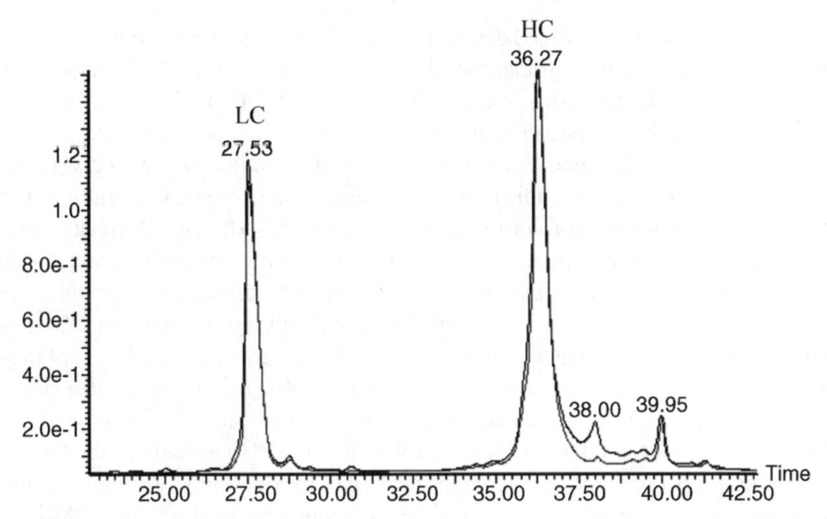

FIGURE 13.11 UV overlay of a reduced and alkylated control antibody and after incubation at 37°C for 1 month at pH 8.

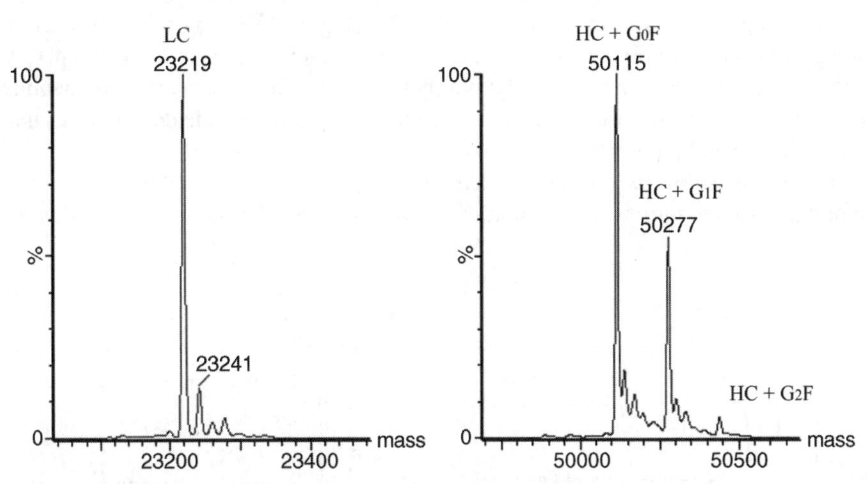

FIGURE 13.12 Deconvoluted MS spectrum of reduced and alkylated antibody showing, in order of retention time, LC and HC manifesting carbohydrate heterogeneity.

13.3.3 Methods for Evaluating Peptides for Impurities and Degradations

13.3.3.1 Reduced and Alkylated Peptide Mapping

The peptide map is the most comprehensive method used to detect and quantitate degradations and *N*- or *C*-terminal variations. It should be remembered that this method cleaves the protein into discrete segments and as such it is not capable of detecting the abundance of multiple cooperative degradations, such as the amount of double oxidation at disparate sites. This method is useful for detecting any manner of chromatographically resolved degradations or impurities, and can detect, via mass spectrometric detection, degradations that are coeluting with other peaks. Minor degradations that do not result in chromatographic resolution when using an intact antibody or MAb fragment can frequently be resolved in a peptide map.

The reduced and alkylated MAb must be desalted (Method 7 in Appendix) prior to digestion (method 8 in Appendix) [90] to remove the chaotrope. Generally, a 1 : 50 ratio of trypsin is ample for complete digestion in 4 h [90]. The chromatographic separation is a critical component to ensure that the peptide map will be most useful for degradant and impurity analysis [90]. A rapid chromatographic gradient will readily enable identification of most peptides but will not resolve, nor will MS detect, many degradants. Thus, longer methods are used to deliver higher resolution of major and many minor species (method 9 in Appendix) [90]. It is no guarantee that all degradations will be detected as resolved peaks; however, the minor peaks are relatively easily observed as discrete and well-resolved UV signals in the baseline between major peptide peaks. The major peaks can be identified by employing most proteomic software applications using either accurate mass analysis or MS/MS data.

With a highly resolving method in place, it can be overwhelming to begin to characterize all the observed degradations and impurities. The first analysis should focus on degradations that are stability indicating (Figure 13.13). Forced-degradation samples can be generated using a 4 week, 37°C incubation of the protein at both pH 5.0 and pH 8.5. By preparing and analyzing these samples in the same run, the peptide maps can be overlaid, resulting in a clear demonstration of peptide degradations that are both pH- and temperature-sensitive.

Given the pathways for various degradations, the pH 5.0 samples will identify aspartate isomerizations and oxidations, while the pH 8.5 samples will identify

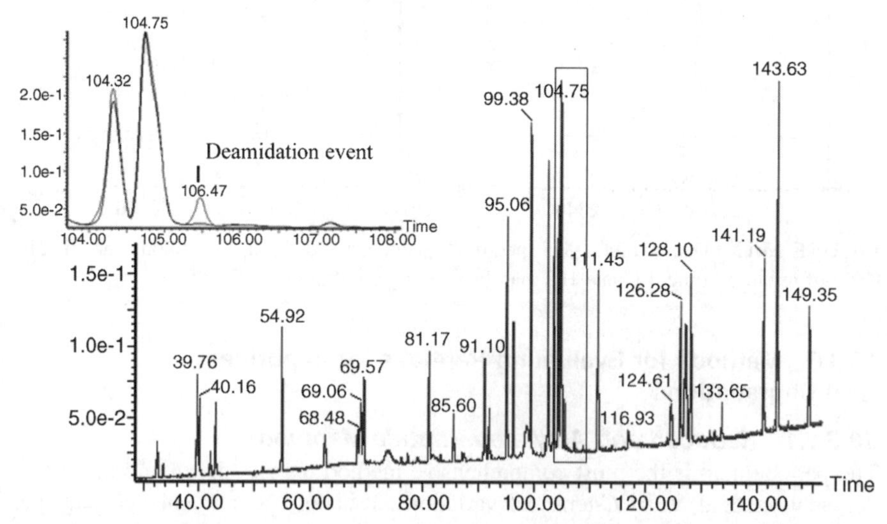

FIGURE 13.13 UV overlay of a tryptic map from an antibody before and after incubation at 37°C for 1 month at pH 8. The region that is boxed has been amplified in the inset to demonstrate a deamidation event that grows after incubation.

TABLE 13.1 Characteristic Masses of Various Degradations

Modification	Accurate Mass (Da)	Average Mass (Da)
Deamidation	0.9840	1.0
Oxidation	15.9949	16.0
Isomerization	0.0000	0.0
Deamidated succinimide	−17.0265	−17.0
Asp/Glu succinimide	−18.0106	−18.0
Glycation	162.0258	162.0
Galactose	162.0258	162.0
Fucose	146.0579	146.0

deamidations and oxidations. To identify the site of a specific modification, first determine the mass of the stability indicating degradation. Most observed degradations have a retention time that is similar to that of the unmodified peptide, but the mass shift is characteristic (Table 13.1).

The minor degradation peak is usually associated with one of the nearby eluting major peaks. By comparing the calculated mass of the degradation, it should be apparent which major peak has been degraded, and the mass shift indicates the nature of the degradation (Figure 13.14). The UV peak height may be used to further confirm the association of the major peak and the degradation since growth in the degraded peak should be associated with a loss of peak height from the proposed unmodified peak—although very minor changes may not result in an observable shift in peak area (Figure 13.15).

As mentioned earlier, deamidations of asparagine result in three possible end-products: aspartate, isoaspartate, and the succinimide intermediate. Isoaspartate is

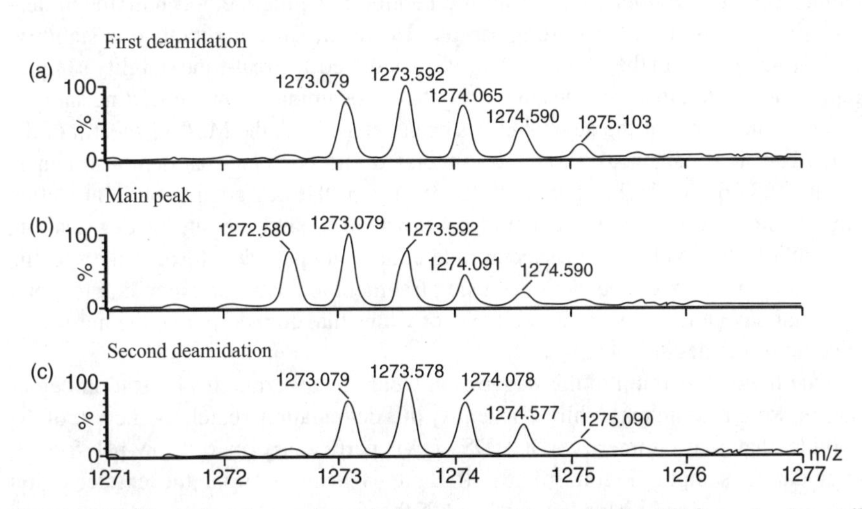

FIGURE 13.14 MS of the two deamidations representing the IsoD and D forms (a, c) as well as the main peak (b).

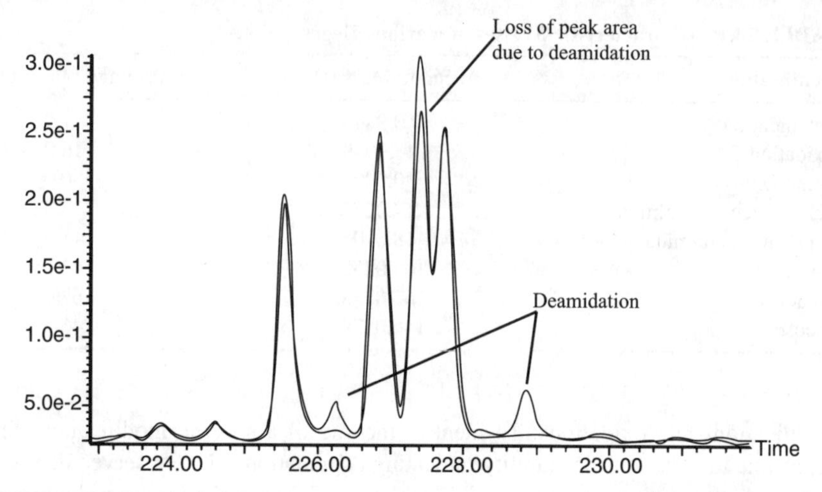

FIGURE 13.15 UV overlay of a tryptic map from an antibody before and after incubation at 37°C for 1 month at pH 8 showing a loss of one peptide, and growth of the two possible products after deamidation: IsoD and D.

frequently observed at earlier retention times and, baring structural constraints in the protein, isoaspartate is often observed at a 3 : 1 ratio compared to the later-eluting aspartate endproduct [60]. The succinimide is observed to elute even later than the aspartate degradation. One particularly vexing issue with deamidation concerns the sample preparation. The rate of deamidation is accelerated at higher pH values, and deamidation is generated as a result of sample preparation. It can be readily demonstrated that longer sample preparation results in more deamidation, but one cannot conclusively determine from the peptide mapping method how much deamidation is present in the starting sample. However, since the method is stability-indicating, growth in the deamidation peak in stressed and real-time stability samples represents quantitative changes in the amount of deamidation in the starting sample.

Once the nature of the degradant has been determined, the MS/MS spectra of the degradant and main peak should be annotated and compared as demonstrated in Figures 13.16 and 13.17. The MS/MS spectra of the degraded peptide are often similar; however, differences should be expected around the site of degradation, particularly for oxidations or succinimide-containing peptides. Ideally, the specific site of degradation will be identified using fragment ion data corresponding to *b* or *y* ions that have no mass shift, as well as *b* or *y* ions that do incorporate the mass shift specific to the degradation.

One major exception to this suggestion would be aspartate isomerization degradation, which has no mass shift. To identify this degradation, search for the *m/z* of the peptide that may isomerize (DG, DS, DA), particularly in the low-pH forced-degradation sample (Figure 13.18). If there is isomerization, the extracted ion chromatogram should have two peaks with the same exact mass or same mass and MS/MS spectra yet slightly different retention times (Figure 13.19). Additionally, the

367 377 387
GF YPSDIAVE WESNGQPENN YK
[Y3] 424.2191m

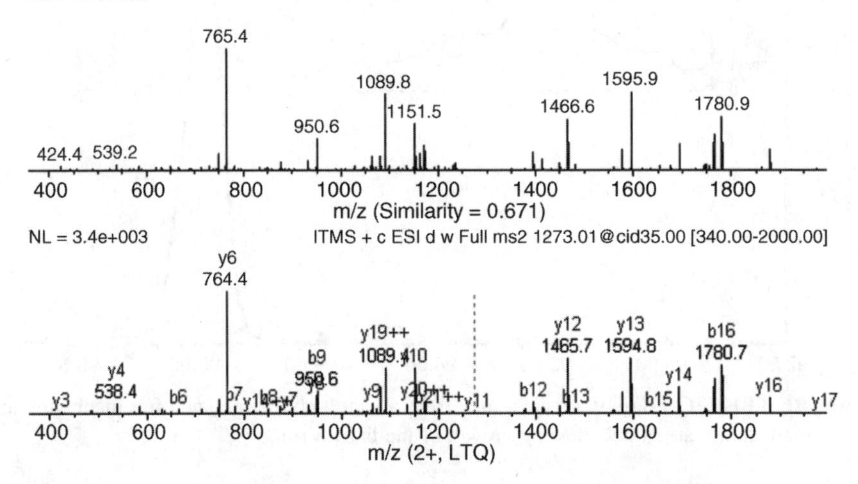

FIGURE 13.16 MS/MS overlays of the second deamidation peptide compared to the main peak demonstrating that y_3 (m/z 424) is unmodified and y_4 is $+1$; hence deamidation at N385.

367 377 387
GFYPSDIAVE WESNGQPENN YK
[y10]1150.5123m,1151.17a

Succinimide form

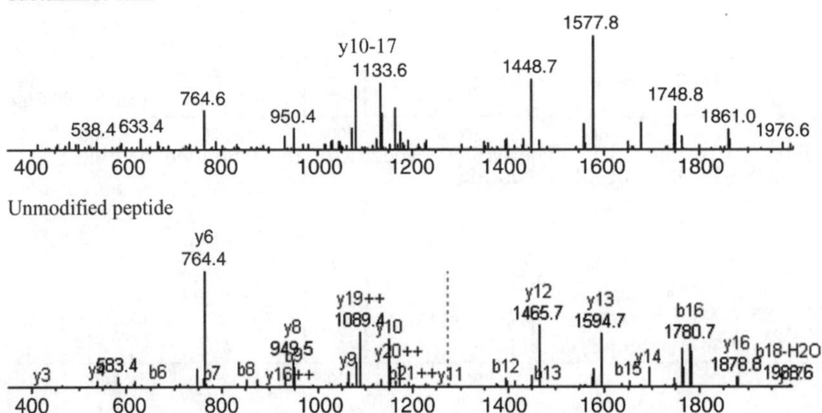

FIGURE 13.17 MS/MS overlays of the succinimide deamidation peptide compared to the unmodified peptide demonstrating that y_6 is not deamidated, while y_{10} (m/z 1150) shows a mass loss of 17 Da, indicating that the succinimide is located at N380.

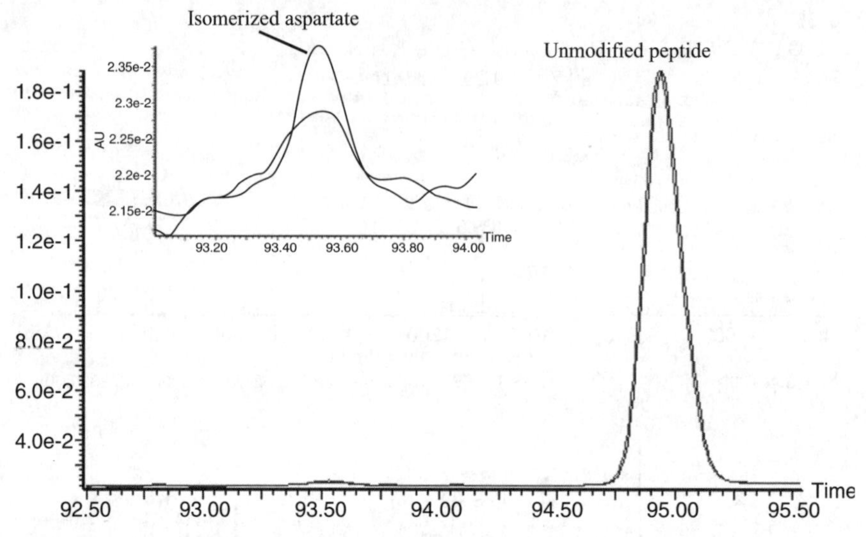

FIGURE 13.18 Overlay of a tryptic map from an antibody before and after incubation at 37°C for 1 month at pH 5.2 showing growth of the IsoD form.

271 281
FNWYVDGVEV HNAK

MS/MS of isomerized peptide

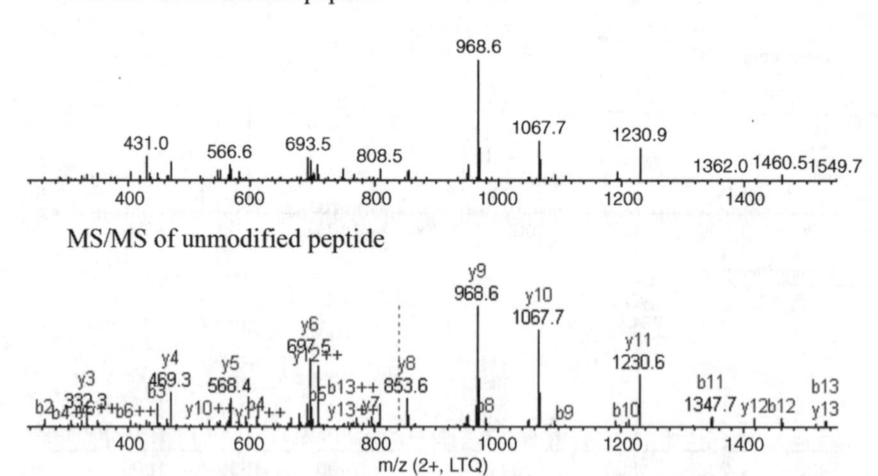

FIGURE 13.19 MS/MS overlays of the *m/z* 839.5 ion eluting at 93.5 min compared to the *m/z* 839.5 ion (unmodified peptide) eluting at 95.0 min, demonstrating that they are the same peptide.

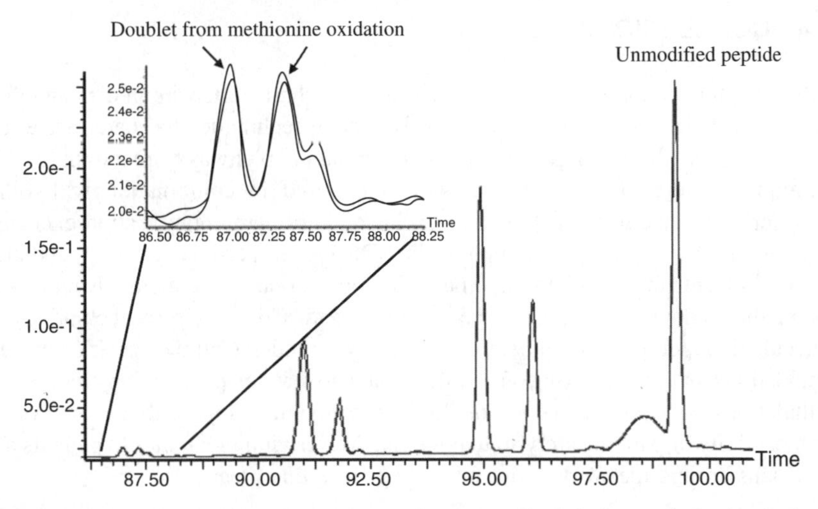

FIGURE 13.20 UV overlays of a tryptic map from an antibody before and after incubation at 37°C for 1 month at pH 8 showing the small growth in the doublet corresponding to methionine oxidation. The parent peptide also loses some peak intensity, accounting for a 0.7% increase in oxidation.

isomerized peptide is most likely to have a low-intensity peak, presuming that the isomerization reaction is slow.

Oxidations of methionine residues are frequently observed as doublet peaks (Figure 13.20). This arises because the sulfur atom has sp^3 geometry in both the sulfide state (which is not chiral) and sulfoxide state (which is chiral) [107]. In this reaction, it is expected that both *S* and *R* states would be formed equally, resulting in diastereomers that are resolved with high-resolution chromatography. Therefore, it is fairly straightforward to identify a methionine oxidation when a growing doublet is observed. However, two peaks are not always observed for methionine oxidation, so the MS/MS spectra for these oxidations should be verified to confirm that no other aromatic residues are oxidized. In the example shown in Figure 13.20, the areas under the curve for the two methionine peaks have been summed and divided by the total peptide area under the curve. In the control sample, the methionine oxidation is 4.0% of the total peptide, while after 1 month at pH 8 and 37°C the methionine oxidation accounts for 4.7% of the total peptide.

Once these degradations have been identified, other minor peaks in the UV chromatogram can be characterized. These may include low-level tryptophan oxidations, deamidations, isomerizations of aspartate or glutamate residues, *N*-terminal and *C*-terminal variants, mutations, backbone cleavages, and most likely some masses that cannot be identified. Since these peaks are not observed to change during formulation, they are formed during the protein production or purification and may be relevant indicators of the drug substance or drug product quality. Careful monitoring of these species during candidate selection and development enables superior product quality.

13.4 CONCLUSIONS

In this chapter, we have covered many methods to characterize antibodies and their degradants. It is critical to use an extensive set of techniques to characterize any antibody, given that a large number of degradation pathways are possible. The techniques covered in this chapter should be used in combination with other characterization methods such as size exclusion chromatography, cation exchange chromatography, capillary electrophoresis, SDS-PAGE, particle counting, and analytical ultracentrifugation, among others. Of course, one of the main advantages of any method coupled with mass spectrometry, and a driving force behind large-molecule therapeutic development, is our ability to understand the specific covalent degradations that are occurring on a molecule and to develop purification processes or formulations that stabilize, reduce, or eliminate the degradations without compromising bioactivity or patient safety so that we can deliver vital medicines to patients and physicians around the world—our ultimate goal in this great endeavor.

APPENDIX

Method 1: Short-Column, Low-Mass Method

Varian Pursuit diphenyl column: 1.0 mm × 50 mm

Mobile phase A is 0.1% trifluoroacetic acid.

Mobile phase B is 90% acetonitrile in 0.086% TFA.

Column set at 80°C; flow rate = 50 μL/min.

Approximately 4 μg of antibody is loaded onto the column in salt-, guanidine-, or urea-containing buffers.

The MS data should be acquired from 1000 to 4000 *m/z* for antibodies, although the mass range will have to be adjusted somewhat for different analytes.

Gradient:

Time (min)	*B%*
0.0	5
2.0	5
3.0	35
28.0	46
30.0	95
32.0	95
33.0	5
40.0	5

Method 2: Disulfide Bond Variants in IgG2

Zorbax 300SB-C8: 2.1 × 150 mm, 5 μm, 300 Å.

Mobile phase A is 0.1% trifluoroacetic acid.

Mobile phase B is 70% isopropanol, 20% acetonitrile, and aqueous 0.1% TFA.

Column set at 75°C; flow rate = 500 μL/min.

Approximately 20 μg of antibody is loaded onto the column.

The MS data should be acquired from 1000 to 4000 *m/z* for antibodies, although the mass range will have to be adjusted somewhat for different analytes.

Gradient:

Time (min)	*B*%
0.0	10
2.0	19
24.0	43.2
25.0	95
30.0	95
31.0	10
36.0	10

Method 3: Limited Lys-C Digestion of an IgG1. Dilute the antibody to 2 mg/mL in 100 mM Tris-HCl, pH 8.0. Reconstitute Lys-C in water to 0.25 mg/mL. Add a 1 : 400 mass ratio of reconstituted Lys-C and incubate at 37°C for 30 min. Quench the reaction with the addition of a 0.25 volume (25 μL is added to 100 μL) of 125 mM NH_4OAc, pH 4.7.

Method 4: Long Column, Higher Resolution

Varian Pursuit diphenyl column: 2.1 mm × 150 mm.

Mobile phase A is 0.1% trifluoroacetic acid.

Mobile phase B is 90% acetonitrile in 0.086% TFA.

Column set at 80°C; flow rate = 200 μL/min.

Approximately 20 μg of antibody is loaded onto the column in salt-, guanidine-, or urea-containing buffers.

The MS data should be acquired from 1000 to 4000 *m/z* for antibodies, although the mass range will have to be adjusted somewhat for different analytes.

Gradient:

Time (min)	*B*%
0.0	5
2.0	5
7.0	35
47.0	46
52.0	95
55.0	95
57.0	5
70.0	5

Method 5: Denaturation and Reduction of IgGs. For this method, the protein is diluted to 2 mg/mL in 7.5 M GndHCl, 250 mM Tris-HCl, pH 7.75 and 2.5 mM EDTA. Dithiothreitol at a stock concentration of 0.5 M in the same dilution buffer is added to a final concentration of 5 mM. This solution is incubated at 45°C for 45 min to ensure complete reduction of all interchain and intrachain disulfide bonds.

Method 6: Alkylation of IgGs. To alkylate the reduced mAb, let the reduced solution sit at room temperature for 5 min. Prepare a fresh stock solution of 0.5 M iodoacetic acid in denaturing buffer. Add IAA to a final concentration of 19 mM. Incubate the samples in the dark and at room temperature for 40 min. The reaction can be quenched with 24 μL of 0.5 M DTT solution.

Method 7: Desalting of IgGs. Desalting is performed using a NAP5 column packed with G25 Sephadex beads that prevent proteins with a MW of >5 kDa from entering the pores. This requires 500 μL of 1 mg/mL reduced and alkylated antibody and a NAP5 column from GE Biosciences equilibrated with 10 mL of 50 mM Tris-HCl, pH 7.75. The protein sample is added to the column and allowed to fully enter the resin. While the protocol calls for a 1 mL elution volume, 800 μL elution is preferred for peptide mapping as it recovers 87% of the protein, while achieving near-quantitative removal of the guanidine salts. The 800 μL of eluate has an approximate concentration of 0.5 mg/mL and can now be digested with trypsin, Lys-C, or other proteases as required.

Method 8: Digestion of IgGs. Antibody digestion is carried out by adding 100 μL of desalted antibody to an Eppendorf tube. Proteomics-grade trypsin can be solubilized in the manufacturer's recommended buffer and added to a final 1 : 50 ratio. Incubation at 37°C for 4 h generally results in complete digestion. The reaction is quenched by adding 3 μL of 10% trifluoracetic acid.

Method 9: Chromatographic Separation of Peptide Maps

Waters Acquity BEH C18: 2.1 mm × 150 mm, 130 Å.

Mobile phase A is 0.1% trifluoroacetic acid.

Mobile phase B is 90% acetonitrile in 0.086% TFA.

Column set at 50°C; flow rate = 200 μL/min.

Approximately 20 μg of antibody is loaded onto the column.

The MS data should be acquired from 200 to 2500 *m/z*.

Gradient:

Time (min)	B%
0.0	0
5.0	0
185.0	50
186.0	95
190.0	95
194.0	0
220.0	0

REFERENCES

1. Hughes, B. (2008), 2007 FDA drug approvals: A year of flux, *Nat. Rev. Drug Discov. 7*, 107–109.
2. Waldmann, T. A. (2003), Immunotherapy: Past, present and future, *Nat. Med. 9*, 269–277.
3. Roque, A. C.; Lowe, C. R.; Taipa, M. A. (2004), Antibodies and genetically engineered related molecules: Production and purification, *Biotechnol. Progress 20*, 639–654.
4. Lobo, E. D.; Hansen, R. J.; Balthasar, J. P. (2004), Antibody pharmacokinetics and pharmacodynamics, *J. Pharm. Sci. 93*, 2645–2668.
5. Wang, W.; Singh, S.; Zeng, D. L.; King, K.; Nema, S. (2007), Antibody structure, instability, and formulation, *J. Pharm. Sci. 96*, 1–26.
6. Wypych, J.; Li, M.; Guo, A.; Zhang, Z.; Martinez, T.; Allen, M. J.; Fodor, S.; Kelner, D. N.; Flynn, G. C.; Liu, Y. D.; Bondarenko, P. V.; Ricci, M. S., Dillon, T. M., and Balland, A. (2008), Human IgG2 antibodies display disulfide-mediated structural isoforms, *J. Biol. Chem. 283*, 16194–16205.
7. van de Winkel, J. G.; Anderson, C. L. (1991), Biology of human immunoglobulin G Fc receptors, *J. Leukocyte Biol. 49*, 511–524.
8. Carter, P. J. (2006), Potent antibody therapeutics by design, *Nat. Rev. 6*, 343–357.
9. Jefferis, R.; Lund, J. (2002), Interaction sites on human IgG-Fc for FcgammaR: Current models, *Immunol. Lett. 82*, 57–65.
10. Morrison, S. L.; Smith, R. I. F.; Wright, A. (1994), Structural determinants of human IgG function, *Immunologist 2*, 119–124.
11. Sandlie, I.; Michaelsen, T. E. (1996), Choosing and manipulating effector functions, *Antibody Eng. 1*, 187–202.
12. Bradley, J. (1974), Immunoglobulins, *J. Med. Genet. 11*, 80–90.
13. Edelman, G. M.; Gall, W. E. (1969), The antibody problem, *Annu. Rev. Biochem. 38*, 415–466.
14. Edelman, G. M.; Cunningham, B. A.; Gall, W. E.; Gottlieb, P. D.; Rutishauser, U.; Waxdal, M. J. (1969), The covalent structure of an entire gammaG immunoglobulin molecule, *Proc. Nat. Acad. Sci. USA 63*, 78–85.
15. Porter, R. R. (1959), The hydrolysis of rabbit y-globulin and antibodies with crystalline papain, *Biochem. J. 73*, 119–126.
16. Huber, R.; Deisenhofer, J.; Colman, P. M.; Matsushima, M.; Palm, W. (1976), Crystallographic structure studies of an IgG molecule and an Fc fragment, *Nature 264*, 415–420.
17. Harris, L. J.; Larson, S. B.; Hasel, K. W.; Day, J.; Greenwood, A.; McPherson, A. (1992), The three-dimensional structure of an intact monoclonal antibody for canine lymphoma, *Nature 360*, 369–372.
18. van der Neut Kolfschoten, M.; Schuurman, J.; Losen, M.; Bleeker, W. K.; Martinez-Martinez, P.; Vermeulen, E.; den Bleker, T. H.; Wiegman, L.; Vink, T.; et al. (2007), Anti-inflammatory activity of human IgG4 antibodies by dynamic Fab arm exchange, *Science 317*, 1554–1557.
19. Kabat, E. A.; Wu, T. T.; Perry, H. M.; Gottesman, K. S.; Foeller, C. (1992), *Sequences of Proteins of Immunological Interest*, DIANE Publishing, Darby, PA.

20. Shukla, A. A.; Hubbard, B.; Tressel, T.; Guhan, S.; Low, D. (2007), Downstream processing of monoclonal antibodies–application of platform approaches, *J. Chromatogr. B (Anal. Technol. Biomed. Life Sci.) 848*, 28–39.
21. Fahrner, R. L.; Knudsen, H. L.; Basey, C. D.; Galan, W.; Feuerhelm, D.; Vanderlaan, M.; Blank, G. S. (2001), Industrial purification of pharmaceutical antibodies: Development, operation, and validation of chromatography processes, *Biotechnol. Genet. Eng. Rev. 18*, 301–327.
22. Trexler-Schmidt, M.; Sze-Khoo, S.; Cothran, A. R.; Thai, B. Q.; Sargis, S.; Lebreton, B.; Kelley, B.; Blank, G. S. (2009), Purification strategies to process 5 g/L titers of monoclonal antibodies, *Bio-Pharm. Int. 8*, 10–15.
23. Mehta, A.; Tse, M. L.; Fogle, J.; Len, A.; Shrestha, R.; Fontes, N.; Lebreton, B.; Wolk, B.; van Reis, R. (2008), Purifying therapeutic monoclonal antibodies, *Chem. Eng. Progress 104*, S14–S20.
24. Tugcu, N.; Roush, D. J.; Goklen, K. E. (2008), Maximizing productivity of chromatography steps for purification of monoclonal antibodies, *Biotechnol. Bioeng. 99*, 599–613.
25. Kelley, B. (2007), Very large scale monoclonal antibody purification: The case for conventional unit operations, *Biotechnol. Progress 23*, 995–1008.
26. Gadgil, H. S.; Pipes, G. D.; Dillon, T. M.; Treuheit, M. J.; Bondarenko, P. V. (2006), Improving mass accuracy of high performance liquid chromatography/electrospray ionization time-of-flight mass spectrometry of intact antibodies, *J. Am. Soc. Mass Spectrom. 17*, 867–872.
27. Gadgil, H. S.; Bondarenko, P. V.; Pipes, G.; Rehder, D.; McAuley, A.; Perico, N.; Dillon, T.; Ricci, M.; Treuheit, M. (2007), The LC/MS analysis of glycation of IgG molecules in sucrose containing formulations, *J. Pharm. Sci. 96*, 2607–2621.
28. Gadgil, H. S.; Bondarenko, P. V.; Pipes, G. D.; Dillon, T. M.; Banks, D.; Abel, J.; Kleemann, G. R.; Treuheit, M. J. (2006), Identification of cysteinylation of a free cysteine in the Fab region of a recombinant monoclonal IgG1 antibody using Lys-C limited proteolysis coupled with LC/MS analysis, *Anal. Biochem. 355*, 165–174.
29. Kleemann, G. R.; Beierle, J.; Nichols, A. C.; Dillon, T. M.; Pipes, G. D.; Bondarenko, P. V. (2008), Characterization of IgG1 immunoglobulins and peptide-Fc fusion proteins by limited proteolysis in conjunction with LC-MS, *Anal. Chem. 80*, 2001–2009.
30. Ren, D.; Pipes, G. D.; Hambly, D. M.; Bondarenko, P. V.; Treuheit, M. J.; Brems, D. N.; Gadgil, H. S. (2007), Reversed-phase liquid chromatography of immunoglobulin G molecules and their fragments with the diphenyl column, *J. Chromatogr. 1175*, 63–68.
31. Liu, H.; Gaza-Bulseco, G.; Faldu, D.; Chumsae, C.; Sun, J. (2008), Heterogeneity of monoclonal antibodies, *J. Pharm. Sci. 97*, 2426–2447.
32. Rehder, D. S.; Chelius, D.; McAuley, A.; Dillon, T. M.; Xiao, G.; Crouse-Zeineddini, J.; Vardanyan, L.; Perico, N.; Mukku, V.; Brems, D. N.; Matsumura, M.; Bondarenko, P. V. (2008), Isomerization of a single aspartyl residue of anti-epidermal growth factor receptor immunoglobulin gamma2 antibody highlights the role avidity plays in antibody activity, *Biochemistry 47*, 2518–2530.
33. Robinson, N. E.; Robinson, A. B. (2001), Molecular clocks, *Proc. Natl. Acad. Sci. USA 98*, 944–949.
34. Ikeda, K.; Higo, J. (2003), Free-energy landscape of a chameleon sequence in explicit water and its inherent alpha/beta bifacial property, *Protein Sci. 12*, 2542–2548.

35. Van Buren, N.; Rehder, D.; Gadgil, H.; Matsumura, M.; Jacob, J. (2009), Elucidation of two major aggregation pathways in an IgG2 antibody, *J. Pharm. Sci. 98*, 3013–3030.
36. Chu, J. W.; Trout, B. L. (2004), On the mechanisms of oxidation of organic sulfides by H2O2 in aqueous solutions, *J. Am. Chem. Soc. 126*, 900–908.
37. Levine, R. L.; Berlett, B. S.; Moskovitz, J.; Mosoni, L.; Stadtman, E. R. (1999), Methionine residues may protect proteins from critical oxidative damage, *Mech. Ageing Devel. 107*, 323–332.
38. Levine, R. L.; Mosoni, L.; Berlett, B. S.; Stadtman, E. R. (1996), Methionine residues as endogenous antioxidants in proteins, *Proc. Natl. Acad. Sci. USA 93*, 15036–15040.
39. Luo, S.; Levine, R. L. (2009), Methionine in proteins defends against oxidative stress, *FASEB J. 23*, 464–472.
40. Liu, H.; Gaza-Bulseco, G.; Xiang, T.; Chumsae, C. (2008), Structural effect of deglycosylation and methionine oxidation on a recombinant monoclonal antibody, *Molec. Immunol. 45*, 701–708.
41. Gaza-Bulseco, G.; Faldu, S.; Hurkmans, K.; Chumsae, C.; Liu, H. (2008), Effect of methionine oxidation of a recombinant monoclonal antibody on the binding affinity to protein A and protein G, *J. Chromatogr. B (Anal. Technol. Biomed. Life Sci.) 870*, 55–62.
42. Nimmerjahn, F.; Ravetch, J. V. (2008), Analyzing antibody-Fc-receptor interactions, *Meth. Molec. Biol. 415*, 151–162.
43. Presta, L. G.; Shields, R. L.; Namenuk, A. K.; Hong, K.; Meng, Y. G. (2002), Engineering therapeutic antibodies for improved function, *Biochem. Soc. Transact. 30*, 487–490.
44. Ottaviano, F. G.; Handy, D. E.; Loscalzo, J. (2008), Redox regulation in the extracellular environment, *Circ. J. 72*, 1–16.
45. Saphire, E. O.; Parren, P. W.; Pantophlet, R.; Zwick, M. B.; Morris, G. M.; Rudd, P. M.; Dwek, R. A.; Stanfield, R. L.; Burton, D. R.; Wilson, I. A. (2001), Crystal structure of a neutralizing human IGG against HIV-1: A template for vaccine design, *Science 293*, 1155–1159.
46. Zhang, W.; Marzilli, L. A.; Rouse, J. C.; Czupryn, M. J. (2002), Complete disulfide bond assignment of a recombinant immunoglobulin G4 monoclonal antibody, *Anal. Biochem. 311*, 1–9.
47. Burton, D. R. (1985), Immunoglobulin G: Functional sites, *Molec. Immunol. 22*, 161–206.
48. Brekke, O. H.; Bremnes, B.; Sandin, R.; Aase, A.; Michaelsen, T. E.; Sandlie, I. (1993), Human IgG3 can adopt the disulfide bond pattern characteristic for IgG1 without resembling it in complement mediated cell lysis, *Molec. Immunol. 30*, 1419–1425.
49. Lacy, E. R.; Baker, M.; Brigham-Burke, M. (2008), Free sulfhydryl measurement as an indicator of antibody stability, *Anal. Biochem. 382*, 66–68.
50. Zhang, W.; Czupryn, M. J. (2002), Free sulfhydryl in recombinant monoclonal antibodies, *Biotechnol. Progress 18*, 509–513.
51. Dillon, T. M.; Ricci, M. S.; Vezina, C.; Flynn, G. C.; Liu, Y. D.; Rehder, D. S.; Plant, M.; Henkle, B.; Li, Y.; Deechongkit, S.; Varnum, B.; Wypych, J.; Balland, A.; Bondarenko, P. V. (2008), Structural and functional characterization of disulfide isoforms of the human IgG2 subclass, *J. Biol. Chem. 283*, 16206–16215.

52. Liu, Y. D.; Chen, X.; Enk, J. Z.; Plant, M.; Dillon, T. M.; Flynn, G. C. (2008), Human IgG2 antibody disulfide rearrangement in vivo, *J. Biol. Chem. 283*, 29266–29272.
53. Banks, D. D.; Gadgil, H. S.; Pipes, G. D.; Bondarenko, P. V.; Hobbs, V.; Scavezze, J. L.; Kim, J.; Jiang, X. R.; Mukku, V.; Dillon, T. M. (2008), Removal of cysteinylation from an unpaired sulfhydryl in the variable region of a recombinant monoclonal IgG1 antibody improves homogeneity, stability, and biological activity *J. Pharm. Sci. 97*, 775–790.
54. Li, B.; Gorman, E. M.; Moore, K. D.; Williams, T.; Schowen, R. L.; Topp, E. M.; Borchardt, R. T. (2005), Effects of acidic N + 1 residues on asparagine deamidation rates in solution and in the solid state, *J. Pharm. Sci. 94*, 666–675.
55. Patel, K.; Borchardt, R. T. (1990), Chemical pathways of peptide degradation. II. Kinetics of deamidation of an asparaginyl residue in a model hexapeptide, *Pharm. Res. 7*, 703–711.
56. Patel, K.; Borchardt, R. T. (1990), Chemical pathways of peptide degradation. III. Effect of primary sequence on the pathways of deamidation of asparaginyl residues in hexapeptides, *Pharm. Res. 7*, 787–793.
57. Rehder, D. S.; Dillon, T. M.; Pipes, G. D.; Bondarenko, P. V. (2006), Reversed-phase liquid chromatography/mass spectrometry analysis of reduced monoclonal antibodies in pharmaceutics, *J. Chromatogr. A 1102*, 164–175.
58. Xiao, G.; Bondarenko, P. V.; Jacob, J.; Chu, G. C.; Chelius, D. (2007), 18O labeling method for identification and quantification of succinimide in proteins, *Anal. Chem. 79*, 2714–2721.
59. Geiger, T.; Clarke, S. (1987), Deamidation, isomerization, and racemization at asparaginyl and aspartyl residues in peptides. Succinimide-linked reactions that contribute to protein degradation *J. Biol. Chem. 262*, 785–794.
60. Chelius, D.; Rehder, D. S.; Bondarenko, P. V. (2005), Identification and characterization of deamidation sites in the conserved regions of human immunoglobulin gamma antibodies, *Anal. Chem. 77*, 6004–6011.
61. Stephenson, R. C.; Clarke, S. (1989), Succinimide formation from aspartyl and asparaginyl peptides as a model for the spontaneous degradation of proteins, *J. Biol. Chem. 264*, 6164–6170.
62. Capasso, S.; Mazzarella, L.; Sica, F.; Zagari, A. (1991), First evidence of spontaneous deamidation of a glutamine residue via a cyclic imide to an alpha- and gamma-glutamic residue under physiological conditions, *J. Chem. Soc., Chem. Commun. 1991*, 1667–1668.
63. Robinson, A. B.; Scotchler, J. W.; McKerrow, J. H. (1973), Rates of nonenzymatic deamidation of glutaminyl and asparaginyl residues in pentapeptides, *J. Am. Chem. Soc. 95*, 8156–8159.
64. Robinson, N. E.; Robinson, A. B. (2004), Prediction of primary structure deamidation rates of asparaginyl and glutaminyl peptides through steric and catalytic effects, *J. Peptide Res. 63*, 437–448.
65. Robinson, N. E.; Robinson, Z. W.; Robinson, B. R.; Robinson, A. L.; Robinson, J. A.; Robinson, M. L.; Robinson, A. B. (2004), Structure-dependent nonenzymatic deamidation of glutaminyl and asparaginyl pentapeptides, *J. Peptide Res. 63*, 426–436.
66. Chelius, D.; Jing, K.; Lueras, A.; Rehder, D. S.; Dillon, T. M.; Vizel, A.; Rajan, R. S.; Li, T.; Treuheit, M. J.; Bondarenko, P. V. (2006), Formation of pyroglutamic acid from

N-terminal glutamic acid in immunoglobulin gamma antibodies, *Anal. Chem. 78*, 2370–2376.

67. Wakankar, A. A.; Borchardt, R. T. (2006), Formulation considerations for proteins susceptible to asparagine deamidation and aspartate isomerization, *J. Pharm. Sci. 95*, 2321–2336.
68. Wakankar, A. A.; Borchardt, R. T.; Eigenbrot, C.; Shia, S.; Wang, Y. J.; Shire, S. J.; Liu, J. L. (2007), Aspartate isomerization in the complementarity-determining regions of two closely related monoclonal antibodies, *Biochemistry 46*, 1534–1544.
69. Chu, G. C.; Chelius, D.; Xiao, G.; Khor, H. K.; Coulibaly, S.; Bondarenko, P. V. (2007), Accumulation of succinimide in a recombinant monoclonal antibody in mildly acidic buffers under elevated temperatures, *Pharm. Res. 24*, 1145–1156.
70. Capasso, S.; Di Cerbo, P. (2000), Formation of an RNase A derivative containing an aminosuccinyl residue in place of asparagine 67, *Biopolymers 56*, 14–19.
71. Kosky, A. A.; Razzaq, U. O.; Treuheit, M. J.; Brems, D. N. (1999), The effects of alpha-helix on the stability of Asn residues: Deamidation rates in peptides of varying helicity, *Protein Sci 8*, 2519–2523.
72. Qi, P.; Volkin, D. B.; Zhao, H.; Nedved, M. L.; Hughes, R.; Bass, R.; Yi, S. C.; Panek, M. E.; Wang, D.; Dalmonte, P.; Bond, M. D. (2009), Characterization of the photodegradation of a human IgG1 monoclonal antibody formulated as a high-concentration liquid dosage form, *J. Pharm. Sci. 98*, 3117–3130.
73. Cordoba, A. J.; Shyong, B. J.; Breen, D.; Harris, R. J. (2005), Non-enzymatic hinge region fragmentation of antibodies in solution, *J. Chromatogr. B (Anal. Technol. Biomed. Life Sci.) 818*, 115–121.
74. Smith, M. A.; Easton, M.; Everett, P.; Lewis, G.; Payne, M.; Riveros-Moreno, V.; Allen, G. (1996), Specific cleavage of immunoglobulin G by copper ions, *Int. J. Peptide Protein Res. 48*, 48–55.
75. Yashiro, M.; Sonobe, Y.; Yamamura, A.; Takarada, T.; Komiyama, M.; Fujii, Y. (2003), Metal-ion-assisted hydrolysis of dipeptides involving a serine residue in a neutral aqueous solution, *Org. Biomolec. Chem. 1*, 629–632.
76. Skribanek, Z.; Mezo, G.; Mak, M.; Hudecz, F. (2002), Mass spectrometric and chemical stability of the Asp-Pro bond in herpes simplex virus epitope peptides compared with X-Pro bonds of related sequences, *J. Peptide Sci. 8*, 398–406.
77. Liu, H.; Gaza-Bulseco, G.; Lundell, E. (2008), Assessment of antibody fragmentation by reversed-phase liquid chromatography and mass spectrometry, *J. Chromatogr. B (Anal. Technol. Biomed. Life Sci.) 876*, 13–23.
78. Davagnino, J.; Wong, C.; Shelton, L.; Mankarious, S. (1995), Acid hydrolysis of monoclonal antibodies, *J. Immunol. Meth. 185*, 177–180.
79. Li, A.; Sowder, R. C.; Henderson, L. E.; Moore, S. P.; Garfinkel, D. J.; Fisher, R. J. (2001), Chemical cleavage at aspartyl residues for protein identification, *Anal. Chem. 73*, 5395–5402.
80. Vrdoljak, A.; Trescec, A.; Benko, B.; Hecimovic, D.; Simic, M. (2004), In vitro glycation of human immunoglobulin G, *Clin. Chim. Acta 345*, 105–111.
81. Lyubarskaya, Y.; Houde, D.; Woodard, J.; Murphy, D.; Mhatre, R. (2006), Analysis of recombinant monoclonal antibody isoforms by electrospray ionization mass spectrometry as a strategy for streamlining characterization of recombinant monoclonal antibody charge heterogeneity, *Anal. Biochem. 348*, 24–39.

82. Harris, R. J. (2005), Heterogeneity of recombinant antibodies: Linking structure to function, *Devel. Biol. 122*, 117–127.
83. Gadgil, H. S.; Bondarenko, P. V.; Treuheit, M. J.; Ren, D. (2007), Screening and sequencing of glycated proteins by neutral loss scan LC/MS/MS method, *Anal. Chem. 79*, 5991–5999.
84. Ahmad, W.; Li, L.; Deng, Y. (2008), Identification of AGE-precursors and AGE formation in glycation-induced BSA peptides, *BMB Rep. 41*, 516–522
85. Frolov, A.; Hoffmann, R. (2008), Separation of Amadori peptides from their unmodified analogs by ion-pairing RP-HPLC with heptafluorobutyric acid as ion-pair reagent, *Anal. Bioanal. Chem. 392*, 1209–1214.
86. Haus, J. M.; Carrithers, J. A.; Trappe, S. W.; Trappe, T. A. (2007), Collagen, cross-linking, and advanced glycation end products in aging human skeletal muscle, *J. Appl. Physiol. 103*, 2068–2076.
87. Bailey, A. J. (2001), Molecular mechanisms of ageing in connective tissues, *Mech. Ageing Devel. 122*, 735–755.
88. Grillo, M. A.; Colombatto, S. (2008), Advanced glycation end-products (AGEs): Involvement in aging and in neurodegenerative diseases, *Amino Acids 35*, 29–36.
89. Bleckwenn, N. A.; Shiloach, J. (2004), Large-scale cell culture, in Coligan, J. E. et al., eds., *Current Protocols in Immunology*, Appendix 1U, John Wiley & Sons, Inc.
90. Banks, D. D.; Hambly, D. M.; Scavezze, J. L.; Siska, C. C.; Stackhouse, N. L.; Gadgil, H. S. (2009), The effect of sucrose hydrolysis on the stability of protein therapeutics during accelerated formulation studies, *J. Pharm. Sci. 98*, 4501–4510..
91. Zhang, Z.; Henzel, W. J. (2004), Signal peptide prediction based on analysis of experimentally verified cleavage sites, *Protein Sci. 13*, 2819–2824.
92. Antes, B.; Amon, S.; Rizzi, A.; Wiederkum, S.; Kainer, M.; Szolar, O.; Fido, M.; Kircheis, R.; Nechansky, A. (2007), Analysis of lysine clipping of a humanized Lewis-Y specific IgG antibody and its relation to Fc-mediated effector function, *J. Chromatogr. B (Anal. Technol. Biomed. Life Sci.) 852*, 250–256.
93. Dick, L. W., Jr.; Qiu, D.; Mahon, D.; Adamo, M.; Cheng, K. C. (2008), C-terminal lysine variants in fully human monoclonal antibodies: Investigation of test methods and possible causes, *Biotechnol. Bioeng. 100*, 1132–1143.
94. Johnson, K. A.; Paisley-Flango, K.; Tangarone, B. S.; Porter, T. J.; Rouse, J. C. (2007), Cation exchange-HPLC and mass spectrometry reveal C-terminal amidation of an IgG1 heavy chain, *Anal. Biochem. 360*, 75–83.
95. Jefferis, R. (2005), Glycosylation of recombinant antibody therapeutics, *Biotechnol. Progress 21*, 11–16.
96. Wagner-Rousset, E.; Bednarczyk, A.; Bussat, M. C.; Colas, O.; Corvaia, N.; Schaeffer, C.; Van Dorsselaer, A.; Beck, A. (2008), The way forward, enhanced characterization of therapeutic antibody glycosylation: Comparison of three level mass spectrometry-based strategies, *J. Chromatogr. B (Anal. Technol. Biomed. Life Sci.) 872*, 23–37.
97. Beck, A.; Wagner-Rousset, E.; Bussat, M. C.; Lokteff, M.; Klinguer-Hamour, C.; Haeuw, J. F.; Goetsch, L.; Wurch, T.; Van Dorsselaer, A.; Corvaia, N. (2008), Trends in glycosylation, glycoanalysis and glycoengineering of therapeutic antibodies and Fc-fusion proteins, *Curr. Pharm. Biotechnol. 9*, 482–501.
98. Jefferis, R. (2009), Glycosylation as a strategy to improve antibody-based therapeutics, *Nat. Rev. Drug Discov. 8*, 226–234.

99. Prater, B. D.; Connelly, H. M.; Qin, Q.; Cockrill, S. L. (2009), High-throughput immunoglobulin G N-glycan characterization using rapid resolution reverse-phase chromatography tandem mass spectrometry, *Anal. Biochem. 385*, 69–79.

100. Martinez, T.; Pace, D.; Brady, L.; Gerhart, M.; Balland, A. (2007), Characterization of a novel modification on IgG2 light chain. Evidence for the presence of O-linked mannosylation, *J. Chromatogr. 1156*, 183–187.

101. Kim, H.; Yamaguchi, Y.; Masuda, K.; Matsunaga, C.; Yamamoto, K.; Irimura, T.; Takahashi, N.; Kato, K.; Arata, Y. (1994), O-glycosylation in hinge region of mouse immunoglobulin G2b, *J. Biol. Chem. 269*, 12345–12350.

102. Chen, F. T.; Evangelista, R. A. (1998), Profiling glycoprotein n-linked oligosaccharide by capillary electrophoresis, *Electrophoresis 19*, 2639–2644.

103. Chang, L. L.; Shepherd, D.; Sun, J.; Ouellette, D.; Grant, K. L.; Tang, X. C.; Pikal, M. J. (2005), Mechanism of protein stabilization by sugars during freeze-drying and storage: Native structure preservation, specific interaction, and/or immobilization in a glassy matrix? *J. Pharm. Sci. 94*, 1427–1444.

104. Ren, D.; Pipes, G.; Xiao, G.; Kleemann, G. R.; Bondarenko, P. V.; Treuheit, M. J.; Gadgil, H. S. (2008), Reversed-phase liquid chromatography-mass spectrometry of site-specific chemical modifications in intact immunoglobulin molecules and their fragments, *J. Chromatogr. 1179*, 198–204.

105. Ren, D.; Pipes, G. D.; Hambly, D.; Bondarenko, P. V.; Treuheit, M. J.; Gadgil, H. S. (2009), Top-down N-terminal sequencing of Immunoglobulin subunits with electrospray ionization time of flight mass spectrometry, *Anal. Biochem. 384*, 42 48.

106. Sinha, S.; Pipes, G.; Topp, E. M.; Bondarenko, P. V.; Treuheit, M. J.; Gadgil, H. S. (2008), Comparison of LC and LC/MS methods for quantifying N-glycosylation in recombinant IgGs, *J. Am. Soc. Mass Spectrom. 19*, 1643–1654.

107. Akazome, M.; Hirabayashi, A.; Takaoka, K.; Nomura, S.; Ogura, K. (2005), Molecular recognition of l-leucyl-l-alanine: Enantioselective inclusion of alkyl methyl sulfoxides, *Tetrahedron 61*, 1107–1113.

INDEX

Characterization of Impurities and Degradants Using Mass Spectrometry, First Edition.
Edited by Birendra N. Pramanik, Mike S. Lee, and Guodong Chen.